Random Processes with Applications to Circuits and Communications

Bernard C. Levy

Random Processes with Applications to Circuits and Communications

 Springer

Bernard C. Levy
University of California
Davis, CA, USA

Additional material to this book can be downloaded from http://springer.com.

ISBN 978-3-030-22299-4   ISBN 978-3-030-22297-0   (eBook)
https://doi.org/10.1007/978-3-030-22297-0

This Springer imprint is published by the registered company Springer Nature Switzerland AG.
The registered company address is: Gewerbestrasse 11, 6330 Cham, Switzerland

It is likely that unlikely things should happen.

Aristotle

Preface

This textbook is based on teaching materials (lecture notes, problem sets, exams) generated over the last 20 years while teaching a first-year graduate course on random processes in the Department of Electrical and Computer Engineering (ECE), UC Davis. Since the ECE Department is a medium-size department, the course has been designed to appeal to a constituency extending beyond signal processing, communications, control, and networking and including in particular circuits, RF, and optics graduate students. So compared to recent textbooks which tend to focus on communications and networking applications, this book represents in part a return to the more physically oriented focus of early ECE random processes textbooks by Davenport and Root, Papoulis, or Helstrom. Another reason for this focus is that with the relentless shrinking of electronic devices over the last 20 years and the accompanying decrease of voltages to reduce power consumption, noise has become progressively a more severe source of performance limitation, and in this context, circuit designers need often to understand noise modeling issues when evaluating circuit architectures. So, while covering classical material on Brownian motion, Poisson processes, and power spectral densities, I have attempted to insert discussions of thermal noise, shot noise, quantization noise, and oscillator phase noise. At the same time, techniques used to analyze modulated communications and radar signals, such as the baseband representation of bandpass random signals or the computation of power spectral densities of a wide variety of modulated signals, are presented.

Although the early chapters of this book include a review of probability theory, random variables convergence, and of asymptotic results such as the law of large numbers and the central limit theorem, it is expected that all the readers of this book will have some training at an undergraduate level in probability and statistics. Ideally, some prior exposure to random processes, in particular the computation of their mean and autocorrelation and the effect of linear filtering operations, would be beneficial, but is not required as these concepts are derived from scratch. Also, while a rigorous presentation of random processes requires the use of measure theory, the technological focus of modern ECE curricula makes it unrealistic to expect that first- or even second-year ECE graduate students should have taken a course on measure theory and Lebesgue integration prior to being exposed to random processes. So, like most ECE graduate textbooks on random processes, measure theory will not be used openly, but since measure theory concepts like sigma fields are unavoidable, they will be lurking in the background, giving rise to unaddressed technical details. A reader already familiar with measure theory and Lebesgue integration should find it easy to identify and fill in gaps in technical derivations when needed. On the other hand, ECE graduate students have typically received a strong training in transform theory (Fourier, Laplace, and z-transforms) and linear algebra, and this experience will be leveraged throughout this book.

Another important feature of stochastic and noise phenomena arising in electrical and computer engineering systems is that they often require a significant modeling effort prior to any form of analysis. Thus, a second goal of this book is to provide some modeling training, primarily through the inclusion of long problems at the end of each chapter, where starting from a description of the operation of a system, a model is constructed and then analyzed. Some of these problems are rather arduous and will test the patience of students. But they reflect my view that learning is often best achieved by working through challenging problems. It is easy to be lulled in a false sense of confidence when reading course notes, where neatly organized answers are provided to each question posed. But the true test of understanding arises when solving messy problems extracted from real-world situations.

I would like to express my gratitude to the educators, mentors, and colleagues who helped me gain a better understanding of random processes at various stages of my career. In particular, I thank Philippe Formery, Georges Matheron, Jean Jacod, and Pierre Faurre at Mines ParisTech, and Kai Lai Chung, Donald Iglehart, and Thomas Kailath, my thesis adviser, at Stanford University. In addition, I would like to acknowledge Sanjoy Mitter, Alan Willsky at MIT, Albert Benveniste at INRIA, and Art Krener and Bill Gardner at UC Davis for their influence and help in clarifying various aspects of stochastic analysis. I would also like to thank all the students who took my EEC260 random processes course during the last 20 years for their interesting questions. Many of the problems appearing in this book originate in fact from student discussions during office hours.

Finally, I would like to express my deep appreciation to my wife, Chuc Thanh, for her support and indulgence in allowing me to spend so much time on this book project.

Davis, CA, USA   Bernard C. Levy

Contents

1	**Introduction**	1
	1.1 Book Organization	2
	1.2 Course Management and Curriculum Organization	6
	References	8

Part I Background

2	**Probability and Random Variables**	13
	2.1 Introduction	13
	2.2 Probability Space	13
	2.3 Conditional Probability	16
	2.4 Random Variables	19
	2.4.1 Discrete Random Variables	19
	2.4.2 Continuous Random Variables	21
	2.5 Transformation of Random Variables	27
	2.5.1 One-to-One Transformation	27
	2.5.2 Several-to-One Transformation	30
	2.6 Expectation and Characteristic Function	36
	2.6.1 Moments	36
	2.6.2 Characteristic Function	37
	2.6.3 Moment Generating Function	40
	2.7 Joint Distribution of Random Variables	41
	2.7.1 Joint PMF, PDF, and CDF	41
	2.7.2 Conditional Distributions	45
	2.7.3 Joint Transformations	47
	2.8 Sums of Independent Random Variables	51
	2.9 Joint and Conditional Expectations	53
	2.9.1 Joint Moments and Characteristic Function	54
	2.9.2 Characterization of Independence and Uncorrelatedness	56
	2.9.3 Conditional Expectation	57
	2.10 Random Vectors	59
	2.11 Problems	64
	References	77

3	**Convergence and Limit Theorems**	79
	3.1 Introduction	79
	3.2 Inequalities	79

3.3   Modes of Convergence 86
3.4   Law of Large Numbers 94
3.5   Regular, Moderate, and Large Deviations 97
3.6   Stable Distributions 102
3.7   Bibliographical Notes 104
3.8   Problems 104
References 113

Part II   Main Topics

4   Specification of Random Processes 117
4.1   Introduction 117
4.2   Specification of Random Processes 117
4.3   Mean and Autocorrelation 122
        4.3.1   Examples 122
        4.3.2   Autocorrelation and Autocovariance Properties 130
4.4   Cross-Correlation and Cross-Covariance of Random Processes 134
4.5   Mean-Square Continuity, Differentiation, and Integration 135
4.6   Classes of Random Processes 139
        4.6.1   Gaussian Processes 139
        4.6.2   Markov Processes 140
        4.6.3   Independent Increments Processes 143
4.7   Bibliographical Notes 147
4.8   Problems 147
References 163

5   Discrete-Time Finite Markov Chains 165
5.1   Introduction 165
5.2   Transition Matrices and Probability Distribution 165
5.3   Classification of States 170
5.4   Convergence 173
5.5   First-Step Analysis 180
        5.5.1   Number of Visits to Transient States 181
        5.5.2   Absorption Probabilities 183
5.6   Markov Chain Modulation 186
5.7   Bibliographical Notes 188
5.8   Problems 189
References 205

6   Wiener Process and White Gaussian Noise 207
6.1   Introduction 207
6.2   Definition and Basic Properties of the Wiener Process 208
6.3   Constructions of the Wiener Process 209
        6.3.1   Scaled Random Walk Model 209
        6.3.2   Paul Lévy's Construction 210
6.4   Sample Path Properties 214
        6.4.1   Hölder Continuity 214
        6.4.2   Non-differentiability 214
        6.4.3   Quadratic and Total Variation 215

6.5   Wiener Integrals and White Gaussian Noise 216
6.6   Thermal Noise 219
6.7   Bibliographical Notes 222
6.8   Problems 223
References 233

7   Poisson Process and Shot Noise 235
7.1   Introduction 235
7.2   Poisson Process Properties 235
7.3   Residual Waiting Time and Elapsed Time 241
7.4   Poisson Process Asymptotics 242
7.5   Merging and Splitting of Poisson Processes 244
7.6   Shot Noise 246
7.7   Bibliographical Notes 251
7.8   Problems 251
References 258

8   Processing and Frequency Analysis of Random Signals 259
8.1   Introduction 259
8.2   Random Signals Through Linear Systems 259
8.3   Power Spectral Density 263
8.4   Nyquist–Johnson Model of RLC Circuits 271
8.5   Power Spectral Densities of Modulated Signals 276
      8.5.1   Pulse Amplitude Modulated Signals 276
      8.5.2   Markov Chain Modulated Signals 278
8.6   Sampling of Bandlimited WSS Signals 281
8.7   Rice's Model of Bandpass WSS Random Signals 283
8.8   Bibliographical Notes 287
8.9   Problems 287
References 307

Part III   Advanced Topics

9   Ergodicity 311
9.1   Introduction 311
9.2   Finite Markov Chains 312
      9.2.1   Mean Duration of State Excursions 313
      9.2.2   Convergence of Empirical Distributions 314
      9.2.3   Convergence of Joint Empirical Distributions 314
9.3   Mean-Square Ergodicity 316
      9.3.1   Mean-Square Ergodicity Criterion 316
      9.3.2   Mean-Square Ergodicity of the Autocorrelation 319
      9.3.3   Mean-Square Ergodicity of the CDF 321
9.4   Bibliographical Notes 323
9.5   Problems 323
References 328

10   Scalar Markov Diffusions and Ito Calculus 331
   10.1   Introduction 331
   10.2   Diffusion Processes 332
          10.2.1   Diffusion Parametrization 332
          10.2.2   Fokker–Planck Equation 334
          10.2.3   Backward Kolmogorov Equation 341
   10.3   Ito Calculus 345
          10.3.1   Ito Integral 345
          10.3.2   The Ito Formula 350
   10.4   Stochastic Differential Equations 352
   10.5   Bibliographical Notes 356
   10.6   Problems 357
   References 362

Part IV   Applications

11   Wiener Filtering 365
   11.1   Introduction 365
   11.2   Linear Least-Squares Estimation 366
          11.2.1   Orthogonality Property of Linear Least-Squares Estimates 367
          11.2.2   Finite Dimensional Case 369
          11.2.3   Linear Observation Model 372
   11.3   Noncausal Wiener Filter 374
   11.4   Spectral Factorization 381
   11.5   Causal Wiener Filter 384
   11.6   Special Cases 388
          11.6.1   Signal in White Noise 388
          11.6.2 k-Steps Ahead Prediction 391
          11.6.3   Fixed-Lag Smoothing 393
   11.7   CT Wiener Filtering 395
   11.8   Bibliographical Notes 398
   11.9   Problems 399
   References 405

12   Quantization Noise and Dithering 407
   12.1   Introduction 407
   12.2   Quantization 408
          12.2.1   Single-Stage Quantizer 408
          12.2.2   Multistage Quantizer 411
   12.3   Quantization Noise Statistics 412
   12.4   Random Process Quantization 421
   12.5   Dithering 425
   12.6   Bibliographical Notes 430
   12.7   Problems 430
   References 435

13   Phase Noise in Autonomous Oscillators 437
   13.1   Introduction 437
   13.2   Oscillator Characteristics 438
       13.2.1   Limit Cycles 438
       13.2.2   Floquet Theory 441
       13.2.3   Isochrons 447
   13.3   Phase Noise Model 450
       13.3.1   General Model 451
       13.3.2   Slow Time-Scale Model 453
   13.4   Oscillator Autocorrelation and Spectrum 455
   13.5   Bibliographical Notes 458
   13.6   Problems 458
   References 463

Introduction

Random processes have a wide range of applications outside mathematics to fields as different as physics and chemistry, engineering, biology, or economics and mathematical finance. When addressed at an audience in one of its fields of applications, random processes take a slightly different flavor since each discipline tends to use tools which are best adapted to the class of problems it seeks to analyze. The goal of this book is to present random processes techniques applicable to the analysis of electrical and computer engineering systems. In this context, random process analysis has traditionally played an important role in several areas: the analysis of noise in electronic, radio-frequency (RF) and optical devices, the study of communications signals in the presence of noise, and the evaluation of the performance of computer and networking systems.

Interest in the effect of noise on electrical circuits goes back to the pioneering work of Schottky [1] on shot noise and of Nyquist and Johnson [2, 3] on thermal noise. However, with the advent of transistors and integrated circuits, researchers identified other types of noise arising commonly in electronic circuits, such as $1/f$ noise [4], switching noise created by the substrate transmission of digital buffer switching [5,6], or oscillator phase noise [7,8]. In addition to noise originating from the physics of circuits, other sources of noise such as quantization noise [9] or dither [10] arise whenever analog signals are converted to digital form, or when synthesized noise is introduced deliberately to mask imperfections in digital audio systems, or to facilitate certain signal processing tasks, such as device or sensor calibration. With the progressive reduction of voltage ranges in integrated circuits to minimize power consumption, and the shrinking of cell sizes, signal to noise ratio has been reduced, making it more important to model noise sources accurately, and to develop strategies to mitigate their effects.

Analytical studies of noise in communications systems started with Rice's work [11–14] on the statistical performance of analog communications systems, such as FM receivers. Subsequently, as digital communications replaced analog communications, Shannon's paper [15] characterizing the capacity of communications channels highlighted the limitation placed by noise on such systems. Over time, a number of considerations have become important in the study of communications systems. These include the bandwidth occupied by communications signals, since for wireless channels, bandwidth efficiency has become crucial in order to accommodate as many users as possible over a limited resource. For channels subject to interference, devising modulation and demodulation schemes (such as spread spectrum) that can mitigate the effect of interferers and jammers is important. For channels that are shared either in coordinated or uncoordinated manner, developing sensing and access schemes that maximize the use of the common resource by users is desirable. It is also worth noting that the nature of communication channels has evolved rapidly over the last 30 years,

B. C. Levy, *Random Processes with Applications to Circuits and Communications*, https://doi.org/10.1007/978-3-030-22297-0_1

since twisted pairs (phone lines and DSL), Ethernet cables, coax (cable services), wireless channels, magnetic storage read and write devices, solid state memories, power lines, and optical cables are all used to exchange or store information. All these media have very different characteristics, yet communications over all these channels can be analyzed by using techniques which in one way or another are based on the study of random processes.

Finally, random processes are commonly used to analyze the performance of computer networks and systems. Very early in the development of computer networks, Kleinrock [16, 17] realized that the behavior of switches and routers could be modeled by using queuing theory techniques. Unfortunately, as computer networks increased in size and traffic control protocols became more complicated, it became clear that exact analytical techniques would become untractable, but queueing theory insights are still commonly used to develop network modeling approximations.

Since the scope of potential applications of random processes techniques to electrical and computer engineering systems is unusually large, most textbooks tend to favor a certain class of applications. Whereas early random processes textbooks for electrical engineers by Davenport and Root [18], Papoulis and Pillai [19], and Helstrom [20] had a strong physics flavor, recent texts, such as those by Gallager [21] and Hajek [22], while significantly more rigorous than their predecessors, tend to be geared primarily towards communications and networking applications. As explained in the preface, owing both to the composition of the audience for the random processes course I have been teaching, and to increased role of noise considerations in circuit design, the present textbook is focused more on circuits and communications applications and less on computer networking applications.

1.1   Book Organization

Although the Accreditation Board for Engineering and Technology (ABET) mandates that electrical and computer engineering (ECE) undergraduates should be exposed to probability and statistics in the context of electrical and computer engineering applications, the level of probabilistic training received by students prior to their first year of graduate studies varies widely, ranging from elementary one-quarter presentations of basic probability and random variables material to comprehensive two-quarter sequences including coverage of convergence concepts and Markov chains. Typical introductory probability and random processes textbooks or lecture notes used by ECE Departments include [23–26]. To bring diversely prepared students to an adequate level of preparation, two review chapters are included. Chapter 2 covers probability spaces, random variables, joint and conditional distributions, transformations of random variables, and characteristic functions. Chapter 3 reviews inequalities, such as the Markov and Chebyshev inequalities, the various concepts of random variable convergence (almost sure convergence, mean-square convergence, convergence in probability and distribution), as well as their relative implications. Fundamental theorems of statistics, such as the weak and strong laws of large numbers and the central limit theorem, are also presented. More advanced material on Chernoff's inequality and large deviations, and on stable random distributions, is also included. This material can easily be skipped and is included only to provide material for further study to more inquisitive students who sometimes ask questions related to these topics. Chapters 2 and 3 are intended strictly to serve as a review of probability and statistics, as well as to introduce the notations used in the remainder of the book. The coverage is fast-paced and concise, and is not indented to serve as a first-time introduction to probability theory, except possibly for students with a strong mathematical training. Like most graduate random processes textbooks addressed at electrical and computer engineers or physicists, the presentation does not rely explicitly on measure theory, since the increasing technological focus of modern undergraduate and first-year graduate electrical engineering curricula leaves little space for advanced mathematics topics.

The core material of the book is presented in the next five chapters. Chapter 4 describes the specification of random processes in terms of the joint distributions of their samples at a finite number of times. In this context, the non-reliance on measure theory implies that all expectations need to be expressed with respect to the finite joint distributions of the process instead of an integration with respect to the measure of the underlying probability space. Although this approach is more concrete and intuitive, it is rather clumsy and cumbersome and represents one of the main drawbacks of non-measure theoretic presentations of random processes. Then, important properties such as strict and wide sense stationarity are introduced. Classes of random processes, such as Gaussian and Markov processes, for which the amount of information needed to describe the process is reduced, are also presented. In addition, since many electrical and computer engineering applications tend to rely primarily on the first- and second-order statistics of random processes, the properties of autocorrelation and covariance functions of random processes are examined in detail. Chapter 5 studies homogeneous discrete-time finite Markov chains. Although the scope of this study is narrow, it represents the simplest class of random processes that can be analyzed completely with elementary analytical tools, such as linear algebra. The classification of states in transient and recurrent classes is presented, and the convergence of aperiodic irreducible chains is established by using a contraction approach. The Perron–Frobenius theorem is a consequence of this analysis. Exit probabilities and mean exit times from transient states are also evaluated by using a first-step step analysis approach. Finally, since finite Markov chains are often used as an engine to generate modulated signals, an analysis of the autocorrelation function of Markov chains modulated sequences is presented.

The next two chapters examine two important Markov processes, which are often used as building blocks in the construction of stochastic models. Chapter 6 examines the Wiener process which is often referred to as Brownian motion, since it represents a mathematical idealization of the Brownian motion of suspended particles in a fluid. This process has a rich set of properties, since it has independent increments, and is thus Markov, and is Gaussian. However, its most striking feature is that its sample paths are continuous almost surely, but are also nowhere differentiable almost surely. In other words. this process is highly erratic and constantly changes direction, so that its sample paths are fractal curves [27]. The earliest construction of Brownian motion as a limit of a random walk was proposed by Bachelier [28] who sought to model fluctuations of financial instruments. This construction illustrates nicely the independent increments feature of the process, but an alternative construction which exhibits the continuity of sample paths was later proposed by the French mathematician Paul Lévy. Both of these constructions are described. In engineering applications, the derivative of the Wiener process, which is called white Gaussian noise, in spite of being mathematically undefined, is often used when performing autocorrelation calculations. It is shown that such calculations can be given a rigorous interpretation by introducing the Wiener integral, which is defined as the integral of a deterministic (nonrandom) square-integrable function with respect to the increments of the Wiener process. There exists an isometry between the second moments of Wiener integrals and the square integral of the functions used in their construction. Since all solutions of linear continuous-time dynamical systems can be expressed as Wiener integrals, this allows a rigorous justification of all white noise calculations performed by engineers. To illustrate this approach, the Nyquist–Johnson model of thermal noise in a noisy resistor is introduced and is used to analyze an RC circuit. It is shown that in thermal equilibrium, the capacitor voltage obeys an Ornstein–Uhlenbeck process, i.e., a first-order wide-sense stationary Markov process. In this context it is also observed that the linear relation existing between the intensity of the noise fluctuations and the resistance of the resistor under consideration is a manifestation of the fluctuation dissipation theorem of statistical physics.

Chapter 7 examines the Poisson process, which also plays a major role in stochastic modeling. Like the Wiener process, it has stationary independent increments and is therefore Markov, but it is integer valued and exhibits unit jumps at successive random times. It is the prototype of counting processes and is used to model discrete phenomena, like photons hitting a detector, or communications packets arriving at a router. This process admits several equivalent definitions. On one hand, it can be constructed by using the independence and Poisson distribution of its increments. On the other hand, it also has the feature that the time intervals (called interarrival times) between successive jumps of the process are independent and exponentially distributed. As such it represents the simplest among the class of renewal processes. The chapter establishes the equivalence between the different definitions of the Poisson process and, while doing so, shows that conditioned on the value of the Poisson process at a terminal time, the jump times (also called epochs of the process) are independent and uniformly distributed. This feature provides a simple mechanism for simulating Poisson processes. It is also shown that merging and splitting (with a Bernoulli switch) operations preserve the Poisson property of Poisson processes. In electrical engineering applications, Poisson processes are often used to model shot noise, which represents the response of an electrical circuit to discrete excitations occurring at the epochs of the Poisson process. Shot noise was first investigated by Walter Schottky, and the autocorrelation of shot noise is evaluated, as well as its distribution at a fixed time.

Chapter 8 focuses on the effect of signal processing operations and the Fourier analysis of random signals. Linear filtering operations are commonly used in electrical engineering to improve the conditioning of random signals, or extract information they contain. Expressions for the autocorrelation of the output and the cross-correlations of the input and output of a linear filter driven by a random input are derived. In particular it is shown that if a wide-sense stationary (WSS) signal (a signal with constant mean whose autocorrelation is invariant under time shifts) is applied as input to a linear time-invariant filter, the input and output are jointly WSS. Although it is possible to apply Fourier analysis techniques to the larger class of harmonizable processes [29, 30], for simplicity we restrict our attention to WSS signals. The power spectral density (PSD) of a WSS signal is defined as the Fourier transform of its autocorrelation. It has the property of being nonnegative definite, and if we consider a small frequency band centered around a nominal frequency, the PSD at this frequency is proportional to the signal power contained in the band. The PSD of a signal provides a convenient tool for visualizing the amount of power a signal contains in different frequency bands, and thus extends to finite-power signals the usual Fourier transform used for finite energy signals. In this context, filtering operations can be analyzed easily in terms of the filter frequency response, and a further justification of white noise calculations is provided by observing that the output PSD of a filter driven by white noise is indistinguishable from the PSD produced by bandlimited input white noise as long as the noise bandwidth is larger than the filter bandwidth. An extension of the Nyquist/Shannon sampling theorem to bandlimited WSS signals is presented. Finally, Rice's representation of bandpass WSS signals in terms of WSS baseband in-phase and quadrature components is derived. This representation simplifies greatly the analysis of modulated communications or radar signals in the presence of noise by reducing it to the analysis of an equivalent complex baseband problem.

Whereas the material extending from Chaps. 4 to 8 can be viewed as essential, the five remaining chapters cover advanced topics and applications, and thus can be studied selectively, depending on the interests of the reader. Chapters 9 and 10 cover advanced topics. Specifically, Chap. 9 examines conditions under which a random process is ergodic, which means that some of its ensemble average statistics can be evaluated as time averages. To ensure that a random process is ergodic in an appropriate sense, a stationarity property (in either the strict- or wide-sense) is required in order to guarantee the convergence of time-averages. Several ergodicity results are presented. For stationary irreducible aperiodic finite Markov chains, it is shown that the time averages of a function of several consecutive states converge almost surely to the ensemble average of this function. This result plays a key role in the implementation of Markov chain Monte Carlo methods. For continuous-valued random processes, conditions ensuring ergodicity of the mean, autocorrelation, or CDF are presented. Note that as the class of statistics for which ergodicity is required becomes larger, ergodicity conditions become stricter. As a general rule of thumb, ergodicity is ensured when certain mixing conditions are satisfied. These conditions guarantee that events expressed in terms of random variables separated in time become asymptotically independent at their time separation increases. Intuitively, this means that by waiting sufficiently long enough, samples of a mixing random process can be treated as if they were independent, and thus subject to the weak and strong laws of large numbers. Chapter 10 studies scalar Markov diffusions and the rules of Ito calculus. Like the Wiener process, Markov diffusions have sample paths which are continuous almost surely, but nowhere differentiable. However, they are far more general than the Wiener process in the sense that they are not homogeneous in time and space. They are described by two functions, the drift and diffusion parameters, which, together with the initial distribution, are sufficient to completely characterize the process. It is shown that the PDF of the process obeys a forward propagating hyperbolic equation called the Fokker–Planck equation. Likewise, the expected value of a future functional of the process conditioned on the present position satisfies a backward propagating equation called the backwards Kolmogorov equation. Another important feature of Markov diffusions is that they can be described as the solution of stochastic differential equations. In integrated form, the solution can be expressed in terms of an Ito integral which can be viewed as an extension of the Wiener integral to the case where the integrand is random, but nonanticipative. Since Markov diffusions, like the Wiener process, have the feature that their squared increments vary like dt, they require a modification to the standard rules of calculus, such as the chain rule of differentiation, which is referred to as Ito calculus. Then, under appropriate Lipschitz conditions, it is shown that the familiar Picard iteration technique of ordinary differential equations can be used to establish the existence of solutions to Ito stochastic differential equations (SDEs). From this perspective, SDEs can be viewed as an efficient mechanism for constructing Markov diffusions with certain desired properties.

The final three chapters of the book focus on selected application areas. Chapter 11 examines noncausal and causal Wiener (optimal in the linear least-squares sense) filtering of WSS random processes. It is shown that the problem of finding the linear least-squares estimate of a random variable given a set, possibly infinite, of observations can be formulated as projection problem in the Hilbert space formed by random variables with finite second moments. From this perspective, the optimum linear least-squares estimate is characterized by an orthonormality property which is used to construct all optimum filtering estimates considered in this chapter. The problem of finding the optimum noncausal filter for estimating a process of interest given observations which are jointly WSS with it is first considered and is used to solve a noisy deconvolution problem. The causal Wiener filtering problem is then examined and solved by introducing the minimum phase spectral factorization of the observations process and interpreting it in terms of the corresponding innovations process. The optimum causal Wiener filter is applied to several special cases, such as the k-step ahead prediction problem and the fixed lag smoothing problem. In particular it is shown that as the fixed lag k of the smoothing estimate tends to infinity, the optimal noncausal smoother is recovered.

Chapter 12 studies the properties of the quantization noise produced by analog-to-digital converters which, in addition to sampling a random input signal, approximates it with a finite number of discrete levels. Building on some empirical observations of Bennett, Widrow proposed in the 1950s some sufficient conditions for the quantization noise to be uniformly distributed and white. By assuming that the input signal PDF has a finite support, and using a Fourier series representation of the PDF, it is shown that quantization is equivalent to performing a downsampling operation in the Fourier coefficient domain. This insight is used to derive necessary and sufficient conditions which ensure that the quantization noise is uniformly distributed and white. It is also shown that when the Fourier series representing the input PDF is summable, the quantization operation is a contraction for which the uniform distribution is a fixed point. Thus if the quantizer is implemented in stages, as is the case for a pipelined ADC, the quantization residuals converge to a uniform distribution. An unfortunate aspect of quantization is that the quantization noise is not independent of the input signal, which results in unsatisfactory perceptual artifacts in audio or visual signal processing systems. This limitation can be overcome by adding a small amplitude dither signal to the input and removing it from the quantized output, and it is shown that when the dithered PDF satisfies certain conditions, the quantization error becomes independent of the input signal.

Finally, Chap. 13 presents a model of phase noise in autonomous oscillators developed approximately 20 years ago by Kaertner [31] and Demir et al. [32]. A key feature of this model is that it accounts properly for the fact that under the effect of noise, the oscillator spectrum becomes Lorentzian in the vicinity of each harmonic. The key insight used in this analysis is that when a stable oscillator is subjected to random perturbations, the perturbation component colinear to the tangent to the oscillator orbit is unperturbed, whereas the components along the isochron surface are attenuated, since the oscillator is stable. After reviewing the Floquet theory of periodic oscillators, and introducing the concept of isochron surface, it is shown that when the noise is small and isochron surfaces are nearly flat at points on the oscillator trajectory, the phase noise is a scalar diffusion process. Furthermore on a slow time scale (compared to the rate of oscillation of the oscillator), the phase noise can be modeled as a scaled Wiener process. This model is used to derive the Lorentzian shape of the oscillator PSD in the vicinity of each of its harmonics.

1.2   Course Management and Curriculum Organization

This textbook represents an expansion of lecture notes which were written for a quarter-length course. In this context, with some judicious skipping of nonessential material, it is possible to cover the background material of Chaps. 2 and 3 in 2 1/2 to 3 weeks, and then cover the core topics extending from Chaps. 4 to 8 in 6 weeks, which leaves about a week for covering either the ergodicity material of Chap. 9 or Wiener filtering in Chap. 11, depending on the interests of the audience and of the instructor. For a semester long course, it would be possible to include the presentation of Markov diffusions of Chap. 10 and their application to the study of phase noise in oscillators described in Chap. 13. The logical dependence of the different book chapters is shown in Fig. 1.1.

To provide some perspective on the choice of applications material included in this textbook, it should be mentioned that at UC Davis, the random processes course serves as prerequisite to three quarter-length courses covering optimal and adaptive filtering, signal detection and parameter estimation, and digital communications, respectively. The optimal and adaptive filtering course assumes some familiarity with state space models and digital signal processing. It starts with FIR and IIR Wiener filtering, followed by Kalman filtering, including filter convergence, and state-space smoothing. It then proceeds to adaptive filtering algorithms and their convergence using the ODE method. This course can therefore be viewed as a highly compressed version of material appearing in [33, 34]. The course on signal detection is based on [35], but [36] contains equivalent material. It covers binary and M-ary hypothesis testing, parameter estimation, composite hypothesis testing, and the detection of signals, possibly with unknown parameters in white or colored noise. Material on sequential hypothesis testing, the asymptotics of tests, and robust detection is also discussed in [35] and can be included at the instructor's discretion. Finally, the course on digital communications is relatively standard and its material is covered in [37, 38] or [39]. The assumed prerequisites for the above-mentioned courses are depicted in Fig. 1.2.

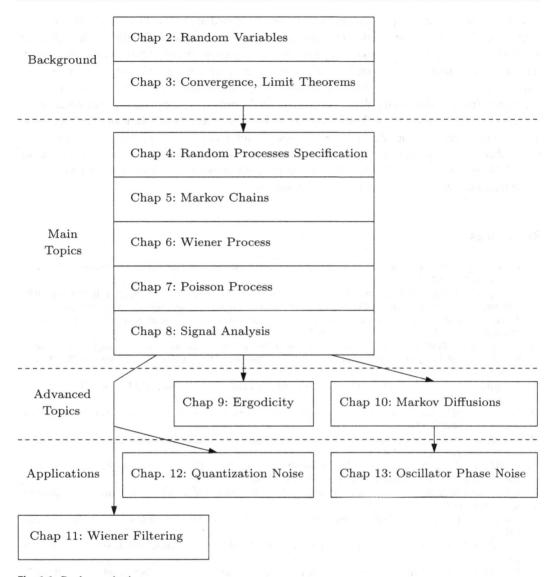

Fig. 1.1 Book organization

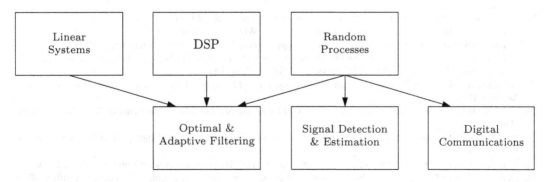

Fig. 1.2 Curriculum organization

In this context, while random processes textbooks addressed at electrical engineers often include brief discussions of Kalman filtering or signal detection and parameter estimation, these topics are deferred to full-length courses which can cover these subjects in greater depth. On the other hand, since the early chapters of this book include studies of thermal noise and shot noise, as well as brief discussions of $1/f$ noise and switching noise in problem sections, the inclusion of chapters on quantization noise and oscillator phase noise completes the overall picture of noise issues which are of interest in modern circuit design. In this respect, while some observers may be under the impression that random processes in electrical engineering systems form a mature, if not stale, object of inquiry, it should be pointed out that the oscillator phase noise analysis proposed by Kaertner [31], and Demir et al. [32], and further extended by other researchers, which is described in Chap. 13 represents a major recent advance which deserves wider recognition.

References

1. W. Schottky, "Über spontane stromschwankungen in verschiedenen elektrizitätsleitern," *Annalen der Physik*, vol. 57, pp. 541–567, 1918.
2. H. Nyquist, "Thermal agitation of electric charge in conductors," *Physical Review*, vol. 32, pp. 110–113, July 1928.
3. J. B. Johnson, "Thermal agitation of electricity in conductors," *Physical Review*, vol. 32, pp. 97–109, July 1928.
4. A. Van Der Ziel, "Unified presentation of $1/f$ noise in electronic devices: fundamental $1/f$ noise sources," *Proceedings of the IEEE*, vol. 76, pp. 233–258, Mar. 1988.
5. A. Demir and P. Feldman, "Modeling and simulation of the interference due to digital switching in mixed-signal ICs," in *Proc. of the 1999 IEEE/ACM Internat. Conf. on Computer-aided Design (ICCAD '99)*, pp. 70–75, Nov. 1999.
6. P. Heydari, "Analysis of the PLL jitter due to power/ground and substrate noise," *IEEE Trans. Circuits Syst. I*, vol. 51, Dec. 2004.
7. A. Hajimiri and T. H. Lee, "A general theory of phase noise in electrical oscillators," *IEEE J. Solid State Circuits*, vol. 33, pp. 179–194, 1998.
8. E. Pankratz and E. Sanchez-Sinencio, "Survey of integrated-circuit oscillator phase-noise analysis," *Internat. J. Circuit Theory and Appl.*, vol. 42, pp. 871–938, 2014.
9. B. Widrow and I. Kollar, *Quantization Noise–Roundoff Error in Digital Computation, Signal Processing, Control, and Communications*. Cambridge, U. K: Cambridge University Press, 2008.
10. S. P. Lipshitz, R. A. Wannamaker, and I. Vanderkooy, "Quantization and dither: a theoretical survey," *J. Audio Eng. Soc.*, vol. 5, pp. 355–375, May 1992.
11. S. O. Rice, "Mathematical analysis of random noise, parts I and II," *Bell System Tech J.*, vol. 23, pp. 282–332, July 1944.
12. S. O. Rice, "Mathematical analysis of random noise, parts III and IV," *Bell System Tech J.*, vol. 24, pp. 46–156, Jan. 1945.
13. S. O. Rice, "Statistical properties of a sine wave plus random noise," *Bell System Tech J.*, vol. 27, pp. 109–157, Jan. 1948.
14. S. O. Rice, "Noise in FM receivers," in *Time Series Analysis* (M. Rosenblatt, ed.), pp. 395–422, New York: John Wiley, 1963.
15. C. E. Shannon, "Communication in the presence of noise," *Proc. of the IRE*, vol. 37, pp. 10–21, Jan. 1949.
16. L. Kleinrock, *Queueing Systems. Vol I: Theory*. New York: Wiley-Interscience, 1975.
17. L. Kleinrock, *Queueing Systems. Vol 2: Computer Applications*. New York: Wiley-Interscience, 1975.
18. J. W. B. Davenport and W. L. Root, *An Introduction to the Theory of Random Signals and Noise*. New York, NY: McGraw-Hill, 1958. Reprinted by IEEE Press, New York, NY, in 1987.
19. A. Papoulis and S. U. Pillai, *Probability, Random Variables and Stochastic Processes, 4th edition*. New York, NY: McGraw-Hill, 2001.
20. C. W. Helstrom, *Probability and Stochastic Processes for Engineers, 2nd edition*. Englewood Cliffs, NJ: Prentice-Hall, 1991.
21. R. G. Gallager, *Stochastic Processes: Theory for Applications*. Cambridge, United Kingdom: Cambridge University Press, 2014.
22. B. Hajek, *Random Processes for Engineers*. Cambridge, United Kingdom: Cambridge University Press, 2015.
23. D. P. Bertsekas and J. N. Tsitsiklis, *Introduction to Probability, 2nd edition*. Belmont, MA: Athena Scientific, 2008.

24. R. D. Yates and D. J. Goodman, *Probability and Stochastic Processes: A Friendly Introduction for Electrical and Computer Engineers, 3rd edition.* Hoboken, NJ: J. Wiley & Sons, 2014.
25. J. Walrand, *Probability in Electrical Engineering and Computer Science– An Application-Driven Course.* Quoi?, Feb. 2014.
26. B. Hajek, "Probability with engineering applications." Notes for Course ECE 313 at University of Illinois at Urbana-Champaign, June 2015.
27. B. Mandelbrot, "The variation of certain speculative prices," *J. of Business*, vol. 36, pp. 394–419, Oct. 1963.
28. L. Bachelier, *Théorie de la Spéculation.* Paris, France: Gauthier-Villars, 1900.
29. M. Loève, *Probability Theory, 3rd edition.* Princeton, NJ: Van Nostrand, 1963.
30. H. Cramér and M. R. Leadbetter, *Stationary and Related Stochastic Processes: Sample Function Properties and Their Applications.* New York, NY: John Wiley & Sons, 1967. Reprinted by Dover Publications, Mineola NY, 2004.
31. F. X. Kaertner, "Analysis of white and $f^{-\alpha}$ noise in oscillators," *Int. J. Circuit theory and Applications*, vol. 18, pp. 485–519, 1990.
32. A. Demir, A. Mehrotra, and J. Roychowdhury, "Phase noise in oscillators: a unifying theory and numerical methods for characterization," *IEEE Trans. Circuits Syst. I*, pp. 655–674, May 2000.
33. T. Kailath, A. H. Sayed, and B. Hassibi, *Linear Estimation.* Upper Saddle River, NJ: Prentice-Hall, 2000.
34. A. H. Sayed, *Adaptive Filters.* Hoboken, NJ: J. Wiley– IEEE Press, 2008.
35. B. C. Levy, *Principles of Signal Detection and Parameter Estimation.* New York, NY: Springer Verlag, 2008.
36. H. V. Poor, *An Introduction to Signal Detection and Estimation, 2nd edition.* New York: Springer Verlag, 1994.
37. G. A. Pavliotis and A. M. Stuart, *Multiscale Methods–Averaging and Homogenization.* New York: Springer, 2008.
38. R. G. Gallager, *Principles of Digital Communication.* Cambridge, United Kingdom: Cambridge University Press, 2008.
39. U. Madhow, *Fundamentals of Digital Communication.* Cambridge, United Kingdom: Cambridge University Press, 2008.

Part I

Background

Probability and Random Variables

2

2.1 Introduction

This chapter reviews the concepts of probability space, random variable, jointly distributed random variables, and random vectors, as well as transformations of random variables and vectors, expectations, moments, conditional expectations, and characteristic and moment generating functions. The presentation is fast paced and is not intended as a first-time introduction to the above topics. Readers are referred to [1–4] for deliberately paced undergraduate level presentations of probability and statistics addressed at electrical engineers and computer scientists.

2.2 Probability Space

A probability space is a triple $(\Omega, \mathcal{F}, P)$ formed by a set of outcomes Ω, a set of events $\mathcal{F}$, and a probability law P. The set of outcomes Ω, which is also called the sample space, is formed by all the possible outcomes ω of the experiment we conduct. For example, if we roll a single die, since a die has six faces marked by a number of dots ranging between 1 and 6, the set representing all possible outcomes of the roll of a die is $\Omega = \{1, 2, 3, 4, 5, 6\}$. Similarly, if we exclude the possibility that a coin might land on its side, the toss of a coin produces two outcomes, heads or tails, so $\Omega = \{H, T\}$. In many situations, the set of outcomes is not finite, and may not even be countable. If, for example, we measure an analog voltage between $-1\,\mathrm{V}$ and $1\,\mathrm{V}$ with an infinite precision voltmeter, in this case each outcome ω is a real number in interval $\Omega = [-1, 1]$ corresponding to the measured voltage value.

The set of events $\mathcal{F}$ is formed by a family of subsets of Ω, to which a probability can be assigned. The set of events $\mathcal{F}$ must satisfy three axioms.

(i) The entire sample space Ω must belong to $\mathcal{F}$.
(ii) If A belongs to $\mathcal{F}$, then its complement A^c must belong to $\mathcal{F}$. We recall that the complement A^c of set A is defined as

$$A^c = \{\omega \in \Omega \,:\, \omega \notin A\}\,.$$

In other words A^c is formed by all outcomes of Ω not in A. Together A and A^c form a nonoverlapping partition of Ω, since

© Springer Nature Switzerland AG 2020
B. C. Levy, *Random Processes with Applications to Circuits and Communications*,
https://doi.org/10.1007/978-3-030-22297-0_2

$$\Omega = A \cup A^c \text{ and } A \cap A^c = \emptyset,$$

where $\emptyset$ denotes the empty set.

(iii) If A_1 and A_2 are two events of $\mathcal{F}$, their union $A_1 \cup A_1$ must be in $\mathcal{F}$. More generally, if we consider an infinite sequence $\{A_k, k \geq 1\}$ of events of $\mathcal{F}$, their union $\bigcup_{k=1}^{\infty} A_k$ must belong to $\mathcal{F}$. Recall that ω belongs to $\cup_{k=1}^{\infty} A_k$ if ω belongs to at least one of the A_ks.

The above axioms have several important implications. First, by Axioms (i) and (ii), since the complement of the whole set Ω is the empty set $\emptyset$, $\emptyset$ must belong to $\mathcal{F}$.

Next, recall De Morgan's laws of set theory:

$$(A_1 \cap A_2)^c = A_1^c \cup A_2^c \tag{2.1}$$

$$(A_1 \cup A_2)^c = A_1^c \cap A_2^c. \tag{2.2}$$

These laws, in combination with Axioms (ii) and (iii), imply that if A_1 and A_2 belong to $\mathcal{F}$, their intersection $A_1 \cap A_2$ must be in $\mathcal{F}$. More generally, if $\{A_k, k \geq 1\}$ is an infinite sequence of events of $\mathcal{F}$, their intersection $\cap_{k=1}^{\infty} A_k$ must be in $\mathcal{F}$. Accordingly, we conclude that the set $\mathcal{F}$ of events is closed under all elementary set operations, such as union, intersection, and complementation.

For example, if we consider the sample space $\Omega = \{1, 2, 3, 4, 5, 6\}$ corresponding to the roll of a die, a set of events meeting all the above axioms is given by

$$\mathcal{F} = \{\Omega, \emptyset, \{1, 2, 3, 4, 5\}, \{6\}, \{1, 2\}, \{3, 4, 5, 6\}, \{3, 4, 5\}, \{1, 2, 6\}\}.$$

This illustrates the fact that not all possible subsets of Ω need to be contained in $\mathcal{F}$. For example, all single outcomes, except $\{6\}$, are not contained in $\mathcal{F}$.

The third element of a probability space is the probability law P, which assigns to each event A of $\mathcal{F}$ a probability $P(A)$. This law must satisfy three axioms.

1) $0 \leq P(A)$ for each $A \in \mathcal{F}$.
2) $P(\Omega) = 1$.
3) If A_1 and A_2 are two events of $\mathcal{F}$ which are disjoint, in the sense that $A_1 \cap A_2 = \emptyset$, then

$$P(A_1 \cup A_2) = P(A_1) + P(A_2). \tag{2.3}$$

More generally, if we consider a sequence $\{A_k, k \geq 1\}$ of disjoint events ($A_k \cap A_\ell = \emptyset$ for any $k \neq \ell$) of $\mathcal{F}$, then

$$P(\cup_{k=1}^{\infty} A_k) = \sum_{k=1}^{\infty} P(A_k). \tag{2.4}$$

The above axioms have a number of direct consequences. First, since $A \cup A^c = \Omega$ forms a partition of the sample space Ω, by Axioms 2 and 3 we deduce that

$$P(A) + P(A^c) = P(\Omega) = 1,$$

so $P(A^c) = 1 - P(A)$. Also, since $P(A^c)$ is necessarily nonnegative (Axiom 1), we conclude that for any set A, $0 \leq P(A) \leq 1$, i.e., the probability of an event is necessarily between 0 and 1.

Next, since the identity (2.3) defines the probability of the union $A_1 \bigcup A_2$ of events A_1 and A_2 only when the events are disjoint, we consider the case when A_1 and A_2 are not disjoint. Let $B = A_1 \cap A_2$ denote the intersection of A_1 and A_2, and let

$$C_1 = A_1 - B \ , \ \ C_2 = A_2 - B$$

denote the complements of B inside A_1 and inside A_2, respectively. In this case,

$$A_1 = C_1 \cup B \ , \ \ A_2 = C_2 \cup B \ , \ \ A_1 \cup A_2 = C_1 \cup B \cup C_2$$

form three nonoverlapping partitions of A_1, A_2, and $A_1 \cup A_2$, respectively. Therefore by Axiom 3

$$P(A_1) = P(C_1) + P(B) \tag{2.5}$$

$$P(A_2) = P(C_2) + P(B) \tag{2.6}$$

$$P(A_1 \cup A_2) = P(C_1) + P(B) + P(C_2) \,, \tag{2.7}$$

and by using (2.5) and (2.6) to eliminate $P(C_1)$ and $P(C_2)$ in (2.7), we find

$$P(A_1 \cup A_2) = P(A_1) + P(A_2) - P(A_1 \cap A_2) \,. \tag{2.8}$$

An intuitive interpretation of (2.8) is that $P(A)$ is essentially a measure of the volume of set A. So, when measuring the volume of set $A_1 \cup A_2$ in terms of the volumes of its components A_1 and A_2, we must ensure not to double count the volume of the overlapping component $A_1 \cap A_2$, which must therefore be subtracted from the sum of the volumes of A_1 and A_2, as indicated by (2.8).

The counterpart of formula (2.8) for three nondisjoint events A_1, A_2 and A_3 is given by

$$P(A_1 \cup A_2 \cup A_3) = P(A_1) + P(A_2) + P(A_3) - P(A_1 \cap A_2) - P(A_1 \cap A_3)$$
$$-P(A_2 \cap A_3) + P(A_1 \cap A_2 \cap A_3) \,,$$

which indicates that keeping track of overlapping volumes of multiple sets gets progressively more complicated as more sets are introduced.

The axioms satisfied by P imply also that if we consider a monotone increasing sequence $A_1 \subset A_2 \cdots A_k \subset A_{k+1} \cdots$ of events of $\mathcal{F}$ with $A = \lim_{k \to \infty} A_k$, then

$$\lim_{k \to \infty} P(A_k) = P(A) \,. \tag{2.9}$$

Note that since A can be expressed as $A = \cup_{\ell=k}^{\infty} A_\ell$ for any k, it necessarily belongs to $\mathcal{F}$. By observing that if $D_k = A_{k+1} - A_k$ for $k \geq 1$, A_k and A admit the disjoint partitions

$$A_k = A_1 \cup_{\ell=1}^{k-1} D_\ell \ , \ \ A = A_1 \cup_{\ell=1}^{\infty} D_\ell$$

for $k \geq 2$, so

$$P(A_k) = P(A_1) + \sum_{\ell=1}^{k-1} P(D_\ell) \ , \ \ P(A) = P(A_1) + \sum_{\ell=1}^{\infty} P(D_\ell) \,,$$

and thus $\lim_{k \to \infty} P(A_k) = P(A)$.

Similarly, if we consider a monotone decreasing sequence $A_1 \supset A_2 \cdots A_k \supset A_{k+1} \cdots$ of events of $\mathcal{F}$, with $A = \lim_{k \to \infty} A_k = \cap_{\ell \geq k} A_\ell$ for any k, then

$$\lim_{k \to \infty} P(A_k) = P(A) \,. \tag{2.10}$$

This result can be proved by noting that the sequence $\{A_k^c, \, k \geq 1\}$ is monotone increasing whenever the sequence $\{A_k, \, k \geq 1\}$ is monotone decreasing, and then using the property (2.9) of sequence $\{A_k^c, \, k \geq 1\}$.

2.3   Conditional Probability

If we consider two events A and B of $\mathcal{F}$, the conditional probability of A given B (i.e., the probability of event A given that event B has taken place) is defined by

$$P(A|B) = \frac{P(A \cap B)}{P(B)} \,, \tag{2.11}$$

and the events A and B are *independent* if

$$P(A|B) = P(A) \,,$$

or equivalently

$$P(A \cap B) = P(A)P(B) \,. \tag{2.12}$$

Roughly speaking, A and B are independent if the knowledge that B has occurred does not affect the probability of A. To illustrate the concept of independence, consider three events $A =$ it rains in Davis in the afternoon, $B =$ it rains in Woodland in the afternoon (Davis and Woodland are two Northern California cities located about 10 miles apart), and $C =$ the stock market goes up today. Assuming that we are considering a day in December, a rough guess would be $P(A) = 0.15$, but $P(A|B) = 0.95$. Because of the proximity of Davis and Woodland, it is safe to assume that if it rains in one of these two cities, it rains in the other. On the other hand, since there is no relation between stock market behavior and weather in California, it is reasonable to guess that $P(A|C) = P(A) = 0.15$. In other words, A and C are independent, but A and B are definitively not independent, and a person leaving from Woodland in the rain would be well advised to take an umbrella when traveling to Davis.

More generally, if we consider events $\{A_k, \, k \in K\}$, where index set K is either finite or infinite, the events A_k are mutually independent if for any $m \geq 2$ and any choice of $\ell_1 < \ell_2 \cdots < \ell_m$, we have

$$P(\cap_{j=1}^m A_{\ell_j}) = \prod_{j=1}^m P(A_{\ell_j}) \,.$$

From this definition, we see that pairwise independence of events in a collection of events does not imply mutual independence of all events.

By using the two symmetric expressions

$$P(A|B)P(B) = P(A \cap B) = P(B|A)P(A)$$

for $P(A \cap B)$, we obtain *Bayes' rule*

$$P(B|A) = \frac{P(A|B)P(B)}{P(A)}, \tag{2.13}$$

which can be used to reverse conditioning operations between two events.

A convenient trick to evaluate the probability of a complex event A of $\mathcal{F}$ consists of decomposing it into simple disjoint sub-events whose probabilities can be evaluated easily. Specifically, consider a partition

$$\Omega = \cup_{k \in K} B_k$$

of the sample space where the events $B_k \in \mathcal{F}$ are disjoint, i.e., $B_k \cap B_\ell = \emptyset$ for $k \neq \ell$. Then any event A of $\mathcal{F}$ can be decomposed as

$$A = A \cap \Omega = \cup_{k \in K} A \cap B_k \,,$$

where the sets $A \cap B_k$ are disjoint, so

$$P(A) = \sum_{k \in K} P(A \cap B_k) = \sum_{k \in K} P(A|B_k)P(B_k) \,. \tag{2.14}$$

The identity (2.14) is called the *principle of total probability* since it can be used to evaluate the probability of event A by summing the probabilities of sub-events $A \cap B_k$.

Next, we introduce the limit superior and limit inferior of a sequence $\{A_k \,, k \geq 1\}$ of events of $\mathcal{F}$. They are defined as

$$\limsup_{k \to \infty} A_k = \cap_{\ell=1}^{\infty} \cup_{k=\ell}^{\infty} A_k \tag{2.15}$$

$$\liminf_{k \to \infty} A_k = \cup_{\ell=1}^{\infty} \cap_{k=\ell}^{\infty} A_k \,. \tag{2.16}$$

Note that

$$\liminf_{k \to \infty} A_k = \left(\limsup_{k \to \infty} A_k^c \right)^c \,. \tag{2.17}$$

An outcome ω belongs to $\limsup_{k \to \infty} A_k$ if and only if it belongs to infinitely many events of the sequence $\{A_k \,, k \geq 1\}$. Similarly, an outcome ω belongs to $\liminf_{k \to \infty} A_k$ if and only if it belongs to all A_k's past a certain index ℓ. To verify the first of the two above assertions, note that if ω belongs to infinitely many A_k's, it must belong to

$$B_\ell = \cup_{k=\ell}^{\infty} A_k \tag{2.18}$$

for all ℓ, and thus it belongs to

$$\cap_{\ell=1}^{\infty} B_\ell = \limsup_{k \to \infty} A_k \,.$$

The assertion concerning $\liminf_{k \to \infty} A_k$ follows from (2.17).

We have found therefore that

$$P(\limsup_{k \to \infty} A_k) = P(A_k \text{ i. o. }) \,,$$

where "i.o." stands for infinitely often. We are now in a position to establish the following result, which will prove useful in studying almost sure convergence in the next chapter.

Borel–Cantelli Lemma

a) If a sequence of events A_k with $k \geq 1$ of $\mathcal{F}$ is such that

$$\sum_{k=1}^{\infty} P(A_k) < \infty , \qquad (2.19)$$

then $P(A_k \text{ i. o.}) = 0$.

b) If a sequence of independent events A_k of $\mathcal{F}$ is such that

$$\sum_{k=1}^{\infty} P(A_k) = \infty , \qquad (2.20)$$

then $P(A_k \text{ i. o.}) = 1$.

Proof To derive a), note that if the event B_ℓ is defined as in (2.18), we have

$$P(B_\ell) \leq \sum_{\ell=k}^{\infty} P(A_k),$$

where the right-hand side of the above inequality tends to zero as $\ell \to \infty$ when (2.19) holds, so $P(B_\ell) \to 0$ as $\ell \to \infty$. Since the events B_ℓ form a monotone decreasing sequence with limit $B = \limsup_{k \to \infty} A_k$, the continuity property (2.10) implies

$$P(B) = \lim_{\ell \to \infty} P(B_\ell) = 0 .$$

To prove b), we consider

$$C_\ell = \cap_{k \geq \ell} A_k^c ,$$

and note that the sequence $\{C_\ell, \ell \geq 1\}$ is monotone increasing with limit

$$C = \cup_{\ell=1}^{\infty} C_\ell = \liminf_{k \to \infty} A_k^c .$$

The continuity property (2.9) implies therefore

$$P(\liminf_{k \to \infty} A_k^c) = \lim_{\ell \to \infty} P(C_\ell) .$$

By observing that the events A_k^c are independent if the events A_k are independent, we have

$$P(C_\ell) = \lim_{m \to \infty} P(\cap_{k=\ell}^{m} A_k^c)$$

$$= \lim_{m \to \infty} \prod_{k=\ell}^{m} P(A_k^c) = \lim_{m \to \infty} \prod_{k=\ell}^{m} (1 - P(A_k)) . \qquad (2.21)$$

Observing that $1 - x \leq \exp(-x)$ for all $x \geq 0$, we deduce that

$$\prod_{k=\ell}^{m}(1 - P(A_k)) \leq \exp\left(-\sum_{k=\ell}^{m} P(A_k)\right),$$

which tends to zero as m tends to infinity, since the sum (2.20) is infinite. The identity (2.21) implies therefore that $P(C_\ell) = 0$, so

$$0 = P(\liminf_{k\to\infty} A_k^c) = P((\limsup_{k\to\infty} A_k)^c),$$

which proves b).   □

Remark Since the second part of the Borel–Cantelli lemma requires the independence of events A_k, it is less convenient to use than the first part. This lemma is an example of zero-one law, whereby when a condition is satisfied (here the convergence or divergence of the series $\sum_{k=1}^{\infty} P(A_k)$), an event will occur either with zero probability or probability one. The Kolmogorov zero-one law [5, Thm. 8.1.2] governing tail events of sequences of independent random variables is of the same type.

2.4   Random Variables

A random variable X is a function from Ω to the real line $\mathbb{R}$ which associates with each outcome ω a real value $X(\omega)$, such that for every real number x

$$X^{-1}((-\infty, x]) = \{\omega : X(\omega) \leq x\}$$

belongs to $\mathcal{F}$. In equivalent words, a random variable X is a function from Ω to $\mathbb{R}$ which is measurable with respect to the Borel field of $\mathbb{R}$. The Borel field $\mathcal{B}$ is the set of subsets of $\mathbb{R}$ obtained by union, intersection, and complementation of sets of the form $(-\infty, x]$ for all x. Then X is measurable with respect to the Borel field if for any set B of $\mathcal{B}$, its pre-image $A = X^{-1}(B) = \{\omega : X(\omega) \in B\}$ belongs to $\mathcal{F}$.

2.4.1   Discrete Random Variables

A random variable X is *discrete-valued* if it takes only values x_k, $1 \leq k \leq K$, with k integer where K is either finite or infinite. Then the measurability condition of X requires that the sets $A_k = \{\omega : X(\omega) = x_k\}$ belong to $\mathcal{F}$, and X can be characterized entirely by its *probability mass function* (PMF)

$$p_k = P(A_k) = P(X = x_k) \tag{2.22}$$

for all k. Since

$$\Omega = \cup_{1 \leq k \leq K} A_k$$

forms a nonoverlapping partition of Ω (if outcome ω belonged to both A_k and A_ℓ with $k \neq \ell$, this would mean that $X(\omega)$ would take simultaneously the values x_k and x_ℓ, which is impossible), we have

$$1 = P(\Omega) = \sum_{k=1}^{K} p_k, \tag{2.23}$$

so that the total probability mass must add up to one. Together with the positivity of the p_ks, the constraint (2.23) is the only one that needs to be satisfied by a PMF. A few examples are considered below.

Bernoulli Random Variable If X takes the values 1 or 0 with probabilities

$$P(X = 1) = p \ , \quad P(X = 0) = q \ ,$$

where $p + q = 1$, X is Bernoulli. X can represent for example the success or failure of a data packet transmission by a user when other users use the same channel and can create a collision and thus a loss of data packet if they decide to transmit at the same time.

Binomial Random Variable Let X denote the number of successful transmissions out of n transmissions, when each transmission is independent from other ones, with probability of success p. Then X can take any integer value k between 0 and n, and since there are $\binom{n}{k}$ ways of selecting k successful transmission slots out of n possible slots, the PMF of X is given by the binomial distribution

$$P(X = k) = \binom{n}{k} p^k q^{n-k}$$

for $0 \leq k \leq n$. The binomial series expansion

$$[p + q]^n = \sum_{k=0}^{n} \binom{n}{k} p^k q^{n-k}$$

ensures that the total probability mass is one.

Geometric Random Variable If we consider again the problem of transmitting a data packet over a shared channel, suppose that an automatic repeat request (ARQ) protocol is employed, so that if a transmission fails, another one is automatically attempted, until ultimately a transmission is successful. Let X denote the total number of transmissions until a successful transmission occurs, so that $X = k$ with $k \geq 1$ if $k - 1$ transmissions failed but the k-th transmission was successful. In this case

$$p_k = P(X = k) = pq^{k-1}$$

for $k \geq 1$ is a geometric distribution (of Type 1 since k starts at 1) and the total probability mass is one since

$$\sum_{k=1}^{\infty} p_k = p \sum_{k'=0}^{\infty} q^{k'} = \frac{p}{1 - q} = 1 \ ,$$

where the second equality is obtained by setting $k' = k - 1$.

Poisson Random Variable The number of photons hitting a detector per second (or the number of packets arriving per second at a router in a network) can be modeled by an integer valued Poisson random variable X with PMF

$$p_k = P(X = k) = \frac{\lambda^k}{k!} \exp(-\lambda)$$

with $k = 0, 1, 2, \cdots$, where the parameter λ denotes the average arrival rate per second. The power series expansion

$$\exp(\lambda) = \sum_{k=0}^{\infty} \frac{\lambda^k}{k!}$$

which is valid for any λ ensures that the total probability mass is one.

2.4.2 Continuous Random Variables

When a random variable X takes continuous values, it is convenient to describe it in terms of this *cumulative distribution function* (CDF)

$$F_X(x) = P(X \le x).$$

The measurability of X with respect to the Borel field ensures the existence of this function.

CDF Properties The function $F_X(x)$ has the following properties.

i) $F_X(-\infty) = 0$ and $F_X(\infty) = 1$.
ii) $F_X(x)$ is monotone nondecreasing, since $F_X(x_1) \le F_X(x_2)$ whenever $x_1 < x_2$.
iii) $F_X(x)$ is right continuous, since for $\epsilon > 0$,

$$\lim_{\epsilon \to 0} F_X(x + \epsilon) = F_X(x)$$

for all x.
iv) Define the left limit of $F_X(x)$ as

$$F_X(x-) = \lim_{\epsilon \to 0} F_X(x - \epsilon)$$

for $\epsilon > 0$. Then X takes the discrete value x_0 with nonzero probability if and only if $F_X(x)$ is discontinuous at x_0, in which case

$$P(X = x_0) = F_X(x_0) - F_X(x_0-).$$

In other words, whenever a jump occurs in the CDF at, say, point x_0, the size of the jump is the probability that X equals x_0.

Proof The property i) is due to the fact that

$$\{\omega : X(\omega) \le -\infty\} = \emptyset , \quad \{\omega : X(\omega) \le \infty\} = \Omega,$$

since a random variable X takes finite real values (it cannot be equal to $\pm\infty$). Thus

$$F_X(-\infty) = P(\emptyset) = 0 \text{ and } F_X(\infty) = P(\Omega) = 1.$$

To prove ii), for $x_1 < x_2$ consider the pre-images

$$A_i = X^{-1}((-\infty, x_i]) = \{\omega : X(\omega) \le x_i)$$

for $i = 1, 2$. The set $A_1 \subset A_2$, and if we consider the event

$$C = X^{-1}((x_1, x_2]) = \{\omega : x_1 < X(\omega) \le x_2\},$$

C is the complement of A_1 inside A_2, i.e.,

$$A_2 = A_1 \cup C,$$

with $A_1 \cap C = \emptyset$. This implies

$$F_X(x_2) = P(A_2) = P(A_1) + P(C), \qquad (2.24)$$

where $P(C) \ge 0$ and $P(A_1) = F_X(x_1)$, so that $F_X(x_2) \ge F_X(x_1)$.

To prove iii) consider the events

$$B_n = X^{-1}((-\infty, x + 1/n]).$$

As n increases, these sets form a monotone decreasing sequence with limit

$$B = \cap_{n \ge m}^{\infty} B_n = X^{-1}((-\infty, x]).$$

Since $P(B_n) = F_X(x + 1/n)$ and $P(B) = F_X(x)$, the continuity property (2.10) implies

$$\lim_{n \to \infty} F_X(x + 1/n) = F_X(x),$$

so that $F_X(x)$ is right continuous.

To prove iv), consider the set

$$D = X^{-1}((-\infty, x_0)).$$

By observing that D is the limit of the monotone increasing sequence of sets

$$D_n = X^{-1}((-\infty, x_0 - 1/n]),$$

the continuity property (2.9) implies

$$F_X(x_0-) = \lim_{n \to \infty} F_X(x - 1/n) = \lim_{n \to \infty} P(D_n) = P(D) = P(X < x_0).$$

Since the event $E = \{\omega : X(\omega) = x_0\}$ is the complement of D inside $X^{-1}(-\infty, x_0])$ we conclude that

$$P(X = x_0) = F_X(x_0) - F_X(x_0-),$$

so that X has a nonzero probability of taking the values x_0 if and only if $F_X(x)$ exhibits a nonzero jump of size $F_X(x_0) - F_X(x_0-)$ at x_0. $\diamond$

A consequence of property iv) of CDFs is that if X is a discrete valued random variable which assumes the values $\{x_k, k \in K\}$ with PMF $p_k = P(X = x_k)$, it admits a staircase CDF of the form

$$F_X(x) = \sum_{1 \le k \le K} p_k u(x - x_k),$$

where

$$u(x) = \begin{cases} 1 & x \ge 0 \\ 0 & x < 0 \end{cases}$$

is the unit step function. The rises of the staircase function are the probabilities p_k of the discrete values x_k, and the runs represent the separation $x_{k+1} - x_k$ between successive values of X. For example, the CDF of a discrete valued random variable X uniformly distributed over $\{0, 1, 2, 3, 4\}$ with PMF

$$P(X = k) = 1/5 \text{ for } 0 \le k \le 4$$

is the uniform staircase function depicted in Fig. 2.1.

At this point, it is worth pointing out that the axioms of probability spaces and the measurability of X ensure that the CDF can be used to evaluate the probability that X belongs to an arbitrary set of the Borel field. For example, since

$$\Omega = X^{-1}((-\infty, x]) \cup X^{-1}((x, \infty))$$

forms a nonoverlapping partition of Ω, we have

$$1 = P(\Omega) = P(X \le x) + P(X > x),$$

so that

$$P(X > x) = 1 - F_X(x).$$

Similarly, if we consider an arbitrary interval (x_1, x_2) of $\mathbb{R}$, the identity (2.24) implies

$$P(x_1 < X \le x_2) = F_X(x_2) - F_X(x_1). \tag{2.25}$$

Fig. 2.1 CDF of a discrete valued random variable uniformly distributed over integers $\{0, 1, 2, 3, 4\}$

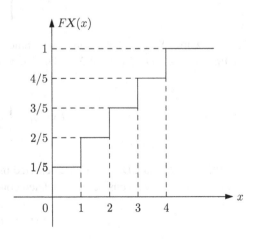

A subclass of random variables of particular interest is those for which the CDF $F_X(x)$ is *absolutely continuous*, in which case it can be represented as

$$F_X(x) = \int_{-\infty}^{x} f_X(u)du \tag{2.26}$$

for all real numbers x, where

$$f_X(x) = \frac{dF_X}{dx}(x)$$

is called the *probability density function* (PDF) of X. Since $F_X(x)$ is monotone nondecreasing and $f_X(x)$ represents its slope at point x, we deduce that the PDF $f_X(x)$ must be *nonnegative*, i.e., $f_X(x) \geq 0$. Also, setting $x = \infty$ inside (2.26), we find that

$$1 = F_X(\infty) = \int_{-\infty}^{\infty} f_X(u)du, \tag{2.27}$$

which expresses that the *total probability mass* of a PDF must be equal to one.

Combining (2.25) and (2.26), we find that for an arbitrary interval $(a, b]$

$$P(a < X \leq b) = \int_{a}^{b} f_X(u)du. \tag{2.28}$$

In particular, for an infinitesimal interval $(x, x + h]$, this ensures that

$$P(x < X \leq x + h) \approx f_X(x)h ,$$

so that intuitively, $f_X(x)$ can be interpreted as a measure of how likely the random variable X will be in the vicinity of x. In other words, the larger $f_X(x)$, the more likely X is to achieve the value x. To illustrate the above discussion, we consider several important classes of random variables.

Uniform Random Variable A random variable X is uniform over interval $[a, b]$ if its density

$$f_X(x) = \begin{cases} \dfrac{1}{b-a} & a \leq x \leq b \\ 0 & \text{otherwise} . \end{cases}$$

In other words X is as likely to take any value as another over interval $[a, b]$. Integrating the density, we find that the CDF $F_X(x)$ of X is the piecewise linear function

$$F_X(x) = \begin{cases} 0 & x < a \\ \dfrac{x-a}{b-a} & a \leq x \leq b \\ 1 & x > b . \end{cases}$$

The PDF $f_X(x)$ and CDF $F_X(x)$ are depicted in Fig. 2.2.

For example, if we consider a modulated communications signal

$$X(t) = a \cos(\omega_c t + \Theta)$$

Fig. 2.2 PDF and CDF of a uniformly distributed random variable over $[a, b]$

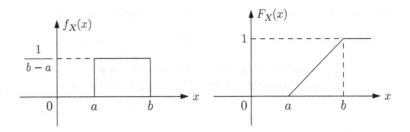

with amplitude A and carrier frequency ω_c, in the absence of synchronization information it is customary to assume that the phase Θ is random and uniformly distributed over $[0, 2\pi)$, since any phase value is as likely to occur as another.

Exponential Random Variable A random variable X is exponential with parameter $\lambda > 0$ if its CDF is given by

$$F_X(x) = (1 - \exp(-\lambda x))u(x),$$

where $u(x)$ is the unit step function. By differentiation, we find its PDF is given by

$$f_X(x) = \lambda \exp(-\lambda x)u(x).$$

Exponential distributions are commonly used to model the lifetime of elementary electrical components, such as capacitors, resistors, or incandescent lightbulbs. In this context, $1/\lambda$ corresponds to the average lifetime of a device.

Gaussian Random Variable A Gaussian/normal random variable X with mean/median m and standard deviation σ (or variance σ^2) has the exponential quadratic PDF

$$f_X(x) = \frac{1}{(2\pi)^{1/2}\sigma} \exp\left(-\frac{(x-m)^2}{2\sigma^2}\right). \tag{2.29}$$

It reaches its peak at $x = m$, and the parameter σ measures the width of the distribution about $x = m$. Specifically, for $x = m \pm \sigma$, we have

$$f_X(m \pm \sigma) = \exp(-1/2)f_X(m).$$

In other words, if we move away from m by σ on either side, the density decreases by a factor $\exp(-1/2)$. A random variable X with density (2.29) is usually said to be $N(m, \sigma^2)$ distributed. The CDF of X cannot be expressed in closed-form, but it can be evaluated rapidly by using two tabulated functions: the function

$$\Phi(x) = \frac{1}{(2\pi)^{1/2}} \int_{-\infty}^{x} exp(-u^2/2)du$$

corresponding to the CDF of a $N(0, 1)$ random variable, and its complement, also called the "Q function"

$$Q(x) = 1 - \Phi(x) = \frac{1}{(2\pi)^{1/2}} \int_{x}^{\infty} \exp(-u^2/2)du,$$

which is commonly used in signal detection applications [6,7]. Note that since

$$Q(-x) = 1 - Q(x)$$

for $x \geq 0$, the Q function needs only to be tabulated for positive values of x. In this context, the CDF of a $N(m, \sigma^2)$ distributed random variable can be expressed as

$$F_X(x) = \Phi((x - m)/\sigma) = 1 - Q((x - m)/\sigma) .$$

Cauchy Random Variable The PDFs considered thus far have the feature that they have no tail, like the uniform distribution which is nonzero only over a finite interval, or exponentially decaying tails like the exponential and Gaussian distributions. A Cauchy random variable X has a PDF of the form

$$f_X(x) = \frac{1}{\pi\sigma} \frac{1}{1 + \left(\frac{x - \mu}{\sigma}\right)^2},$$

which decays very slowly as x becomes large. In fact its $1/x^2$ decay rate is so slow that it admits no moment of any order. The location parameter μ is not the mean of the distribution, but its median. The scale parameter σ is used to adjust the width of the distribution. The corresponding CDF is given by

$$F_X(x) = \frac{1}{\pi} \arctan\left(\frac{x - \mu}{\sigma}\right) + \frac{1}{2} .$$

Median The above discussion makes reference to the median μ of a probability distribution. A median μ is defined as a real number such that

$$F_X(\mu) = P(X \leq \mu) \geq 1/2$$

and

$$1 - F_X(\mu-) = P(X \geq \mu) \geq 1/2 .$$

Thus, at least half of the probability mass must be located to the left of or at μ, and at least half must be located to the right of or at μ. When the distribution of X is absolutely continuous, the above definition can be rewritten as

$$F_X(\mu) = 1 - F_X(\mu) = \int_{-\infty}^{\mu} f_X(u)du = 1/2 .$$

Note that a median is not necessarily unique. for example, if we consider a random variable X uniformly distributed over the union of intervals $[-1, -1/2]$ and $[1/2, 1]$, so that

$$f_X(x) = \begin{cases} 1 & 1/2 \leq |x| \leq 1 \\ 0 & \text{otherwise}, \end{cases}$$

then any μ in interval $[-1/2, 1/2]$ is a median. When there is only one median, it is referred to as the median.

Distributions Conditioned on an Event It is sometimes required to compute the CDF or PDF of a random variable X conditioned on an event E depending on X. To do so, one needs only to apply the conditional probability of two events defined in (2.11) to the CDF and PDF definition, respectively. Specifically, the conditional CDF $F_X(x|E)$ is defined as

$$F_X(x|E) = \frac{P(\{X \leq x\} \cap E)}{P(E)},$$

and the conditional PDF

$$f_X(x|E) = \frac{d}{dx} F_X(x|E)$$

satisfies

$$f_X(x|E)h \approx \frac{P(\{x < X \leq x + h\} \cap E)}{P(E)}$$

for small h.

Example 2.1 Consider the residual lifetime $Z = X - x$ of an exponential random variable with parameter λ conditioned on the event $E = \{X > x\}$ (the lifetime is greater than x). The probability of event $\{Z > z\}$ with $z \geq 0$ conditioned on E is

$$P(Z > z|E) = P(X > x + z|X > x) = \frac{P(\{X > x + z\} \cap \{X > x\})}{P(X > x)}$$

$$= \frac{P(X > x + z)}{P(X > x)} = \frac{\exp(-\lambda(z + x))}{\exp(-\lambda x)} = \exp(-\lambda z),$$

where the first equality on the second line is due to the fact that the event $\{X > x + z\}$ is contained in E. Thus the CDF of Z conditioned on E is given by

$$F_Z(z|E) = 1 - P(Z > z|E) = 1 - \exp(-\lambda z),$$

so that Z conditioned on E is also exponential with parameter λ. This result is known as the *memoryless property* of the exponential distribution. It has important consequences. We have seen earlier that lightbulbs have an exponential lifetime. Suppose that after living a few years in your house, you wonder if your working lightbulbs should be replaced. The answer is no! Any working lightbulb is functionally equivalent to a brand new one, no matter how long ago it was installed.

2.5   Transformation of Random Variables

Given a random variable X with CDF $F_X(x)$ and, if it is absolutely continuous, PDF $f_X(x)$, a common problem is to find the CDF or PDF of a transformed random variable $Y = g(X)$.

2.5.1   One-to-One Transformation

To solve this problem, we first consider the case where g is *one-to-one*. In this case, g admits an inverse function $X = h(Y)$. It is also useful to observe that if g is continuous over an interval (a, b), it is necessarily strictly monotone increasing or decreasing over this interval, and accordingly its inverse h is also continuous and strictly monotone increasing or decreasing. Finally, if g is differentiable over interval (a, b), it inverse is differentiable over the image of (a, b) under g.

Suppose now that g is one-to-one and continuous over $\mathbb{R}$. Then both g and h are either monotone increasing or monotone decreasing over the entire real line. When g is monotone increasing, we have

$$F_Y(y) = P(Y = g(X) \le y) = P(X \le h(y)) = F_X(h(y)) , \tag{2.30}$$

and when g is monotone decreasing, we have

$$F_Y(y) = P(Y = g(X) \le y) = P(X \ge h(y)) = 1 - F_X(h(y)) . \tag{2.31}$$

Next, assume that g is one-to-one and $F_X(x)$ is absolutely continuous. Assume also that g is differentiable over an infinitesimal interval $(x, x + dx]$. In this case g and its inverse are either monotone increasing or monotone decreasing over this interval and h is differentiable over the image $(y, y + dy] = g((x, x + dx])$ of interval $(x, x + dx]$. In the monotone increasing case, we have

$$f_Y(y)dy = P(Y = g(X) \in (y, y + dy]) = P(X \in (x, x + dx])$$
$$= f_X(h(y))\frac{dh}{dy}dy, \tag{2.32}$$

where the last equality is due to the fact that $x = h(y)$ and

$$dx = \frac{dh}{dy}dy .$$

Comparing the left and right sides of (2.32), we find

$$f_Y(y) = \frac{dh}{dy} f_X(h(y)) . \tag{2.33}$$

In the case when g is monotone decreasing over $(x, x + dx]$, the only change in the analysis appearing in (2.32) is that

$$dx = -\frac{dh}{dy}dy .$$

Accordingly, the expression (2.33) can be replaced by the more general expression

$$f_Y(y) = J(y)f_X(h(y)), \tag{2.34}$$

where

$$J(y) = \left| \frac{dh}{dy} \right| = \left| \frac{1}{\frac{dg}{dx}(h(y))} \right| \tag{2.35}$$

denotes the Jacobian of the transformation $Y = g(X)$, and where the second expression is obtained by using the chain rule of differentiation in identity

$$g(h(y)) = y .$$

In this respect, it does not matter which one of the two expressions is used to evaluate the Jacobian in (2.34). The differentiation which is easiest to perform should always be selected.

As final verification, note that if g is differentiable over the entire real line, then it is monotone increasing or decreasing over $\mathbb{R}$, and either expression (2.30) or (2.31) holds. Differentiating (2.30) and (2.31) yields (2.34), so all transformation rules described up to this point are consistent.

Example 2.2 Suppose X is a Cauchy random variable with $\mu = 0$ and scale parameter σ, so its PDF

$$f_X(x) = \frac{1}{\pi\sigma(1 + (x/\sigma)^2)}.$$

Then consider the transformation $Y = g(X) = 1/X$. It is one-to-one and differentiable over $\mathbb{R}$ except at 0. The inverse transformation is $X = h(Y) = 1/Y$ and since $dh/dy = -1/y^2$, the Jacobian of the transformation is $J(y) = 1/y^2$. The PDF of Y is therefore given by

$$f_Y(y) = J(y)f_X(h(y)) = \frac{1}{y^2}\frac{1}{\pi\sigma(1 + (1/\sigma y)^2)}$$

$$= \frac{\sigma}{\pi}\frac{1}{1 + (\sigma y)^2},$$

which is Cauchy with location parameter $\mu = 0$ and scale parameter $1/\sigma$.

Simulation of Continuous Random Variables One important issue arising in simulating probabilistic models is how to generate random variables with a specific distribution. MATLAB can only generate random variables uniformly distributed over [0, 1] and $N(0, 1)$ random variables with the rand and randn commands, respectively. So we need to figure out how to generate random variables with other distributions. Since the CDF of Gaussian random variables does not admit a closed form, it is more convenient in practice to start from a uniformly distributed random variable. The *transformation method* uses a one-to-one transformation $Y = g(X)$ to generate a random variable Y with CDF $F_Y(y)$ from a uniformly distributed random variable X over [0, 1]. The random variable Y is required to have a CDF $F_Y(y)$ which is monotone increasing over an interval (a, b) and such that $F_Y(a) = 0$ and $F_Y(b) = 1$. This condition rules out discrete valued random variables, since their PDF is piecewise constant. In this case, the inverse function F_Y^{-1} is uniquely defined and has image (a, b). The CDF inversion method generates the random variable $Y = F_Y^{-1}(X)$ as shown in Fig. 2.3. To verify that Y has CDF $F_Y(y)$, note that

$$P(Y \leq y) = P(F_Y^{-1}(X) \leq y) = P(X \leq F_Y(y)) = F_X(F_Y(y)), \tag{2.36}$$

where $0 \leq F_Y(y) \leq 1$. But since the CDF of uniform random variable X is such that $F_X(x) = x$ for $0 \leq x \leq 1$, the identity (2.36) yields

Fig. 2.3 Random variable generation by the CDF inversion method

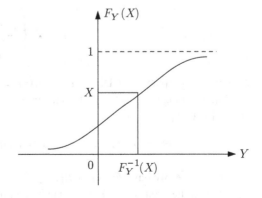

$$P(Y \leq y) = F_Y(y)$$

as desired.

Exponential Random Variable Generation To generate an exponential random variable Y with parameter λ, we invert

$$X = F_Y(Y) = 1 - \exp(-\lambda Y),$$

which yields

$$Y = F_Y^{-1}(X) = \frac{1}{\lambda} \ln \left(\frac{1}{1 - X} \right).$$

Cauchy Random Variable Generation Similarly, to generate a Cauchy random variable with parameters (μ, σ), we invert

$$X = F_Y(Y) = \frac{1}{\pi} \arctan \left(\frac{Y - \mu}{\sigma} \right) + \frac{1}{2},$$

which yields

$$Y = \mu + \sigma \tan(\pi(X - 1/2)).$$

2.5.2   Several-to-One Transformation

Next, we consider the case where the transformation $Y = g(X)$ is not one-to-one, but *p-to-one* with $p \geq 2$. When p remains constant over the entire range of g, we can construct p inverse functions $X = h_i(Y)$, with $1 \leq i \leq p$ which can be used to compute the CDF or PDF of Y. To illustrate how this works, consider the following example.

Example 2.3 Let Θ be a uniformly distributed random variable over $[-\pi, \pi)$, with density

$$f_\Theta(\theta) = \begin{cases} \dfrac{1}{2\pi} & -\pi \leq \theta < \pi \\ 0 & \text{otherwise}, \end{cases}$$

and CDF

$$F_\Theta(\theta) = \begin{cases} 0 & \theta < -\pi \\ \dfrac{\theta + \pi}{2\pi} & -\pi \leq \theta \leq \pi \\ 1 & \theta > \pi. \end{cases}$$

We seek to compute the CDF of $Y = g(\Theta) = A \cos(\Theta)$, whose graph over $[-\pi, \pi)$ is shown in Fig. 2.4. The range of g is $[-A, A]$, and since $g(\Theta)$ is even, it admits two inverse functions for values of $Y \in [-A, A]$:

$$\Theta_+ = h_+(Y) = \arccos(Y/A) \text{ and } \Theta_- = h_-(Y) = -\arccos(Y/A).$$

For $y < -A$, we have $F_Y(y) = P[Y \leq y] = 0$, since Y cannot be less than $-A$. For $-A \leq y \leq A$, we note that the event $E = \{Y \leq y\}$ can be composed as

Fig. 2.4 Function $g(\Theta) = A\cos(\Theta)$ and its two inverses

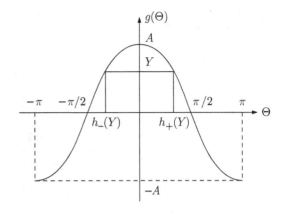

$$E = E_- \cup E_+,$$

where

$$E_- = \{\Theta : -\pi \le \Theta \le h_-(y)\}$$
$$E_+ = \{\Theta : h_+(y) \le \Theta < \pi\}.$$

Since the events E_+ and E_- are disjoint, we conclude that

$$F_Y(y) = P(E) = P(E_+) + P(E_-) = F_\Theta(h_-(y)) + 1 - F_\Theta(h_+(y))$$
$$= \frac{1}{2\pi}(\pi - \arccos(y/A)) + 1 - \frac{1}{2\pi}(\pi + \arccos(y/A))$$
$$= 1 - \frac{1}{\pi}\arccos(y/A) \tag{2.37}$$

for $-A \le y \le A$, and $F_Y(y) = P(Y \le y) = 1$ for $y \ge A$. The CDF is plotted in Fig. 2.5 for $A = 1$. Differentiating $F_Y(y)$ yields

$$f_Y(y) = -\frac{1}{\pi}\frac{d}{dy}\arccos(y/A)$$
$$= \frac{1}{A\pi}\frac{1}{(1-(y/A)^2)^{1/2}}$$

for $|y| \le A$ and $f_Y(y) = 0$ otherwise, which is plotted in Fig. 2.6 for the case when $A = 1$. Note that the density $f_Y(y)$ tends to ∞ as y approaches $\pm A$. This reflects the fact that as Θ varies between $-\pi$ and π, the amplitudes of the cosine function $g(\Theta) = A\cos(\theta)$ stay in the vicinity of $\pm A$ for long intervals close to $\pm\pi$ (for $-A$) and close to 0 (for A), due to the fact that the derivative of $g(\theta)$ is zero at $\pm\pi$ and 0.

In the example above, the PDF of Y was obtained by first evaluating the CDF $F_Y(y)$, and then differentiating it. When the distribution of X is absolutely continuous, and the inverse functions $h_i(y)$ are differentiable, it is also possible to obtain an identity for $f_Y(y)$ which generalizes the Jacobian expression (2.35). Specifically, assume that the $h_i(y)$s are differentiable over an infinitesimal interval $(y, y+dy]$. Then the image of $(y, y+dy]$ under h_i is $(x_i, x_i+dx_i]$ with $x_i = h_i(y)$ and $dx_i = J_i(y)dy$, where

Fig. 2.5 CDF of
$Y = A \cos(\Theta)$ for $A = 1$

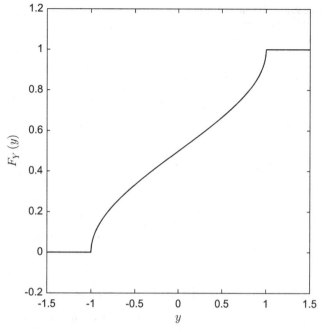

Fig. 2.6 PDF of
$Y = A \cos(\Theta)$ for $A = 1$

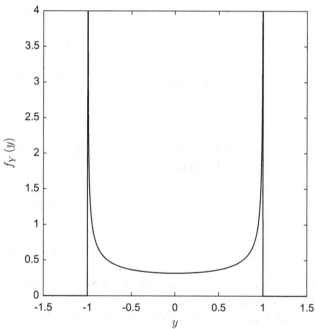

$$J_i(y) = |\frac{dh_i}{dy}| = \frac{1}{|\frac{dg}{dx}(h_i(y))|} \tag{2.38}$$

denotes the Jacobian of the i-th inverse function. By observing that the event $E = \{Y \in (y, y + dy]\}$
admits the decomposition

$$E = \cup_{i=1}^{p} E_i,$$

where the events $E_i = \{X \in (x_i, x_i + dx_i]\}$ are disjoint, we deduce that

$$f_Y(y)dy = P(E) = \sum_{i=1}^{p} P(E_i) = \sum_{i=1}^{p} f_X(h_i(y))J_i(y)dy,$$

which yields expression

$$f_Y(y) = \sum_{i=1}^{p} J_i(y)f_X(h_i(y)),\qquad (2.39)$$

which generalizes the formula (2.34) for the one-to-one case. Once again, it is worth noting that it does not matter which of the two identities is used to evaluate the Jacobian in (2.38).

Example 2.3 (Continued) Instead of using the inverse functions $h_\pm(y)$ of $y = g(\theta) = A\cos(\theta)$ to evaluate the Jacobian $J_\pm(y)$ of $h_\pm$, it is more convenient to evaluate

$$\frac{dg}{d\theta} = -A\sin(\theta).$$

We have

$$\left| \frac{1}{dg/d\theta} \right| = \frac{1}{A(1 - \cos^2(\theta))^{1/2}},$$

so

$$J_\pm(y) = \frac{1}{A(1 - (y/A)^2)^{1/2}}.$$

In this case, for $-A \le y \le A$, the identity (2.39) yields

$$f_Y(y) = J_+(y)f_\Theta(h_+(y)) + J_-(y)f_\Theta(h_-(y)),$$

where

$$f_\Theta(h_\pm(y)) = \frac{1}{2\pi},$$

since $-\pi \le h_\pm(y) = \pm\arccos(y/A) < \pi$. Accordingly, we find

$$f_Y(y) = \frac{1}{A\pi(1 - (y/A)^2)^{1/2}},$$

which coincides with the expression obtained earlier by differentiating the CDF.

Example 2.4 Assume that X is Laplace distributed random variable with PDF

$$f_X(x) = \frac{1}{2}\exp(-|x|).$$

Then consider the asymmetric rectifying transformation

$$Y = g(X) = \begin{cases} X & X \ge 0 \\ -X/2 & X < 0, \end{cases}$$

which is two-to-one and has range $\mathbb{R}^+$ (the positive real line).

Fig. 2.7 Two-to-one asymmetric rectifier transformation and its inverses

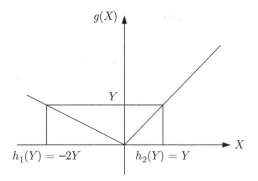

As shown in Fig. 2.7, it admits two inverses

$$h_1(Y) = -2Y \text{ and } h_2(Y) = Y$$

for $Y \in \mathbb{R}^+$, which have for Jacobian

$$J_1(y) = |\frac{dh_1}{dy}| = 2 \text{ and } J_2(y) = |\frac{dh_2}{dy}| = 1 \,,$$

respectively. From (2.29), we find therefore that the density of Y is given by

$$f_Y(y) = J_1(y) f_X(h_1(y)) + J_2(y) f_X(h_2(y))$$

$$= \exp(-2y) + \frac{1}{2} \exp(-y) \qquad (2.40)$$

for $y \geq 0$ and $f_Y(y) = 0$ for $y < 0$, since Y cannot take negative values. The density Y can be interpreted as a mixture of two exponential densities, since that it can be viewed as obtained by selecting an exponential random variable with parameter $\lambda = 1$ with probability $1/2$ and an exponential random variable with parameter $\lambda = 2$ with probability $1/2$.

Example 2.5 When the CDF $F_X(x)$ or its derivatives exhibit discontinuities, and the transformation $Y = g(X)$ is several-to-one, a good strategy to evaluate the CDF of Y is to divide the range of g into subintervals formed by the images under g of the x points where F_X or its derivatives are discontinuous. For example, consider the CDF

$$F_X(x) = \begin{cases} 0 & x \leq -2 \\ (x+2)/3 & -2 \leq x < -1 \\ 1/3 & -1 \leq x < 1 \\ 2/3 + (x-1)/3 & 1 \leq x < 2 \\ 1 & x \geq 2 \,, \end{cases}$$

which is plotted in Fig. 2.8. It can be interpreted as a mixture

$$F_X(x) = \frac{1}{3} F_{X_1}(x) + \frac{1}{3} F_{X_2}(X) + \frac{1}{3} F_{X_3}(x) \,,$$

Fig. 2.8 CDF of mixture random variable X

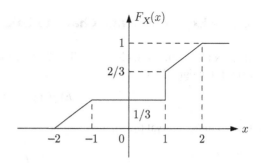

Fig. 2.9 CDF of the transformed random variable Y

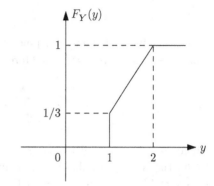

where the random variables X_1, X_2 are uniform over $[-2, -1]$ and over $[1, 2]$, respectively, and $X_3 = 1$ is deterministic. Thus X is obtained by selecting with probability $1/3$ one of the three random variables X_i, $1 \leq i \leq 3$.

The transformation $Y = g(X) = |X|$ is two-to-one and admits as inverses

$$X_{\pm} = h_{\pm}(Y) = \pm Y.$$

Since Y is necessarily nonnegative $F_Y(y) = 0$ for $y < 0$. For $y \geq 0$, we have

$$F_Y(y) = P[Y \leq y] = P[-y \leq X \leq y] = F_X(y) - F_X(-y_-).$$

The discontinuity points of F_X and its derivatives are $x = \pm 1$ and $x = \pm 2$, which are mapped into $y = 1$ and $y = 2$, so we find

$$F_Y(y) = 0$$

for $y < 1$, $F_Y(1) = 1/3$, and

$$F_Y(y) = \frac{2}{3} + \frac{y-1}{3} - \frac{(-y+2)}{3} = \frac{2y-1}{3}$$

for $1 \leq y \leq 2$, and $F_Y(y) = 1$ for $y > 2$. The CDF $F_Y(y)$ is plotted in Fig. 2.9.

2.6   Expectation and Characteristic Function

If X is a random variable with CDF $F_X(x)$, the expectation of a function $g(X)$ is defined as the Stieltjes integral [8, Chap. 6]

$$E[g(X)] = \int_{-\infty}^{\infty} g(x) d F_X(x) . \tag{2.41}$$

Over a finite interval $[a, b]$, the integral

$$\int_{a}^{b} g(x) d F_X(x)$$

is evaluated by subdividing $[a, b]$ into N subintervals $[x_{i-1}, x_i]$, $1 \leq i \leq N$ with $x_0 = a$, $x_N = b$ of width $x_i - x_{i-1} = (b - a)/N$ and forming the sum

$$\sum_{i=1}^{N} g(c_i)(F_X(x_i) - F_X(x_{i-1})),$$

where $x_{i-1} < c_i < x_i$. Then the limit of this sum as N tends to infinity is the Stieltjes integral over $[a, b]$. The integral over the entire real line is obtained by evaluating the limit of the integral as a tends to $-\infty$ and b tends to ∞. When the CDF $F_X(x)$ is absolutely continuous, the expectation reduces to the familiar Riemann integral

$$E[g(X)] = \int_{-\infty}^{\infty} g(x) f_X(x) dx . \tag{2.42}$$

Similarly, when X is a discrete valued random variable with PMF $p_k = P(X = x_k)$ for $1 \leq k \leq K$, the integral (2.41) reduces to the familiar expression

$$E[g(X)] = \sum_{1 \leq k \leq K} g(x_k) p_k . \tag{2.43}$$

Thus the definition (2.41) of the expectation through the Stieltjes integral is an elegant approach which allows us to handle continuous, discrete, and mixed valued random variables in a unified manner.

2.6.1   Moments

For the special case where $g(X) = X^\ell$ with ℓ integer, $E[X^\ell]$ is called the ℓ-th moment of distribution F_X. For $\ell = 1$

$$m_X = E[X] = \int_{\infty}^{\infty} x d F_x(x) \tag{2.44}$$

is the mean of X, and for $\ell = 2$

$$E[X^2] = \int x^2 d F_X(x)$$

is the second moment. When $g(X) = (X - m_X)^2$, we obtain the centered second moment or *variance*

$$K_X = E[(X - m_x)^2] = \int_\infty^\infty (x - m_X)^2 dF_X(x) \tag{2.45}$$

of X. Its square-root $\sigma_X = (K_X)^{1/2}$ is the *standard deviation* of X. By using the expansion $(X - m_X)^2 = X^2 - 2Xm_X + m_X^2$, we find that the variance can be expressed in terms of the second moment as

$$K_X = E[X^2] - m_X^2 , \tag{2.46}$$

so that given m_X, the second moment and the variance can be obtained from one another.

Uniform Distribution If X is uniformly distributed over $[a, b]$, its mean

$$m_X = \frac{1}{b-a} \int_a^b x dx = \frac{1}{2(b-a)} x^2 \Big|_a^b = \frac{a+b}{2}$$

corresponds to the midpoint of interval $[a, b]$. Its variance

$$K_X = \frac{1}{b-a} \int_a^b \left(x - \left(\frac{a+b}{2}\right) \right)^2 dx$$

$$= \frac{1}{3(b-a)} \left(x - \left(\frac{a+b}{2}\right) \right)^3 \Big|_a^b = \frac{(b-a)^2}{12} .$$

Exponential Distribution If X is exponentially distributed with parameter $\lambda > 0$, by integration by parts, we find

$$m_X = \int_0^\infty x\lambda \exp(-\lambda x) dx = -x \exp(-\lambda x)\Big|_0^\infty + \int_0^\infty \exp(-\lambda x) dx$$

$$= \frac{-1}{\lambda} \exp(-\lambda x)\Big|_0^\infty = \frac{1}{\lambda}$$

so that the average lifetime of X is $1/\lambda$. Similarly, after integration by parts, the second moment

$$E[X^2] = \int_0^\infty x^2 \lambda \exp(-\lambda x) = \frac{2}{\lambda^2} .$$

Using (2.46), we then find that the variance $K_X = 1/\lambda^2$.

It is worth pointing out that expectations and moments don't always exist. For example, it was observed earlier that the Cauchy density $f_X(x)$ decays like $1/x^2$ as x becomes large. Since in order to ensure convergence of the Riemann integral (2.42), $g(x)f_X(x)$ must decay like $1/|x|^\alpha$ with $\alpha > 1$, we conclude that the Cauchy distribution does not admit any moment.

2.6.2 Characteristic Function

To compute all the moments of a random variable X directly without having to perform long calculations, a simple approach consists of introducing the *characteristic function*

$$\Phi_X(u) = E[\exp(juX)] = \int_{-\infty}^{\infty} \exp(jux)dF_X(x) . \tag{2.47}$$

This function always exists since $F_X(x)$ is monotone and bounded. When F_X is absolutely continuous, it reduces to the Fourier transform

$$\Phi_X(u) = \int_{-\infty}^{\infty} f_X(x) \exp(jux)dx \tag{2.48}$$

of the PDF, and $f_X(x)$ can be recovered from the characteristic function by using the inverse Fourier transform

$$f_X(x) = \frac{1}{2\pi} \int_{-\infty}^{\infty} \Phi_X(u) \exp(-jux)du . \tag{2.49}$$

Note that the exponent convention for the Fourier transform used here is the opposite of the one employed by electrical engineers who use $\exp(-jux)$ in the forward transform and $\exp(jux)$ in its inverse. Similarly, when X is discrete and takes only integer values ($x_k = k, k \geq 0$), the characteristic function

$$\Phi_X(u) = \sum_{k=0}^{\infty} p_k \exp(jku) \tag{2.50}$$

can be viewed as a discrete-index Fourier transform, whose inverse is given by

$$p_k = \frac{1}{2\pi} \int_{-\pi}^{\pi} \Phi_X(u) \exp(-jku)du . \tag{2.51}$$

Properties of the Characteristic Function The function $\Phi_X(u)$ has several noteworthy properties:

i) $\Phi_X(u)$ is bounded, i.e., $|\Phi_X(u)| \leq 1 = \Phi_X(0)$ for all u;
ii) $\Phi_X(u)$ is uniformly continuous over $\mathbb{R}$;
iii) $\Phi_X^*(u) = \Phi_X(-u)$;
iv) $\Phi_X(u)$ is nonnegative definite in the sense that for any integer n and any real numbers $u_1 < u_2 < \cdots < u_n$, the $n \times n$ matrix

$$\boldsymbol{\Phi} = (\Phi_X(u_\ell - u_m), \ 1 \leq \ell, m \leq n)$$

is nonnegative, i.e., for any complex vector $\mathbf{v} \in \mathbb{C}^n$, $\mathbf{v}^H \boldsymbol{\Phi} \mathbf{v} \geq 0$ where $\mathbf{v}^H$ denotes the Hermitian transpose of $\mathbf{v}$.
v) the moments of X can be computed by using

$$E[X^\ell] = \frac{1}{j^\ell} \frac{d^\ell \Phi_X}{du^\ell}(0) . \tag{2.52}$$

Property i) is due to the inequality

$$|\Phi_X(u)| \leq \int_{-\infty}^{\infty} |\exp(jux)|dF_X(x) = F_X(\infty) = 1 .$$

To prove property ii), note that for arbitrary real u and h

$$\Phi_X(u+h) - \Phi_X(u) = \int_{-\infty}^{\infty} [\exp(j(u+h)x) - \exp(jux)] d F_X(x) \,,$$

so

$$|\Phi_X(u+h) - \Phi_X(u)| \leq \int_{-\infty}^{\infty} |\exp(jux)| |\exp(jhx) - 1| d F_X(x)$$

$$= \int_{-\infty}^{\infty} |\exp(jhx) - 1| d F_X(x) \,.$$

In this last expression $|\exp(jhx) - 1|$ tends to zero pointwise as h tends to zero, and is bounded by 2, so the integral tends to zero as h tends to zero by the bounded convergence theorem [8, p. 246]. Since the integral does not contain u, the convergence is uniform. Property iii) is due to the fact that the CDF $F_X(x)$ is real. The nonnegative definiteness of $\Phi_X(u)$ relies on the observation that

$$\mathbf{v}^H \mathbf{\Phi} \mathbf{v} = \sum_{\ell=1}^{n} \sum_{m=1}^{n} \Phi_X(u_\ell - u_m) v_\ell^* v_m$$

$$= \int_{-\infty}^{\infty} |\sum_{\ell=1}^{n} v_\ell^* \exp(ju_\ell x)|^2 d F_X(x) = E[|\sum_{\ell=1}^{n} v_\ell^* \exp(ju_\ell X)|^2] \geq 0 \,,$$

where v_ℓ denotes the ℓ-th component of complex vector $\mathbf{v}$. By differentiating $\Phi_X(u)$ ℓ times with respect to u, we obtain

$$\frac{d^\ell \Phi_X}{du^\ell}(u) = j^\ell E[X^\ell \exp(iuX)],$$

which yields (2.52) after setting $u = 0$. Thus, the computation of the characteristic function greatly facilitates the evaluation of the successive moments of X.

Gaussian Distribution To evaluate the characteristic function of Gaussian density (2.29) note that the exponent inside the Fourier transform integral can be written as

$$-\frac{(x-m)^2}{2\sigma^2} + jux = -\frac{(x - j\sigma^2 u - m)^2}{2\sigma^2} + jum - \sigma^2 u^2/2 \,,$$

so that after integration, we find

$$\Phi_X(u) = \exp(jum - \sigma^2 u^2/2) \,.$$

The first and second derivatives of $\Phi_X(u)$ are given respectively by

$$\frac{d\Phi_X}{du} = (jm - \sigma^2 u)\Phi_X(u)$$

$$\frac{d^2\Phi_X}{du^2} = [-\sigma^2 + (jm - \sigma^2 u)^2]\Phi_X(u) \,,$$

so that after setting $u = 0$ in the above expressions, we find

$$E(X) = \frac{1}{j}\frac{d\Phi_X}{du}(0) = m$$

$$E(X^2) = -\frac{d^2\Phi_X}{du}(0) = \sigma^2 + m^2 \,.$$

This shows that the parameter m appearing in the density expression (2.29) is indeed the mean, and the variance $K_X = E[X]^2 - m^2 = \sigma^2$, so the parameter σ represents the standard deviation of X.

Cauchy Distribution By using the space-shift property of the Fourier transform and Fourier transform tables, the characteristic function of a Cauchy density with parameters μ and $\sigma > 0$ can be expressed as

$$\Phi_X(u) = \exp(ju\mu - \sigma|u|) \,.$$

Since $|u|$ is not differentiable at $u = 0$, this function is not differentiable at $u = 0$, so the Cauchy density has no moment.

2.6.3 Moment Generating Function

When X is a discrete-valued random variable taking integer values with PMF $p_k = P(x = k)$ for $0 \le k \le K$, the use of the moment generating function

$$G_X(z) = E[z^X] = \sum_{k=0}^{K} p_k z^k \tag{2.53}$$

is often preferred to the characteristic function to compute the moments of X. It can be viewed as a z transform with the customary z^{-1} replaced by z. Since the sequence p_k is defined for $k \ge 0$, and the total probability mass is one, the region of convergence of $G_X(z)$ includes the unit disk $|z| \le 1$. Then the first two derivatives of $G_X(z)$ with respect to z are given by

$$\frac{dG_X}{dz} = E[Xz^{X-1}]$$

$$\frac{d^2G_X}{dz^2} = E[X(X-1)z^{X-2}] \,,$$

so that by setting $z = 1$ and backsubstitution, we find

$$E[X] = \frac{dG_X}{dz}(1) \tag{2.54}$$

$$E[X^2] = \frac{d^2G_X}{dz^2}(1) + \frac{dG_x}{dz}(1) \,. \tag{2.55}$$

Binomial Distribution For a binomial random variable X with parameters n and p, if $q = 1 - p$, the generating function can be expressed as

$$G_X(z) = \sum_{k=0}^{n} \binom{n}{k} p^k q^{n-k} z^k = (q + pz)^n,$$

where the last equality uses the binomial series expansion. Taking its first two derivatives gives

$$\frac{dG_X}{dz} = np(q + pz)^{n-1}, \quad \frac{d^2 G_X}{dz} = n(n-1)p^2(q + pz)^{n-1},$$

so

$$m_X = np, \quad E(x^2) = n^2 p^2 + npq,$$

and $K_X = E[X^2] - m_X^2 = npq$.

Poisson Distribution If X is Poisson distributed with parameter λ, its generating function is given by

$$G_X(z) = \sum_{k=0}^{\infty} \frac{\lambda^k}{k!} \exp(-\lambda) z^k = \exp(\lambda(z-1)),$$

where we have used the power series expansion

$$\exp(x) = \sum_{k=0}^{\infty} \frac{x^k}{k!}$$

for $x = \lambda z$. Differentiating $G_X(z)$ twice yields

$$\frac{dG_X}{dz} = \lambda G_X(z), \quad \frac{d^2 G_X}{dz^2} = \lambda^2 G_X(z),$$

so $m_X = \lambda$, $E[X^2] = \lambda^2 + \lambda$, and $K_X = E[X^2] - m_X^2 = \lambda$.

2.7 Joint Distribution of Random Variables

Up to this point, we have focused on a single random variable, but many applications involve two (or more) random variables, where say we are interested in random variable X, but we cannot measure it directly, and have only access to random variable Y. To describe how two random variables vary together, we use their joint PMF (when both are discrete valued), their joint CDF, or their joint PDF (when both are continuous valued).

2.7.1 Joint PMF, PDF, and CDF

The *joint probability mass function* of two discrete random variables X and Y taking values $\{x_k, k \in K\}$ and $\{y_\ell, \ell \in L\}$, respectively, is given by

$$p_{XY}(k, \ell) = p(X = x_k, Y = y_\ell).$$

It satisfies

$$\sum_{k \in K} \sum_{\ell \in L} p_{XY}(k, \ell) = 1,$$

since the total probability mass is one.

The *joint cumulative distribution function* of two random variables X and Y is a function of two real variables x and y defined by

$$F_{XY}(x, y) = P(X \le x, Y \le y),\tag{2.56}$$

where the comma appearing in $P(X \le x, Y \le y)$ needs to be interpreted as a logical "and," i.e., if we consider the two events $A = \{X \le x\}$ and $B = \{Y \le y\}$, then

$$P(X \le x, Y \le y) \overset{\triangle}{=} P(A \cap B).$$

Properties of the Joint CDF The joint CDF $F_{XY}(x, y)$ satisfy several properties which extend the properties of the CDF of a single random variable.

i) $F_{XY}(x, -\infty) = F_{XY}(-\infty, y) = 0$ and $F_{XY}(\infty, \infty) = 1$.
ii) $F_{XY}(x, \infty) = F_X(x)$ and $F_{XY}(\infty, y) = F_Y(y)$.
iii) $F_{XY}(x, y)$ is jointly monotone nondecreasing in the two variables x and y, i.e.,

$$F_{XY}(x_1, y_1) \le F_{XY}(x_2, y_2)\tag{2.57}$$

if $x_1 \le x_2$ and $y_1 \le y_2$.

Proof Property i) is due to the fact that $\{X \le -\infty\} = \{Y \le -\infty\} = \emptyset$, and the intersection of the empty set with any set is empty. Similarly, $\{X \le \infty\} = \{Y \le \infty\} = \Omega$, and the intersection of the entire set of outcomes Ω with itself is Ω. Property ii) relies on the observation that

$$F_{XY}(x, \infty) = P(\{X \le x\} \cap \Omega) = P(X \le x) = F_X(x)$$

and the matching result when the roles of X and Y are switched. This property is important since it indicates that the CDFs $F_X(x)$ and $F_Y(y)$ of X alone and Y alone, i.e., the marginal CDFs, are contained in the joint CDF $F_{XY}(x, y)$. To prove (2.57), consider the quarter plane $Q_2 = \{(x, y) : x \le x_2, y \le y_2\}$. As indicated in Fig. 2.10, it can be decomposed as

$$Q_2 = Q_1 \cup H \cup V \cup R,\tag{2.58}$$

Fig. 2.10 Decomposition of quarter plane Q_2

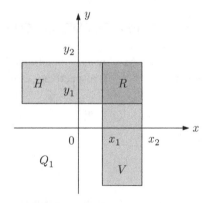

where the quarter plane Q_1, rectangle R, horizontal strip H and vertical strip V are defined as

$$Q_1 = \{(x, y) : x \le x_1, y \le y_1\}$$
$$R = \{(x, y) : x_1 < x \le x_2, y_1 < y \le y_2\}$$
$$H = \{(x, y) : x \le x_1, y_1 < y \le y_2\}$$
$$V = \{(x, y) : x_1 < x \le x_2, y \le y_1\}.$$

Since these subdomains do not overlap, we have

$$F_{XY}(x_2, y_2) = P((X, Y) \in Q_2)$$
$$= P((X, Y) \in Q_1) + P((X, Y) \in R)$$
$$+ P((X, Y) \in H) + P((X, Y) \in V), \tag{2.59}$$

where $P((X, Y) \in Q_1) = F_{X,Y}(x_1, y_1)$. since all probabilities appearing in the decomposition (2.59) are nonnegative, this implies (2.57). $\diamond$

The CDF $F_{XY}(x, y)$ can be used to evaluate the probability that (X, Y) belongs to arbitrary sets of $\mathcal{B} \times \mathcal{B}$. For example, we have

$$F_{XY}(x_1, y_2) = P(X \le x_1, Y \le y_2)$$
$$= P((X, Y) \in H) + P((X, Y) \in Q_1),$$

so

$$P[(X, Y) \in H) = F_{XY}(x_1, y_2) - F_{XY}(x_1, y_1). \tag{2.60}$$

Similarly, by exchanging the roles of X and Y,

$$P[(X, Y) \in V) = F_{XY}(x_2, y_1) - F_{XY}(x_1, y_1), \tag{2.61}$$

and by substituting (2.60) and (2.61) inside (2.59), we find that the probability that the pair (X, Y) belongs to rectangle R is given by

$$P((X, Y) \in R) = F_{XY}(x_2, y_2) + F_{XY}(x_1, y_1)$$
$$- F_{XY}(x_1, y_2) - F_{XY}(x_2, y_1). \tag{2.62}$$

As for the case of a single random variable, the CDF $F_{XY}(x, y)$ is absolutely continuous if it can be represented as

$$F_{XY}(x, y) = \int_{-\infty}^{\infty} f_{XY}(u, v) du dv, \tag{2.63}$$

in which case

$$f_{XY}(x, y) = \frac{\partial^2}{\partial x \partial y} F_{XY}(x, y)$$

is the joint PDF of X and Y. The joint monotone nondecreasing property implies $f_{XY}(x, y) \geq 0$, and by setting $x = y = \infty$ in (2.63), we find

$$1 = F_{XY}(\infty, \infty) = \int_{-\infty}^{\infty} \int_{-\infty}^{\infty} f_{XY}(u, v) du dv ,$$

so the total probability mass is one. By setting $x_1 = x$, $y_1 = y$, $x_2 = x + dx$, $y_2 = y + dy$ with dx and dy infinitesimal in (2.62) and performing Taylor series expansions of $F_{XY}(x + dx, y)$, $F_{XY}(x, y + dy)$, and $F_{XY}(x + dx, y + dy)$ in the vicinity of (x, y), we find

$$P(x \leq X < x + dx, y \leq Y < y + dy) = f_{XY}(x, y) dx dy,$$

so that $f_{XY}(x, y) dx dx y$ can be interpreted as the probability that the pair (X, Y) belongs to an infinitesimal rectangle of area $dx dy$ about point (x, y). By differentiating

$$F_X(x) = F_{XY}(x, \infty) = \int_{-\infty}^{x} \int_{-\infty}^{\infty} f_{XY}(u, v) du dv$$

with respect to x, we find

$$\frac{d}{dx} F_X(x) = f_X(x) = \int_{-\infty}^{\infty} f_X(x, v) dv, \tag{2.64}$$

so that the marginal density $f_X(x)$ is just obtained by integrating away the y variable in the joint density $f_{XY}(x, y)$. Similarly, the marginal density of Y can be evaluated as

$$f_Y(y) = \int_{-\infty}^{\infty} f_{XY}(x, y) dx. \tag{2.65}$$

This shows, as we had already observed for the CDF, that all the information about the separate distributions of X and Y is contained in their joint distribution.

Jointly Gaussian Random Variables The random variables X and Y are jointly Gaussian if, when we form the random vector

$$\mathbf{Z} = \begin{bmatrix} X \\ Y \end{bmatrix}$$

and represent a point (x, y) of $\mathbb{R}^2$ by the vector

$$\mathbf{z} = \begin{bmatrix} x \\ y \end{bmatrix},$$

the joint density of X and Y can be written as

$$f_{X,Y}(x, y) = f_{\mathbf{Z}}(\mathbf{z}) = \frac{1}{2\pi |\mathbf{K}_{\mathbf{Z}}|^{1/2}} \exp\left(-\frac{1}{2}(\mathbf{z} - \mathbf{m}_{\mathbf{Z}})^T \mathbf{K}_{\mathbf{Z}}^{-1}(\mathbf{z} - \mathbf{m}_{\mathbf{Z}}) \right), \tag{2.66}$$

where T denotes the matrix transposition operation. In this expression

$$\mathbf{m}_{\mathbf{Z}} = \begin{bmatrix} m_X \\ m_Y \end{bmatrix}$$

and

$$\mathbf{K_Z} = E[(\mathbf{Z} - \mathbf{m_Z})(\mathbf{Z} - \mathbf{m_Z})^T] = \begin{bmatrix} K_X & K_{XY} \\ K_{YX} & K_Y \end{bmatrix}$$

denote respectively the mean and covariance matrices of random vector $\mathbf{Z}$, where m_X, m_Y, K_X, and K_Y are the means and variances of random variables X and Y, and

$$K_{XY} = E[(X - m_X)(Y - m_Y)] \tag{2.67}$$

denotes the covariance of random variables X and Y. Since the random variables $X - m_X$ and $Y - m_Y$ can be exchanged in the expectation (2.67), $K_{XY} = K_{YX}$, so the matrix $\mathbf{K_z}$ is symmetric. It is sometimes convenient to express K_{XY} as

$$K_{XY} = \rho (K_X K_Y)^{1/2},$$

where ρ represents the *correlation coefficient* of X and Y. In (2.66), we have used the compact notation

$$|\mathbf{K_Z}| = |\det \mathbf{K_Z}| = (1 - \rho^2) K_X K_Y,$$

and we recall that the inverse of covariance matrix $\mathbf{K_Z}$ can be expressed as

$$\mathbf{K_Z}^{-1} = \frac{1}{1 - \rho^2} \begin{bmatrix} K_X^{-1} & -\rho (K_X K_Y)^{-1/2} \\ -\rho (K_X K_Y)^{-1/2} & K_Y^{-1} \end{bmatrix}.$$

So viewed as a function of x and y, the joint density can be written explicitly as

$$f_{XY}(x, y) = \frac{1}{2\pi ((1 - \rho^2) K_X K_Y)^{1/2}} \exp(-Q(x, y)), \tag{2.68}$$

where the quadratic function $Q_{XY}(x, y)$ is given by

$$Q(x, y) = \frac{1}{2(1 - \rho^2)} \left(\frac{(x - m_X)^2}{K_X} - 2 \frac{\rho (x - m_X)(y - m_Y)}{(K_X K_Y)^{1/2}} + \frac{(y - m_Y)^2}{K_Y} \right). \tag{2.69}$$

This shows that the density $f_{XY}(x, y)$ reaches its maximum at (m_X, m_Y), and its level sets are ellipses specified by $Q(x, y) = c$ with c constant.

2.7.2   Conditional Distributions

Given two discrete valued random variables X and Y taking values $\{x_k, k \in K\}$ and $\{y_\ell, \ell \in L\}$, respectively, the conditional PMF of X given Y is defined as

$$p_{X|Y}(k|\ell) = P(X = x_k | Y = y_\ell) = \frac{P(X = x_k, Y = y_\ell)}{P(Y = y_\ell)}$$

$$= \frac{p_{XY}(k, \ell)}{p_Y(\ell)}. \tag{2.70}$$

It is obviously nonnegative, and viewed as a function of k, it is a PMF for any ℓ since

$$\sum_{k \in K} p_{X|Y}(k|\ell) = \frac{1}{p_Y(\ell)} \sum_{k \in K} p_{XY}(k, \ell)$$

$$= \frac{p_Y(\ell)}{p_Y(\ell)} = 1 \, ,$$

where to go from the first to the second line we have used the marginalization property of the joint PMF of X and Y. Then the random variables X and Y are said to be independent if

$$p_{X|Y}(k|\ell) = p_X(k) \, , \tag{2.71}$$

i.e., knowing that $Y = y_\ell$ does not affect the distribution of X. This implies of that the joint PMF

$$p_{XY}(k, \ell) = p_X(k) p_Y(\ell)$$

is the product of the marginal PMFs of X and Y.

Likewise, suppose that X and Y are continuous valued with an absolutely continuous CDF. Then the conditional density of X given Y is defined as

$$f_{X|Y}(x|y) = \frac{f_{XY}(x, y)}{f_Y(y)} \, . \tag{2.72}$$

It is clearly nonnegative, and using the marginalization property of the joint density it can readily be checked that

$$\int_{-\infty}^{\infty} f_{X|Y}(x|y)dx = 1$$

for any real y, so viewed as a function of x, $f_{X|Y}(x|y)$ is a PDF. The conditional PDF can also be interpreted in terms of events $A = \{x < X \le x + dx\}$ and $B = \{y < Y \le y + dy\}$ with dx and dy infinitesimal as

$$f_{X|Y}(x|y)dx = P(A|B) = \frac{P(A \cap B)}{P(B)} = \frac{f_{X,Y}(x, y)dxdy}{f_Y(y)dy} \, .$$

The random variables are X and Y independent if

$$f_{X|Y}(x|y) = f_X(x) \, , \tag{2.73}$$

or equivalently if the joint PDF

$$f_{X,Y}(x, y) = f_X(x) f_Y(y)$$

is the product of the marginal PDFs.

Conditional Gaussian Density When X and Y are jointly Gaussian with PDF (2.68) and (2.69), if we write

$$f_Y(y) = \frac{1}{(2\pi K_Y)^{1/2}} \exp(-Q(y))$$

with $Q(y) = (Y - m_Y)^2/(2K_Y)$, we find after some elementary manipulations

$$f_{X|Y}(x|y) = \frac{f_{X,Y}(x, y)}{f_Y(y)} = \frac{1}{(2\pi K_{X|Y})^{1/2}} \exp(-Q(x|y)), \qquad (2.74)$$

where

$$K_{X|Y} = (1 - \rho^2)K_X = K_X - K_{XY}^2 K_Y^{-1} \qquad (2.75)$$

and

$$Q(x|y) = Q(x, y) - Q(y) = (x - m_{X|Y})^2/(2K_{X|Y})$$

with

$$m_{X|Y} = m_X + K_{XY} K_Y^{-1}(Y - m_Y). \qquad (2.76)$$

We conclude therefore that conditioned on the knowledge of Y, X remains Gaussian, with conditional mean $m_{X|Y}$ and conditional variance $K_{X|Y}$. It is also interesting to note that $m_{X|Y}$ depends linearly on Y, and $K_{X|Y}$ does not depend on Y, so it can be computed even before Y is observed. Furthermore, the random variables X and Y are independent if and only if

$$f_{X|Y}(x|y) = f_X(x),$$

which requires $m_{X|Y} = m_X$ and $K_{X|Y} = K_X$. From expressions (2.75) and (2.76) we conclude that X and Y are independent if and only if $K_{XY} = 0$, or equivalently $\rho = 0$. In other words, two Gaussian random variables are independent if and only if they are uncorrelated. However, this is not true in general, as the independence property is much stronger than uncorrelatedness, as will be shown in Sect. 2.9.

Bayes Law Given the conditional density $f_{Y|X}(y|x)$ and marginal density $f_X(x)$, it is sometimes desired to reverse the conditioning operation to find $f_{X|Y}(x|y)$. This is accomplished by observing that

$$f_{X,Y}(x, y) = f_{Y|X}(y|x)f_X(x) = f_{X|Y}(x|y)f_Y(y),$$

and noting that the marginal density of Y is given by

$$f_Y(y) = \int_{-\infty}^{\infty} f_{Y|X}(y|x)f_X(x)dx,$$

so that the conditional density of X given Y satisfies *Bayes' law*

$$f_{X|Y}(x|y) = \frac{f_{Y|X}(y|x)f_X(x)}{\int_{-\infty}^{\infty} f_{Y|X}(y|x)f_X(x)dx}, \qquad (2.77)$$

which can be used to reverse the conditioning operation.

2.7.3 Joint Transformations

As in the case of a single random variable, given two random variables (X_1, X_2) with joint PDF $f_{X_1 X_2}(x_1, x_2) = f_{\mathbf{X}}(\mathbf{x})$, where

$$\mathbf{X} = \begin{bmatrix} X_1 \\ X_2 \end{bmatrix} \quad , \quad \mathbf{x} = \begin{bmatrix} x_1 \\ x_2 \end{bmatrix} ,$$

it is often of interest to find the joint density of transformed random variables

$$\mathbf{Y} = \begin{bmatrix} Y_1 \\ Y_2 \end{bmatrix} = \begin{bmatrix} g_1(X_1, X_2) \\ g_2(X_1, X_2) \end{bmatrix} \overset{\triangle}{=} \mathbf{g}(\mathbf{X}) .$$

As in the single variable case, different expressions need to be used depending on whether $\mathbf{g}$ is one-to-one, or several-to-one. We consider first the one-to-one case. Then $\mathbf{Y} = \mathbf{g}(\mathbf{X})$ has for inverse transformation

$$\mathbf{X} = \begin{bmatrix} X_1 \\ X_2 \end{bmatrix} = \begin{bmatrix} h_1(Y_1, Y_2) \\ h_2(Y_1, Y_2) \end{bmatrix} \overset{\triangle}{=} \mathbf{h}(\mathbf{Y}) ,$$

whose Jacobian $J(\mathbf{y})$ can be expressed as

$$J(\mathbf{y}) = \left| \det \begin{bmatrix} \dfrac{\partial h_1}{\partial y_1} & \dfrac{\partial h_1}{\partial y_2} \\ \dfrac{\partial h_2}{\partial y_1} & \dfrac{\partial h_2}{\partial y_2} \end{bmatrix} \right| , \tag{2.78}$$

or equivalently, in terms of the forwards transformation as

$$J(\mathbf{y}) = 1/\left| \det \begin{bmatrix} \dfrac{\partial g_1}{\partial x_1} & \dfrac{\partial g_1}{\partial x_2} \\ \dfrac{\partial g_2}{\partial x_1} & \dfrac{\partial g_2}{\partial x_2} \end{bmatrix} \right|_{\mathbf{x}=\mathbf{h}(\mathbf{y})} . \tag{2.79}$$

Then the PDF of the transformed vector $\mathbf{Y}$ is given by

$$f_{\mathbf{Y}}(\mathbf{y}) = J(\mathbf{y}) f_{\mathbf{X}}(\mathbf{h}(\mathbf{y})) . \tag{2.80}$$

Example 2.6 Consider two independent $N(0, \sigma^2)$ distributed Gaussian random variables X_1 and X_2 with joint density

$$f_{X_1, X_2}(x_1, x_2) = f_{X_1}(x_1) f_{X_2}(x_2) = \frac{1}{2\pi\sigma^2} \exp\left(-\frac{(x_1^2 + x_2^2)}{2\sigma^2} \right).$$

A point represented by (X_1, X_2) in rectangular coordinates can also be represented in polar coordinates as (R, Θ) as shown in Fig. 2.11, where

$$R = g_1(X_1, X_2) = (X_1^2 + X_2^2)^{1/2} \quad , \quad \Theta = g_2(X_1, X_2) = \arctan(X_2/X_1) .$$

The ranges of the polar variables are $0 \leq R < \infty$ and $0 \leq \Theta < 2\pi$, and the inverse transformation from polar to rectangular coordinates is given by

$$X_1 = h_1(R, \Theta) = R\cos(\Theta) \quad , \quad X_2 = h_2(R, \Theta) = R\sin(\Theta) .$$

Fig. 2.11 Rectangular to polar coordinates transformation

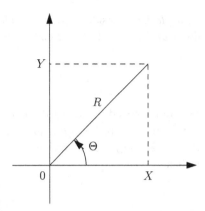

To find the Jacobian of the transformation, it is more convenient to use (2.78) since it is easier to differentiate sines and cosines. We find

$$J(r, \theta) = |\det \begin{bmatrix} \frac{\partial h_1}{\partial r} & \frac{\partial h_1}{\partial \theta} \\ \frac{\partial h_2}{\partial r} & \frac{\partial h_2}{\partial \theta} \end{bmatrix}| = |\det \begin{bmatrix} \cos(\theta) & -r\sin(\theta) \\ \sin(\theta) & r\cos(\theta) \end{bmatrix}| = r,$$

so the joint PDF of R and Θ is given by

$$f_{R\Theta}(r, \theta) = r f_{X_1, X_2}(r\cos(\theta), r\sin(\theta))$$

$$= \frac{r}{2\pi\sigma^2} \exp(-r^2/(2\sigma^2))$$

for $0 \leq r < \infty$ and $0 \leq \theta < 2\pi$, and $f_{R\Theta}(r, \theta) = 0$ otherwise. Since the joint density does not depend on θ, the marginal density of R is

$$f_R(r) = \int_0^{2\pi} f_{R\Theta}(r, \theta) d\theta = \frac{r}{\sigma^2} \exp(-r^2/(2\sigma^2)) u(r),$$

which corresponds to a Rayleigh distribution with parameter σ^2. The conditional density of Θ given R is

$$f_{\Theta|R}(\theta|r) = \frac{1}{2\pi}$$

for $0 \leq \theta < 2\pi$, which does not depend on r, so R and Θ are *independent* and Θ is uniformly distributed over $[0, 2\pi)$.

Remark The result derived above implies that if we consider a quadrature modulated signal

$$X(t) = X_c \cos(\omega_c t) - X_s \sin(\omega_c t),$$

where X_c and X_s are independent $N(0, \sigma^2)$ distributed, the envelope representation

$$X(t) = R \cos(\omega_c t + \Theta)$$

is such that the amplitude R is Rayleigh distributed and independent of the phase Θ, which is uniformly distributed.

Example 2.7 Consider two independent random variables X_1 and X_2, such that X_i is $N(0, \sigma_i^2)$ distributed for $i = 1, 2$. The joint PDF of X_1 and X_2 is therefore given by

$$f_{X_1 X_2}(x_1, x_2) = \frac{1}{2\pi\sigma_1\sigma_2} \exp\left(-\frac{1}{2}((x_1/\sigma_1)^2 + (x_2/\sigma_2)^2)\right).$$

We are interested in finding the PDF of $Y = g_1(X_1, X_2) = X_1/X_2$ but since the transformation procedure described above requires two transformed variables, we introduce the auxiliary variable $Z = g_2(X_1, X_2) = X_2$, which will be eliminated later by marginalization. The inverse transformation is then

$$\begin{bmatrix} X_1 \\ X_2 \end{bmatrix} = \begin{bmatrix} h_1(Y, Z) \\ h_2(Y, Z) \end{bmatrix} = \begin{bmatrix} YZ \\ Z \end{bmatrix},$$

whose Jacobian is given by

$$J(y, z) = \left| \det \begin{bmatrix} \dfrac{\partial h_1}{\partial y} & \dfrac{\partial h_1}{\partial z} \\ \dfrac{\partial h_2}{\partial y} & \dfrac{\partial h_2}{\partial z} \end{bmatrix} \right| = \left| \det \begin{bmatrix} z & y \\ 0 & 1 \end{bmatrix} \right| = |z|.$$

The joint PDF of Y and Z is therefore

$$f_{YZ}(y, z) = J(y, z) f_{X_1 X_2}(h_1(y, z), h_2(y, z))$$

$$= \frac{|z|}{2\pi\sigma_1\sigma_2} \exp\left(-((y/\sigma_1)^2 + 1/\sigma_2^2)z^2/2\right).$$

However, we are really interested in the density of Y, and by marginalizing we find

$$f_Y(y) = \int_{-\infty}^{\infty} f_{YZ}(y, z)dz = 2\int_0^{\infty} f_{YZ}(y, z)dz$$

$$= \frac{1}{\pi\sigma_1\sigma_2} \int_0^{\infty} \exp(-((y/\sigma_1)^2 + 1/\sigma_2^2)u)du$$

$$= \frac{1}{\pi\gamma} \frac{1}{1 + (y/\gamma)^2}$$

with $\gamma = \sigma_1/\sigma_2$, where the second equality of the first line uses the fact that $f_{YZ}(y, z)$ is even in z, and the first equality of the second line relies on the change of variable $u = z^2/2$. This shows that the ratio $Y = X_1/X_2$ of two zero mean Gaussian random variables is Cauchy distributed with location parameter $\mu = 0$ and scale parameter $\gamma = \sigma_1/\sigma_2$ (the ratio of the standard deviations of the Gaussian variables).

More generally, if we consider a p-to-one transformation $\mathbf{Y} = g(\mathbf{X})$ with p inverses of the form $\mathbf{x} = \mathbf{h}_i(\mathbf{Y})$, the PDF of $\mathbf{Y}$ can be expressed as

$$f_{\mathbf{Y}}(\mathbf{y}) = \sum_{i=1}^{p} J_i(\mathbf{y}) f_{\mathbf{X}}(\mathbf{h}_i(\mathbf{y})), \tag{2.81}$$

where $J_i(\mathbf{y})$ denotes the Jacobian of $h_i(\mathbf{y})$.

2.8   Sums of Independent Random Variables

The idea of introducing an auxiliary variable to form a vector transformation, and then marginalizing to evaluate the PDF of a variable of interest can be used to find the PDF of the sum

$$Y = X_1 + X_2$$

of two independent random variables with PDFs $f_{X_i}(x_i)$ for $i = 1, 2$. We introduce the variable $Z = X_2$, which allows us to construct the inverse transformation

$$\begin{bmatrix} X_1 \\ X_2 \end{bmatrix} = \begin{bmatrix} h_1(Y, Z) \\ h_2(Y, Z) \end{bmatrix} = \begin{bmatrix} Y - Z \\ Z \end{bmatrix} .$$

The Jacobian of this transformation is

$$J(y, z) = \left| \det \begin{bmatrix} \dfrac{\partial h_1}{\partial y} & \dfrac{\partial h_1}{\partial z} \\ \dfrac{\partial h_2}{\partial y} & \dfrac{\partial h_2}{\partial z} \end{bmatrix} \right|$$

$$= \left| \det \begin{bmatrix} 1 & -1 \\ 0 & 1 \end{bmatrix} \right| = 1 ,$$

so the joint density of Y and Z is

$$f_{Y,Z}(y, z) = f_{X_1}(y - z) f_{X_2}(z) .$$

The marginal density of Y is therefore

$$f_Y(y) = \int_{-\infty}^{\infty} f_{YZ}(y, z) dz$$

$$= \int_{-\infty}^{\infty} f_{X_1}(y - z) f_{X_2}(z) dz = f_{X_1}(y) * f_{X_2}(y), \tag{2.82}$$

where $*$ denotes the convolution operation. Thus the PDF of the sum of two independent random variables is the convolution of their PDFs. Since the convolution in the space domain becomes a product in the Fourier domain, the characteristic function of Y can be expressed in terms of the characteristic functions $\Phi_{X_i}(u)$, $i = 1, 2$ of X_1 and X_2 as

$$\Phi_Y(u) = \Phi_{X_1}(u)\Phi_{X_2}(u) . \tag{2.83}$$

In particular if the random variables X_i are $N(m_i, \sigma_i^2)$ distributed for $i = 1, 2$, by forming the product (2.81), we find that Y has for characteristic function

$$\Phi_Y(u) = \exp(jum_Y - \sigma_Y^2 u^2/2)$$

with

$$m_Y = m_1 + m_2 \ , \quad \sigma_Y^2 = \sigma_1^2 + \sigma_2^2 ,$$

so that Y is $N(m_Y, \sigma_Y^2)$ distributed. Similarly, if the random variables X_i are Cauchy distributed with location parameter μ_i and scale parameter σ_i for $i = 1, 2$, their sum Y admits the characteristic function

$$\Phi_Y(u) = \exp(ju\mu_Y - \sigma_Y|u|)$$

with

$$\mu_Y = \mu_1 + \mu_2 \ , \quad \sigma_Y = \sigma_1 + \sigma_2 \ ,$$

so that Y is also Cauchy distributed with location parameter μ_Y and scale parameter σ_Y. Thus the Gaussian and Cauchy distributions are stable under the addition of random variables. However, this property is quite rare, and most random variable distributions are not stable under the addition operation.

Example 2.8 Let X_1 and X_2 be two independent exponential random variables with parameter λ. Since their characteristic function is

$$\Phi_{X_i}(u) = \frac{\lambda}{\lambda - ju}$$

with $i = 1, 2$, the characteristic function of the sum $y = X_1 + X_2$ is

$$\Phi_Y(u) = \left(\frac{\lambda}{\lambda - ju}\right)^2,$$

whose inverse Fourier transform is

$$f_Y(y) = \lambda^2 y \exp(-\lambda y) u(y),$$

which is the PDF of an Erlang random variable with parameters 2 and λ.

Up to this point, we have restricted our attention to absolutely continuous distributions. Similar results hold for discrete valued random variables. For simplicity, assume that independent discrete random variables X_1 and X_2 are integer valued with PMF $p_{X_i}(k), k \geq 0$ for $i = 1, 2$. To evaluate the PMF of Y, we use the principle of total probability, so

$$p_Y(k) = \sum_{\ell=0}^{\infty} P(Y = k|X_2 = \ell) p_{X_2}(\ell),$$

where

$$P(Y = k|X_2 = \ell) = P(X_1 = k - \ell|X_2 = \ell) = P(X_1 = k - \ell) = p_{X_1}(k - \ell) \ ,$$

where the first equality relies on the observation that $X_1 = Y - X_2 = Y - \ell$ if $X_2 = \ell$, and the second equality uses the independence of X_1 and X_2. We conclude therefore that the PMF of Y

$$p_Y(k) = \sum_{\ell=0}^{\infty} p_{X_1}(k - \ell) p_{X_2}(\ell) = p_{X_1}(k) * p_{X_2}(k) \tag{2.84}$$

is the discrete convolution of the PMFs of X_1 and X_2. Since the z transform converts a convolution into a product in the z-domain, we deduce that the generating function $G_Y(z)$ of Y can be expressed in terms of the generating functions $G_{X_i}(z)$ of random variables X_i with $i = 1, 2$ as

$$G_Y(z) = G_{X_1}(z)G_{X_2}(z). \tag{2.85}$$

Suppose that X_1 and X_2 are two independent Poisson distributed random variables with arrival rates λ_1 and λ_2. Then by forming the product (2.85), we find that the generating function of $Y = X_1 + X_2$ can be expressed as

$$G_Y(z) = \exp(\lambda_Y(z - 1))$$

with $\lambda_Y = \lambda_1 + \lambda_2$, so Y is also Poisson distributed, but its rate is the sum of the rates of X_1 and X_2.

2.9   Joint and Conditional Expectations

As for a single random variable, if X and Y are two random variables with joint CDF $F_{XY}(x, y)$, the expectation of $g(X, Y)$ is defined as the Stieltjes integral

$$E[g(X, Y)] = \int_{\mathbb{R}^2} g(x, y)dF_{X,Y}(x, y), \tag{2.86}$$

where the Stieltjes integral over a rectangle $R = [a, b] \times [c, d]$ is obtained by dividing R in elementary rectangles $R_{ij} = [x_{i-1}, x_i] \times [y_{j-1}, y_j]$ of horizontal and vertical width $x_i - x_{i-1} = (b - a)/N$ and $y_j - y_{j-1} = (d - c)/N$ with $0 \le i \le N, 0 \le j \le M, x_0 = a, x_N = b, y_0 = c, y_M = d$ and then taking the limit as N and M tend to infinity of the sum

$$\sum_{i=1}^{N}\sum_{j=1}^{M} g(c_{ij})dF_{YX}(R_{ij}),$$

where c_{ij} is a point located inside rectangle R_{ij} and

$$dF_{XY}(R_{ij}) = F_{XY}(x_i, y_j) + F_{XY}(x_{i-1}, y_{j-1})$$
$$-F_{XY}(x_i, y_{j-1}) - F_{XY}(x_{i-1}, y_j).$$

Then the integral over $\mathbb{R}^2$ is obtained by letting a and c tend to $-\infty$ and b and d to ∞.

When F_{XY} is absolutely continuous with PDF f_{XY}, the expectation becomes

$$E[g(X, Y)] = \int_{\mathbb{R}^2} g(x, y)f_{X,Y}(x, y)dxdy, \tag{2.87}$$

and when X and Y are discrete valued with PMF $p_{XY}(k, \ell)$, the definition (2.86) yields

$$E[g(X, Y)] = \sum_{k \in K}\sum_{\ell \in L} g(x_k, y_\ell)p_{XY}(k, \ell). \tag{2.88}$$

2.9.1 Joint Moments and Characteristic Function

When $g(X, Y) = X^k Y^\ell$, $E[X^k Y^\ell]$ is called the joint moment of order (k, ℓ) of X and Y. In particular for $k = \ell = 1$, $E[XY]$ is the joint moment of order $(1, 1)$ and the centered cross-moment

$$K_{XY} = E[(X - m_X)(Y - m_Y) = E[XY] - m_X m_Y \qquad (2.89)$$

is called the covariance of X and Y. If X and Y have finite variances K_X and K_Y, the correlation coefficient ρ of X and Y is defined as

$$\rho \overset{\triangle}{=} \frac{K_{XY}}{(K_X K_Y)^{1/2}} \,. \qquad (2.90)$$

Correlation Coefficient Property The correlation coefficient ρ satisfies $|\rho| \leq 1$ with $|\rho| = 1$ if and only if $Y - m_Y = \alpha(X - m_X)$, i.e., if the centered random variables $Y - m_Y$ and $X - m_X$ are proportional to each other.

Proof Consider the random variable

$$Z = (Y - m_Y) - z(X - m_X),$$

where z is an arbitrary real number. The function $h(z) = E[Z^2]$ is nonnegative for all z, with

$$h(z) = E[((Y - m_Y) - z(X - m_X))^2] = K_Y - 2z K_{XY} + z^2 K_X.$$

The quadratic function $h(z)$ becomes infinite as z tends to infinity and reaches its minimum when

$$\frac{dh}{dz} = 2(z K_X - K_{XY}),$$

i.e., for $z_{\min} = K_{XY}/K_X$. For this value, we have

$$0 \leq h(z_{\min}) = K_Y - (K_{XY})^2/K_X \,,$$

which implies $|\rho| \leq 1$. Furthermore, $|\rho| = 1$ if and only if $h(z_{\min}) = 0$, which implies $E[Z^2] = 0$ and thus $Z = 0$ for $z = z_{\min}$, Thus $Y - m_Y = \alpha(X - m_X)$ with $\alpha = z_{\min}$. Finally, note that $\rho = 1$ whenever $\alpha > 0$ and $\rho = -1$ whenever $\alpha < 0$.

Joint Characteristic Function Direct computation of the mixed moments of order (k, ℓ) is often tedious, so as in the single variable case it is usually preferable to evaluate the moments by differentiating the joint characteristic function

$$\Phi_{XY}(u, v) = E[\exp(j(uX + vY))] = \int_{\mathbb{R}^2} \exp(j(ux + vy)) dF_{XY}(x, y) \,. \qquad (2.91)$$

Its properties are similar to those of characteristic functions of a single random variable. In particular, when F_{XY} is absolutely continuous with PDf f_{XY}, the characteristic function is the 2-D Fourier transform of the PDF, i.e.,

$$\Phi_{XY}(u, v) = \int_{-\infty}^{\infty} \int_{-\infty}^{\infty} \exp(j(ux + vy)) f_{XY}(x, y) dx dy , \tag{2.92}$$

so the PDF

$$f_{XY}(x, y) = \frac{1}{(2\pi)^2} \int_{\infty}^{\infty} \int_{-\infty}^{\infty} \exp(-j(ux + vy) \Phi_{XY}(u, v) du dv \tag{2.93}$$

is the inverse Fourier transform of the characteristic function. In other words, the PDF and characteristic function contain exactly the same information, but presented differently. As in the 1-D case, the characteristic function has the differentiation property

$$E[X^k Y^\ell] = \frac{1}{j^{k+\ell}} \frac{\partial^{k+\ell}}{\partial_u^k \partial_v^\ell} \Phi_{XY}(u, v) \,|_{(u,v)=(0,0)}, \tag{2.94}$$

which can be derived in a straightforward manner by taking the mixed partial derivative of order (k, ℓ) of $\Phi_{XY}(u, v)$ and setting $(u, v) = (0, 0)$ in the resulting expression.

Jointly Gaussian Distribution To illustrate the computation of cross-moments of X and Y, consider the case where X and Y are jointly Gaussian with density (2.68) and (2.69). In this case, by completing the square in the Fourier transform integral (2.92), we find that the characteristic function $\Phi_{XY}(u, v)$ is also Gaussian, since

$$\Phi_{X,Y}(u, v) = \exp(j(um_X + vm_Y)) \exp(-\Psi(u, v))$$

with

$$\Psi(u, v) = \begin{bmatrix} u & v \end{bmatrix} \mathbf{K_Z} \begin{bmatrix} u \\ v \end{bmatrix} / 2$$

$$= (K_X u^2 + 2K_{XY} uv + K_Y v^2)/2 .$$

Then

$$\frac{\partial^2}{\partial u \partial v} \Phi_{XY}(u, v) = [(jm_X - (K_X u + K_{XY} v))(jm_Y - (K_Y + K_{XY} u))$$

$$- K_{XY}]\Phi_{XY}(u, v) ,$$

and by setting $u = v = 0$, we find

$$E[XY] = -\frac{\partial^2}{\partial u \partial v} \Phi_{XY}(u, v) \,|_{(u,v)=(0,0)} = K_{XY} + m_X m_Y ,$$

so that the parameter K_{XY} appearing in the definition of the joint PDF (2.68) and (2.69) corresponds indeed to the cross-variance of X and Y.

2.9.2   Characterization of Independence and Uncorrelatedness

The independence of two random variables X and Y can be characterized in several equivalent ways. For simplicity, we assume that their joint CDF $F_{XY}(x, y)$ is absolutely continuous. Then X and Y are independent if any one of the following three properties hold:

i) Their joint PDF can be factored as

$$f_{XY}(x, y) = f_X(x) f_Y(y),$$

where $f_X(x)$ and $f_Y(y)$ are the marginal PDfs of X and Y;

ii) Their joint characteristic function can be factored as

$$\Phi_{XY}(u, v) = \Phi_X(u) \Phi_Y(v),$$

where $\Phi_X(u)$ and $\Phi_Y(v)$ denote the characteristic functions of X and Y, respectively;

iii) For all functions g and h, the expectation

$$E[g(X)h(Y)] = E[g(X)]E[h(Y)]. \tag{2.95}$$

Proof The equivalence of i) and ii) is due to the fact that the separability property of functions is preserved by the Fourier transform and its inverse. Property i) implies iii) since

$$E[g(X)h(Y)] = \int_{-\infty}^{\infty} \int_{-\infty}^{\infty} g(x)h(y) f_X(x) f_Y(y) dx dy$$

$$= \int_{-\infty}^{\infty} g(x) f_X(x) dx \times \int_{-\infty}^{\infty} h(y) f_Y(y) dy = E[g(X)]E[h(Y)]$$

due to the separability of the integrand on the first line. Finally, by selecting $g(X) = \exp(iuX)$ and $h(Y) = \exp(ivY)$, we conclude that iii) implies ii).

The property (2.95) is quite useful and will be used repeatedly in the following chapters. In contrast, we say that two random variables X and Y are *uncorrelated* if their covariance $K_{XY} = 0$. It is interesting to observe that X and Y are uncorrelated if and only if the property (2.95) holds for all *linear* functions functions $g(X) = aX + b$ and $h(Y) = cY + d$. To prove this result note that if X and Y have means m_X and m_Y respectively

$$E[g(X)] = am_X + b \ , \quad E[h(Y)] = cm_Y + d \ ,$$

and we can rewrite

$$g(X) = a(X - m_X) + E[g(X)] \ , \quad h(Y) = c(Y - m_Y) + E[h(Y)] \ .$$

Then

$$E[g(X)h(Y)] = acK_{XY} + E[g(X)]E[h(Y)] \ ,$$

where since $g(X)$ and $h(Y)$ are arbitrary linear functions, we can assume $ac \neq 0$. Accordingly, the property (2.94) holds if and only if $K_{XY} = 0$. This shows that uncorrelatedness is much weaker than independence, since it ensures that property (2.95) holds only for linear functions g and h. Thus independence implies uncorrelatedness, but the converse is not true in general, except as we saw earlier for jointly Gaussian random variables. Readers may wish to examine Problem 2.12, which constructs an example where two random variables are marginally Gaussian and uncorrelated, but are not jointly Gaussian and independent.

2.9.3 Conditional Expectation

Since conditional PMFs and PDFs satisfy all properties of PMFs and PDFs, they can be used to define conditional expectations. If we consider a function $g(X, Y)$ of two continuous valued random variables X and Y such that the conditional PDF of X given Y is $f_{X|Y}(x|y)$, the conditional expectation of $g(X, Y)$ is defined as

$$E[g(X, Y)|Y] = \int_{-\infty}^{\infty} g(x, Y) f_{X|Y}(x|Y) dx . \qquad (2.96)$$

Similarly, if X and Y are discrete valued with conditional PMF $p_{X|Y}(x_k|y_\ell)$, the conditional expectation of $g(X, Y)$ is expressed as

$$E[g(X, Y)|Y] = \sum_{k \in K} g(x_k|Y) p_{X|Y}(x_k|Y) . \qquad (2.97)$$

An interesting feature of conditional expectations is that since they are functions of the conditioning random variable Y, they are themselves random variables.

Properties of Conditional Expectations

i) Since the conditioning random variable Y is viewed as known,

$$E[g(X)h(Y)|Y] = h(Y)E[g(X)|Y]$$

for arbitrary functions g and h. In other words, since Y is known, $h(Y)$ can be treated as a constant when evaluating a conditional expectation.

ii) It is sometimes convenient to compute expectations in stages and use the *iterated expectation principle*:

$$E[g(X, Y)] = E_Y[E_{X|Y}[g(X, Y)|Y]] , \qquad (2.98)$$

where the inner expectation $E_{X|Y}[\cdot]$ represents a conditional expectation of X given Y and the outer expectation $E_Y[\cdot]$ is taken with respect to Y.

The iterated/repeated expectation property (2.98) is derived for continuous-valued random variables. We have

$$E[g(X, Y)] = \int_{\infty}^{\infty} \int_{-\infty}^{\infty} g(x, y) f_{XY}(x, y)_d x dy$$

$$= \int_{\infty}^{\infty} \left[\int_{-\infty}^{\infty} g(x, y) f_{X|Y}(x|y) dx \right] f_Y(y) dy$$

$$= E_Y[E_{X|Y}[g(X, Y)|Y]] .$$

The idea of conditioning with respect to a random variable, i.e., treating it temporarily as known, is a fruitful approach to analyze difficult probability problems.

Example 2.9 Buffon's needle problem was devised by Buffon, a French naturalist of the 18th century who was the director of what is now known as the Jardin des Plantes in Paris. Like most scientists of his time, Buffon had a wide range of interests and posed the following problem. Consider an infinite table with evenly spaced vertical lines separated by a in the horizontal direction. A needle of length ℓ is dropped at random on the table with an arbitrary orientation. What is the probability that the needle will straddle/hit one of the vertical lines? As shown in Fig. 2.12, the position and orientation of the needle are described by two independent random variables: the location X of its center, and its angle Θ with respect to the horizontal axis. Since the table is infinite, we can assume without loss of generality that X is located between the lines corresponding to $x = 0$ and $x = a$, so X is uniformly distributed over $[0, a)$ with PDF

$$f_X(x) = \begin{cases} 1/a & 0 \le x < a \\ 0 & \text{otherwise} . \end{cases}$$

Since the needle is not oriented, we cannot distinguish its front and back, so the angle Θ is uniformly distributed over $[-\pi/2, \pi/2)$, and its density is

$$f_\Theta(\theta) = \begin{cases} 1/\pi & -\pi/2 \le \theta < \pi/2 \\ 0 & \text{otherwise} . \end{cases}$$

We assume $\ell < a$ to exclude the possibility that the needle might hit two lines at the same time. Then the line $x = 0$ is hit if $X - \ell \cos(\Theta)/2 < 0$ and the line $x = a$ is hit if $X + \ell \cos(\Theta)/2 > a$. Consider the subset of $[0, a) \times [-\pi/2, \pi/2)$ defined by

$$A = \{(x, \theta) : x \le \ell \cos(\theta)/2 \text{ or } x + \ell \cos(\theta)/2 > a\},$$

and let

$$g(X, \Theta) = 1_A(X, \Theta) = \begin{cases} 1 & (X, \Theta) \in A \\ 0 & \text{otherwise} . \end{cases}$$

The function $1_A(\cdot)$ is usually called the indicator function of set A. The probability that the needle hits a line is therefore

$$P_H = P((X, \Theta) \in A) = E[g(X, \Theta)] .$$

To evaluate the expectation of g, it is convenient to assume that the orientation angle Θ of the needle is known. In this case,

Fig. 2.12 Needle position with respect to vertical lines separated by a

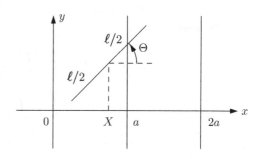

$$E[g(X, \Theta)|\Theta] = P(\{X < \ell \cos(\Theta)/2\} \cup \{X > a - \ell \cos(\Theta)/2\})$$

$$= \frac{\ell}{a} cos(\Theta) .$$

Using the iterated expectation principle, we find

$$P_H = E[g(X, \Theta)] = \frac{\ell}{a} E[\cos(\Theta)]$$

$$= \frac{\ell}{a\pi} \int_{-\pi/2}^{\pi/2} \cos(\theta) d\theta = \frac{2\ell}{a\pi} .$$

Note that the assumption $\ell < a$ ensures that $P_H < 1$.

2.10 Random Vectors

The tools used to analyze joint distributions of random variables can be extended in a straightforward manner to the study of random vectors with more than two components. Consider a random vector

$$\mathbf{Z} = \begin{bmatrix} \mathbf{X} \\ \mathbf{Y} \end{bmatrix} , \tag{2.99}$$

where $\mathbf{X}$ and $\mathbf{Y}$ are random vectors in $\mathbb{R}^n$ and $\mathbb{R}^m$ respectively, so that $\mathbf{Z}$ has $n + m$ components Z_i with $1 \le i \le n + m$. Let

$$\mathbf{z} = \begin{bmatrix} \mathbf{x} \\ \mathbf{y} \end{bmatrix}$$

denote an arbitrary vector of $\mathbb{R}^{n+m}$ partitioned in such a way that $\mathbf{x}$ and $\mathbf{y}$ belong to $\mathbb{R}^n$ and $\mathbb{R}^m$ respectively. Then if $\mathbf{z} \preceq \mathbf{z}'$ denotes the componentwise inequality such that all components z_i and z_i' of $\mathbf{z}$ and $\mathbf{z}'$ satisfy $z_i \le z_i'$ for $1 \le i \le n + m$, the CDF of random vector $\mathbf{Z}$ is defined as

$$F_{\mathbf{Z}}(\mathbf{z}) = P(\mathbf{Z} \preceq \mathbf{z}) = P(\mathbf{X} \preceq \mathbf{x}, \mathbf{Y} \preceq \mathbf{y}) = F_{\mathbf{X}\mathbf{Y}}(\mathbf{x}, \mathbf{y}) . \tag{2.100}$$

It is such that $F_{\mathbf{Z}}(\mathbf{z}) = 0$ if *one* of the components $z_i = -\infty$, and $F_{\mathbf{Z}}(\mathbf{z}) = 1$ if *all* $z_i = \infty$. It is also jointly monotone nondecreasing in the sense that

$$F_{\mathbf{Z}}(\mathbf{z})(\mathbf{z}) \le F_{\mathbf{Z}}(\mathbf{z}') \tag{2.101}$$

if $\mathbf{z} \preceq \mathbf{z}'$. The distribution information for subvectors $\mathbf{X}$ and $\mathbf{Y}$ is contained in the CDF $F_{\mathbf{Z}}(\mathbf{z})$ since

$$F_{\mathbf{X}}(\mathbf{x}) = F_{\mathbf{X}\mathbf{Y}}(\mathbf{x}, \infty) , \quad F_{\mathbf{Y}}(\mathbf{y}) = F_{\mathbf{X}\mathbf{Y}}(\infty, \mathbf{y}) , \tag{2.102}$$

i.e., the marginal CDF $F_{\mathbf{X}}(\mathbf{x})$ is obtained by setting to infinity the entries of $\mathbf{y}$ in $F_{\mathbf{X}\mathbf{Y}}(\mathbf{x}, \mathbf{y})$, and the marginal CDF $F_{\mathbf{Y}}(\mathbf{y})$ is obtained by setting to infinity the entries of $\mathbf{x}$ in the joint CDF.

The CDF $F_{\mathbf{Z}}(\mathbf{z})$ is absolutely continuous if it can be expressed as

$$F_{\mathbf{Z}}(\mathbf{z}) = \int_{D(\mathbf{z})} f_{\mathbf{Z}}(\mathbf{u}) d\mathbf{u}, \tag{2.103}$$

where the domain $D(\mathbf{z})$ is defined by

$$D(\mathbf{z}) = \{\mathbf{u} \in \mathbb{R}^{n+m} : \mathbf{u} \preceq \mathbf{z}\} \, ,$$

in which case

$$f_{\mathbf{Z}}(\mathbf{z}) = \frac{\partial^{n+m}}{\partial z_1 \partial z_2 \cdots \partial z_{n+m}} F_{\mathbf{Z}}(\mathbf{z})$$

is the PDF of $f_{\mathbf{Z}}(\mathbf{z})$. Since $F_{\mathbf{Z}}(\mathbf{z})$ is jointly non-monotone decreasing, the PDF $f_{\mathbf{Z}}(\mathbf{z})$ is nonnegative, and since the CDF is unity when all its entries are infinity, the total probability mass

$$\int_{\mathbb{R}^{n+m}} f_{\mathbf{Z}}(\mathbf{z}) d\mathbf{z} = 1 \, .$$

The joint PDF can be rewritten in partitioned form in terms of the subvectors $\mathbf{X}$ and $\mathbf{Y}$ of $\mathbf{Z}$ as $f_{\mathbf{Z}}(\mathbf{z}) = f_{\mathbf{XY}}(\mathbf{x}, \mathbf{y})$, and the CDF marginalization relations (2.102) imply the corresponding PDF marginalization identities

$$f_{\mathbf{X}}(\mathbf{x}) = \int_{\mathbb{R}^m} f_{\mathbf{XY}}(\mathbf{x}, \mathbf{y}) d\mathbf{y}$$

$$f_{\mathbf{Y}}(\mathbf{y}) = \int_{\mathbb{R}^n} f_{\mathbf{XY}}(\mathbf{x}, \mathbf{y}) d\mathbf{x} \, .$$

Then the conditional PDF of $\mathbf{X}$ given $\mathbf{Y}$ is defined by

$$f_{\mathbf{X}|\mathbf{Y}}(\mathbf{x}|\mathbf{y}) = \frac{f_{\mathbf{XY}}(\mathbf{x}, \mathbf{y})}{f_{\mathbf{Y}}(\mathbf{y})} \, . \tag{2.104}$$

Conditional Gaussian Distribution Consider a Gaussian random vector $\mathbf{Z}$ partitioned as indicated in (2.99), with mean vector and covariance matrix

$$\mathbf{m}_Z = \begin{bmatrix} \mathbf{m}_X \\ \mathbf{m}_Y \end{bmatrix} \, , \quad \mathbf{K}_Z = E[(\mathbf{Z} - \mathbf{m}_Z)(\mathbf{Z} - \mathbf{m}_Z)^T] = \begin{bmatrix} \mathbf{K}_X & \mathbf{K}_{XY} \\ \mathbf{K}_{YX} & \mathbf{K}_Y \end{bmatrix} \, .$$

Its PDF can be expressed compactly as

$$f_{\mathbf{Z}}(\mathbf{z}) = \frac{1}{(2\pi)^{(n+m)/2} |\mathbf{K}_Z|^{1/2}} \exp(-Q_Z(\mathbf{z})) \, ,$$

where $|\mathbf{K}_Z|^{1/2} \stackrel{\triangle}{=} |\det \mathbf{K}_Z|^{1/2}$ and

$$Q_Z(\mathbf{z}) = (\mathbf{z} - \mathbf{m}_Z)^T \mathbf{K}_Z^{-1} (\mathbf{z} - \mathbf{m}_Z)/2 \, .$$

By using the matrix factorization

$$\mathbf{K}_Z = \begin{bmatrix} \mathbf{I} & \mathbf{K}_{XY}\mathbf{K}_Y^{-1} \\ \mathbf{0} & \mathbf{I} \end{bmatrix} \begin{bmatrix} \mathbf{K}_{X|Y} & \mathbf{0} \\ \mathbf{0} & \mathbf{K}_Y \end{bmatrix} \begin{bmatrix} \mathbf{I} & \mathbf{0} \\ \mathbf{K}_Y^{-1}\mathbf{K}_{YX} & \mathbf{I} \end{bmatrix} \, ,$$

where

$$\mathbf{K}_{X|Y} = \mathbf{K}_X - \mathbf{K}_{XY}\mathbf{K}_Y^{-1}\mathbf{K}_{YX} \tag{2.105}$$

denotes the Schur complement of $\mathbf{K}_Y$ inside covariance matrix $\mathbf{K}_Z$, we find

$$|\mathbf{K}_Z| = |\mathbf{K}_{X|Y}||\mathbf{K}_Y|$$

and

$$\mathbf{K}_Z^{-1} = \begin{bmatrix} \mathbf{I} & \mathbf{0} \\ -\mathbf{K}_Y^{-1}\mathbf{K}_{YX} & \mathbf{I} \end{bmatrix} \begin{bmatrix} \mathbf{K}_{X|Y}^{-1} & \mathbf{0} \\ \mathbf{0} & \mathbf{K}_Y^{-1} \end{bmatrix} \begin{bmatrix} \mathbf{I} & -\mathbf{K}_{XY}\mathbf{K}_Y^{-1} \\ \mathbf{0} & \mathbf{I} \end{bmatrix} .$$

These identities can be used to decompose $Q_Z(\mathbf{z})$ as

$$Q_Z(\mathbf{z}) = Q_{X|Y}(\mathbf{x}|\mathbf{y}) + Q_Y(\mathbf{y}),$$

where

$$Q_Y(\mathbf{y}) = (\mathbf{y} - \mathbf{m}_Y)^T \mathbf{K}_Y^{-1} (\mathbf{y} - \mathbf{m}_Y)/2$$
$$Q_{X|Y}(\mathbf{x}|\mathbf{y}) = (\mathbf{x} - \mathbf{m}_{X|Y})^T \mathbf{K}_{X|Y}^{-1} (\mathbf{x} - \mathbf{m}_{X|Y})/2 ,$$

with

$$\mathbf{m}_{X|Y} = \mathbf{m}_X + \mathbf{K}_{XY}\mathbf{K}_Y^{-1}(\mathbf{Y} - \mathbf{m}_Y) , \tag{2.106}$$

so that the conditional density given by (2.103) can be expressed as

$$f_{X|Y}(\mathbf{x}|\mathbf{y}) = \frac{1}{(2\pi)^{n/2}|\mathbf{K}_{X|Y}|^{1/2}} \exp(-Q_{X|Y}(\mathbf{x}|\mathbf{y})) . \tag{2.107}$$

This result generalizes to the vector case the conditional density expression (2.74)–(2.76) for jointly Gaussian scalar random variables X and Y. This shows that jointly Gaussian random vectors remain Gaussian under conditioning.

The expectation of $g(\mathbf{Z})$ is defined as the $n + m$-th dimensional Stieltjes integral

$$E[g(\mathbf{Z})] = \int_{\mathbb{R}^{n+m}} g(\mathbf{z})dF_{\mathbf{Z}}(\mathbf{z}) ,$$

and to evaluate mixed moments of the entries of $\mathbf{Z}$ it is again convenient to introduce the characteristic function

$$\Phi_{\mathbf{Z}}(\mathbf{w}) = E[\exp(j\mathbf{w}^T\mathbf{Z})] = \int_{\mathbb{R}^{n+m}} \exp(j\mathbf{w}^T\mathbf{z})dF_{\mathbf{Z}}(\mathbf{z}) , \tag{2.108}$$

where

$$\mathbf{w} = \begin{bmatrix} w_1 \\ w_2 \\ \vdots \\ w_{m+n} \end{bmatrix} = \begin{bmatrix} \mathbf{u} \\ \mathbf{v} \end{bmatrix}$$

is an arbitrary vector of $\mathbb{R}^{n+m}$, which is partitioned in subvectors $\mathbf{u} \in \mathbb{R}^n$ and $\mathbf{v} \in \mathbb{R}^m$, and where

$$\mathbf{w}^T\mathbf{z} = \sum_{i=1}^{n+m} w_i z_i$$

denotes the inner product of $\mathbb{R}^{n+m}$. When $F_\mathbf{Z}(\mathbf{z})$ is absolutely continuous, $\Phi_\mathbf{Z}(\mathbf{w}) = \Phi_{\mathbf{XY}}(\mathbf{u}, \mathbf{v})$ is just the $n + m$-th dimensional Fourier transform of the PDF $f_\mathbf{Z}(\mathbf{z}) = f_{\mathbf{XY}}(\mathbf{x}, \mathbf{y})$, which can be therefore recovered from the characteristic function by the inverse Fourier transform

$$f_\mathbf{Z}(\mathbf{z}) = \frac{1}{(2\pi)^{(n+m)/2}} \int_{\mathbb{R}^{n+m}} \exp(-j\mathbf{w}^T \mathbf{z}) \Phi_\mathbf{Z}(\mathbf{w}) d\mathbf{w} . \tag{2.109}$$

Then, the random vectors $\mathbf{X}$ and $\mathbf{Y}$ are independent if any one of the following properties holds:

i) $F_\mathbf{Z}(\mathbf{z}) = F_\mathbf{X}(\mathbf{x}) F_\mathbf{Y}(\mathbf{y})$ or equivalently, for absolutely continuous distributions, $f_\mathbf{Z}(\mathbf{z}) = f_\mathbf{x}(\mathbf{x}) f_\mathbf{Y}(\mathbf{y})$;

ii) $\Phi_\mathbf{Z}(\mathbf{w}) = \Phi_\mathbf{X}(\mathbf{u}) \Phi_\mathbf{Y}(\mathbf{v})$;

iii) for all functions g and h we have $E[g(\mathbf{X})h(\mathbf{Y})] = E[g(\mathbf{X})]E[h(\mathbf{Y})]$.

In the discussion above, we have considered the case where $\mathbf{Z}$ is divided into two component vectors $\mathbf{X}$ and $\mathbf{Y}$. But $\mathbf{Z}$ can be partitioned into an arbitrary number of components which are mutually independent if the CDF, PDF, or characteristic function can be factored according to this partition. For example, the components Z_i, $1 \leq i \leq n + m$ of $\mathbf{Z}$ are mutually independent if the CDF or characteristic function can be factored as

$$F_\mathbf{Z}(\mathbf{z}) = \prod_{i=1}^{n+m} F_{Z_i}(z_i)$$

$$\Phi_\mathbf{Z}(\mathbf{w}) = \prod_{i=1}^{n+m} \Phi_{Z_i}(w_i) ,$$

or equivalently if for all functions g_i with $1 \leq i \leq n + m$, the expectation of the product of $g_i(Z_i)$ is the product of the expectations, i.e.,

$$E[\prod_{i=1}^{n+m} g_i(Z_i)] = \prod_{i=1}^{n+m} E[g_i(Z_i)] .$$

If we consider a random vector $\mathbf{Z}$ partitioned as indicated in (2.99) with an absolutely continuous distribution, the conditional expectation $E[\mathbf{g}(\mathbf{X}, \mathbf{Y})|\mathbf{Y}]$ of a vector function of the form

$$\mathbf{g}(\mathbf{X}, \mathbf{Y}) = \begin{bmatrix} g_1(\mathbf{X}, \mathbf{Y}) \\ g_2(\mathbf{X}, \mathbf{Y}) \\ \vdots \\ g_p(\mathbf{X}, \mathbf{Y}) \end{bmatrix}$$

can be defined component-wise as

$$E[g_i(\mathbf{X}, \mathbf{Y}))|\mathbf{Y}] = \int_{\mathbb{R}^n} g_i(\mathbf{x}, \mathbf{Y}) f_{\mathbf{X}|\mathbf{Y}}(\mathbf{x}|\mathbf{Y}) d\mathbf{x}$$

for $1 \leq i \leq p$. Then all properties of scalar conditional expectations carry over to the vector case, except that iterated conditioning can now be performed in multiple stages, instead of just two stages in (2.98).

Random Sums of Independent Random Variables The study of sums of independent random variables with a random number of terms provides a simple illustration of the interplay of independence and iterated conditioning. Suppose we consider a sum

$$Z = \sum_{k=1}^{N} X_k, \tag{2.110}$$

where the random variables X_k, $i \geq 1$ are independent identically distributed (iid) with CDF $F_X(x)$ and characteristic function $\Phi_X(u)$, and N is an integer valued random variable with PMF $p_N(n) = P(N = n)$ for $n \geq 1$. Note that the random variables X_k can be either discrete or continuous. The generating function of N is denoted as

$$G_N(z) = \sum_{n=1}^{\infty} z^n p_n . \tag{2.111}$$

A convenient approach to study the sum (2.110) consists of conditioning with respect to N. Given that $N = n$, the number of terms is fixed and the characteristic function of Z can be expressed as

$$\Phi_{Z|N}(u|n) = E[\exp(ju\sum_{k=1}^{n} X_k)] = E[\prod_{k=1}^{n} \exp(juX_k)]$$

$$= \prod_{k=1}^{n} E[\exp(juX_k)] = \Big(\Phi_X(u)\Big)^n ,$$

where the first equality of the second line is due to the independence of the X_ks. Then, by iterated conditioning, the characteristic function of Z can be expressed as

$$\Phi_Z(u) = E_N[E_{Z|N}[\exp(juZ)|N]] = \sum_{n=1}^{\infty} \Phi_{Z|N}(u|n) p_N(n)$$

$$= \sum_{n=1}^{\infty} \Big(\Phi_Z(u)\Big)^n p_n = G_N(\Phi_Z(u)) , \tag{2.112}$$

where the last equality is obtained by observing that the first expression on the second line is obtained by replacing z by $\Phi_X(u)$ in (2.111).

The expression (2.112) provides a convenient mechanism for evaluating the moments of Z. By using the chain rule of differentiation, we find

$$\frac{d}{du}\Phi_Z(u) = \frac{dG_N}{dz}(\Phi_X(u))\frac{d}{du}\Phi_X(u)$$

$$\frac{d^2}{du^2}\Phi_Z(u) = \frac{d^2G_N}{dz^2}(\Phi_X(u))\Big(\frac{d}{du}\Phi_X(u)\Big)^2$$

$$+ \frac{dG_N}{dz}(\Phi_X(u))\frac{d^2}{dz^2}\Phi_X(u) ,$$

and by setting $u = 0$ and dividing the resulting expressions by i and i^2 respectively, we find

$$E[Z] = E[N]E[X] \tag{2.113}$$

$$E[Z^2] = (E[N^2] - E[N])(E[X])^2 + E[N]E[X^2]$$
$$= E[N^2](E[X])^2 + E[N]K_X, \tag{2.114}$$

so that by squaring (2.113) and subtracting from (2.114), the variance of Z is given by

$$K_Z = K_N(E[X])^2 + E[N]K_X. \tag{2.115}$$

Example 2.10 Assume that in (2.111) the iid random variables X_i are exponential with parameter λ and N is a geometrically distributed random variable with parameter p, so

$$\Phi_X(u) = \frac{\lambda}{\lambda - ju} \quad, \quad G_N(z) = \frac{pz}{1 - (1 - p)z}.$$

Then

$$\Phi_Z(u) = G_N(\Phi_X(u)) = \frac{\dfrac{p\lambda}{\lambda - ju}}{1 - (1 - p)\dfrac{\lambda}{\lambda - ju}} = \frac{p\lambda}{p\lambda - ju},$$

which shows that Z is also exponential, but with parameter $\lambda_Z = p\lambda$. Then expressions (2.113) and (2.115) yield

$$E[Z] = \frac{1}{p} \times \frac{1}{\lambda} = \frac{1}{\lambda_Z}$$

$$K_Z = \frac{(1-p)}{p^2} \times \frac{1}{\lambda^2} + \frac{1}{p} \times \frac{1}{\lambda^2} = \frac{1}{\lambda_Z^2}, \tag{2.116}$$

which match exactly the mean and variance of an exponential random variable with parameter λ_Z.

2.11 Problems

2.1 Let $\{A_k, \; 1 \le k \le n\}$ be a set of mutually independent events of $\mathcal{F}$. Show that the events $\{B_k, \; 1 \le k \le n\}$, where for each k B_k is selected as either A_k or A_k^c, are also mutually independent. When $B_k = A_k^c$ for all k, this implies that if a set of events are independent, their complements are also independent.

2.2 A random variable X has the cumulative probability distribution

$$F_X(x) = (1 - \exp(-2x))u(x),$$

where $u(\cdot)$ denotes the unit-step function.

a) Calculate the probabilities $P(X \le 1)$, $P(X \ge 2)$, and $P(X = 2)$.
b) Find $f_X(x)$, the probability density function of X.

c) Let Y be a random variable obtained from X as follows:

$$Y = \begin{cases} 0 \text{ for } X < 2 \\ 1 \text{ for } X \geq 2 . \end{cases}$$

Find the CDF $F_Y(y)$ of Y.

2.3 The probability density function of the random variable X is

$$f_X(x) = \begin{cases} 1 - |x| \text{ for } |x| < 1 \\ 0 \quad \text{ for } |x| \geq 1 . \end{cases}$$

a) Find and sketch the cumulative probability distribution of X.
b) Calculate the probability that $|X| > 1/2$.
c) Calculate the conditional probability density function $f_X(x \mid |X| > 1/2)$ of the random variable X, given that $|X| > 1/2$, and sketch it.
d) Calculate the conditional CDF

$$F_X(x \mid |X| > 1/2) = P[X < x \mid |X| > 1/2],$$

and sketch it.

2.4 A limiter has the input–output characteristic shown in Fig. 2.13. Specifically, if the limiter input is X, its output Y is given by

$$Y = \begin{cases} -b \quad \text{for} \quad X < -a \\ bX/a \text{ for } -a \leq X \leq a \\ b \quad \text{for} \quad X > a . \end{cases}$$

The input to the limiter is a sine wave

$$X(t) = A \cos(\omega t + \Theta)$$

with $A > a$, where A and ω are known, but the phase Θ is random and uniformly distributed over $[0, 2\pi)$. The output $Y(t)$ of the limiter is sampled at an arbitrary time t to obtain a random variable Y. Find the CDF of the sampled output Y. *Hint*: If Θ is uniformly distributed over $[0, 2\pi)$, so is $\Psi = (\omega t + \Theta)$ modulo 2π, and we can write $X = A \cos(\Psi)$.

Fig. 2.13 Limiter
input–output characteristic

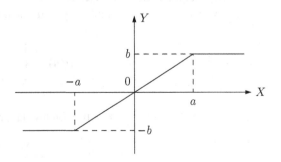

2.5 Let X be a random variable uniformly distributed over interval $[-3, 3]$. Find the cumulative probability distribution function (CDF) of $Y = g(X)$ if

$$g(x) = \begin{cases} 2 & 0 \le |x| \le 1 \\ \dfrac{2}{x^2} & 1 < |x| \le 2 \\ 0 & \text{otherwise} \,. \end{cases}$$

2.6 Let X be a Laplace distributed random variable with probability density

$$f_X(x) = \frac{a}{2} \exp(-a|x|),$$

where the parameter $a > 0$. Consider the transformation $Y = |X|^{1/2}$. Find the PDF of Y. Can you recognize this distribution?

2.7 One difficulty associated with the use of the inversion method to generate a random variable Y with arbitrary CDF $F_Y(y)$ is that it is not always possible to evaluate $F_Y(y)$ or its inverse in closed form. Another widely used random variable generation technique that can be used in such situations is John Von Neumann's *acceptance/rejection method*. Let $f_Y(y)$ denote the PDF of random variable Y. Von Neumann's method generates Y by using another random variable X which is easy to generate, whose density $f_X(x)$ is "close" to $f_Y(x)$, in the sense that the ratio $f_Y(x)/f_X(x)$ must be bounded by a fixed constant c, i.e.,

$$\frac{f_Y(x)}{f_X(x)} \le c$$

for all x. It also uses a second random variable U which is uniformly distributed over $[0, 1]$ and which is used to decide if X needs to be accepted or rejected. The method works as follows.

(i) Generate a random variable X with density $f_X(x)$.
(ii) Generate a random variable U independent of X which is uniformly distributed over $[0, 1]$.
(iii) If

$$U \le \frac{f_Y(X)}{cf_X(X)},$$

accept the random variable X and set $Y = X$. Otherwise, return to step (i).

To verify that the acceptance/rejection method works properly, it is useful to recall that the CDF of a random variable U which is uniformly distributed over $[0, 1]$ is

$$F_U(u) = \begin{cases} 0 & u < 0 \\ u & 0 \le u \le 1 \\ 1 & 1 < u \end{cases}$$

a) Use the CDF of U to show that the conditional probability

$$P\left(U \le \frac{f_Y(X)}{cf_X(X)} \Big| X = x\right) = \frac{f_Y(x)}{cf_X(x)} \,.$$

b) By using the principle of total probability, verify that

$$P(U \leq \frac{f_Y(X)}{cf_X(X)}) = \frac{1}{c} .$$

c) To show that the method works properly, it is necessary to prove that

$$F_Y(y) = P(X \leq y | U \leq \frac{f_Y(X)}{cf_X(X)}) ,$$

i.e., $F_Y(y) = P(A|B)$ where events $A = \{X \leq y\}$ and

$$B = \{U \leq \frac{f_Y(X)}{cf_X(X)}\} .$$

This can be done by using the definition

$$P(A|B) = \frac{P(B \cap A)}{P(B)}$$

of the conditional probability of two events, where $P(B)$ has already been computed in b), and by using again the principle of total probability to evaluate $P(B \cap A)$.

The question b) showed that the probability of acceptance of a pair (X, U) is only $p = 1/c$. Accordingly, the upper bound c should not be selected too large, since increasing c lowers the probability of acceptance. In fact, it is not difficult to verify that the number N of tries needed to generate one accepted pair obeys a geometric distribution of type 1 with parameter p. Accordingly, the expected number of tries needed to generate one accepted pair is

$$E[N] = \frac{1}{p} = c .$$

To illustrate the acceptance/rejection method, write a MATLAB program which generates 10^4 Beta(4,3) random variables Y with the density

$$f_Y(y) = \begin{cases} 60y^3(1 - y)^2 & 0 \leq y \leq 1 \\ 0 & \text{otherwise.} \end{cases}$$

To generate the random variables Y, it is suggested that you should use uniformly distributed random variables X over $[0, 1]$, so that in this case

$$f_X(x) = \begin{cases} 1 & 0 \leq x \leq 1 \\ 0 & \text{otherwise .} \end{cases}$$

d) Verify that

$$\frac{f_Y(x)}{f_X(x)} = f_Y(x) \leq 3$$

so that c can be selected equal to 3 in the acceptance/rejection method.

·

e) To verify that your Matlab program of the acceptance/rejection method works properly, construct a histogram (say, with 20 bins) of the random variables you have generated and compare it to the desired Beta(4,3) density.

f) *Optional:* In your Matlab program, include a counter which keeps track of the total number N_T of pairs (X, U) needed to generate 10^4 MATLAB Beta random variables. Verify that $N_T/10^4$ is approximately $c = 3$.

2.8 To generate discrete random variables, the transformation method described of Sect. 2.5 can be modified as follows. Let X be a uniform random variable over $[0, 1]$, and suppose that we seek to generate a discrete-valued random variable Y taking values y_k for $1 \leq k \leq K$, with PMF $p_k = P(Y = y_k)$.

a) Show that if we select $Y = y_k$ whenever the uniform random variable X satisfies

$$F_Y(y_{k-1}) = \sum_{\ell=1}^{k-1} p_\ell \leq X < F_Y(y_k) = \sum_{\ell=1}^{k} p_\ell ,$$

then Y has the desired PMF. Verify that the function $Y = h(X)$ thus constructed is a staircase function with rises $y_k - y_{k-1}$ and runs p_k, so it can be viewed as an inverse of the staircase function $X = F_Y(Y)$.

b) Consider the case where Y is geometric of Type 1, so that $P(Y = k) = p(1 - p)^{k-1}$ for $k \geq 1$. Verify that the rule of part a) can be implemented as

$$Y = \lfloor \frac{\ln(1 - X)}{\ln(1 - p)} \rfloor + 1 ,$$

where the floor function $\lfloor z \rfloor$ denotes the largest integer smaller than or equal to z.

c) Write a MATLAB program that generates $N = 10^3$ geometric random variables Y_i, $1 \leq i \leq N$ with parameter $p = 0.2$ and also evaluates the empirical probability distribution $q_k = N_k/N$ for $1 \leq k \leq 10$, where N_k denotes the number of Y_is equal to k. Use a stem plot to compare q_k and p_k for $1 \leq k \leq 10$.

2.9 Let X_1 and X_2 be two independent random variables with CDF $F_{X_1}(x_1)$ and $F_{X_2}(x_2)$, respectively. Let

$$Y = \max(X_1, X_2) \text{ and } Z = \min(X_1, X_2).$$

a) Find the CDF of Y.

b) Find the CDF of Z.

c) Suppose that X_1 and X_2 are exponential with parameters λ_1 and λ_2. Can you recognize the distribution satisfied by Z?

2.10 The random variables X, Y, and Z have the joint probability density function

$$f_{XYZ}(x, y, z) = \frac{1}{(2\pi)^{3/2}} \exp(-(x - y)^2/2) \exp(-(y - z)^2/2) \exp(-z^2/2).$$

a) Find the marginal joint density $f_{XY}(x, y)$.
b) Find the means and variances of X and Y and find the covariance $K_{XY} = E[(X - m_X)(Y - m_Y)]$.

2.11 The random variables X, Y, and Z have the joint probability density function

$$f_{X,Y,Z}(x, y, z) = Cz^2 \exp(-z(1 + x + y))u(x)u(y)u(z),$$

where $u(\cdot)$ denotes the unit step function.

a) Calculate the value of C, the marginal density functions $f_Z(z)$, $f_{X,Y}(x, y)$, and $f_X(x)$, and the conditional probability density $f_{X,Y|Z}(x, y \mid z)$.
b) Find the joint probability density of

$$U = Z(X + Y), \qquad V = X + Y, \qquad W = X.$$

From this, calculate the joint probability density $f_{U,V}(u, v)$ and the probability density $f_U(u)$.

2.12 Let X and Y be independent random variables such that X takes the discrete values 1 and -1 with probability $1/2$ each, i.e.

$$P(X = 1) = P(X = -1) = 1/2,$$

and Y is $N(0, 1)$ distributed. Consider the two random variables

$$Z = X + Y \text{ and } W = XY.$$

a) Find the probability density $f_Z(z)$ of Z.
b) Find the conditional probability density functions functions $f_{Z|X}(z|X = -1)$ and $f_{Z|X}(z|X = 1)$.
c) Find the mean values m_Y, m_W, the variances σ_Y^2, σ_W^2, and the covariance K_{YW}. Are Y and W uncorrelated random variables? Are Y and W statistically independent random variables?

2.13 Let X and Y be two independent random variables with Rayleigh densities

$$f_X(x) = \frac{x}{\alpha^2} \exp\left(-\frac{x^2}{2\alpha^2}\right)u(x)$$

$$f_Y(y) = \frac{y}{\beta^2} \exp\left(-\frac{y^2}{2\beta^2}\right)u(y).$$

a) Find the probability density of $Z = Y/X$. To do so you may want to find the joint density of the pair (Z, X) and then evaluate the marginal density of Z. To compute the marginal density, you may need the integral

$$\int_0^\infty r \exp(-cr)dr = \frac{1}{c^2}$$

for $c > 0$.
b) Use the result of part a) to evaluate $P[Y \geq X]$.

The physical motivation for this problem is as follows. In wireless communications over a flat fading channel, the amplitude of the signal received by an antenna is Rayleigh distributed. In switched diversity antenna systems, a receiver uses several antennas, and selects the one with the largest signal amplitude. From this perspective, X and Y represent here the signal amplitudes at two antenna elements, and $P[X \leq Y]$ is the probability that the output of the second antenna is selected when the signal fading coefficients at the two antennas are α^2 and β^2, respectively.

2.14 Let X and Y be two random variables with finite second moments $E[X^2]$ and $E[Y^2]$. By applying the derivation following (2.90) to the random variable

$$Z = |Y| - z|X|$$

with z real, and minimizing $h(z) = E[Z^2]$, prove the *Cauchy–Schwartz inequality*

$$|E[XY]| \leq E[|XY|] \leq (E[X^2]E[Y^2])^{1/2} . \tag{2.117}$$

2.15 The Cauchy–Schwartz inequality is a special case of Hölder's inequality: let p and q be such that $p, q \geq 1$ and

$$\frac{1}{p} + \frac{1}{q} = 1 ,$$

and denote the L_p norm of random variable X as $||X||_p = (E[|X|^p])^{1/p}$. Then if X and Y are two random variables such that $||X||_p$ and $||Y||_q$ are finite, we have

$$|E[XY]| \leq |E[|XY|] \leq ||X||_p ||Y||_q . \tag{2.118}$$

a) The derivation of inequality (2.118) relies on Young's inequality

$$uv \leq \frac{1}{p}u^p + \frac{1}{q}v^q ,$$

which holds for any nonnegative real numbers u and v. Derive this inequality by observing that

$$uv = \exp(\ln(uv)) = \exp(\frac{1}{p}\ln(u^p) + \frac{1}{q}\ln(v^q))$$

and then using the convexity of the exponential function. Recall that a function $f(x)$ is convex if its second derivative is strictly positive and in this case for $x = wx_1 + (1 - w)x_2$ with $0 \leq w \leq 1$ we have

$$f(x) \leq wf(x_1) + (1 - w)f(x_2) .$$

b) By setting $u = |x|/||X||_p$ and $v = |x|/||Y||_q$ in Young's inequality, and taking expectations on both sides of the resulting expression, i.e., performing a Stieltjes integration with respect to the joint CDF $F_{XY}(x, y)$, prove (2.118).

2.16 Show that if a random variable satisfies $E[|X|^n] < \infty$, it has moments of all orders up to n. This result can be proved by using the inequality

$$|x|^k \le |x|^n + 1$$

for $1 \le k \le n - 1$. It can also be derived by using Hölder's inequality.

2.17 Let X, Y, and U be independent random variables where X and Y are exponential with parameter λ, i.e., their density is given by

$$f_X(x) = \lambda \exp(-\lambda x) u(x).$$

Assume that $\lambda > 1$. The random variable U is uniformly distributed over $[0, 1]$. Evaluate

$$E[\exp((X + Y)U)].$$

Hint: Condition with respect to U and use the iterated expectation rule.

2.18 In addition to the transformation and acceptance/rejection methods, which can be used to generate arbitrary random variables, several specialized techniques can be employed to generate Gaussian random variables. The *polar method* [9] uses two independent random variables X and Y uniformly distributed over $[-1, 1]$, whose PDF is given by

$$f_X(u) = f_Y(u) = \begin{cases} \frac{1}{2} & -1 \le u \le 1 \\ 0 & \text{otherwise}. \end{cases}$$

Let $R = (X^2 + Y^2)^{1/2}$ be the amplitude of the pair (X, Y) and let

$$M = \{R \le 1\}$$

denote the event that the pair (X, Y) is located inside the unit disk centered at the origin.

a) If (R, Θ) represents the polar coordinate representation of (X, Y), i.e.

$$X = R\cos(\Theta) \quad Y = R\sin(\Theta),$$

find the joint probability density $f_{R\Theta}(r, \theta | M)$ of R and Θ conditioned on the event M.
b) Given two random variables X and Y uniformly distributed over $[-1, 1]$ with amplitude R, verify that *conditioned on the event M*, the random variables

$$Z = \frac{X}{R}\sqrt{-4\ln R} \quad W = \frac{Y}{R}\sqrt{-4\ln R}$$

are two independent zero-mean Gaussian random variables with unit variance.
c) Use the result of part b) and the `rand` command of MATLAB to generate 10^4 pairs of independent Gaussian random variables (Z, W) with zero mean and unit variance. Construct separate histograms, say with 20 bins, of the Z and W coordinates and compare them to the ideal Gaussian distribution. Evaluate the sampled mean and variances of the pairs you have generated.

2.19 The *Box–Muller* method for generating $N(0, 1)$ distributed random variables uses two independent random variables X and Y uniformly distributed over $[0, 1]$. Consider the transformed random variables

$$U = \left(-2\ln(X)\right)^{1/2}\cos(2\pi Y)$$

$$V = \left(-2\ln(X)\right)^{1/2}\sin(2\pi Y).$$

a) Find the joint probability density of U and V.
b) Verify that U and V are independent and $N(0, 1)$ distributed.
c) Write a MATLAB program to generate 10^4 pairs of independent $N(0, 1)$ distributed random variables (U, V) by using the Box–Muller method. Construct separate histograms, say with 20 bins, for U and V and compare them to the ideal Gaussian distribution.

2.20 Let N be a random variable taking only integer values, with probabilities $p_n = P[N = n]$ for $n \geq 0$. Its moment generating function is defined as

$$G_N(z) = E[z^N] = \sum_{n=0}^{\infty} p_n z^n.$$

a) Show that the moments $E[N]$, $E[N^2]$, ..., $E[N^k]$ of the random variable N can be expressed in terms of the derivatives

$$\frac{dG_N}{dz}(1) \quad , \quad \frac{d^2G_N}{dz^2}(1) \quad , \quad \frac{d^kG_N}{dz^k}(1).$$

b) If N is a Poisson distributed with parameter λ, so that

$$P[N = n] = \frac{\lambda^n}{n!}\exp(-\lambda)$$

for $n \geq 0$, find $G_N(z)$. Find the mean and variance of N.
c) Consider a sequence of independent, identically distributed discrete-valued random variables X_k, with generating function $K(z) = E[z^X]$. Let N be a random variable taking only positive integer values, and with generating function $G_N(z)$. Show that the generating function of

$$S = \sum_{k=1}^{N} X_k$$

satisfies

$$G_S(z) = G_N(K(z)).$$

Hint: Condition with respect to N and use the law of iterated expectations to evaluate

$$E[z^S] = E[E[z^S|N]].$$

d) In part c) assume that N is Poisson distributed with parameter λ, and that the random variables X_k are Bernoulli distributed with

$$P[X_k = 1] = p \ , \quad P[X_k = 0] = 1 - p \ .$$

Show that

$$G_S(z) = \exp(\lambda p(z - 1)) \ .$$

What type of random variables is S? What are its mean and variance?

2.21 In a simple photomultiplier, incident light ejects photoelectrons from a thin cathode, and these, accelerated across a short distance, impinge on a second metallic surface. Each electron striking it knocks out a random number X of secondary electrons, which are collected at an anode and counted. If the k-th primary electron ejects X_k secondaries, and if a total of N primaries are produced by the incident light during a time interval $[0, T]$, the total number of secondaries is

$$S = \sum_{k=1}^{N} X_k \ .$$

The number of primary electrons, however, is a random variable, since their ejection by incident light is a random phenomenon.

We assume here that the number N of primary electrons has the Poisson probability distribution

$$P(N = n) = \frac{\lambda^n}{n!} \exp(-\lambda)$$

with parameter λ. The number X of secondary electrons per primary has also a Poisson distribution

$$P(X = k) = \frac{\mu^k}{k!} \exp(-\mu)$$

but with parameter μ.

a) Find the generating function

$$G_X(z) = E[z^X] = \sum_{k=0}^{\infty} P(X = k)z^k$$

of the number X of electrons generated by a single primary. You may want to use the infinite sum

$$\exp(x) = \sum_{k=0}^{\infty} \frac{x^k}{k!} \ .$$

b) Given that $N = n$, find the generating function $E[z^S \mid N = n]$ of the number of secondary electrons. Assume that the secondaries X_k are independent.

c) Find the generating function

$$G_S(z) = E[z^S]$$

of the number of secondary electrons. Verify that it can be expressed as

$$G_S(z) = G_N(G_X(z)) \,,$$

where

$$G_N(z) = \sum_{n=0}^{\infty} P(N = n)z^n$$

is the generating function of the number N of primary electrons.
d) Evaluate the probability $P[S = 0]$ that no secondary electron is emitted.
e) Find the mean of S.

2.22 The Sacramento Police Department has $n = 420$ patrolmen/women on its payroll. Unfortunately, because of day offs, vacations, and sick leaves, each patrol person has only a probability $p = 4/7$ of being on duty on any given day. Thus the total number of patrol persons on duty on a given day can be modeled as

$$K = \sum_{i=1}^{n} D_i,$$

where the random variables D_i (indicating whether the i-th patrolperson is on duty or not) are independent and Bernoulli with

$$P(D_i = 1) = p \quad , \quad P(D_i = 0) = 1 - p \,.$$

a) Explain why the random variable K has the binomial distribution

$$P(K = k) = \binom{n}{k} p^k (1 - p)^{n-k} \,.$$

Evaluate the generating function $G_K(z) = E[z^K]$ of K. To do so, you may want to use the binomial expansion

$$(x + y)^n = \sum_{k=0}^{n} \binom{n}{k} x^k y^{n-k}$$

for appropriately selected values of x and y. Find the mean m_K and variance K_K of K.
b) Each day, in addition to responding to reports of crime in progress, patrol persons are supposed to monitor traffic violations and write tickets to help the City pay its share of the new Sacramento Kings downtown arena. The number M of tickets that each patrol person writes is uniformly distributed between 2 and 6, so that

$$P(M = m) = \begin{cases} 1/5 & 2 \le m \le 6 \\ 0 & \text{otherwise}. \end{cases}$$

Evaluate the generating function $G_M(z)$ of M. Find the mean m_M and variance K_M of M.

The total number of tickets written by all patrolmen/women on duty on a given day can therefore be written as

$$S = \sum_{\ell=1}^{K} M_\ell,$$

where the numbers M_ℓ of tickets written by each patrolperson are independent and uniformly distributed between 2 and 6 for $1 \le \ell \le K$, and as observed in a), K is random with a binomial distribution.

c) Suppose that the number K of patrol persons on duty is known and equals k. Find the conditional generating function

$$G_{S|K}(z|k) = E[z^S | K = k]$$

of the total number S of traffic tickets.
d) Use the principle of total probability in the form

$$G_S(z) = E[Z^S] = \sum_{k=0}^{n} E[z^S | K = k] P[K = k]$$

to find the generating function of S when K is random and unknown.
e) Evaluate the mean m_S and variance K_S of S. As much as possible try to express the values you obtain in function of $m_K, K_K, m_M,$ and K_M.

2.23 Consider a continuous random variable X with probability density $f_X(x)$ and characteristic function $\Phi_X(u) = E[e^{juX}]$. The random variable X is the input of a roundoff quantizer with step size Δ. The quantizer output is $Z = q(X)$, where the quantizer characteristic $q(X)$ is plotted in Fig. 2.14. The output

$$Z = n\Delta$$

can be represented as an integer multiple of the step size Δ where the integer $n = 0, \pm 1, \pm 2, \ldots$ The quantization error

$$E = X - q(X)$$

satisfies $-\Delta/2 \le E < \Delta/2$.

In digital signal processing, the quantization error E is often modeled as a random variable uniformly distributed over interval $[-\Delta/2, \Delta/2)$. so that its density is expressed as

$$f_E(e) = \begin{cases} \frac{1}{\Delta} & -\Delta/2 \le e < \Delta/2 \\ 0 & \text{otherwise}. \end{cases}$$

Fig. 2.14 Input–output transfer characteristic of a roundoff quantizer

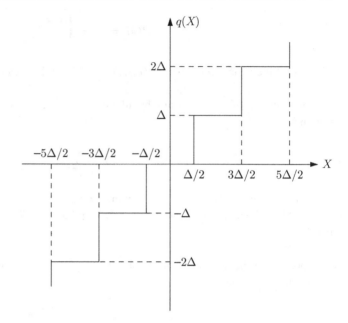

The purpose of this problem is to obtain a precise set of conditions in terms of the characteristic function $\Phi_X(u)$ under which the error density is exactly uniform.

a) Since the error E equals to X modulo $n\Delta$, show that the event $A = \{-\Delta/2 \le E \le e\}$ with $e < \Delta/2$ admits the partition

$$A = \cup_{n=-\infty}^{\infty}\{(n-1/2)\Delta \le X \le n\Delta + e\}.$$

Use this result to express the CDF $F_E(e)$ of E in terms of the CDF $F_X(x)$ of X, and by differentiating the resulting expression, show that the PDF $f_E(e)$ of E can be expressed in terms of the PDF $f_X(x)$ of X as

$$f_E(e) = \begin{cases} \sum_{n=-\infty}^{\infty} f_X(e+n\Delta) & -\Delta/2 \le e < \Delta/2 \\ 0 & \text{otherwise}. \end{cases}$$

b) Consider the function

$$g(e) = \sum_{n=-\infty}^{\infty} f_X(e+n\Delta) \tag{2.119}$$

for $-\Delta/2 \le e < \Delta/2$, and extend it periodically over the entire real line $(-\infty, \infty)$. Because $g(e)$ is periodic with period Δ, it admits a Fourier series of the form

$$g(e) = \sum_{k=-\infty}^{\infty} G_k e^{jk\omega_0 e}, \tag{2.120}$$

where

$$\omega_0 = \frac{2\pi}{\Delta}$$

denotes the fundamental frequency, and

$$G_k = \frac{1}{\Delta} \int_{-\Delta/2}^{\Delta/2} g(e) e^{-jk\omega_0 e} de$$

represents the k-th Fourier coefficient in the Fourier series representation of $g(e)$. By substituting (2.119) inside this expression, show that

$$G_k = \frac{1}{\Delta} \Phi_X(-k\omega_0). \tag{2.121}$$

Verify that $G_0 = 1/\Delta$.

c) By substituting (2.121) inside (2.120) prove that E is uniformly distributed over $[-\Delta/2, \Delta/2)$ if and only if

$$\Phi_X(k\omega_0) = 0 \tag{2.122}$$

for all $k \neq 0$.

d) Consider a PDF $f_X(x)$ which is piecewise constant over intervals corresponding to the quantizer steps, i.e.,

$$f_X(x) = c_n \tag{2.123}$$

for $(n - 1/2)\Delta \leq x < (n + 1/2)\Delta$, where

$$\sum_{n=-\infty}^{\infty} c_n = 1/\Delta$$

to ensure that the total probability mass is one. Evaluate the characteristic function of X and verify that it satisfies (2.122). Let $Z = \sum_{m=1}^{M} X_m$ be the sum of M independent random variables with the property (2.123). Verify that it also satisfies condition (2.122).

e) Assume now that X is $N(m, \sigma^2)$ distributed. Evaluate its characteristic function $\Phi_X(u)$. Please indicate how the step size Δ should be selected in comparison with the standard deviation σ to ensure that condition (2.122) is approximately satisfied.

References

1. D. P. Bertsekas and J. N. Tsitsiklis, *Introduction to Probability, 2nd edition*. Belmont, MA: Athena Scientific, 2008.
2. B. Hajek, "Probability with engineering applications." Notes for Course ECE 313 at University of Illinois at Urbana-Champaign, Jan. 2017.
3. C. Therrien and M. Tummala, *Probability and Random Processes for Electrical and Computer Engineers, 2nd edition*. Boca Raton, FL: CRC Press, 2011.
4. J. Walrand, *Probability in Electrical Engineering and Computer Science– An Application-Driven Course*. Quoi?, Feb. 2014.
5. K. L. Chung, *A Course in Probability Theory, Second Edition*. New York: Academic Press, 1968.
6. S. M. Kay, *Fundamentals of Statistical Signal Processing: Detection Theory*. Prentice-Hall, 1998.
7. B. C. Levy, *Principles of Signal Detection and Parameter Estimation*. New York, NY: Springer Verlag, 2008.
8. W. Rudin, *Principles of Mathematical Analysis, 2nd edition*. New York, NY: McGraw-Hill, 1964.
9. G. Marsaglia and T. A. Bray, "A convenient method for generating normal random variables," *SIAM Review*, vol. 6, pp. 260–264, July 1964.

Convergence and Limit Theorems

3

3.1 Introduction

The construction of models of random phenomena often relies on simplifications and approximations whose validity is typically justified by asymptotic arguments. Such arguments require extending to random variables the familiar convergence analysis of real numbers. We recall that if x_n, $n \geq 1$ is a sequence of real numbers, it converges to a real limit x if for each $\epsilon > 0$, there exists an integer $n(\epsilon)$ such that $|x_n - x| < \epsilon$ for all $n \geq n(\epsilon)$. One drawback of this definition is that it assumes the limit x is known, but given a sequence x_n, $n \geq 1$ it is not always easy to guess the limit. A convenient tool to analyze the convergence of sequences of real numbers when the limit is not yet known is the Cauchy criterion [1, p. 46], which states that the sequence x_n converges if and only if $\lim_{n,m \to \infty} |x_n - x_m| = 0$. The main difficulty in extending these concepts to sequences $X_n(\omega)$, $n \geq 1$ of random variables is that random variable sequences can be viewed as families of real number sequences indexed by the outcome $\omega \in \Omega$. So convergence can be defined in a variety of ways depending on whether we look at each sequence individually, or on whether we allow averaging across outcomes to measure the closeness of X_n to its limit X.

Convergence concepts play an important role in deriving key results of statistics such as the law of large numbers and the central limit theorem. However, these results hold only under specific conditions, and in the case of the central limit theorem, are often misapplied. The second part of this chapter will discuss in detail the derivation of limit theorems and the circumstances under which they can be applied safely.

3.2 Inequalities

It is often difficult to compute exactly probabilities of events of interest, and in such cases, inequalities can provide useful bounds for the magnitude of event probabilities.

Markov Inequality Consider a nonnegative random variable X with mean $E[X]$. Then, for any $z > 0$

$$P(X > z) \leq \frac{E[X]}{z} . \tag{3.1}$$

© Springer Nature Switzerland AG 2020
B. C. Levy, *Random Processes with Applications to Circuits and Communications*,
https://doi.org/10.1007/978-3-030-22297-0_3

Note that since all probabilities are less than or equal to 1, the inequality (3.1) is of interest only if $z > E[X]$. It is derived by observing that

$$E[X] = \int_0^z x \, dF_X(x) + \int_z^\infty x \, dF_X(x),$$
(3.2)

where the first term is nonnegative and the second term satisfies

$$\int_z^\infty x \, dF_X(x) \geq z \int_z^\infty dF_X(x) = z P(X > z).$$
(3.3)

The applicability of the Markov inequality on its own is somewhat limited, since the random variable X is required to be nonnegative, but by applying it to transformed random variables of the form $Y = g(X)$ with $Y \geq 0$ and mean $E[Y]$, where X is no longer assumed to be nonnegative, we can generate several important inequalities.

Chebyshev Inequality Let X be a random variable with mean m_X and variance K_X. Then, by applying the Markov inequality to $Y = (X - m_X)^2 \geq 0$ and observing that $E[Y] = K_X$, we find

$$P(|X - m_X| > z) = P((X - m_X)^2 > z^2) \leq \frac{K_X}{z^2}.$$
(3.4)

Note again that since probabilities are less than 1, the bound (3.4) is informative only if $z > \sigma_X = K_X^{1/2}$, i.e., if we are trying to evaluate the probability that X is at least one standard deviation away from its mean. It is also worth observing that the bound (3.4) is two-sided in the sense that it considers the probability that X belongs either to the left tail $X < m_X - z$ or to the right tail $X > m_X + z$. There exists also a one-sided version of the Chebyshev inequality that considers only one tail at a time.

One-Sided Chebyshev Inequality If X is a random variable with mean m_X and variance K_X, for $z > 0$ the right and left tail probabilities satisfy

$$P(X - m_X > z) \leq \frac{K_X}{K_X + z^2}$$
(3.5)

$$P(X - m_X < -z) \leq \frac{K_X}{K_X + z^2}.$$
(3.6)

By symmetry, we need only to consider the right tail. Let $Z = X - m_X$. Then for any $s > 0$, we have

$$P(Z > z) = P(Z + s > z + s) = P((Z + s)/(z + s) > 1)$$
$$\leq P(((Z + s)/(z + s))^2 > 1)$$
$$\leq E[((Z + s)/(z + s))^2] = \frac{K_X + s^2}{(z + s)^2} \triangleq b(s),$$
(3.7)

where to go from the second to the third line, we have used the Markov inequality for $Y = ((Z + s)/(z + s))^2$. Since s is arbitrary, it can be selected to minimize the bound $b(s)$. Setting the derivative of $b(s)$ equal to zero yields $s_0 = K_X/z$ and the corresponding minimal bound

$$b(s_0) = \frac{K_X}{K_X + z^2} \, ,$$

which proves (3.5).

Chernoff Bound Consider the random variable $Y = \exp(sX)$ where s is real and such that $E[Y] = E[\exp(sX)] = M_X(s)$ exists. Then $Y \geq 0$, and by applying the Markov inequality to Y, we find

$$P(X > z) = P(\exp(sX) > \exp(sz)) \leq \frac{M_X(s)}{\exp(sz)} \tag{3.8}$$

for $s > 0$, and

$$P(z > X) = P(\exp(sX) > \exp(sz)) \leq \frac{M_X(s)}{\exp(sz)} \tag{3.9}$$

for $s < 0$. The function

$$M_X(s) = E[\exp(sX)] = \int_{-\infty}^{\infty} \exp(sx) dF_X(x) \tag{3.10}$$

is the *moment generating function* of X. For the case of an absolutely continuous distribution, it can be expressed as

$$M_X(s) = \int_{-\infty}^{\infty} \exp(sx) f_X(x) dx \, ,$$

which is a Laplace transform, but with the variable s replaced by $-s$. The moment generating function can be used as a replacement for the characteristic function to compute all the moments of X, since it satisfies

$$\frac{d^k M_X}{ds^k}(0) = E[X^k]$$

for $k \geq 1$. One slight disadvantage of relying on this function is the need to keep track of its domain of convergence. Specifically, the characteristic function

$$\Phi_X(u) = M_X(s)\,|_{s=ju}$$

is guaranteed to exist because F_X has a total variation equal to 1. But $M_X(s)$ with s complex typically has only a finite domain of convergence, which includes the $s = ju$ axis since $M_X(s)$ is identical to the characteristic function on this axis. Since the integrand in (3.10) is nonnegative, when the integral diverges, we can set $M_X(s) = +\infty$ by convention, in which case the bounds (3.8) and (3.9) remain valid (but trivial). Since the bound $b(s) = \exp(-sz)M_X(s)$ on the right-hand side of (3.8) and (3.9) is s-dependent, we can make it as tight as possible by minimizing $b(s)$, or equivalently by maximizing

$$\beta(s) = -\ln b(s) = sz - \ln M_X(s) \tag{3.11}$$

over $s \geq 0$ and $s \leq 0$, respectively, where we observe that the function $\Gamma_X(s) = \ln M_X(s)$ is the *cumulant generating function* of X, in the sense that

$$\frac{d^k \Gamma_X}{ds^k}(0) = C_k,$$

where C_k denotes the k-th cumulant of X, i.e., $C_1 = m_X, C_2 = K_X$, etc.

Fig. 3.1 Construction of
the Legendre transform of
$\Gamma_X(s)$

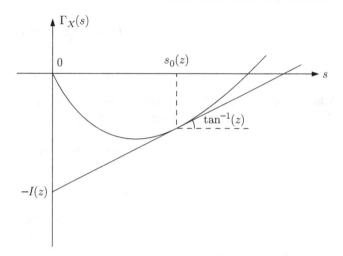

To perform this maximization, it is useful to observe that $\Gamma_X(s)$ is a convex function of s, as proved in Problem 3.2. Then, since $M_X(0) = 1$, we have $\beta(0) = 0$, and for $z > m_X = d\Gamma_X/ds(0)$, we recognize that the maximization of $\beta(s)$ for $s \geq 0$ is equivalent to its maximization over the entire real line, so

$$\sup_{s \geq 0}(sz - \Gamma_X(s)) = \sup_{s \in \mathbb{R}}(sz - \Gamma_X(s)) \overset{\triangle}{=} I(z) \tag{3.12}$$

for $z > m_X$, where sup denotes the supremum (least upper bound). The function $I(z)$ can be recognized at the Legendre transform of $\Gamma_X(s)$. It admits the following geometric interpretation when the maximizing value of $\beta(s)$ is finite. At the maximum $d\beta/ds = z - d\Gamma_X/ds = 0$, so the maximizing point s_0 is such that the slope of Γ_X at s_0 is z. The tangent of $\Gamma_X(s)$ at s_0 admits therefore the equation

$$y = \Gamma_X(s_0) + z(s - s_0) \,,$$

whose intercept with the vertical axis is $\Gamma_X(s_0) - zs_0 = -I(z)$. This shows therefore that $-I(z)$ is the intercept of the tangent to $\Gamma_X(s)$ of slope z with the vertical axis, as shown in Fig. 3.1. The Legendre transform has the property that it preserves convexity, so that $I(z)$ is itself convex, and it serves as its own inverse [2, Sec. 7.1], so that $\Gamma_X(s)$ can be recovered by Legendre transformation of $I(z)$.

Similarly, for the case when $z < m_X$ and the maximization of $\beta(s)$ is performed for $s \leq 0$, we conclude that the maximum for $s \leq 0$ is the same as the maximum over $\mathbb{R}$, which is again equal to $I(z)$. Accordingly, the tightest form of the Chernoff bounds (3.8) and (3.9) is given by

$$P(X > z) \leq \exp(-I(z)) \text{ for } z > m_X \tag{3.13}$$

$$P(X < z) \leq \exp(-I(z)) \text{ for } z < m_X \,. \tag{3.14}$$

The main application of this result will be for evaluating the probability of large deviations of the sampled mean

$$M_n = \frac{1}{n}\sum_{k=1}^{n} X_k$$

of independent identically distributed random variables X_k away from the statistical mean m_X. In this context, $I(z)$ is usually called the *rate function*.

Exponential Distribution Consider the case where X is exponential with parameter λ. Then the CDF and PDF are given respectively by $F(x) = [1 - \exp(-\lambda x)]u(x)$ and $f_X(x) = \lambda \exp(-\lambda x)u(x)$, and X has mean $m_X = 1/\lambda$. The generating function

$$M_X(s) = \int_0^\infty \lambda \exp(s - \lambda) dx = \frac{\lambda}{\lambda - s}$$

has for region of convergence $\Re s < \lambda$. By setting the derivative

$$\frac{d\Gamma_X}{ds} = \frac{1}{\lambda - s}$$

of $\Gamma_X(s) = \ln M_X(s)$ equal to z, we find that the tangency point is located at

$$s_0 = \lambda - 1/z$$

and

$$I(z) = zs_0 - \Gamma_X(s_0) = \lambda z - 1 - \ln(\lambda z) .$$

This expression admits an interesting interpretation in terms of the relative entropy or Kullback–Leibler (KL) divergence [3]

$$D(f_1|f_0) = \int_{-\infty}^\infty \ln(f_1(x)/f_0(x)) f_1(x) dx \tag{3.15}$$

of two PDFs. If f_1 and f_0 are exponential with parameters λ_1 and λ_0, respectively, a simple evaluation shows that the divergence is given by

$$D(f_1|f_0) = \frac{\lambda_0}{\lambda_1} - 1 - \ln(\lambda_0/\lambda_1) .$$

If we assume that the nominal PDF f_0 has parameter λ, and f_1 has parameter $\lambda_1 = 1/z$, we recognize that

$$I(z) = D(f_1|f_0) .$$

The choice of $\lambda_1 = 1/z$ can be explained by observing that when we compare z to $m_X = 1/\lambda$, z can be interpreted as the mean $1/\lambda_1$ of a second exponential distribution f_1 with parameter λ_1.

Since the CDF of X is known, the right and left tail probabilities can be computed exactly and compared to the Chernoff bounds (3.13) and (3.14) to assess their tightness. For the right tail specified by $z > m_X = 1/\lambda$, this gives

$$P(X > z) = \exp(-\lambda z) \leq \lambda z \exp(1 - \lambda z) ,$$

and for the left tail corresponding to $z < 1/\lambda$, we obtain

$$P(X < z) = 1 - \exp(-\lambda z) \leq \lambda z \exp(1 - \lambda z) .$$

Gaussian Distribution If X is $N(m_X, K_X)$ distributed, its moment generating function is given by

$$M_X(s) = \exp(m_X s + K_X s^2/2)$$

and is defined over the entire complex plane. Then by setting to zero the derivative of

$$\beta(s) = zs - \ln M_X(s) = (z - m_X)s - K_X s^2/2,$$

we find that the point where $\Gamma_X(s)$ has slope z is

$$s_0 = (z - m_X)/K_X,$$

and at this point

$$I(z) = \beta(s_0) = \frac{(z - m_X)^2}{2K_X}.$$

As in the exponential distribution case, this expression admits a KL divergence interpretation. The divergence between two $N(m_1, K_1)$ and $N(m_0, K_0)$ Gaussian PDFs f_1 and f_0 is given by [4]

$$D(f_1|f_0) = \frac{1}{2}\left[\frac{K_1}{K_0} - 1 - \ln(K_1/K_0) + (m_1 - m_0)^2/K_0\right].$$

Therefore if $f_0 = f_X$ and f_1 is $N(z, K_X)$ distributed, the rate function $I(z)$ can be expressed as distribution, then

$$I(z) = D(f_1|f_X).$$

The interpretation of the rate function $I(z)$ in terms of a KL divergence is not coincidental, and is a consequence of an elegant characterization of minimum discrimination information distributions due to Kullback [3, Sec. 3.2]. Specifically consider a reference density $f_X(x)$. Then the minimum of $D(f_1|f_X)$ under the constraint

$$E_1[X] = \int_{-\infty}^{\infty} x f_1(x) dx = z \tag{3.16}$$

is $I(z)$, where $E_1[.]$ denotes here the expectation with respect to density f_1. Furthermore, the PDF minimizing $D(f_1|f_X)$ takes the form

$$f_1(x) = \exp(s_0 x) f_X(x)/M_X(s_0), \tag{3.17}$$

where

$$s_0 = \arg\min_{s \in \mathbb{R}} \beta(s)$$

is the point where the tangent of $\Gamma_X(s)$ has slope x.

In this respect it is worth noting that, since the PDF f_1 minimizing the divergence is obtained by applying a multiplicative exponential factor to the nominal density f_X, f_1 will belong to the same family as f_X only if f_X belongs to the canonical exponential class

$$f_X(x|\boldsymbol{\theta}) = g(x)h(\boldsymbol{\theta})\exp(\boldsymbol{\theta}^T \mathbf{S}(x)),$$

Fig. 3.2 Tangent $t(x)$ to convex function $g(x)$ at x_0

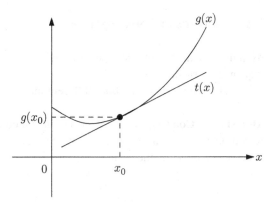

where θ is a parameter vector and $\mathbf{S}(x)$ is a vector statistic. This class includes the exponential and Gaussian distributions considered above. Finally, although the previous discussion has focused on absolutely continuous distributions, the minimum discrimination information interpretation of the rate function can be extended to discrete-valued random variables as shown in Problems 3.3 and 3.4.

Jensen's Inequality If X is a random variable such that $E[|X|]$ is finite, and $g(x)$ is a convex function, then

$$g(E[X]) \leq E[g(X)] . \tag{3.18}$$

For simplicity, assume that $g(x)$ is differentiable, even though the result holds in general [5]. Then if $g(x)$ is convex, its derivative dg/dx is monotone nondecreasing for all x. This implies that for an arbitrary point x_0

$$g(x) \geq g(x_0) + \frac{dg}{dx}(x_0)(x - x_0) = t(x), \tag{3.19}$$

where $t(x)$ denotes the tangent of $g(x)$ at x_0, as shown in Fig. 3.2. To verify (3.19), assume that $x \geq x_0$. Then

$$g(x) = g(x_0) + \int_{x_0}^{x} \frac{dg}{dx}(u)du$$

$$\geq g(x_0) + \frac{dg}{dx}(x_0) \int_{x_0}^{x} du = g(x_0) + \frac{dg}{dx}(x_0)(x - x_0) \stackrel{\triangle}{=} t(x),$$

where to go from the first to the second line, we have used the monotonicity of dg/dx. A similar proof holds for $x \leq x_0$. Suppose now that we select $x_0 = E[X]$. We have

$$t(X) \leq g(X),$$

so by taking the expectation on both sides, and observing that

$$E[t(X)] = t(E[X]) = g(E[X])$$

since $t(x)$ is linear and $x_0 = E[X]$, we obtain (3.18).

3.3   Modes of Convergence

As indicated in the introduction, the main conceptual difficulty in evaluating the convergence of sequences of random variables is that these sequences define families of real number sequences $X_n(\omega)$ indexed by outcome $\omega \in \Omega$. Four different modes of convergence will be examined.

Almost Sure Convergence   The most stringent concept requires that, except for a few outcomes with zero probability, each one of the separate real sequences $X_n(\omega)$ converges to the real limit $X(\omega)$. Specifically, consider the event

$$A = \{\omega : \lim_{n \to \infty} X_n(\omega) = X(\omega).\}$$

Then X_n converges to X almost surely, which is denoted as $X_n \overset{a.s.}{\to} X$ if $P(A) = 1$.

Typically, the best tool for establishing almost sure convergence is the Borel–Cantelli lemma.

Example 3.1   Consider a fair game of roulette where all slots on the wheel are evenly divided between red and black (no slot is reserved for the casino). A gambler arrives at the table with an initial capital X_0 and decides to bet everything on black at each turn until red finally occurs, in which case the gambler stops. If the ball stops on black, the payoff is equal to the money wagered, and if it stops on red, the wager is lost. Let B_k denote a Bernoulli random variable representing the outcome of the k-th spin of the wheel, where $B_k = 1$ or 0 depending on whether the ball falls on black, or red, respectively. The random variables B_k are independent, and since black and red are equally likely

$$P(B_k = 1) = P(B_k = 0) = 1/2 \ .$$

Then if X_n denotes the capital of the gambler after the n-th spin,

$$X_n = (\prod_{k=1}^{n} B_k) 2^n X_0 \ .$$

It is intuitively obvious that the gambler will almost surely leave the casino with empty pockets. This can be demonstrated rigorously by considering for an arbitrary small $\epsilon > 0$ the sequence of events $E_n = \{\omega : X_n(\omega) > \epsilon\}$. In other words, an outcome belongs to E_n if the gambler has not yet gone bust after the n-th spin, which of course means that $B_k = 1$ for $1 \le k \le n$. We find

$$P(E_n) = \prod_{k=1}^{n} P(B_k = 1) = (1/2)^n \ ,$$

and since

$$\sum_{n=1}^{\infty} P(E_n) = \sum_{n=1}^{\infty} (1/2)^n = 1,$$

we conclude that the probability that event E_n will occur infinitively often is zero. This means that the gambler goes bust almost surely. Note, however, that this conclusion relies on the implicit assumption that the casino has infinite capital and cannot itself go bankrupt.

Convergence in the Mean Square Instead of examining separately the distance $|X_n(\omega) - X(\omega)|$ for each ω as n tends to infinity, it is possible to average this distance across all outcomes by considering the mean-square $E[(X_n - X)^2]$. Then X_n converges to X in the mean square, which is denoted as $X_n \overset{m.s.}{\to} X$, if

$$\lim_{n \to \infty} E[(X_n - X)^2] = 0 . \tag{3.20}$$

Convergence in the mean square is not ensured by almost sure convergence, since the squared error $(X_n(\omega) - X(\omega))^2$ may become progressively larger over a set of vanishingly small probability, leading to a nonzero limit for $E[(X_n - X)^2]$.

Example 3.1 (Continued) For the roulette problem examined earlier, we have

$$E[X_n^2] = 2^{2n} X_0^2 \prod_{k=1}^{n} P(B_k = 1) = 2^n X_0^2,$$

which diverges as n tends to infinity, so X_n does not converge to 0 in the mean-square sense, even though it converges to 0 almost surely.

Conversely, mean-square convergence does not imply almost sure convergence either, as can be seen from the following example.

Example 3.2 Consider a sequence $\{X_n, n \geq 1\}$ of independent Bernoulli random variables, where X_n has parameter $1/n$, so that

$$P(X_n = 1) = \frac{1}{n} , \quad P(X_n = 0) = 1 - \frac{1}{n} .$$

Then consider for a small positive ϵ the sequence of events $E_n = \{\omega : X_n(\omega) > \epsilon\}$. Since the random variables are independent, the events E_n are independent and

$$P(E_n) = P(X_n = 1) = \frac{1}{n} .$$

The sum

$$\sum_{n=1}^{\infty} P(E_n) = \sum_{n=1}^{\infty} \frac{1}{n} = \infty ,$$

and by the second part of the Borel Cantelli lemma, we conclude that the probability that E_n occurs infinitely often is one, so X_n does not converge to zero almost surely. On the other hand

$$E[X_n^2] = 1 \times P(X_n = 1) = \frac{1}{n} ,$$

so that X_n converges to 0 in the mean-square sense.

Together Examples 3.1 and 3.2 indicate that almost sure convergence does not imply mean-square convergence and vice versa.

Convergence in Probability The third convergence mode we introduce is similar in concept to mean-square convergence, but it employs a different measure to evaluate deviations of X_n from X across all outcomes. For every $\epsilon > 0$, it uses the deviation metric

$$1_{|X_n - X| > \epsilon} = \begin{cases} 1 & |X_n - X| > \epsilon \\ 0 & |X_n - X|| \le \epsilon \end{cases}$$

Thus it applies the same unit penalty too all deviations $|X_n - X|$ larger than ϵ, no matter how large they may be. Since

$$E[1_{|X_n - x| > \epsilon}] = P(|X_n - X| > \epsilon) \,,$$

this amounts of course to computing the probability that $|X_n - X|$ exceeds ϵ. Then X_n converges in probability to X as n tends to infinity, which is denoted as $X_n \xrightarrow{p} X$ if for every $\epsilon > 0$

$$\lim_{n \to \infty} P(|X_n - X| > \epsilon) = 0 \,. \tag{3.21}$$

It turns out that convergence in probability is implied by both mean-square convergence and almost sure convergence. Since $(X_n - X)^2$ is nonnegative and has a finite mean whenever X_n converges to X in the mean square, the Markov inequality implies

$$P(|X_n - X| > \epsilon) = P((X_n - X)^2 > \epsilon^2) \le \frac{E[(X_n - X)^2]}{\epsilon^2} \,. \tag{3.22}$$

This means that when $E[(X_n - X)^2]$ tends to zero, so does $P(|X_n - X| > \epsilon)$ for all choices of $\epsilon > 0$, and thus mean-square convergence implies convergence in probability. To prove that almost sure convergence implies convergence in probability, consider the event sequence

$$B_n = \cup_{m \ge n}\{\omega : |X_m(\omega) - X(\omega)| > \epsilon\} \tag{3.23}$$

for $n \ge 1$. This sequence is monotone decreasing and converges to $B = \cap_{n \ge 1} B_n$. By the continuity property (2.10), the monotone decreasing sequence $P(B_n)$ has limit $P(B)$. If X_n converges towards X almost surely, for each ω such that $X_n(\omega)$ converges towards $X(\omega)$, and for each $\epsilon > 0$, there exists an integer n_0 such that $|X_n - X| \le \epsilon$ for all $n \ge n_0$. This means ω does not belong to B_{n_0} and thus does not belong to B. In other words, if A is the set of outcomes for which $X_n(\omega)$ converges to $X(\omega)$, B is contained in A^c. Since $P(A) = 1$, this implies $P(B) = 0$. Then the definition (3.23) of B_n implies

$$0 \le P(|X_n - X| > \epsilon) \le P(B_n) \,.$$

Since $P(B_n)$ converges to $P(B) = 0$ this means that $P(|X_n - X| > \epsilon)$ also converges to zero, thus proving convergence in probability.

Conversely, convergence in probability does not imply either almost sure convergence or mean-square convergence. To see this, consider Examples 3.1 and 3.2. In Example 3.1, the gambler's capital converges almost surely to 0, and thus converges in probability to 0, but it does not converge in the mean square to 0, so that convergence in probability does not imply convergence in the mean square. In Example 3.2, X_n converges in the mean square, and thus in probability, to 0 but it does not converge almost surely. This shows that convergence in probability is weaker than almost sure convergence.

Convergence in Distribution The final convergence concept does not even consider sequences of outcomes $X_n(\omega)$, but just the statistical distribution of the random variables X_n and their limit X. If X_n has CDF $F_{X_n}(x)$, X_n is said to converge in distribution to X which is denoted as $X_n \overset{d}{\to} X$ if

$$\lim_{n \to \infty} F_{X_n}(x) = F_X(x) \tag{3.24}$$

for all points x where the CDF $F_X(x)$ of X is continuous.

Example 3.3 To illustrate the intrinsic weakness of the convergence in distribution concept, let X be a $N(0, 1)$ random variable, and consider the sequence

$$X_n = (-1)^n X .$$

Unless $X(\omega) = 0$, this sequence does not converge since it keeps oscillating between $X(\omega)$ for n even and $-X(\omega)$ for n odd, so it does not converge almost surely. But each X_n is $N(0, 1)$ distributed so $X_n \overset{d}{\to} X$. On the other hand, since

$$X_n - X = \begin{cases} 0 & n \text{ even} \\ 2X & n \text{ odd} , \end{cases}$$

we conclude that for n odd

$$P[|X - X_n| > \epsilon] = P[|X| > \epsilon/2] = 2Q(\epsilon/2) ,$$

which does not tend to zero (in fact it is close to 1 for small ϵ) as n tends to infinity.

Convergence in Distribution Criteria It turns out that convergence in distribution, which is also called *weak convergence* can be characterized in a variety of ways which are equivalent to the CDF convergence criterion (3.24). Specifically, $X_n \overset{d}{\to} X$ if and only if any one of the following properties holds.

i) For all bounded continuous functions g,

$$\lim_{n \to \infty} E[g(X_n)] = E[g(X)] . \tag{3.25}$$

ii) If $\Phi_{X_n}(u)$ and $\Phi_X(u)$ denote respectively the characteristic functions of X_n and X

$$\lim_{n \to \infty} \Phi_{X_n}(u) = \Phi_X(u) \tag{3.26}$$

pointwise for all real u.

Proof To prove that (3.24) implies (3.25), note that for any $\epsilon > 0$, $E[g(X_n)]$ and $E[g(X)]$ can be approximated with less than ϵ error by finite Stieltjes sums of the form

$$\sum_{i=1}^{I} g(c_i)[F_{X_n}(x_i) - F_{X_n}(x_{i-1})]$$

and

$$\sum_{i=1}^{I} g(c_i)[F_X(x_i) - F_X(x_{i-1})]$$

with $x_{i-1} \leq c_i < x - i$, where the x_i's are selected to be continuity points of F_X. then since $F_{X_n}(x_i)$ converges to $F_X(x_i)$ for all $1 \leq i \leq I$, and the function g is bounded, we can find $N(\epsilon)$ such that for all $n \geq N(\epsilon)$, the two Stieltjes sums are within ϵ of each other, ensuring that $E[g(X_n)]$ and $E[g(X)]$ are within 3ϵ of one another. The convergence property of the characteristic functions follows from (3.25) by selecting $g(x) = \exp(jux)$ which is continuous and bounded by 1. The difficult part of the result consists in proving that (3.26) implies (3.24). This is a consequence of the Lévy continuity theorem discussed below, which is proved in [5, Sec. 6.3].

Example 3.4 Consider a sequence of binomial random variables X_n, such that X_n has parameters n and $p_n = \lambda/n$, so that its PMF is given by

$$p_{X_n}(k) = \binom{n}{k}(p/n)^k(1 - p/n)^{(n-k)}$$

for $0 \leq k \leq n$. As n becomes large, it is customary to approximate X_n by a Poisson distributed random variable with parameter λ, so that its PMF is given by

$$p_X(k) = \frac{\lambda^k}{k!}\exp(-\lambda)$$

for $k \geq 0$. To prove the convergence in distribution of X_n to X, consider the characteristic functions

$$\Phi_{X_n}(u) = \sum_{k=0}^{n} p_{X_n}(k)\exp(juk) = \left[1 + \frac{p}{n}(\exp(ju) - 1)\right]^n$$

$$\Phi_X(u) = \sum_{k=0}^{\infty} p_X(k)\exp(juk) = \exp(\lambda(\exp(ju) - 1)),$$

and note that

$$\lambda = \frac{1}{j}\frac{d}{du}\Phi_{X_n}(u)\mid_{u=0} = \frac{1}{j}\frac{d}{du}\Phi_X(u)\mid_{u=0},$$

so that λ is the mean of both X_n and X. Then, by using the approximation

$$(1 + a/n)^n \approx \exp(a)$$

for large n with $a = p(\exp(ju) - 1)$, we deduce that

$$\lim_{n \to \infty} \Phi_{X_n}(u) = \Phi_X(u)$$

pointwise, which establishes the convergence in distribution of X_n to X as n tends to infinity.

Lévy's Continuity Theorem One aspect of the convergence criterion (3.26) that can create some difficulty is that even if the characteristic functions $\Phi_{X_n}(u)$ admit a pointwise limit $\Phi(u)$, it is not always easy to determine if $\Phi(u)$ is the characteristic function of a valid CDF. This question was

Fig. 3.3 Relationship
between the four modes of
convergence

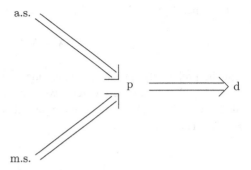

answered by the French mathematician Paul Lévy, who showed that if the limit $\Phi(u)$ is continuous at $u = 0$, it is the characteristic function of a CDF $F_X(x)$, and $X_n \overset{d}{\to} X$ (see [5]). The surprising part of this result is not that a continuity requirement is imposed, since characteristic functions are continuous over $\mathbb{R}$, but that continuity at $u = 0$ is sufficient.

The relationship existing in general between the four modes of convergence is depicted in Fig. 3.3. Unfortunately, the relation between modes of convergence is slightly more complicated than suggested by this figure, since there exists partial converses to two of the implications appearing in Fig. 3.3.

Convergence to a Constant If X_n converges in distribution to a constant c, it also converges in probability to c. To prove this, observe that the CDF of $X = c$ is the step function

$$F_X(x) = u(x - c) = \begin{cases} 0 & x < c \\ 1 & x \geq c. \end{cases}$$

The only point where $F_X(x)$ is discontinuous is $x = c$, so when X_n converges in distribution to $X = c$, we have

$$\lim_{n \to \infty} F_n(x) = \begin{cases} 0 & x < c \\ 1 & x > c. \end{cases}$$

Then for $\epsilon > 0$, the event $A_n = \{|X_n - c| > \epsilon\}$ admits the nonoverlapping decomposition

$$A_n = \{X_n > c + \epsilon\} \cup \{X_n < c - \epsilon\},$$

so

$$P(|X_n - c| > \epsilon) = 1 - F_{X_n}(c + \epsilon) + F_{X_n}(c - \epsilon),$$

where the terms $1 - F_{X_n}(c + \epsilon)$ and $F_{X_n}(x - \epsilon)$ both converge to zero as n tends to infinity. This implies

$$\lim_{n \to \infty} P(|X_n - c| > \epsilon) = 0,$$

which shows that X_n converges to $X = c$ in probability.

Uniformly Bounded Random Variables If the random variables X_n satisfy $P(|X_n| < C) = 1$ where C denotes a positive constant, then if X_n converges to X in probability, it converges to X in the mean square. We show first that if X_n converges in probability to X, then X is bounded by C almost surely. The inequality

$$|X| \leq |X - X_n| + |X_n|$$

implies that if $|X_n - X| \leq \epsilon$ with $\epsilon > 0$ and $X_n \leq C$, then $|X| \leq C + \epsilon$. Thus

$$P(|X - X_n| \leq \epsilon) \leq P(|X| \leq C + \epsilon). \tag{3.27}$$

But the probability on the left-hand side of (3.27) tends to 1 as n tends to infinity, so $P(|X| \leq C + \epsilon) = 1$. Letting ϵ go to zero, we conclude that $P(|X| \leq C) = 1$. Then let $Z_n = |X - X_n|$. Since $|X - X_n| \leq |X| + |X_n|$, this implies that the CDF of Z_n satisfies $F_{Z_n}(z) = P(Z_n \leq z) = 1$ for $z \geq 2C$. Thus for $\epsilon > 0$

$$E[(X_n - x)^2] = \int_0^\epsilon z^2 dF_{Z_n}(z) + \int_\epsilon^{2C} z^2 dF_{Z_n}(z)$$
$$\leq \epsilon^2 + (2C)^2 P(|X_n - X| > \epsilon), \tag{3.28}$$

where $P(|X_n - X| > \epsilon)$ tends to zero as n tends to infinity, since X_n converges to X in probability. This shows

$$\lim_{n \to \infty} E[(X_n - X)^2] \leq \epsilon^2$$

and by letting ϵ go to zero, we conclude that X_n converges to X in the mean square.

Example 3.5 To illustrate simultaneously the two partial converses we just derived, consider a sequence of random variables $X_k, k \geq 1$, which are independent and uniformly distributed over $[0, 1]$. Let $Z_n = \max_{1 \leq k \leq n} X_k$ denote the maximum of the first n random variables. We would normally expect that as n becomes large, the maximum will approach 1, since as more samples are drawn at random, it becomes increasingly likely that samples very close to 1 will be selected. To prove that X_n converges to 1 in distribution, we recall that the distribution of the X_ks is given by

$$F_X(x) = \begin{cases} 0 & x < 0 \\ x & 0 \leq x < 1 \\ 1 & 1 \leq x. \end{cases}$$

To evaluate the CDF of Z_n, we observe that in order for the maximum Z_n to to be less than or equal to z, each of the random variables $X_k, 1 \leq k \leq n$ must be less than or equal to z, so

$$F_{Z_n}(z) = P(Z_n \leq z) = \prod_{k=1}^n P(X_k \leq z)$$

$$= (F_X(z))^n = \begin{cases} 0 & z < 0 \\ z^n & 0 \leq z < 1 \\ 1 & z \geq 1. \end{cases}$$

But z^n tends to zero for all $0 \leq z < 1$, so

$$\lim_{n \to \infty} F_{Z_n}(z) = u(z - 1)$$

pointwise, where we recognize that the unit step $u(z - 1)$ is the CDF of the constant random variable $Z = 1$. This shows that Z_n converges in distribution to the constant 1. But this implies it also converges to 1 in probability, and since the random variables $|Z_n|$ are bounded by 1, Z_n converges to 1 in the mean square!

Sometimes the convergence of a sequence can be established indirectly, by observing that the sequence of interest is obtained by transforming or combining sequences which are known to be convergent. This can be accomplished by using the following two results.

Continuous Transformations If $g(x)$ is a continuous function on $\mathbb{R}$, then

i) $g(X_n) \xrightarrow{a.s.} g(X)$ if $X_n \xrightarrow{a.s.} X$;

ii) $g(X_n) \xrightarrow{p} g(X)$ if $X_n \xrightarrow{p} X$;

iii) $g(X_n) \xrightarrow{d} g(X)$ if $X_n \xrightarrow{d} X$.

Proof i) is a consequence of the fact that if the real number sequence $X_n(\omega)$ converges to $X(\omega)$ for a given outcome ω, the continuity of g implies

$$\lim_{n\to\infty} g(X_n(\omega)) = g(\lim_{n\to\infty} X_n(\omega)) = g(X(\omega)),$$

so the set A of outcomes for which $X_n(\omega)$ and $g(X_n(\omega))$ converge to $X(\omega)$ and $g(X(\omega))$, respectively is the same, and $P(A) = 1$.

To prove ii), let B_h be the set of real numbers x such that there exist y with $|x - y| < h$, but $|g(x) - g(y)| > \epsilon$. If $X(\omega) \notin B_h$, and $|g(X_n(\omega)) - g(X(\omega))| > \epsilon$, then $|X_n(\omega) - X(\omega)| \geq h$. Thus

$$P(|g(X_n) - g(X)| > \epsilon) \leq P(X \in B_h) + P(|X_n - X| > h).$$

For a fixed $h > 0$, the second term tends to zero as n tends to infinity, since X_n converges to X in probability. Because of the continuity of g, the first term tends to zero as h tends to zero.

To prove iii), let h denote an arbitrary bounded continuous function. Then the function $h \circ g$ defined by $h \circ g(x) = h(g(x))$ is also bounded and continuous, so

$$\lim_{n\to\infty} E[h \circ g(X_n)] = E[h \circ g(X)],$$

since $X_n \xrightarrow{d} X$. But conversely, since h is an arbitrary continuous bounded function, this implies that $g(X_n) \xrightarrow{d} g(X)$.   ◇

Slutsky's Theorem If $g(x, y)$ is a continuous function over $\mathbb{R}^2$, $X_n \xrightarrow{d} X$, and $Y_n \xrightarrow{d} c$, where c denotes a constant, then $g(X_n, Y_n) \xrightarrow{d} g(X, c)$.

The derivation goes along the following lines: first note that if Y_n converges to c in distribution, it also converges to c in probability. Then, one can show that the joint distribution of (X_n, Y_n) converges in distribution to the joint distribution of (X, c), so

$$(X_n, Y_n) \xrightarrow{d} (X, c).$$

Then by using a two variable version of the continuous mapping theorem proved earlier, we conclude that $g(X_n, Y_n)$ converges in distribution to $g(X, c)$. This result implies in particular that if $X_n \xrightarrow{d} X$ and $Y_n \xrightarrow{d} c$, then as n tends to infinity

i) $X_n + Y_n \overset{d}{\to} X + c$;

ii) $X_n Y_n \overset{d}{\to} cX$;

iii) $X_n / Y_n \overset{d}{\to} X/c$ if $c \neq 0$,

which is sometimes referred to as Slutsky's theorem.

3.4   Law of Large Numbers

Even if we consider random variables whose distribution admit a mean, there is no reason to assume that the mean is known. In practice, this mean must be discovered by performing statistical surveys, or in the case of natural phenomena, such as weather patterns, by collecting data for an extended period of time. The underpinning of such statistical methods is the *law of large numbers*. This law relates the sampled mean

$$M_n = \frac{1}{n} \sum_{k=1}^{n} X_k \tag{3.29}$$

of a set of identically distributed random variables $\{X_k, k \geq 1\}$ to its statistical mean m_X. Depending on the strength of assumptions placed on X_k, there are several different flavors to the basic result which states that M_n converges in some sense to m_X.

Mean-Square Convergence If the random variables X_k are uncorrelated with finite mean m_X and variance K_X, the sample mean M_n converges in the mean square to the statistical mean m_X.

Proof We have

$$M_n - m_X = \frac{1}{n} \sum_{k=1}^{n} (X_k - m_X) \,,$$

so

$$E[(M_n - m_X)^2] = \frac{1}{n^2} \sum_{k=1}^{n} \sum_{\ell=1}^{n} E[((X_k - m_X)(X_\ell - m_X)] \,. \tag{3.30}$$

Then by using the fact that

$$E[(X_k - m_X)(X_\ell - m_X)] = K_X \delta(k - \ell) \,,$$

where

$$\delta(m) = \begin{cases} 1 & m = 0 \\ 0 & m \neq 0 \end{cases}$$

denotes the Kronecker delta function, the expression (3.30) yields

$$E[(M_n - m_X)^2] = \frac{K_X}{n} \,,$$

which tends to zero as n tends to infinity, so $Z_n \overset{m.s.}{\to} m_X$.   ◇

Since mean-square convergence implies convergence in probability, we deduce that when the X_k's are uncorrelated and have finite mean and variance, M_n converges also to m_X in probability. However, it is also possible to establish convergence in probability under different assumptions.

Weak Law of Large Numbers (WLLN) If the X_k's are iid with mean m_X, M_n converges in probability to m_X. The main difference between this result and the mean-square convergence result described above is that the X_k's are no longer required to have a finite variance, but they need to be independent, instead of being merely uncorrelated.

Proof We prove that M_n converges in distribution to m_X, which then implies convergence in probability since m_X is a constant. By using the independence of the X_k's, the characteristic function of Z_n can be expressed as

$$\Phi_{M_n}(u) = E[\exp(juZ_n)] = E[\prod_{k=1}^{n} \exp(juX_k/n)]$$

$$= \prod_{k=1}^{n} E[\exp(juX_k/n))] = (\Phi_X(u/n))^n, \qquad (3.31)$$

where $\Phi_X(u)$ denotes the characteristic function of the X_k's. For a fixed u, the behavior of $\Phi_X(u/n)$ as n becomes large is governed by the behavior of $\Phi_X(v)$ in the vicinity of $v = 0$. Since $\Phi_X(0) = 1$ and X has mean m_X, $\Phi_X(v)$ admits a Taylor series expansion of the form

$$\Phi_X(v) = 1 + jm_X v + o(v),$$

where $o(v)$ denotes a term that decays to zero faster than v as v tends to zero. But

$$\Phi_{M_n}(u) = (1 + jum_X/n + o(u/n))^n$$

tends to $\Phi_M(u) = \exp(ium_X)$ pointwise as n tends to infinity, where $\Phi_M(u)$ is the characteristic function of the constant random variable $M = m_X$. This proves that M_n converges in distribution, and thus in probability to m_X.

Strong Law of Large Numbers (SLLN) If the random variables X_k are iid and such that $E[|X_1|] < \infty$, then M_n converges to m_X almost surely. The proof of the SLLN under the minimal assumption $E[|X_1|] < \infty$ is rather complicated (see [6, Sec. 7.5]), so we derive the result under the stronger assumption that X_1 has a finite 4-th order moment, i.e., $E[X_1^4] < \infty$. Note that this implies that X_1 has lower order moments as well, as shown in Problem 2.16. By replacing the X_ks by the centered random variables $X_k - m_X$, we can assume $m_X = 0$, without loss of generality. Then consider

$$E[M_n^4] = \frac{1}{n^4} \sum_{j=1}^{n} \sum_{k=1}^{n} \sum_{\ell=1}^{n} \sum_{m=1}^{n} E[X_j X_k X_\ell X_m].$$

Depending on the choice of indices j, k, ℓ, and m, the terms on the right-hand side take the form

$$E[X_j^4] \;, \quad E[X_j^3 X_k] = E[X_j^3]E[X_k] = 0 \;, \quad E[X_j^2 X_k^2] = E[X_j^2]E[X_k^2]$$

$$E[X_j^2 X_k X_\ell] = E[X_j^2]E[X_k]E[X_\ell] = 0 \;,$$

$$E[X_j X_k X_\ell X_m] = E[X_j]E[X_k]E[X_\ell]E[X_m] = 0 \;,$$

where all indices j, k, ℓ, and m are different. There are n terms of the form $E[X_j^4]$ and $3n(n-1)$ terms of the form $E[X_j^2]E[X_k^2]$, so we find

$$E[M_n^4] = \frac{1}{n^4}\left(nE[X_1^4] + 3n(n-1)(E[X_1^2])^2\right).$$

Since the right-hand side is a summable sequence, we have

$$\sum_{n=1}^{\infty} E[M_n^4] < \infty. \tag{3.32}$$

But for any $\epsilon > 0$, the Markov inequality implies

$$P(M_n^4 > \epsilon) \leq E[M_n^4]/\epsilon,$$

which in light of (3.32) yields

$$\sum_{n=1}^{\infty} P(M_n^4 > \epsilon) < \infty.$$

The Borel Cantelli lemma implies that the probability that $M_n^4 > \epsilon$ infinitely often is zero, so M_n^4 and thus M_n converge almost surely to 0.

Remark As expressed, the law of large numbers gives the impression that it can only be used to estimate the means of random variables. However, given a set of iid random variables X_k, the law of large numbers can be applied to any transformed sequence $Y_k = g(X_k)$, which is iid since the X_ks are iid, and used to estimate $m_Y = E[g(X_k)]$ as long as the required moment conditions are satisfied by Y_1.

i) For example, suppose that the iid sequence X_k with $k \geq 1$ has a known mean m_X, but its variance K_X is unknown and needs to be estimated. Then the random variables $Y_k = (X_k - m_X)^2$ have mean $m_Y = E[(X_k - m_X)^2] = K_X$, so by the SLLN, the sampled variance

$$V_n = \frac{1}{n}\sum_{k=1}^{n} Y_k = \frac{1}{n}\sum_{k=1}^{n}(X_k - m_X)^2$$

converges almost surely to K_X. On the other hand, if mean-square convergence of V_n to K_X is required, Y_k must have second-order moments, and thus X_k must have fourth-order moments.

ii) Likewise, suppose we seek to estimate the CDF $F_X(x)$ of a sequence of iid random variables X_k for a fixed value of x. Consider the iid random variables $Y_k = u(x - X_k)$. We have

$$E[Y_k] = 1 \times P(X_k \leq x) = F_X(x),$$

so the random variables Y_k have the desired mean. Then the empirical CDF

$$\hat{F}_n(x) = \frac{1}{n}\sum_{k=1}^{n} Y_k = \frac{1}{n}\sum_{k=1}^{n} u(x - X_k)$$

converges almost surely to $F_X(x)$, where among the first n random variables X_k, $1 \leq k \leq n$, $\hat{F}_n(x)$ denotes the fraction of those which are less than or equal to to x. Since $Y_k^2 = Y_k$, the second moment of the Y_ks is $F_X(x)$, so the empirical CDF $\hat{F}_n(x)$ converges also in the mean square to $F_X(x)$.

Given an iid sequence X_k, $k \geq 1$, it is also possible to apply the SLLN to several different statistics of X_k to establish the almost sure convergence of a function of these statistics. For example, suppose the X_ks have a finite second moment. When the mean m_X is unknown, the variance is often estimated by

$$V_n = \frac{1}{n-1} \sum_{k=1}^{n} (X_k - M_n)^2 ,$$

where

$$M_n = \frac{1}{n} \sum_{k=1}^{n} X_k$$

is the sample mean of the X_k's. The SLLN implies that M_n converges almost surely to m_X, and that the sampled power

$$P_n = \frac{1}{n} \sum_{k=1}^{n} X_k^2$$

converges almost surely to $E[X_1^2]$. Then, since

$$V_n = \frac{n}{n-1}(P_n - (M_n)^2),$$

where $n/(n-1)$ converges to 1 independently of the outcome ω, then

$$V_n \overset{a.s.}{\rightarrow} K_X = E[X_1^2] - m_X^2 .$$

3.5 Regular, Moderate, and Large Deviations

Section 3.4 was concerned with showing that the sample mean M_n converges to the statistical mean m_X under appropriate conditions. In this section, we examine the fluctuations of M_n about m_X as n becomes large.

Central Limit Theorem (CLT) The first result of interest examines the convergence of

$$Y_n = \frac{\sum_{k=1}^{n}(X_k - m_X)}{n^{1/2}} , \tag{3.33}$$

when the random variables X_k are iid, with mean m_X and finite variance K_X. The finite variance assumption is essential and cannot be omitted, as will be seen later in Sect. 3.6. Then as n tends to infinity, we have

$$Y_n \overset{d}{\rightarrow} Y, \tag{3.34}$$

where Y is $N(0, K_X)$ distributed.

Proof Let $\check{X}_k = X_k - m_X$. The random variables $\check{X}_k$s are iid with zero mean, variance K_X, and their characteristic function can be denoted as $\Phi_{\check{X}}(u)$. The characteristic function of Y_n can be expressed as

$$\Phi_{Y_n}(u) = E[\prod_{k=1}^{n} \exp(j\check{X}_k u/n^{1/2})]$$

$$= \prod_{k=1}^{n} E[\exp(j\check{X}_k u/n^{1/2})] = \left(\Phi_{\check{X}}(u/n^{1/2})\right)^n, \tag{3.35}$$

where the first equality on the second line is due to the independence of the $\check{X}_k$s. Because $\check{X}_k$ has zero mean and variance K_X,

$$\Phi_{\check{X}}(v) = 1 - K_X v^2/2 + o(v^2)$$

in the vicinity of $v = 0$. Using again the approximation

$$(1 + a/n)^n \approx \exp(a)$$

for large n with $a = -K_X u^2/2$, we find

$$\lim_{n\to\infty} \Phi_{Y_n}(u) = \exp(-K_X u^2/2) \overset{\triangle}{=} \Phi_Y(u)$$

pointwise, where we recognize that the limit $\Phi_Y(u)$ is the characteristic function of a $N(0, K_X)$ random variable Y. This proves (3.34). $\diamond$

Regular Deviations Interpretation Since Y_n can be expressed in terms of the sampled mean M_n as

$$Y_n = n^{1/2}(M_n - m_X)$$

for $y > 0$, the CLT can be recast as

$$Q(y/K_X^{1/2}) = \lim_{n\to\infty} P(Y_n > y) = \lim_{n\to\infty} P(M_n > m_X + y/n^{1/2}). \tag{3.36}$$

Similarly for $y < 0$, the CLT takes the form

$$Q(-y/K_X^{1/2}) = \lim_{n\to\infty} P(Y_n < y) = \lim_{n\to\infty} P(M_n < m_X + y/n^{1/2}). \tag{3.37}$$

The identities (3.36) and (3.37) show that the CLT can be interpreted as a tool for evaluating the probability of deviations proportional to $|y|/n^{1/2}$ of the sampled mean M_n away from its limit m_X as n becomes large. Deviations of this type are the most frequent and are therefore called regular deviations.

Unfortunately, the CLT is often applied mistakenly not just to regular deviations, but also to large deviations, which often leads to incorrect results, as we shall see below. In this respect, it is useful to recall that for large z, $Q(z)$ admits the approximation

$$Q(z) \approx \frac{1}{(2\pi)^{1/2} z} \exp(-z^2/2). \tag{3.38}$$

Large Deviations Let X_k, $k \geq 1$ denote a sequence of iid random variables with mean m_X, generating function $M_X(s)$, and cumulant generating function $\Gamma_X(s) = \ln M_X(s)$. Then the sum $S_n = \sum_{k=1}^{n} X_k$ has generating function

$$M_{S_n}(s) = E[\exp(s \sum_{k=1}^{n} X_k)] = M_X(s)^n$$

and cumulant generating function

$$\Gamma_{S_n}(s) = \ln M_{S_n}(s) = n \ln \Gamma_X(s) .$$

Then if $z > m_X$, by applying the bound (3.8), we find that the sample mean $M_n = S_n/n$ satisfies

$$P(M_n > z) = P(S_n > nz) \leq \exp(-n(zs - \ln(\Gamma_X(s))), \tag{3.39}$$

so that if $I(z)$ denotes the rate function defined in (3.12), the Chernoff bound takes the form

$$P(M_n > z) \leq \exp(-nI(z)) . \tag{3.40}$$

Similarly for $z < m_X$, we obtain

$$P(M_n < z) \leq \exp(-nI(z)) . \tag{3.41}$$

Amazingly, it turns out that the $I(z)$ decay exponent of the Chernoff upper bound is the correct exponent of large deviation probabilities.

Cramer's Theorem If

$$M_n = \frac{1}{n} \sum_{k=1}^{n} X_k$$

where the random variables X_k are iid with finite mean m_X, then for $z > m_X$

$$\lim_{n \to \infty} \frac{1}{n} \ln P(M_n > z) = -I(z) . \tag{3.42}$$

Similarly, for $z < m_X$

$$\lim_{n \to \infty} \frac{1}{n} \ln P(M_n < z) = -I(z) . \tag{3.43}$$

This result, which is due to the Swedish statistician Harald Cramér [7], is proved in [8, Sec. 1.3] (see also [9, Chap. 3]). At this point, it is of interest to compare the exponents (3.42) and (3.43) with those predicted by an incorrect application of the CLT. If we set $y = n^{1/2}(z - m_X)$ in (3.37) and use the asymptotic approximation of the $Q(\cdot)$ function, the CLT would predict

$$\frac{1}{n} \ln P(M_n > z) \approx -\frac{(z - m_X)^2}{2K_X} , \tag{3.44}$$

which is the correct rate function $I(z)$ when the X_ks are independent $N(m_x, K_X)$ distributed, but is wrong when the X_ks are not Gaussian.

Fig. 3.4 Comparison of $I(z)$ to CLT estimate (3.42) for exponential random variables with parameter $\lambda = 1$

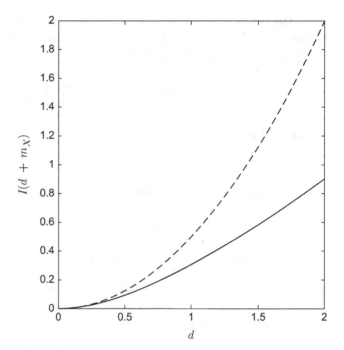

Example 3.6 Assume that the random variables X_k are exponential with parameter λ. Recall that the mean $m_X = 1/\lambda$ and variance $K_X = 1/\lambda^2$, and that it was shown in Sect. 3.2 that

$$I(z) = \lambda z - 1 - \ln(\lambda z) .$$

To compare the decay rates predicted by Cramer's theorem and by the CLT estimate (3.42), we assume $\lambda = 1$, so $m_X = K_X = 1$. The decay rate is plotted as a function of $d = z - m_X$ in Fig. 3.4, where the solid line corresponds to the rate function $I(z)$, and the broken line corresponds to the CLT estimate (3.42). The plots clearly show that the CLT predicts a much higher decay rate of the probability of large deviations than the actual value, which indicates that the CLT should not be used for estimating probabilities of large deviations of the sample mean.

Moderate Deviations Let us consider now moderate deviations of the form $\gamma_n y$, where the sequence γ_n satisfies

$$\lim_{n \to \infty} \gamma_n = 0 , \quad \lim_{n \to \infty} \gamma_n n^{1/2} = \infty .$$

In other words $1/n^{1/2} \ll \gamma_n \ll 1$, so that the deviations we consider are in between regular and large deviations. Then under the assumptions of Cramer theorem, for all $y > 0$, we have

$$\lim_{n \to \infty} \frac{1}{n\gamma_n^2} \ln P(M_n > m_X + \gamma_n y) = -\frac{y^2}{2K_X} , \tag{3.45}$$

which indicates that the decay exponent of the CLT is correct for intermediate deviations.

This result appears in [10, p. 553]. Following [11], the upper bound part can be derived as follows. By observing that the random variables X_k can be replaced by their centered version $\check{X}_k = X_k - m_X$, the mean m_X can be set equal to zero, without loss of generality. Then for $s > 0$, the Markov inequality yields

$$P(M_n \geq \gamma_n y) = P(\exp(M_n s n \gamma_n) \geq \exp(s n \gamma_n^2 y))$$

$$\leq \exp(-s n \gamma_n^2 y) \exp(n \Gamma_X(s \gamma_n)) .$$

But since γ_n approaches zero for large n

$$\Gamma_X(s \gamma_n) \approx (\gamma_n s)^2 K_X / 2 ,$$

so

$$\lim_{n \to \infty} \frac{1}{n \gamma_n^2} \ln P(M_n \geq \gamma_n y) \leq -sy + K_X s^2 / 2.$$

The bound holds for any $s > 0$, and by selecting the minimizing value $s = y / K_X$, we obtain

$$\lim_{n \to \infty} \frac{1}{n \gamma_n^2} \ln P(M_n \geq \gamma_n y) \leq -\frac{y^2}{2 K_X} .$$

Remark The moderate deviations result (3.45) indicates only that the CLT models accurately the asymptotic rate of decay of moderate deviations probabilities. This does not mean that the CLT approximation itself is accurate, since the nonexponential part of the approximation (the $1/((2\pi)^{1/2} z)$ term in (3.38)), may be inaccurate. In fact, it is shown in [10, p. 549] that

$$\lim_{n \to \infty} \frac{P(M_n > m_X + y/n^{1/3})}{Q(n^{1/6} y / K_X^{1/2})} = 1 , \tag{3.46}$$

so that the CLT is accurate for deviations of order $n^{-1/3}$ or less, the CLT decay exponent is valid for deviations of order less than 1, and the rate function $I(z)$ predicts the rate of decay of deviation probabilities of order 1, depicted by Fig. 3.5, which delineates precisely the zones of validity of CLT-based approximations, where d denotes the deviation with respect to the mean m_X. In this respect, it is also worth pointing out that the discussion has focused here on *relative errors* in the CLT approximation. It turns out that for iid random variables X_k such that $E[|X_k|^3]$ is finite, it was shown by Berry and Esseen that the absolute error of the CLT approximation admits a uniform bound proportional to $n^{-1/2}$ [10]. This result is often invoked to justify the use of the CLT approximation, but when evaluating rare events, the CLT has the potential to be wildly inaccurate, so careful use is recommended.

Fig. 3.5 Zones of validity of the CLT, moderate, and large deviations approximations

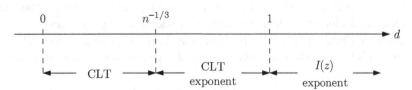

3.6   Stable Distributions

Given iid random variables X_k, with $k \geq 1$, the assumptions that the X_ks have a finite mean, and have a finite variance, play an essential role in the validity of the law of large numbers, and of the CLT, respectively. To see why this is the case, assume that the X_ks are Cauchy distributed with median μ and scale parameter σ. Then the sample mean

$$M_n = \frac{1}{n} \sum_{k=1}^{n} X_k$$

has for characteristic function

$$\Phi_{M_n}(u) = E[\exp(i \sum_{k=1}^{n} X_k u/n)] = (\Phi_X(u/n))^n$$

$$= (\exp(i\mu u/n - \sigma|u/n|)^n = \exp(iu\mu - \sigma|u|),$$

where the second equality on the first line uses the independence of the X_k's. Quite surprisingly, the sample mean M_n is itself Cauchy with median μ and scale parameter σ, so it obviously does not converge to a constant as n tends to infinity, and thus does not satisfy the law of large numbers, as expected since the x_k's do not have a mean. Likewise, $n^{1/2} M_n$ is not asymptotically Gaussian and thus does not satisfy the CLT.

Another interesting feature of the above example is that the sample mean M_n has exactly the same distribution as the iid random variables X_k. This constitutes an example of what is called a *stable distribution*. A CDF $F_X(x)$ with characteristic function $\Phi_X(u)$ is said to be stable if given a set of independent $F_X(x)$ distributed random variables X_k with, $k \geq 1$, there exists scaling and translation constants a_n and b_n such that

$$Z_n = \frac{\sum_{k=1}^{n} X_k - b_n}{a_n} \tag{3.47}$$

is also $F_X(x)$ distributed. In terms of characteristic functions, this implies that $\Phi_X(u)$ must satisfy

$$\Phi_X(u) = E[\exp(iuZ_n)] = E[\exp(i \sum_{k=1}^{n} X_k u/a_n)] \exp(-iub_n/a_n)$$

$$= \left(\Phi_X(u/a_n)\right)^n \exp(-iub_n/a_n) \tag{3.48}$$

for all integers n.

Based on the previous discussion, it is clear that Cauchy random variables are stable with $a_n = n$ and $b_n = 0$. It is also not difficult to verify that if the X_ks are $N(m_X, K_X)$, so will Z_n, as long as we select $a_n = n^{1/2}$ and $b_n = (n - a_n)m_X$, so Gaussian distributions are also stable. The characterization of stable distributions was undertaken by Paul Lévy, who was later joined in this study by Khinchin [12, 13], who found (see [14] for a comprehensive study of stable laws) that if $\Phi_X(u)$ is the characteristic function of a stable distribution, its logarithm can be expressed as

$$\ln \Phi_X(u) = i\mu u - c|u|^{\alpha}[1 - i\beta \text{sgn}(u)\gamma(u, \alpha)], \tag{3.49}$$

where

$$\gamma(u, \alpha) = \begin{cases} \tan(\alpha\pi/2) & \text{if } \alpha \neq 1 \\ -(2/\pi)\ln(|u|) & \text{if } \alpha = 1, \end{cases} \tag{3.50}$$

and

$$\text{sgn}(u) = \begin{cases} 1 & u > 0 \\ 0 & u = 0 \\ -1 & u < 0 \end{cases}$$

denotes the usual sign function. The parametrization (3.49) and (3.50) depends on four parameters: the exponent α with $0 < \alpha \leq 2$ controls the decay rate of the distribution tails, μ is a location parameter controlling the position of the PDF on the real axis, $c > 0$ is a scale parameter controlling the dispersion of the probability mass, and β with $-1 \leq \beta \leq 1$ is a skewness parameter controlling the right-handedness ($\beta > 0$) or left-handedness ($\beta < 0$) of the distribution. When $\beta = 0$, the PDF is symmetric about μ, which is the median of the distribution.

The case $\alpha = 2$, $\beta = 0$ corresponds to $N(\mu, 2c)$ distributions, and Cauchy random variables with median μ and scale parameter $\sigma = c$ are obtained by setting $\alpha = 1$ and $\beta = 0$. Finally, note that for $0 < \alpha < 2$, the CDF of a stable distribution satisfies [14, Chap. 7]

$$\lim_{x \to \infty} x^\alpha F_X(-x) = C_- \tag{3.51}$$

$$\lim_{x \to \infty} x^\alpha (1 - F_X(x)) = C_+ \tag{3.52}$$

with $C_- \geq 0$, $C_+ \geq 0$, and $C_- + C_+ > 0$, so that at least one of its tails decays like $1/|x|^\alpha$. This indicates that, except for the Gaussian case, stable distributions have fat tails, and in fact, because of property (3.51) and (3.52), they are often called Paretian distributions since Pareto distributions with $0 < \alpha < 2$ have similar tail properties. Note in particular that for $1 < \alpha < 2$, stable distributions have a mean, but no second-order moment, and for $0 < \alpha \leq 1$, they have no moment. Finally, note that by taking the logarithm of (3.48), substituting (3.49) and (3.50), and matching coefficients on both sides of the resulting identity, it is easy to verify that the scaling and translation constants are $a_n = n^{1/\alpha}$, and

$$b_n = \begin{cases} (n - a_n)(\mu + c\beta \tan(\alpha\pi/2)) & \alpha \neq 1 \\ \dfrac{2c\beta}{\pi} n \ln(n) & \alpha = 1. \end{cases}$$

Domain of Attraction A CDF $F_X(x)$ belongs to the domain of attraction of a stable distribution $G(x)$ if given a sequence of independent random variables X_k with CDF $F_X(x)$, there exists scaling and translation sequences a_n and b_n such that

$$Z_n = \frac{\sum_{k=1}^n X_k - b_n}{a_n}$$

converges in distribution to a $G(x)$ distributed random variable Z.

From this definition it is clear that by construction, all stable distributions belong to their own zone of attraction, and the CLT indicates that all random variables with finite variance belong to the domain of attraction of Gaussian distributions, which are stable with $\alpha = 2$ and $\beta = 0$. It turns out that it is possible to characterize exactly the laws which are in the domain of attraction of a stable law with $0 < \alpha < 2$. To do so, we need to introduce the concept of *slowly varying* function. A function $L(x)$ defined for $x > 0$ is slowly varying if for all $c > 0$

$$\lim_{x \to \infty} \frac{L(cx)}{L(x)} = 1 . \tag{3.53}$$

For example, $L(x) = \ln(x)$ is slowly varying since

$$\frac{\ln(cx)}{\ln(x)} = \frac{\ln(x) + \ln(c)}{\ln(x)} \to 1$$

as x tends to infinity. Then it is possible to extend the CLT to stable distributions as follows [14].

Generalized Central Limit Theorem A CDF $F_X(x)$ belongs to the domain of attraction of a stable distribution with parameter $0 < \alpha < 2$ if and only if there exists nonnegative constants C_-, C_+ with $C_- + C_+ > 0$ and a slow varying function $L(x)$ such that

$$x^\alpha F_X(-x) = (C_- + o(1))L(x) \tag{3.54}$$

$$x^\alpha (1 - F_X(x)) = (C_+ + o(1))L(x) \tag{3.55}$$

as x tends to infinity.

This result represents a further indication that the CLT should not be employed loosely, since with proper scaling and translation, sums of iid random variables with fat tails converge to non-Gaussian stable distributions. This fact has important consequences in a variety of fields, such as mathematical finance. Indeed, many software tools developed by the finance industry to evaluate risk assume implicitly the validity of the CLT. Yet, it was pointed out by Mandelbrot [15] that the long-term empirical CDF of cotton price changes over the past 200 years suggests that it obeys a Paretian law with α around 1.7, and a further study of stock price changes by Fama [16] also suggests they obey a Paretian distribution with α below 2.

3.7   Bibliographical Notes

The material presented in this chapter is quite standard, except perhaps for the part referring to the theory of large deviations, and the discussion of stable laws. Van den Hollander's book [8] and Morters' course notes [11] constitute excellent elementary introductions to the theory of large deviations. Likewise, the classic probability treatises of Breiman [17] and Feller [10, 18] contain extensive discussions of the CLT and its ramifications, and of stable laws. Finally, the law of large numbers and the CLT constitute only two of the better known results in the much wider field of asymptotic statistics, and readers may wish to consult [19] for a comprehensive yet elementary presentation of key ideas in this field.

3.8   Problems

3.1 Use the one-sided Chebyshev inequalities (3.5) and (3.6) to prove that if a random variable X has mean m_X, median μ_X, and standard deviation $\sigma_X = K_X^{1/2}$, the mean and median cannot be more than one standard deviation away from each other, i.e.,

$$|m_X - \mu_X| \le \sigma_X .$$

3.2 The proof of the convexity of the cumulant generating function $\Gamma_X(s) = \ln M_X(s)$ relies on Hölder's inequality derived in Problem 2.15. Specifically, consider

$$M_X(ws_1 + (1 - w)s_2) = E[\exp((ws_1 + (1 - w)s_2)X)]$$

with $0 \leq w \leq 1$. By applying Hölder's inequality with $p = w^{-1}, q = (1 - w)^{-1}$ to random variables $Y = \exp(ws_1 X)$ and $Z = \exp((1 - w)s_2 X)$, verify that

$$M_X(ws_1 + (1 - w)s_2) \leq M_X(s_1)^w M_X(s_2)^{1-w}$$

and then conclude that $\Gamma_X(s)$ is convex by taking logarithms on both sides of the above inequality (since the function $\ln(x)$ is monotone increasing, it preserves order, i.e., $\ln(x_1) \leq \ln(x_2)$ whenever $x_1 \leq x_2$).

3.3 For an integer valued random variable X with PMF $p_X(k) = P(X = k), k \geq 0$, the moment generating function can be expressed as

$$M_X(s) = E[\exp(sX)] = \sum_{k=0}^{\infty} p_X(k) \exp(sk),$$

which coincides with $G_X(\exp(s))$, where $G_X(z) = E[z^K]$ is the discrete generating function introduced in (2.53). The Chernoff bounds (3.13) and (3.14) remain valid, with the rate function $I(z)$ specified by (3.12).

a) If X is a Bernoulli random variable with

$$P(X = 1) = p \ , \quad P(X = 0) = 1 - p \ ,$$

evaluate its generating function $M_X(s)$ and verify that $E[X] = p$.
b) Evaluate the rate function $I(z)$.

To provide a minimum discrimination interpretation of $I(z)$, note that the relative entropy of two PMFs $p_1(k)$ and $p_0(k)$ with $k \geq 0$ is defined by

$$D(p_1|p_0) = \sum_{k=0}^{\infty} p_1(k) \ln \left(\frac{p_1(k)}{p_0(k)} \right).$$

c) Let p_1 be the PMF of a Bernoulli random variable with parameter z. Verify that the rate function $I(z)$ of part b) can be expressed as $I(z) = D(p_1|p_X)$, where p_X is the Bernoulli PMF with parameter p.

3.4 Consider now a discrete Poisson random variable X with parameter λ, so that its PMF is

$$p_X(k) = \frac{\lambda^k}{k!} \exp(-\lambda)$$

for $k \geq 0$.

a) Evaluate the moment generating function $M_X(s)$ and verify that $E[X] = \lambda$.
b) Compute the rate function $I(z)$.
c) If $p_1(k)$ is the PMF of a Poisson random variable with parameter z, verify that $D(p_1|p_X) = I(z)$.

3.5 The random variables X_k with $k \geq 1$ are independent and $N(0, \sigma^2)$ distributed.

a) Show that as $n \to \infty$, the sequence of random variables

$$Z_n = \frac{1}{n} \sum_{k=1}^{n} |X_{2k} - X_{2k-1}|$$

converges almost surely to a constant c. What is the value of c?
b) Evaluate $E[(Z_n - c)^2]$ and show that Z_n converges to c in the mean-square sense.

3.6 Consider two sequences X_n, $n \geq 1$ and Y_n, $n \geq 1$ which converge in the mean square to X and Y respectively as n tends to infinity.

a) Show that $X_n + Y_n \overset{m.s.}{\to} X + Y$. *Hint:* You may want to use the inequality $(a+b)^2 \leq 2(a^2 + b^2)$.
b) Show that $\lim_{n\to\infty} E[X_n Y_n] = E[XY]$.
c) Show that $\lim_{n\to\infty} E[X_n] = E[X]$.

3.7 To show that a random variable sequence X_n with $n \geq 1$ converges in the mean square towards a limit X without knowing a priori the random variable X, it is necessary to use a Cauchy sequence criterion, whereby X_n converges in the mean square towards the unknown limit X if and only if $E[(X_n - X_m)^2] \to 0$ as n and m converge independently to infinity. Show that $X_n \overset{m.s.}{\to} X$ if and only if

$$E[X_n X_m] \to c$$

as n and m converge independently to infinity. In this case the finite constant $c = E[X^2]$. This convergence criterion is due to Loève [20, Chap. IV].

3.8 Let $\{X_k, k \geq 1\}$ be an infinite sequence of independent random variables uniformly distributed over $[0, 1]$, and let

$$Z_n = \left(\prod_{k=1}^{n} X_k \right)^{1/n}$$

denote the geometrical mean of the first n random variables.

a) Show that Z_n converges almost surely to a constant c and indicate the value of c. *Hint:* consider the logarithm of Z_n.
b) Does Z_n converge to c in the mean-square sense?

3.9 Consider a sequence X_n, $n \geq 1$ of random variables satisfying the recursion

$$X_{n+1} = U_n X_n + 1 - U_n, \tag{3.56}$$

where the random variables U_n are independent, uniformly distributed over $[0, 1]$, and independent of X_1. The recursion (3.56) indicates that X_{n+1} is obtained by averaging X_n and 1 with a random weight U_n. In the analysis of this recursion it is useful to observe that

$$X_{n+1} - 1 = U_n(X_n - 1).$$

Assume in the following that $E[X_1^2]$ is finite.

a) Find a solution of recursion (3.56) expressing X_n in terms of the initial condition X_1 and the averaging random variables U_k for $1 \leq k \leq n - 1$.
b) Evaluate $E[X_n]$ and show that it converges to a finite limit as $n \to \infty$.
c) Does the sequence X_n converge in the mean-square sense? If so, what is its limit?

3.10 Consider the sequence of random variables given by $X_n = nY_n$ where the random variables Y_n are independent and take the values 0 or 1 with probabilities

$$P(Y_n = 1) = p_n \quad , \quad P(Y_n = 0) = 1 - p_n.$$

a) Show that X_n converges in distribution to 0 as long as p_n tends to zero as $n \to \infty$. This is the case in particular for $p_n = n^{-\alpha}$ with $\alpha > 0$.
b) If $p_n = n^{-2}$, use the Borel–Cantelli lemma with events $E_n = \{|X_n| > \epsilon\}$ and $\epsilon > 0$ to show that X_n converges almost surely to 0 as $n \to \infty$.
c) For $p_n = n^{-2}$, does X_n converge to 0 in the mean-square sense?
d) For $p_n = n^{-3}$, does X_n converge to 0 in the mean-square sense?

3.11 Let $\{X_k, k \geq 1\}$ be an infinite sequence of independent random variables uniformly distributed over $[0, 1]$, and let $Z_n = \max_{1 \leq k \leq n} X_k$ denote the maximum of the first n variables. In Example 3.5, it was shown that Z_n converges to 1 in distribution, in probability and in the mean square. Prove that Z_n converges to 1 almost surely. *Hint:* for any $\epsilon > 0$, use the Borel–Cantelli lemma for the events $E_n = \{\omega : Z_n(\omega) < 1 - \epsilon\}$.

3.12 Consider a sequence of iid random variables X_k with $k \geq 1$, with zero mean and variance K_X. Show that

$$R_n = \frac{(\sum_{k=1}^n X_k^2)^{1/2}}{n^{1/2}}$$

converges almost surely to $K_X^{1/2}$.

3.13 Use Jensen's inequality to show that if X is a random variable such that $E[|X|]$ and $E[|X|^{-1}]$ exist, then

$$\frac{1}{E[X^{-1}]} \leq \exp(E[\ln(X)]) \leq E[X].$$

Use this result to show that if a sequence of iid random variables X_k with $k \geq 1$ is such that $E[X_1]$, $E[\ln(X_1)]$ and $E[X_1^{-1}]$ all exist, then as n tends to infinity the arithmetic mean

$$A_n = \frac{1}{n}\sum_{k=1}^n X_k,$$

the geometric mean

$$G_n = (\prod_{k=1}^{n} X_k)^{1/n} ,$$

and the harmonic mean

$$H_n = (\frac{1}{n} \sum_{k=1}^{n} X_k^{-1})^{-1}$$

converge in probability to some constants a, g, and h, respectively, such that

$$h \le g \le a .$$

3.14 Consider an investor with X_k in investible capital in year k. During each successive year, this investor divides his/her capital in a fraction $f X_k$ which is invested in an SP500 index fund whose return varies randomly depending on whether the stock market goes up or down during year k. The remaining fraction $(1 - f)X_k$ is invested in a riskless security, such as Treasury Bills with a fixed but low return $r_0 > 1$. For the stock portion of the investment portfolio, a simple model of the random return R_k takes the form

$$P(R_k = r_U) = p , \quad P(R_k = r_D) = 1 - p ,$$

where p denotes the probability of an up year for the stock market, and r_U denotes the average rate of return during such a year, and r_D denotes the average rate of return during a down year. We assume that $r_U > r_0 > r_D$ and $p r_U + (1 - p)r_D > r_0$. In fact the difference

$$p r_U + (1 - p)r_D - r_0 > 0$$

represents the *risk premium* for assuming the larger risk of a stock investment compared to a riskless asset. The change in the investor's portfolio from 1 year to the next is given by

$$X_{k+1} = M_k X_k ,$$

where

$$M_k = f R_k + (1 - f)r_0 .$$

a) Explain why the long-term growth rate of the portfolio is described by the infinite horizon logarithmic average

$$L = \lim_{n \to \infty} \frac{1}{n} \sum_{k=1}^{n} \ln(M_k).$$

b) Suppose that the investor uses a fixed investment strategy with the same f in each successive year. What is the value of the logarithmic average $L(f)$ corresponding to this strategy?

c) Find the value of f that maximizes $L(f)$, and observe that it is proportional to the risk premium. This result is known as Kelly's criterion [21]. For an interesting discussion of its applications to blackjack, betting on horse races, and investing, see [22].

3.15 Let $X_k, k \geq 1$ denote a sequence of random variables with mean m_X and variance K_X. Denote by

$$M_n = \frac{1}{n} \sum_{k=1}^{n} X_k$$

and

$$V_n = \frac{1}{n-1} \sum_{k=1}^{n} (X_k - M_n)^2$$

the sampled mean and estimated variance. Use the CLT in combination with Slutsky's theorem to show that

$$Z_n = \frac{n^{1/2}(M_n - m_X)}{V_n^{1/2}}$$

converges in distribution to a $N(0, 1)$ distributed random variable Z.

3.16 Consider a sequence of iid random variables $X_k, k \geq 1$, with CDF $F_X(x)$. At the end of Sect. 3.4 it was shown that for a fixed x the empirical CDF

$$\hat{F}_n(x) = \frac{1}{n} \sum_{k=1}^{n} u(x - X_k)$$

converges almost surely to $F_X(x)$. Show that at n tends to infinity, for a fixed x the random variables

$$Z_n = n^{1/2}(\hat{F}_n(x) - F_X(x))$$

converge in distribution to a random variable Z, and identify completely its distribution.

3.17 In some circuit applications, it is sometimes necessary to synthesize digitally a random analog dither signal X which is uniformly distributed over some range, say $[0, 1]$.

a) If X is uniformly distributed over $[0, 1]$, evaluate its characteristic function

$$\Phi_X(u) = E[e^{juX}].$$

The synthesis of X is often achieved approximately by using a digital to analog converter (DAC). If a binary-weighted DAC is used, a signal of the form

$$X_n = \sum_{k=1}^{n} 2^{-k} B_k \tag{3.57}$$

is generated by using a sequence of independent bits B_k with $1 \leq k \leq n$ which take the binary values $\{0, 1\}$ with probability $1/2$, i.e.,

$$P(B_k = 0) = P(B_k = 1) = \frac{1}{2}.$$

b) Evaluate the characteristic function of X_n.

c) Show that as n tends to infinity, X_n converges in distribution to a random variable X which is uniformly distributed over $[0, 1]$.

For part c), you may want to use the product of cosines identity

$$\frac{\sin(v)}{v} = \prod_{k=1}^{\infty} \cos(\frac{v}{2^k}).$$

3.18 To study the almost sure convergence of the random dithers $X_n(\omega)$ given by (3.57), it is useful to observe that for each outcome ω, the sequence $X_n(\omega)$ converges to the real number $X(\omega)$ between 0 and 1 whose infinite binary expansion is given by

$$X(\omega) = \sum_{k=1}^{\infty} 2^{-k} B_k.$$

a) Consider the sequence of events $E_n = \{|X_n - X| > \epsilon\}$ with $\epsilon > 0$. Show that there exists an integer $n_0(\epsilon)$ such that $P(E_n) = 0$ for all $n \geq n_0(\epsilon)$. Conclude that X_n converges to X almost surely.

b) Prove that X_n converges to X in the mean-square sense.

3.19 Consider a sequence of iid random variables X_k, $k \geq 1$ with PDF

$$f_X(x) = \frac{1 - \cos(x)}{\pi x^2}.$$

a) Show that the characteristic function of X_k is

$$\Phi_X(u) = \begin{cases} 1 - |u| & 0 \leq |u| < 1 \\ 0 & \text{otherwise}. \end{cases}$$

Hint: Observe that

$$\Phi_X(u) = G(u) * G(u)$$

where

$$G(u) = \begin{cases} 1 & 0 \leq |u| < 1/2 \\ 0 & \text{otherwise}, \end{cases}$$

and $*$ denotes the convolution operation. Then use the modulation theorem

$$g_1(x)g_2(x) \leftrightarrow \frac{1}{2\pi} G_1(u) * G_2(u)$$

of the Fourier transform.

b) Show that X_k has no moment, and in particular no mean and no variance.

c) Show that as n tends to infinity,

$$Z_n = \frac{1}{n} \sum_{k=1}^{n} X_k$$

converges in distribution to a Cauchy random variable Z with characteristic function $\Phi_Z(u) = \exp(-|u|)$. This shows that the distribution $f_X(x)$ lies in the domain of attraction of the Cauchy distribution.

3.20 Consider a sequence X_k, $k \geq 1$ of independent random variables uniformly distributed over $[-1, 1]$, and let $Y_k = X_k^{-1}$.

a) Find the PDF and CDF of Y_k for $k \geq 1$, and verify that for $y > 1$

$$P(Y_k \leq -y) = P(Y_k \geq y) = \frac{1}{2y},$$

so that the Y_ks satisfy the tail conditions (3.54) and (3.55) with $\alpha = 1$. Do the Y_ks have any moment?

b) By integration by parts, show that the characteristic function of the Y_ks can be expressed as

$$\Phi_Y(u) = \cos(u) + |u|\text{si}(|u|),$$

where the sine integral function is defined as

$$si(u) = - \int_u^\infty \frac{sin(v)}{v} dv$$

for $u \geq 0$ and satisfies $si(0) = -\pi/2$.

c) Show that

$$Z_n = \frac{1}{n} \sum_{k=1}^{n} Y_k$$

converges in distribution to a Cauchy random variable Z with median $\mu = 0$ and scale parameter $\sigma = \pi/2$. This shows that the CDF F_Y of the Y_ks lies in the domain of attraction of the Cauchy distribution F_Z.

3.21 Consider a sequence X_k, $k \geq 1$ of independent exponential random variables with parameter $\lambda > 0$. Recall that the CDF of X_k is

$$F_X(x) = [1 - \exp(-\lambda x)]u(x),$$

where $u(\cdot)$ denotes the unit step function. Let

$$Y_n = \max_{1 \leq k \leq n} X_k \quad \text{and} \quad Z_n = \min_{1 \leq k \leq n} X_k$$

denote the minimum and maximum of the first n X_k's.

a) Evaluate the CDF of Y_n and the CDF of Z_n. Can you recognize the CDF of z_n? *Hint:* note that $Y_n \leq y$ if and only if all the X_k's are less than or equal to y. Similarly $Z_n \geq z$ if and only if all the X_k's are greater or equal to z.

b) Show that the random variables

$$A_n = \frac{Y_n - \lambda^{-1} \ln(n)}{\lambda^{-1}}$$

converge in distribution to a random variable A. What is the CDF of A? In your derivation, you may want to use the fact that

$$\lim_{n \to \infty} \left(1 - \frac{a}{n}\right)^n = e^{-a}$$

for $a > 0$.

c) Show that the random variables

$$B_n = \lambda n Z_n$$

converge in distribution to a random variable B. Can you recognize the distribution of B?

3.22 Consider a sequence $X_k, k \geq 1$ of independent Cauchy distributed random variables with scale parameter a, so that the CDF of each X_k is

$$F_X(x) = \frac{1}{2} + \frac{1}{\pi} \arctan(x/a) .$$

Let $Z_n = \max_{1 \leq k \leq n} X_k$ denote the maximum of the first n elements of this sequence. Evaluate the CDF of Z_n and show that the scaled random variables

$$V_n = \frac{\pi Z_n}{an}$$

converge in distribution to a random variable V as as $n \to \infty$. Specify the CDF $G(v)$ of V. To solve this problem, you may want to use the identity

$$\frac{\pi}{2} = \arctan(z) + \arctan(1/z)$$

for $z > 0$ and

$$\frac{-\pi}{2} = \arctan(z) + \arctan(1/z)$$

for $z < 0$ which can be used to rewrite $F_X(x)$ as

$$F_X(x) = \begin{cases} 1 - \frac{1}{\pi} \arctan(a/x) & x > 0 \\ -\frac{1}{\pi} \arctan(a/x) & x < 0 . \end{cases}$$

Recall also that for small z

$$\arctan(z) \approx z .$$

References

1. W. Rudin, *Principles of Mathematical Analysis, 2nd edition*. New York, NY: McGraw-Hill, 1964.
2. D. Bertsekas, A. Nedic, and A. E. Ozdaglar, *Convex Analysis and Optimization*. Belmont, MA: Athena Scientific, 2003.
3. S. Kullback, *Information Theory and Statistics*. New York: J. Wiley & Sons, 1959. Reprinted by Dover Publ., Mineola, NY, 1968.
4. M. Basseville, "Information: Entropies, divergences et moyennes," Tech. Rep. 1020, Institut de Recherche en Informatique et Systèmes Aléatoires, Rennes, France, May 1996.
5. K. L. Chung, *A Course in Probability Theory, Second Edition*. New York: Academic Press, 1968.
6. G. R. Grimmett and D. R. Stirzaker, *Probability and Random Processes, 3rd edition*. Oxford, United Kingdom: Oxford University Press, 2001.
7. H. Cramér, "Sur un nouveau théorème-limite de la théorie des probabilités," in *Actalités Scientifiques et Industrielles*, vol. 736, Paris: Hermann, 1938.
8. F. den Hollander, *Large Deviations*. Providence, RI: American Mathematical Soc., 2000.
9. B. C. Levy, *Principles of Signal Detection and Parameter Estimation*. New York, NY: Springer Verlag, 2008.
10. W. Feller, *An Introduction to Probability Theory and its Applications, 2nd edition*, vol. 2. New York, NY: J. Wiley, 1970.
11. P. Morters, "Large deviation theory and applications." Lecture notes on large deviation theory, University of Bath, Math. Sciences Dept., Nov. 2008.
12. A. Y. Khintchine and P. Lévy, "Sur les lois stables," *Comptes Rendus Académie des Sciences, Paris*, vol. 202, pp. 374–376, 1936.
13. P. Lévy, *Theorie de l'addition des Variables Aléatoires, 2ème edition*. Paris, France: Gauthier-Villars, 1954.
14. B. V. Gnedenko and A. N. Kolmogorov, *Limit Distributions for Sums of Independent Random Variables*. Cambridge, MA: Addison-Wesley, 1954.
15. B. Mandelbrot, "The variation of certain speculative prices," *J. of Business*, vol. 36, pp. 394–419, Oct. 1963.
16. E. F. Fama, "The behavior of stock-market prices," *J. of Business*, vol. 38, pp. 34–105, Jan. 1965.
17. L. Breiman, *Probability*. Reading, MA: Addison-Wesley, 1968. Republished in 1992 by SIAM, Philadelphia, PA.
18. W. Feller, *An Introduction to Probability Theory and its Applications, 3rd edition*, vol. 1. New York, NY: J. Wiley, 1968.
19. T. S. Ferguson, *A Course in Large Sample Theory*. Boca Raton, FL: CRC Press, 1996.
20. M. Loève, "Sur les fonctions aléatoires stationnaires du second ordre," *Revue Scientifique*, vol. 83, pp. 297–310, 1945.
21. J. J. L. Kelly, "A new interpretation of information rate," *Bell System Technical J.*, pp. 917–926, July 1956.
22. E. O. Thorp, "The Kelly criterion in blackjack, sports betting, and the stock market," in *Handbook of Asset and Liability Management, Vol. 1* (S. A. Zenios and W. T. Ziemba, eds.), Elsevier, 2006.

Part II
Main Topics

Specification of Random Processes

4

4.1 Introduction

The purpose of this chapter is to describe random processes and analyze their specification in terms of finite joint distributions for ordered time samples of the process. Particular emphasis is given to the computation of the second-order statistics of random processes, such as the autocorrelation, since these are heavily used by electrical engineers for signal analysis purposes, even though they provide only a very incomplete characterization of the stochastic behavior of processes, except in the Gaussian case. Several categories of random processes, such as Gaussian and Markov processes, which admit simple parametrizations, are introduced, as well as the class of stationary independent increments processes. This class contains the Wiener and Poisson processes, which play a major role in modeling random phenomena arising in electrical engineering systems.

4.2 Specification of Random Processes

A random process $X(t)$ can be viewed as a collection of random variables indexed by time, where the time t belongs to index set $\mathcal{T}$. For discrete-time random processes, $\mathcal{T}$ will be the set of signed integers $\mathbb{Z}$ or a subset thereof, such as $\mathbb{N}$, the set of nonnegative integers. Likewise, in the continuous time $\mathcal{T}$ will be the set $\mathbb{R}$ of real numbers or a subset of it, such as for example the finite interval $[a, b]$. Based on this definition, we see that effectively a random process $X(t, \omega)$ is a function of two variables: time $t \in \mathcal{T}$ and outcome $\omega \in \Omega$. If we freeze the outcome to say ω_1, the time trajectory $X(t, \omega_1)$ forms what is called a sample path of the process. In other words, to each fixed outcome $\omega \in \Omega$ corresponds a sample path describing the time evolution of physical variable X for this specific outcome. Conversely, if we freeze the time t to, say t_1, $X(t_1, \omega)$ is just an ordinary random variable.

To illustrate the concept of random process, we start by considering a few simple examples.

Example 4.1 Consider the sinusoidal signal

$$X(t, \omega) = A \cos(\omega_c t + \Theta(\omega)),$$

where the amplitude A and frequency ω_c are known, but the phase Θ is random and uniformly distributed over $[-\pi, \pi)$, so that its PDF is given by

© Springer Nature Switzerland AG 2020
B. C. Levy, *Random Processes with Applications to Circuits and Communications*,
https://doi.org/10.1007/978-3-030-22297-0_4

$$f_\Theta(\theta) = \begin{cases} \frac{1}{2\pi} & -\pi \le \theta < \pi \\ \\ 0 & \text{otherwise.} \end{cases}$$

Thus the amplitude A and period $T_c = 2\pi/\omega_c$ of sinusoidal signal $X(t, \omega)$ are fixed, and by observing that it can be rewritten as

$$X(t, \omega) = A\cos(\omega_c(t - D(\omega))),$$

where the delay $D(\omega) = -\Theta/\omega_c$ is uniformly distributed over $(-T_c/2, T_c/2]$, we conclude that the sample paths differ only by the value of the random delay $D(\omega)$ which is selected to shift the waveform $A\cos(\omega_c t)$.

Example 4.2 Likewise, let $p(t)$ be a square wave with amplitude A and period T, as shown in Fig. 4.1. Then consider the random process

$$X(t, \omega) = p(t - D(\omega))$$

obtained by applying a random delay $D(\omega)$ to $p(t)$. It is assumed that D is uniformly distributed over $[0, T)$, so that its PDF is given by

$$f_D(d) = \begin{cases} \frac{1}{T} & 0 \le d < T \\ \\ 0 & \text{otherwise.} \end{cases}$$

So sample paths differ again from one another by the time shift applied to the waveform $p(t)$.

The random processes of Examples 4.1 and 4.2 depend on a single random phase or delay parameter, and as such they are not truly representative of the random processes encountered in engineering applications which are typically produced by random effects occurring consistently across time. Note indeed that once the parameters Θ and D are estimated from observations, the waveforms $X(t, \omega)$ of Examples 4.1 and 4.2 are completely known. In general, the goal of random process analysis is to investigate the statistical properties of processes across time, i.e., at multiple times instead of a single time. Thus, the specification of a random process $X(t)$ requires the knowledge of the joint distributions of $X(t_1)$, $X(t_2)$, ... and $X(t_N)$ for all choices of times $t_1 < t_2 < \cdots < t_N$ and for all N. These distributions must of course be internally consistent in the sense that given the distribution of $X(t_1)$, ... $X(t_{k-1})$, $X(t_k)$, $X(t_{k+1})$, ..., $X(t_N)$ where the index k satisfies $1 < k < N$, the joint distribution of the $N - 1$ random variables $X(t_1)$, ..., $X(t_{k-1})$, $X(t_{k+1})$, ..., $X(t_N)$ obtained by removing t_k from the set of times of interest must of course be the marginal of the joint distribution including $X(t_k)$. For the case of a process whose finite joint densities are absolutely continuous, this means that if

$$f_{X(t_1)\cdots X(t_k)\cdots X(t_N)}(x_1, t_1; \cdots; x_k, t_k; \cdots; x_N, t_N)$$

Fig. 4.1 Square wave $p(t)$ (broken line) and sample path $X(t, \omega) = p(t - D(\omega))$ (solid line)

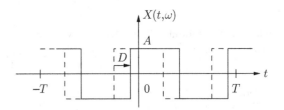

denotes the joint PDF of the $X(t_k)$s for $1 \leq k \leq N$, the joint PDF of the $N-1$ random variables $X(t_1), \ldots, X(t_{k-1}), X(t_{k+1}), \ldots, X(t_N)$ obtained by dropping time $X(t_k)$ is

$$f_{X(t_1)\cdots X(t_{k-1})X(t_{k+1})\cdots X(t_N)}(x_1, t_1; \cdots; x_{k-1}, t_{k-1}; x_{k+1}, t_{k+1}; x_N, t_N) \tag{4.1}$$

$$= \int f_{X(t_1)\cdots X(t_{k-1})X(t_k)X(t_{k+1})\cdots X(t_N)}(x_1, t_1; \cdots; x_k, t_k; \cdots; x_N, t_N) dx_k.$$

Given the finite joint distributions of $X(t)$, it was shown by Kolmogorov [1] that there exists a probability measure P defined on the set of events formed by countable union, intersection, and complementation of events involving only a finite number of time values of $X(t)$. This result is known as the Kolmogorov extension theorem.

The need to specify all finite joint densities of $X(t_k)$, with $1 \leq k \leq N$ for all choices of times $t_1 < \ldots < t_k < \ldots t_N$ and all N represents a formidable task, which typically can be performed only for very simple processes, or processes endowed with additional properties.

Example 4.1 (Continued) If we consider the process

$$X(t) = A\cos(\omega_c t + \Theta),$$

where the random phase Θ is uniformly distributed over $[-\pi, \pi)$, by observing that for a fixed t, the random variable

$$\Psi = \omega_c t + \Theta \mod 2\pi$$

is also uniformly distributed over $[-\pi, \pi)$, we find that

$$X(t) = A\cos(\Psi)$$

is in the form of the nonlinear transformation studied in Example 2.3, so its CDF is given by $F_{X(t)}(x) = 0$ for $x < -A$,

$$F_{X(t)}(x) = 1 - \frac{1}{\pi}\arccos(x/A) \tag{4.2}$$

for $-A \leq x \leq A$ and $F_{X(t)}(x) = 1$ for $x > A$. Its PDF is also given by

$$f_{X(t)}(x) = \frac{1}{A\pi}\frac{1}{(1-(x/A)^2)^{1/2}}$$

for $|x| \leq A$ and $f_{X(t)}(x) = 0$ otherwise. This specifies the distribution of $X(t)$ at *one time only*. An interesting feature of this distribution is that it does not depend on t. Suppose now that $N = 2$ and we seek to find the joint distribution of $X(s)$ and $X(t)$ for $s < t$. Since we already know the distribution of $X(s)$, all what is needed is finding the conditional distribution of $X(t)$ given $X(s)$. To do so, observe that

$$X(t) = A\cos(\omega_c(t - s + s) + \Theta)$$

$$= A[\cos(\omega_c(t-s))\cos(\omega_c s + \Theta) - \sin(\omega_c(t-s))\sin(\omega_c s + \Theta)]. \tag{4.3}$$

In this expression we recognize that $X(s) = A\cos(\omega_c s + \Theta)$ and

$$A\sin(\omega_c s + \Theta) = \epsilon(A^2 - X^2(s))^{1/2},$$

where

$$\epsilon = \begin{cases} 1 \text{ if } & 0 \le \omega_c s + \Theta \quad \text{mod } 2\pi < \pi \\ -1 \text{ if } & -\pi \le \omega_c s + \Theta \quad \text{mod } 2\pi \le 0 \end{cases}$$

The random variable ϵ is independent of $X(s)$ and since $\Psi = \omega_c s + \Theta \mod 2\pi$ is uniformly distributed over $[-\pi, \pi)$,

$$P[\epsilon = 1] = P[\epsilon = -1] = 1/2.$$

By substitution inside (4.3) we find therefore that given $X(s) = x_s$,

$$X(t) = \cos(\omega_c(t - s))x_s - \epsilon \sin(\omega_c(t - s))(A^2 - x_s^2)^{1/2}. \tag{4.4}$$

So if $p_+(x_s, t - s)$ and $p_-(x_s, t - s)$ represent the two possible locations of $X(t)$ obtained by setting $\epsilon = 1$ and $\epsilon = -1$ on the right-hand side of (4.4), we deduce that given $X(s) = x_s$, $X(t)$ is discrete valued and has for conditional CDF

$$F_{X(t)|X(s)}(x_t, t|x_s, s) = \frac{1}{2}[u(x_t - p_+(x_s, t - s)) + u(x_t - p_-(x_s, t - s))]. \tag{4.5}$$

An interesting feature of this CDF is that it depends only on the difference $t - s$ between times t and s. Note also that when the time difference $t - s = mT_c/2$ with m integer, the two possible positions of $X(t)$ merge and $X(t) = X(s)$ for $m = 2\ell$ with ℓ integer, $X(t) = -X(s)$ for $m = 2\ell + 1$. This just reflects that since the signal is periodic, $X(t)$ returns to the same position every period, and takes an opposite position after half a period.

Up to this point we have considered the finite distributions of $X(t)$ for $N = 1$ and $N = 2$ points. If we now consider three points $r < s < t$ and $N = 3$, it is not difficult to observe that unless $s - r$ is an integer multiple of $T_c/2$, $X(t)$ can be computed exactly from the knowledge of $X(s) = x_s$ and $X(r) = x_r$. Specifically, we have

$$X(t) = A \begin{bmatrix} \cos(\omega_c t) & \sin(\omega_c t) \end{bmatrix} \begin{bmatrix} \cos(\Theta) \\ -\sin(\Theta) \end{bmatrix}.$$

This implies

$$\begin{bmatrix} X(s) \\ X(r) \end{bmatrix} = AM(s, r) \begin{bmatrix} \cos(\Theta) \\ -\sin(\Theta), \end{bmatrix}$$

where

$$M(s, r) \triangleq \begin{bmatrix} \cos(\omega_c s) & \sin(\omega_c s) \\ \cos(\omega_c r) & \sin(\omega_c r) \end{bmatrix}.$$

When $s - r \ne mT_c/2$ with m integer, $M(s, r)$ is invertible and

$$M^{-1}(s, r) = \frac{1}{\sin(\omega_s(s - r))} \begin{bmatrix} -\sin(\omega_c r) & \sin(\omega_c s) \\ \cos(\omega_c r) & -\cos(\omega_c s) \end{bmatrix},$$

so that

$$X(t) = \left[\cos(\omega_c t) \; \sin(\omega_c t) \right] M^{-1}(s, r) \begin{bmatrix} X(r) \\ X(s) \end{bmatrix}$$

$$= \frac{1}{\sin(\omega_c(s - r))} [\sin(\omega_c(t - r))X(s) + \sin(\omega_c(s - t))X(r)]. \tag{4.6}$$

Thus $X(t)$ is known exactly if $X(s)$ and $X(r)$ are known. This should not come as a complete surprise since $X(t)$ obeys the second-order differential equation

$$\frac{d^2 X}{dt^2} + \omega_c^2 X(t) = 0$$

which can be solved for $t \geq s$ from two initial conditions. Instead of selecting $X(s)$ and $(dX/dt)(s)$ as initial conditions, $X(r)$ is selected here as the second condition. Since $X(t)$ is completely known from $X(s)$ and $X(r)$ (as long as s and r are not separated my an integer multiple of a half period), there is no point in examining joint distributions for $N = 4$ points or more, since conditioning with respect to additional values of X in the past will not improve our prediction of $X(t)$ based on $X(s)$ and $X(r)$. In fact, as will be seen later, the process $X(t)$ is what is called a second-order Markov process.

In the previous example, we have encountered a situation where the joint CDF

$$F_{X(t_1)\cdots X(t_k)\cdots X(t_N)}(x_1, t_1; \ldots; x_k, t_k; \ldots, x_N, t_N)$$

of random variables $X(t_1), .., X(t_k), \ldots X(t_N)$ does not depend explicitly on all times t_k with $1 \leq k \leq N$. We say that a random process $X(t)$ is *strict-sense stationary* of order N if its finite joint densities of order N are invariant under a change of time origin, i.e., under time-translations by an arbitrary constant h. In other words for $t_1 < t_2 < \ldots < t_N$ and for all $h \in \mathbb{R}$ in the continuous-time case and all $h \in \mathbb{Z}$ in the discrete-time case

$$F_{X(t_1+h)\cdots X(t_k+h)\cdots X(t_N+h)}(x_1, t_1 + h; \ldots; x_k, t_k + h; \ldots, x_N, t_N + h)$$

$$= F_{X(t_1)\cdots X(t_k)\cdots X(t_N)}(x_1, t_1; \ldots; x_k, t_k; \ldots, x_N, t_N). \tag{4.7}$$

By selecting $h = -t_1$, this implies that the joint CDF of $X(t_1)$, $\ldots$, $X(t_k)$, $\ldots$, $X(t_N)$ will depend only $t_k - t_1$ for $2 \leq k \leq N$. When $N = 1$, the invariance of the CDF under time translation implies that $F_{X(t)}(x)$ does not depend on t, and for $N = 2$, for $s < t$ $F_{X(s)X(t)}(x_s, x_t, t - s)$ depends only on $t - s$. Since the CDFs of lower order are contained in the CDF of order N, for example

$$F_{X(t_1)\cdots X(t_{N-1})}(x_1, 0; \ldots; x_{N-1}, t_{N-1} - t_1)$$

$$= F_{X(t_1)\cdots X(t_{N-1})X(t_N)}(x_1, 0; \ldots; x_{N-1}, t_{N-1} - t_1; \infty, t_N - t_1), \tag{4.8}$$

strict-sense stationarity of order N implies strict-sense stationarity of order $N - 1$. A process $X(t)$ is said to be strict-sense stationary (SSS) if it is strict sense stationary of order N for all N.

Example 4.1 (Continued) If we consider the random phase sinusoidal signal $X(t) = A \cos(\omega_c t + \Theta)$, its first order CDF $F_{X(t)}(x)$ given by (4.2) is independent of t, its second-order CDF $F_{X(s)X(t)}(x_s, x_t, t - s)$ given by (4.5) depends only on $t - s$, and the coefficients of identity (4.6) expressing $X(t)$ in terms of $X(s)$ and $X(r)$ for $t > s > r$ depend only on differences between times t, s, and r, so the process $X(t)$ is SSS, i.e., strict sense stationary of all orders.

Of course, for the case of absolutely continuous random processes, i.e., processes whose finite joint distributions of all orders admit PDFs, translation invariance for the CDFs and PDFs is equivalent, so if a random process $X(t)$ satisfies the equivalent of identity (4.7) for its joint PDFs of order N, it is strict-sense stationary of order N.

4.3   Mean and Autocorrelation

Because of the huge amount of information needed to specify the finite joint distributions of a random process $X(t)$, for some applications, electrical engineers focus their attention on the first- and second-order statistics of random processes, which often provide all the information needed for certain signal processing tasks. However, this should be done cautiously, since we will encounter later examples of random processes with identical second-order statistics, but extremely different sample paths and stochastic behavior.

The mean $m_X(t)$ of a random process $X(t)$ is the statistical mean of random variable $X(t)$, which is given by

$$m_X(t) = E[X(t)] = \int_\infty^\infty x\, dF_{X(t)}(x, t). \tag{4.9}$$

The autocorrelation $R_X(t, s)$ of $X(t)$ is just the cross-correlation of random variables $X(t)$ and $X(s)$ at two different times t and s, so

$$R_X(t, s) = E[X(t)X(s)] = \int_{\mathbb{R}^2} x_t x_s\, dF_{X(s),X(t)}(x_s, s; x_t, t). \tag{4.10}$$

Finally, the autocovariance $K_X(t, s)$ is the autocorrelation of the centered process $X(t) - m_X(t)$, which can be expressed as

$$K_X(t, s) = E[(X(t) - m_X(t))(X(s) - m_X(s))] = R_X(t, s) - m_X(t)m_X(s). \tag{4.11}$$

The expression (4.11) indicates that the autocovariance $K_X(t, s)$ can be expressed in terms of the mean $m_X(t)$ and autocorrelation $R_X(t, s)$. Conversely, given $m_X(t)$ and $K_X(t, s)$, the autocorrelation $R_X(t, s)$ can be obtained from

$$R_X(t, s) = K_X(t, s) + m_X(t)m_X(s).$$

Note also that since

$$R_X(t, s) = E[X(t)X(s)] = E[X(s)X(t)] = R_X(s, t),$$

the autocorrelation and autocovariance functions are symmetric.

4.3.1   Examples

Example 4.1 (Continued)   Consider again the sinusoidal process

$$X(t) = A\cos(\omega_c t + \Theta).$$

To find its mean, instead of using the distribution of $X(t)$, it is more convenient to use that of Θ, so

$$m_X(t) = E[X(t)] = A \int \cos(\omega_c t + \theta) f_\Theta(\theta) d\theta$$

$$= \frac{A}{2\pi} \int_{-\pi}^{\pi} \cos(\omega_c t + \theta) d\theta. \tag{4.12}$$

Viewed as a function of θ, $\cos(\omega_c t + \theta)$ is periodic with period 2π, and since the expression (4.12) integrates this function over one period, the average is zero, so

$$m_X(t) = 0.$$

By using the trigonometric identity

$$\cos(a)\cos(b) = \frac{1}{2}[\cos(a+b) + \cos(a-b)],$$

we also find that

$$X(t)X(s) = \frac{A^2}{2}[\cos(\omega_c(t-s)) + \cos(\omega_c(t+s) + 2\Theta)],$$

where the first term is not random so

$$R_X(t,s) = \frac{A^2}{2}[\cos(\omega_c(t-s)) + E[\cos(\omega_c(t+s) + 2\Theta)]],$$

and

$$E[\cos(\omega_c(t+s) + 2\Theta)] = \int \cos(\omega_c(t+s) + 2\theta) f_\Theta(\theta) d\theta$$

$$= \frac{1}{2\pi} \int_{-\pi}^{\pi} \cos(\omega_c(t+s) + 2\theta) d\theta = 0$$

since when viewed as a function of θ, $\cos(\omega_c(t+s) + 2\theta)$ is periodic with period π. This shows that the autocorrelation

$$R_X(t,s) = \frac{A^2}{2}\cos(\omega_c(t-s)) \tag{4.13}$$

is a function of $\tau = t - s$ only. Viewed as a function of τ, $R_X(\tau)$ is such that $R_X(0) = A^2/2$, so the correlation coefficient between $X(t)$ and $X(t+\tau)$ is

$$\rho(\tau) = \frac{R_X(\tau)}{R_X(0)} = \cos(\omega_c \tau).$$

As expected $-1 \le \rho(\tau) \le 1$ with $\rho(\tau) = 1$ if $\tau = mT_c$ and $\rho(\tau) = -1$ for $\tau = (2m+1)T_c/2$ where $T_c = 2\pi/\omega_c$ denotes the period of $X(t)$. This just indicates that $X(t)$ is periodic with period T_c and is out of phase by $180°$ after half a period.

Random processes such as the one considered in Example 4.1 which have a constant mean $m_X(t) = m_X$ and an autocorrelation R_X which depends only on $\tau = t - s$ are said to be *wide-sense stationary* (WSS), since their first- and second-order statistics are invariant under a change of time origin, i.e., under a translation of the time axis where t is replaced by $t + h$, with $h \in \mathbb{R}$ in the continuous-time case and $h \in \mathbb{Z}$ in the discrete-time case. It is not difficult to see that WSS is a much weaker property than SSS. In fact, SSS or order 2 implies WSS, since when $F_{X(t)}(x)$ is independent of t, the mean m_X obtained from (4.9) is also independent of t, and when the joint CDF $F_{X(s)X(t)}(x_s, x_t, t - s)$ depends on $t - s$ only, the autocorrelation R_X given by (4.10) depends only on $t - s$. The converse is generally not true, except for Gaussian processes considered in Sect. 4.6, for which WSS implies SSS.

Example 4.3 Consider the discrete-time binomial process

$$Z_n = \sum_{k=1}^{n} X_k,$$

where the random variables X_k with $k \geq 1$ are independent and Bernoulli distributed with

$$P[X_k = 1] = p \ , \quad P[X_k = 0] = q = 1 - p.$$

Since it is the sum of n independent Bernoulli variables with parameter p, Z_n has the binomial distribution

$$P(Z_n = \ell) = \binom{n}{\ell} p^{\ell} q^{n-\ell}$$

for $0 \leq \ell \leq n$. Because $X_k^2 = X_k$, the first and second moments of X_k are given by

$$E[X_k] = E[X_k^2] = p,$$

and the mean of Z_n is given by

$$m_Z(n) = \sum_{k=1}^{n} E[X_k] = np.$$

To find the autocorrelation of Z_n, we first evaluate the second moment of Z_n. We have

$$E[Z_n^2] = E[(\sum_{k=1}^{n} X_k)(\sum_{\ell=1}^{n} X_\ell)]$$

$$= \sum_{k=1}^{n} \sum_{\ell=1}^{n} E[X_k X_\ell].$$

The independence of the X_ks implies

$$E[X_k X_\ell] = \begin{cases} E[X_k]E[X_\ell] = p^2 & k \neq \ell \\ E[X_k^2] = p & k = \ell, \end{cases}$$

and since over the square grid formed by $1 \leq k, \ell \leq n$ there are n points located on the diagonal $k = \ell$, and there are $n^2 - n$ off-diagonal points, we obtain

$$E[Z_n^2] = np + (n^2 - n)p^2. \tag{4.14}$$

Then to compute autocorrelation of Z_n, we observe that for $m \geq n$, Z_m can be decomposed as

$$Z_m = Z_m - Z_n + Z_n,$$

where

$$Z_m - Z_n = \sum_{n+1}^{m} X_k$$

is independent of Z_n since the random variables X_k over $n + 1 \leq k \leq m$ are independent of the X_ks over $1 \leq k \leq n$. This implies

$$R_Z(m, n) = E[Z_m Z_n] = E[(Z_m - Z_n)Z_n] + E[Z_n^2],$$

where

$$E[(Z_m - Z_n)Z_n] = E[Z_m - Z_n]E[Z_n] = (m - n)np^2. \tag{4.15}$$

Combining (4.14) and (4.15), we find therefore

$$R_Z(m, n) = E[Z_m Z_n] = np(1 - p) + mnp^2$$

for $m \leq n$. Since the autocorrelation needs to be symmetric in m and n, this gives

$$R_Z(m, n) = \min(m, n)p(1 - p) + mnp^2 \tag{4.16}$$

for all m and n. This implies that the autocovariance

$$K_Z(m, n) = R_Z(m, n) - m_Z(m)m_Z(n) = \min(m, n)p(1 - p).$$

Since the mean $m_Z(n)$ is linear in n and $R_Z(m, n)$ depends on both m and n, Z_n is not WSS.

Example 4.4 The random telegraph wave is a continuous-time process $X(t)$ defined for $t \geq 0$ which takes only the values ± 1. The probability distribution of the initial state is given by

$$P(X(0) = 1) = P(X(0) = -1) = \frac{1}{2},$$

and for $t \geq s$, the conditional distribution of $X(t)$ given $X(s)$ is specified by

$$P(X(t) = 1 \mid X(s) = 1) = P(X(t) = -1 \mid X(s) = -1)$$
$$= \frac{1}{2}[1 + \exp(-2a(t - s))]$$
$$P(X(t) = 1 \mid X(s) = -1) = P(X(t) = -1 \mid X(s) = 1)$$
$$= \frac{1}{2}[1 - \exp(-2a(t - s))],$$

where $a > 0$ denotes the transition rate between the states $+1$ and -1. As indicated by Fig. 4.2, which depicts a sample path of $X(t)$, $X(t)$ switches between states 1 and -1 at random times. It will be seen later in Chap. 7 that the time interval between switches is exponentially distributed with parameter a.

Fig. 4.2 Sample path of the random telegraph wave

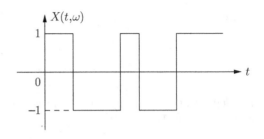

Based on this specification, we find

$$P(X(t) = 1) = P(X(t) = 1|X(0) = 1)P(X(0) = 1)$$
$$+P(X(t) = 1|X(0) = -1)P(X(0) = -1)$$
$$= \frac{1}{4}(1 + \exp(-2at)) + \frac{1}{4}(1 - \exp(-2at)) = \frac{1}{2} \qquad (4.17)$$

and

$$P(X(t) = -1) = 1 - P(X(t) = 1) = \frac{1}{2},$$

so that the distribution of $X(t)$ is the same as that of $X(0)$ and the mean of $X(t)$ is

$$m_X(t) = E[X(t)] = 1 \times \frac{1}{2} - 1 \times \frac{1}{2} = 0.$$

To evaluate the joint distribution of $X(t)$ and $X(s)$, note that for $t \geq s$

$$P(X(t) = 1, X(s) = 1) = P(X(t) = 1|X(s) = 1)P(X(s) = 1)$$
$$= \frac{1}{4}[1 + \exp(-2a(t - s))],$$

and proceeding in a similar manner we find

$$P(X(t) = -1, X(s) = -1) = \frac{1}{4}[1 + \exp(-2a(t - s))]$$
$$P(X(t) = 1, X(s) = -1) = P(X(t) = -1, X(s) = 1)$$
$$= \frac{1}{4}[1 - \exp(-2a(t - s))].$$

The autocorrelation of $X(t)$ is therefore given by

$$R_X(t, s) = E[X(t)X(s)]$$
$$= 1 \times (P(X(t) = 1, X(s) = 1) + P(X(t) = -1, X(s) = -1))$$
$$-1 \times (P(X(t) = -1, X(s) = 1) + P(X(t) = 1, X(s) = -1))$$
$$= \exp(-2a(t - s))$$

for $t \geq s$, and by symmetry

$$R_X(t, s) = \exp(-2a|t - s|) \qquad (4.18)$$

for all t and s. Since m_X is constant and the autocorrelation depends only on $\tau = t - s$, $X(t)$ is WSS. Furthermore, $R_X(\tau_c) = \exp(-1)$ for $\tau_c = 1/(2a)$, where τ_c denotes the coherence time of $X(t)$. Thus $X(t)$ and $X(t + \tau)$ are approximately uncorrelated as long as $\tau \gg \tau_c$. This property just illustrates the fact that, unlike the random phase process of Example 4.1 where the random phase is selected at the start, the random telegraph wave is affected by random effects throughout time, so that the knowledge that $X(s) = 1$ or -1 becomes progressively less useful to guess the state of $X(t)$ whenever $X(t)$ has undergone one or more state transitions after time s.

Example 4.5 Consider the pulse amplitude modulated (PAM) signal

$$Z(t) = \sum_{k=-\infty}^{\infty} X(k) p(t - kT - D), \qquad (4.19)$$

where the transmitted discrete-time random signal $X(k)$ is assumed to be WSS with mean $m_X = E[X(t)]$ and autocorrelation

$$R_X(m) = E[X(k + m)X(k)].$$

In (4.19), T denotes the signaling or baud period, and $p(t)$ is the signaling pulse used for transmission. D denotes a random synchronization delay which is assumed to be independent of signal $X(k)$ and uniformly distributed over $[0, T]$, so its PDF is given by

$$f_D(d) = \begin{cases} \frac{1}{T} & 0 \le d < T \\ 0 & \text{otherwise} . \end{cases}$$

The model (4.19) provides a convenient framework for analyzing communication signals arising in a wide range of applications, such as pulse-coded-modulation (PCM) of digital telephony, optical local area networks, or magnetic recording [2].

To evaluate the mean of $Z(t)$, note that since $X(k)$ is independent from D

$$m_Z(t) = E[Z(t)] = \sum_{k=-\infty}^{\infty} E[X(k) p(t - kT - D)]$$

$$= \sum_{k=-\infty}^{\infty} E[X(k)] E[p(t - kT - D)]$$

$$= \frac{m_X}{T} \sum_{k=-\infty}^{\infty} \int_0^T p(t - kT - u) du.$$

By performing the change of variable $v = t - kT - u$, we find

$$\int_0^T p(t - kT - u) du = \int_{t-(k+1)T}^{t-kT} p(v) dv,$$

and observing that

$$\sum_{k=-\infty}^{\infty} \int_{t-(k+1)T}^{t-kT} p(v) dv = \int_{-\infty}^{\infty} p(v) dv,$$

we obtain

$$m_Z(t) = \frac{m_X}{T} \int_{-\infty}^{\infty} p(v)dv, \tag{4.20}$$

which does not depend on t, so $Z(t)$ has a constant mean. m_Z is zero whenever $X(k)$ has zero mean, or when the total area under pulse $p(t)$ is zero. If

$$P(j\omega) = \int_{-\infty}^{\infty} p(t)\exp(-j\omega t)dt$$

denotes the Fourier transform of $p(t)$, the area of pulse $p(t)$ coincides with the DC value $P(j0)$ of the Fourier transform.

To evaluate the autocorrelation of $Z(t)$, we use again the independence between $X(k)$ and D to write

$$R_Z(t, s) = E[Z(t)Z(s)]$$

$$= \sum_{k=-\infty}^{\infty} \sum_{\ell=-\infty}^{\infty} E[X(k)X(\ell)p(t - kT - D)p(s - \ell T - D)]$$

$$= \sum_{k=-\infty}^{\infty} \sum_{\ell=-\infty}^{\infty} E[X(k)X(\ell)]E[p((t - kT - D)p(s - \ell T - D)]$$

$$= \frac{1}{T} \sum_{k=-\infty}^{\infty} \sum_{\ell=-\infty}^{\infty} R_X(k - \ell) \int_0^T p(t - kT - u)p(s - \ell T - u)du. \tag{4.21}$$

Replacing the sum over k by a sum over $m = k - \ell$ and performing the change of variable $v = s - \ell T - u$, expression (4.21) can be rewritten as

$$R_Z(t, s) = \frac{1}{T} \sum_{m=-\infty}^{\infty} R_X(m) \Big[\sum_{\ell=-\infty}^{\infty} \int_{s-(\ell+1)T}^{s-\ell T} p(t - s - mT + v)p(v)dv \Big].$$

But we recognize that

$$\sum_{\ell=-\infty}^{\infty} \int_{s-(\ell+1)T}^{s-\ell T} p(t - s - mT + v)p(v)dv = \int_{-\infty}^{\infty} p(t - s - mT + v)p(v)dv.$$

The above expressions indicate that R_Z depends only on $\tau = t - s$, so $Z(t)$ is WSS. Furthermore, if

$$r_p(\tau) \overset{\triangle}{=} p(\tau) * p(-\tau) = \int_{-\infty}^{\infty} p(\tau + v)p(v)dv \tag{4.22}$$

denotes the deterministic autocorrelation of pulse $p(\tau)$, where $*$ represents the convolution operation, and

$$R_S(\tau) \overset{\triangle}{=} \sum_{m=-\infty}^{\infty} R_X(m)\delta(\tau - mT), \tag{4.23}$$

where $\delta(\tau)$ denotes the Dirac delta function, we obtain the compact expression

$$R_Z(\tau) = \frac{1}{T} R_S(\tau) * r_p(\tau). \tag{4.24}$$

In this identity, the sampling function $R_S(\tau)$ depends only on the autocorrelation of $X(k)$, and $r_p(\tau)$ depends only on the shape of pulse $p(t)$, so the two contributing factors to $R_Z(\tau)$ are separated.

For example, let $p(t)$ be the rectangular pulse

$$p(t) = \begin{cases} A & 0 \le |\tau| \le T/2 \\ 0 & \text{otherwise}. \end{cases}$$

As shown in Fig. 4.3, its autocorrelation function $r_p(\tau)$ is triangular. Furthermore, assume that

$$X(k) = Y(k) - Y(k-1),$$

where $Y(k)$ a zero-mean uncorrelated WSS sequence, so that

$$R_Y(m) = E[Y(k+m)Y(k)] = P_Y \delta(m),$$

where $\delta(m)$ denotes the Kronecker delta function, i.e.,

$$\delta(m) = \begin{cases} 1 & m = 0 \\ 0 & m \ne 0. \end{cases}$$

Then

$$R_X(m) = E[X(k+m)X(k)] = E[(Y(k+m) - Y(k+m-1))(Y(k) - Y(k-1))]$$
$$= P_Y[2\delta(m) - \delta(m-1) - \delta(m+1)],$$

so

$$R_S(\tau) = \frac{P_Y}{T}[2\delta(\tau) - \delta(\tau - T) - \delta(\tau + T),$$

and

$$R_Z(\tau) = R_S(\tau) * r_p(\tau)$$
$$= \frac{P_Y}{T}[2r_p(\tau) - r_p(\tau - T) - r_p(\tau + T)]$$

is plotted in Fig. 4.4.

Fig. 4.3 Pulse $p(t)$ and its autocorrelation $r_p(\tau)$

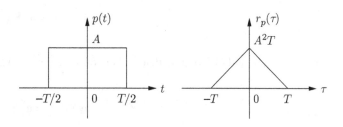

Fig. 4.4 Autocorrelation
$R_X(\tau)$ of $Z(t)$

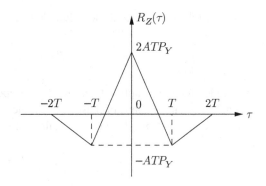

4.3.2 Autocorrelation and Autocovariance Properties

We have already seen that the autocorrelation and autocovariance functions $R_X(t, s)$ and $K_X(t, s)$ are symmetric in t and s. In the WSS case, this implies that if $\tau = t - s$, $R_X(\tau)$, and $K_X(\tau)$ are even, i.e.,

$$R_X(\tau) = R_X(-\tau) \text{ and } K_X(\tau) = K_X(-\tau).$$

Another important property is that $R_X(t, s)$ and $K_X(t, s)$ are nonnegative definite functions. Specifically for any N integer and $t_1 < \ldots < t_k < \ldots < t_N$ consider the N dimensional random vector

$$\mathbf{X} = \begin{bmatrix} X(t_1) \cdots X(t_k) \cdots X(t_N) \end{bmatrix}^T$$

and its mean

$$\mathbf{m}_X = \begin{bmatrix} m_X(t_1) \cdots m_X(t_k) \cdots m_X(t_N) \end{bmatrix}^T.$$

Then consider the $N \times N$ autocorrelation and autocovariance matrices

$$\mathbf{R}_X = E[\mathbf{X}\mathbf{X}^T] \text{ and } \mathbf{K}_X = E[(\mathbf{X} - \mathbf{m}_X)(\mathbf{X} - \mathbf{m}_X)^T].$$

These matrices are *nonnegative definite* in the sense that for all vectors

$$\mathbf{a} = \begin{bmatrix} a_1 \ldots a_k \ldots a_n \end{bmatrix}^T \in \mathbb{R}^N,$$

we have

$$\sum_{k=1}^{N} \sum_{\ell=1}^{N} a_k a_\ell R_X(t_k, t_\ell) = \mathbf{a}^T \mathbf{R}_X \mathbf{a} \geq 0 \tag{4.25}$$

$$\sum_{k=1}^{N} \sum_{\ell=1}^{N} a_k a_\ell K_X(t_k, t_\ell) = \mathbf{a}^T \mathbf{K}_X \mathbf{a} \geq 0. \tag{4.26}$$

To verify (4.25) and (4.26), note that if Z denotes the random variable

$$Z = \sum_{i=1}^{N} a_k X(t_k) = \mathbf{a}^T \mathbf{X},$$

we have

$$m_Z = E[Z] = \mathbf{a}^T \mathbf{m}_X$$

and

$$Z - m_Z = \mathbf{a}^T (\mathbf{X} - \mathbf{m}_X).$$

This implies

$$\mathbf{a}^T \mathbf{R}_X \mathbf{a} = E[Z^2] \geq 0$$

and

$$\mathbf{a}^T \mathbf{K}_X \mathbf{a} = E[(Z - m_Z)^2] = K_Z \geq 0,$$

which shows that both $\mathbf{R}_X$ and $\mathbf{K}_X$ are nonnegative definite matrices. In this respect, note that since

$$\mathbf{a}^T \mathbf{R}_X \mathbf{a} = E[Z^2] = K_Z + m_Z^2$$
$$= \mathbf{a}^T \mathbf{K}_X \mathbf{a} + (\mathbf{a}^T \mathbf{m}_X)^2,$$

the nonnegativeness of $\mathbf{K}_X$ implies that of $\mathbf{R}_X$. When the process $X(t)$ is WSS, the nonnegativeness condition (4.19) can be expressed as

$$\sum_{k=1}^{N} \sum_{\ell=1}^{M} a_k a_\ell R_X(t_k - t_\ell) \geq 0, \tag{4.27}$$

which must hold for all N, all $t_1 < \ldots < t_k < \ldots < t_N$, and all real a_ks, $1 \leq k \leq N$. This condition is similar to the one satisfied by the characteristic function of a random variable. In the case of absolutely continuous distributions, the characteristic function could be viewed as the Fourier transform of a pointwise nonnegative PDF. Similarly, it will be shown in Chap. 8 that the autocorrelation of a WSS process admits a Fourier transform, called the power spectral density (PSD), which is also pointwise nonnegative.

Properties of Nonnegative Definite Matrices At this point it is worth reviewing briefly the properties of real symmetric and nonnegative matrices. More details can be found in [3, 4]. If a real matrix $\mathbf{M}$ of dimension $N \times N$ is symmetric, all its eigenvalues λ_i, $1 \leq i \leq N$ are necessarily real, and there exists an orthonormal matrix $\mathbf{P}$ such that

$$\mathbf{M} = \mathbf{P} \mathbf{\Lambda} \mathbf{P}^T, \tag{4.28}$$

where

$$\mathbf{\Lambda} = \text{diag}\{\lambda_i, \ 1 \leq i \leq N\}$$

is a diagonal eigenvalue matrix. Recall that $\mathbf{P}$ is orthonormal if

$$\mathbf{P}^T \mathbf{P} = \mathbf{P} \mathbf{P}^T = \mathbf{I}_N,$$

where $\mathbf{I}_N$ denotes the identity matrix of size N. By observing that (4.28) is equivalent to

$$\mathbf{M} \mathbf{P} = \mathbf{P} \mathbf{\Lambda}, \tag{4.29}$$

and if $\mathbf{p}_i$ denotes the i-th column of

$$\mathbf{P} = \begin{bmatrix} \mathbf{p}_1 \cdots \mathbf{p}_i \cdots \mathbf{p}_N \end{bmatrix},$$

we recognize that (4.29), and thus (4.28) are just obtained by rewriting the eigenvalue-eigenvector relations

$$\mathbf{M}\mathbf{p}_i = \lambda_i \mathbf{p}_i$$

for all $1 \leq i \leq n$ as a single matrix identity. Hence (4.28) just expresses the fact that it is possible to construct an orthonormal basis of $\mathbb{R}^N$ formed by eigenvectors of $\mathbf{M}$. It is easy to verify that the symmetry of $\mathbf{M}$ ensures that eigenvectors associated with distinct eigenvalues are necessarily orthogonal. Thus all what is required to construct the basis $\{\mathbf{p}_i, \ 1 \leq i \leq n\}$ is to use Gram–Schmidt orthonormalization to orthonormalize the eigenvectors of $\mathbf{M}$ corresponding to repeated eigenvalues.

Given a real symmetric matrix $\mathbf{M}$ with eigenvalue/eigenvector representation (4.28), $\mathbf{M}$ is nonnegative if any one of the following three properties holds:

i) $\mathbf{a}^T\mathbf{M}\mathbf{a} \geq 0$ for all $\mathbf{a} \in \mathbb{R}^N$;
ii) the eigenvalues $\lambda_i \geq 0$ for $1 \leq i \leq N$;
iii) the principal minors of $\mathbf{M}$ are nonnegative.

We recall that a principal minor of order k of $\mathbf{M}$ is the determinant of the submatrix obtained by deleting $N - k$ rows and $N - k$ columns with the same indices.

In the discussion above, it is important to observe that to verify that an autocorrelation or autocovariance function is nonnegative, the nonnegativeness of matrix $\mathbf{R}_X$ or $\mathbf{K}_X$ needs to be tested for all N and all $t_1 < \ldots < t_k < \ldots < t_N$. It is possible to obtain some necessary conditions by testing only for some N and some t_ks, but these conditions are typically not sufficient. Consider for example a zero-mean WSS process $X(t)$ with autocorrelation $R_X(\tau)$. Then for $N = 2$ and $\tau = t - s$, the autocorrelation matrix corresponding to

$$\mathbf{X} = \begin{bmatrix} X(s) \\ X(t) \end{bmatrix}^T$$

is given by

$$\mathbf{R}_X = E[\mathbf{X}\mathbf{X}^T] = \begin{bmatrix} R_X(0) & R_X(\tau) \\ R_X(\tau) & R_X(0), \end{bmatrix}$$

where we have used the fact that $R_X(-\tau) = R_X(\tau)$ since the autocorrelation is symmetric in t and s. According to the principal minor test, we must have both

$$\det \mathbf{R}_X = R_X^2(0) - R_X^2(\tau) \geq 0 \tag{4.30}$$

and $R_X(0) = E[X^2(t)] = P_X \geq 0$, where P_X denotes the power of $X(t)$. The condition (4.30) is equivalent to

$$|R_X(\tau)| \leq R_X(0) = P_X, \tag{4.31}$$

so that the absolute value of autocorrelation function $R_X(\tau)$ must never exceed its value $R_X(0) = P_X$ at $\tau = 0$. The autocorrelation functions of Examples 4.1 and 4.4 clearly satisfy this property. For Example 4.1

$$R_X(\tau) = \frac{A^2}{2} \cos(\omega_c \tau)$$

oscillates between $P_X = A^2/2$ and P_X reaching these values every period or half-period $T_c = 2\pi/\omega_c$ of the waveform $X(t)$, which just reflects the periodicity this signal. In contrast for Example 4.4, $R_X(\tau) = \exp(-2a|\tau|)$ decreases steadily as $|\tau|$ increases.

However, the condition (4.31) is not sufficient to ensure that $R_X(\tau)$ is a valid autocorrelation function, as can be verified by considering the following counter-example.

Example 4.6 Consider the candidate autocorrelation function

$$R_X(\tau) = \begin{cases} 1 - |\tau|/2 & |\tau| \le 1 \\ 0 & |\tau| > 1, \end{cases}$$

which is plotted in Fig. 4.5. This function is symmetric and satisfies condition (4.31). Yet, it is not a valid autocorrelation function. Suppose that $X(t)$ is a WSS process with this autocorrelation. Then the autocorrelation matrix of the vector

$$\mathbf{X} = \begin{bmatrix} X(t) & X(t + 15/16) & X(t + 17/16) \end{bmatrix}^T$$

would be given by

$$\mathbf{R}_X = \begin{bmatrix} R_X(0) & R_X(15/16) & R_X(17/16) \\ R_X(15/16) & R_X(0) & R_X(1/8) \\ R_X(17/16) & R_X(1/8) & R_X(0) \end{bmatrix} = \begin{bmatrix} 1 & 15/32 & 0 \\ 15/32 & 1 & 15/16 \\ 0 & 15/16 & 1 \end{bmatrix}.$$

The determinant of $\mathbf{R}_X$ is given by

$$\det \mathbf{R}_X = 1 - (15/32)^2 - (15/16)^2 = -101/(32)^2 < 0,$$

so $R_X(\tau)$ is not a valid autocorrelation function.

Fig. 4.5 Candidate autocorrelation function $R_X(\tau)$

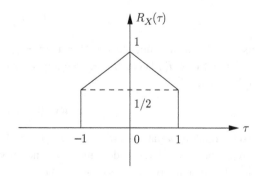

4.4   Cross-Correlation and Cross-Covariance of Random Processes

It is sometimes necessary to examine the joint second-order statistics of two different random processes, say $X(t)$ and $Y(t)$. The cross-correlation of these two processes is defined as

$$R_{XY}(t, s) = E[X(t)Y(s)],$$

and the cross-covariance is the cross-correlation of the centered processes $X(t) - m_X(t)$ and $Y(t) - m_Y(t)$, i.e.,

$$K_{XY}(t, s) = E[(X(t) - m_X(t))(Y(s) - m_Y(s))] = R_{XY}(t, s) - m_X(t)m_Y(s).$$

As was already observed for autocorrelations and autocovariances, assuming that the means $m_X(t)$ and $m_Y(t)$ are known, the autocorrelation and autocovariance can be obtained from each other. Unlike the autocorrelation and autocovariance cases, the cross-correlations and cross-covariance are not symmetric in t and s, but

$$R_{YX}(t, s) = E[Y(t)X(s)] = E[X(s)Y(t)] = R_{XY}(s, t).$$

Likewise $K_{YX}(t, s) = K_{XY}(s, t)$.

Two processes X and Y are said to be uncorrelated if

$$K_{XY}(t, s) = 0$$

for all t and s. It is a common mistake to declare X and Y uncorrelated if $E[(X(t) - m_X(t))(Y(t) - m_Y(t))] = K_{XY}(t, t) = 0$, but this is wrong! The uncorrelatedness of the *processes* X and Y for all times is a much stronger requirement than the uncorrelatedness of the *random variables* $X(t)$ and $Y(t)$ at the same time.

Example 4.1 (Continued) For example, consider the sinusoidal signal with a uniformly distributed random phase

$$X(t) = A \cos(\omega_c t + \Theta)$$

with period $T_c = 2\pi/\omega_c$, and consider the quadrature signal $Y(t) = X(t - T_c/4)$ obtained by delaying $X(t)$ by one-quarter of a period. Then

$$R_{XY}(\tau) = E[X(t + \tau)Y(t)] = \frac{A^2}{2} \cos(\omega_c(\tau - T_c/4))$$

is not identically equal to zero for all $\tau = t - s$, so $X(t)$ and $Y(t)$ are cross-correlated. In fact $X(t) = Y(t + T_c/4)$ can be recovered exactly by advancing $Y(t)$ by one quarter of a period. On the other hand

$$R_{XY}(0) = E[X(t)Y(t)] = 0,$$

so the random variables $X(t)$ and $Y(t)$ obtained by sampling the processes X and Y at the same time are uncorrelated. Thus, to determine whether processes X and Y are uncorrelated, all the relative time lags between the two processes need to be examined.

The processes $X(t)$ and $Y(t)$ examined above have the feature that their cross-correlation depends only on $\tau = t - s$. Two processes $X(t)$ and $Y(t)$ are said to be *jointly WSS* if the vector process

$$\mathbf{Z}(t) = \begin{bmatrix} X(t) \\ Y(t) \end{bmatrix}$$

is WSS, i.e., its mean

$$\mathbf{m}_Z = \begin{bmatrix} m_X \\ m_Y \end{bmatrix}$$

is constant and its 2×2 autocorrelation matrix

$$\mathbf{R}_Z(\tau) = E[\mathbf{Z}(t+\tau)\mathbf{Z}^T(t)] = E[\begin{bmatrix} X(t+\tau) \\ Y(t+\tau) \end{bmatrix} [\, X(t) \; Y(t)\,]]$$

$$= \begin{bmatrix} R_X(\tau) & R_{XY}(\tau) \\ R_{YX}(\tau) & R_Y(\tau) \end{bmatrix}$$

depends on τ only. In other words, for $X(t)$ and $Y(t)$ to be jointly WSS, i) $X(t)$ and $Y(t)$ must be separately WSS and ii) the cross-correlation $R_{XY}(t, s)$ of $X(t)$ and $Y(t)$ must depend on $\tau = t - s$ only. Note that since $R_{YX}(t, s) = R_{XY}(s, t)$, we must have

$$R_{XY}(\tau) = R_{YX}(-\tau).$$

4.5 Mean-Square Continuity, Differentiation, and Integration

The second-order statistics of $X(t)$ are sufficient to study the continuity, differentiability, and integrability of continuous-time random processes in the mean-square sense. A continuous-time process $X(t)$ is said to be mean-square continuous at point t if

$$\lim_{h \to 0} E[(X(t+h) - X(t))^2] = 0,$$

and it is mean-square continuous over interval $I = [a, b]$ if it is continuous at each point of I.

By observing that

$$E[(X(t+h) - X(t))^2] = R_X(t+h, t+h) + R_X(t, t) - 2R_X(t+h, t), \qquad (4.32)$$

we conclude that if $R_X(t, s)$ is continuous at point (t, t), i.e., if

$$\lim_{h \to 0} R_X(t+h, t+h) = \lim_{h \to 0} R_X(t+h, t) = R_X(t, t),$$

then $X(t)$ is mean-square continuous at time t. This means of course that if $R_X(t, s)$ is continuous on the diagonal (t, t) of I^2, it is mean-square continuous over interval I.

In fact, it is possible to prove a much stronger result: $X(t)$ is mean-square continuous over I if and only if $R_X(t, s)$ is continuous over I^2. We have just seen that continuity of R_X on the diagonal implies continuity of $X(t)$ in the mean-square sense over I. But if $X(t)$ is continuous in the mean square sense over I, if $t_n, n \geq 1$ and $s_n, n \geq 1$ are two time sequences of I converging to t and s

respectively, the sequences $X(t_n)$ and $X(s_n)$ converge to $X(t)$ and $X(s)$ in the mean-square sense. By Problem 3.6 of Chap. 3, this implies that

$$\lim_{n\to\infty} E[X(t_n)X(s_n)] = E[X(t)X(s)] = R_X(t,s),$$

so $\lim_{n\to\infty} R_X(t_n, s_n) = R_X(t, s)$, which indicates that R_X is continuous at an arbitrary point (t, s) of I^2.

For the case of WSS processes, the above result indicates that continuity of $R_X(\tau)$ at $\tau = 0$ ensures that $X(t)$ is mean-square-continuous for all t, and ensures as well that $R_X(\tau)$ is continuous for all τ. If we consider the random phase process of Example 4.1, its autocorrelation

$$R_X(\tau) = \frac{A^2}{2}\cos(\omega_c\tau)$$

is continuous at $\tau = 0$, so $X(t)$ is mean-square continuous, which was expected since the sample paths $X(t, \omega)$ are all continuous. Surprisingly, if we consider the random telegraph wave of Example 4.4, we find that its autocorrelation

$$R_X(\tau) = \exp(-2a|\tau|)$$

is continuous at $\tau = 0$, so $X(t)$ is mean-square continuous for all t, even though the sample paths $X(t, \omega)$ are discontinuous with probability 1, since they contain switches between 1 and -1. The mean-square continuity is due to the fact that for a fixed time t, the probability that a switch will occur at this exact time is zero.

Let us turn now to mean-square differentiation. The process $X(t)$ is mean-square differentiable at time t if

$$d_h X(t) = \frac{X(t+h) - X(t)}{h}$$

converges in the mean-square sense to a random variable $\dot{X}(t)$ (the mean-square derivative of $X(t)$ at t) as h tends to zero. Since $\dot{X}(t)$ is not known a priori, to establish the mean-square convergence, we need to rely on a Loève's mean-square convergence criterion of Problem 3.7 which relies on two independent approximations $d_{h_1}X(t)$ and $d_{h_2}X(t)$ of the unknown derivative $\dot{X}(t)$. Then $d_h X(t)$ converges in the mean-square sense as h tends to zero if and only if

$$E[d_{h_1}X(t)d_{h_2}X(t)] = \frac{1}{h_1 h_2} E[(X(t+h_1) - X(t))(X(t+h_2) - X(t))]$$

$$= \frac{1}{h_1 h_2}[R_X(t+h_1, t+h_2) + R_X(t,t) - R_X(t+h_1, t) - R_X(t, t+h_2)] \tag{4.33}$$

admits a finite limit as h_1 and h_2 tend to zero independently. But if we introduce the difference approximations

$$\partial_{t,h_1} R_X(t,s) = \frac{R_X(t+h_1, s) - R_X(t,s)}{h_1}$$

$$\partial_{s,h_2} R_X(t,s) = \frac{R_X(t, s+h_2) - R_X(t,s)}{h_1}$$

of the first order derivatives of R_X with respect to t and s respectively, we recognize that (4.33) can be rewritten as

$$\partial_{s,h_2}\partial_{t,h_1} R_X(t,s)\,|_{t=s},$$

so if the limit of (4.33) exists, it is given by

$$\frac{\partial^2}{\partial t \partial s} R_X(t, t).$$

This shows therefore that $X(t)$ is mean-square differentiable if and only if the second order cross-derivative in t and s of $R_X(t, s)$ exists at $t = s$, and $X(t)$ is mean-square differentiable over $I = [a, b]$ if the cross-derivative exists along the diagonal $t = s$ of I^2. By observing that $d_{h_1} X(t)$ and $d_{h_2} X(s)$ converge in the mean-square sense to $\dot{X}(t)$ and $\dot{X}(s)$, we also conclude by using the result of Problem 3.6 that

$$\lim_{h_1, h_2 \to 0} E[d_{h_1} X(t) d_{h_2} X(s)] = \frac{\partial^2}{\partial t \partial s} R_X(t, s) = E[\dot{X}(t)\dot{X}(s)] \tag{4.34}$$

exists over I^2. So the existence of the cross-derivative along the diagonal $t = s$ of I^2 guarantees its existence over I^2.

By invoking again the result of Problem 3.6 (the correlation of mean-square limits of sequences is the limit of the correlation of the sequences), we deduce that the cross-correlation of processes $\dot{X}$ and X is given by

$$\lim_{h \to 0} E[d_h X(t) X(s)] = \frac{\partial}{\partial t} R_X(t, s) = E[\dot{X}(t) X(s)] = R_{\dot{X}X}(t, s). \tag{4.35}$$

By symmetry we have of course

$$R_{X\dot{X}}(t, s) = E[X(t)\dot{X}(s)] = \frac{\partial}{\partial s} R_X(t, s). \tag{4.36}$$

Together, identities (4.34)–(4.36) specify the auto- and cross-correlations of processes $\dot{X}$ and X.

In the case when $X(t)$ is WSS, $X(t)$ is mean-square differentiable for all t if

$$\frac{\partial^2}{\partial t \partial s} R_X(t - s) = -\frac{d^2}{d\tau^2} R_X(\tau)$$

exists at $\tau = 0$. In this case $\dot{X}(t)$ is jointly WSS with $X(t)$ with

$$E[\dot{X}(t + \tau)\dot{X}(t)] = -\frac{d^2}{d\tau^2} R_X(\tau)$$

$$E[\dot{X}(t + \tau)X(t)] = \frac{d}{d\tau} R_X(\tau)$$

$$E[X(t + \tau)\dot{X}(t)] = -\frac{d}{d\tau} R_X(\tau).$$

If we consider the random phase process of Example 4.1, by observing that

$$\frac{d^2}{d\tau^2} R_X(\tau) = \frac{\omega_c^2 A^2}{2} \cos(\omega_c \tau),$$

we conclude that $X(t)$ is mean-square differentiable. This should not come as a surprise, since each sample path is differentiable with

$$\frac{d}{dt}X(t, \omega) = -\omega_c A \sin(\omega_c t + \Theta(\omega)).$$

On the other hand, the random telegraph wave process of Example 4.4 is not mean-square differentiable since the first derivative

$$\frac{d}{d\tau}R_X(\tau) = \begin{cases} 2a \exp(2a\tau) & \tau < 0 \\ -2a \exp(-2a\tau) & \tau > 0 \end{cases}$$

of the autocorrelation is discontinuous at $\tau = 0$. Note that the sample paths are almost surely not differentiable since with probability one, they include jumps between 1 and -1.

Let us now turn to mean-square integration. Let $h(t)$ be a piecewise continuous function over $I = [a, b]$ and let $X(t)$ be piecewise mean-square continuous over I. Recall that if a function $h(t)$ is piecewise continuous over I, there exists at most a finite number of points t_i of I where $h(t)$ is discontinuous, and at such points $h(t)$ admits finite left and right limits. Then, let $h_n = (b-a)/n$ and consider points $t_k^n = a + kh_n$ for $0 \leq k \leq n$. The integral

$$H = \int_a^b h(t)X(t)dt \tag{4.37}$$

is defined as the mean-square limit of

$$H_n = \sum_{k=0}^n h(t_k^n)X(t_k^n)h_n$$

as n tends to infinity. Applying the mean-square convergence criterion of Problem 3.7, we conclude H_n converges in the mean-square if and only if

$$E[H_n H_m] = \sum_{k=0}^n \sum_{\ell=0}^m h(t_k^n)R_X(t_k^n, t_\ell^m)h(t_\ell^m)h_n h_m \tag{4.38}$$

converges to the Riemann integral

$$\int_a^b \int_a^b h(t)R_X(t, s)h(s)dtds$$

as n and m tend to infinity. But the convergence of the Riemann sum (4.38) is guaranteed as n and m tend to infinity if h is piecewise continuous and R_X is piecewise continuous over I^2. If H denotes the mean-square limit of H_n, applying again the result of Problem 3.6 yields

$$\lim_{n \to \infty} E[H_n] = E[H] = \int_a^b h(t)m_X(t)dt, \tag{4.39}$$

$$\lim_{n \to \infty} E[H_n^2] = E[H^2] = \int_a^b \int_a^b h(t)R_X(t, s)h(s)dtds. \tag{4.40}$$

Finally, by letting a or b tend to $-\infty$ or ∞, it is also possible to define the mean-square integral (4.37) over a half-line or the entire real line, but in this case $h(t)R_X(t, s)h(s)$ needs to integrable over I^2.

4.6 Classes of Random Processes

We now return to the specification of random processes in terms of their finite joint distributions. The amount of information needed to accomplish this task is very large, but it is possible to identify several random processes classes for which the needed information can be reduced to a reasonable level.

4.6.1 Gaussian Processes

A process $X(t)$ is said to be Gaussian if all its finite joint densities are Gaussian. In this case $X(t)$ is specified entirely by its mean $m_X(t)$ and its autocorrelation $R_X(t, s)$ or its autocovariance $K_X(t, s)$. Specifically for $t_1 < \ldots < t_k < \ldots < t_N$, consider the random vector

$$\mathbf{X} = \left[\, X(t_1) \ldots X(t_k) \ldots X(t_N)\,\right]^T,$$

and denote by

$$\mathbf{m}_X = \left[\, m_X(t_1) \ldots m_X(t_k) \ldots m_X(t_N)\,\right]^T$$

and

$$\mathbf{K}_X = (K_X(t_k, t_\ell))_{1 \le k, \ell \le N}$$

its mean vector and covariance matrix. Then if

$$\mathbf{x} = \left[\, x_1 \ldots x_k \ldots x_N\,\right]^T,$$

$X(t)$ is Gaussian if for all integers N and all choices of $t_1 < \ldots < t_k < \ldots < t_N$, $\mathbf{X}$ admits the Gaussian PDF

$$f_{\mathbf{X}}(\mathbf{x}) = \frac{1}{(2\pi)^{N/2}|\mathbf{K}_X|^{1/2}} \exp\left(-(\mathbf{x} - \mathbf{m}_X)^T \mathbf{K}_X^{-1}(\mathbf{x} - \mathbf{m}_X)/2\right). \tag{4.41}$$

Thus for a process $X(t)$ to be Gaussian it is not enough that the random variable $X(t)$ at time t alone should be Gaussian, but all joint samples at different times need to be Gaussian together.

The reduction in the information needed to specify a Gaussian process is considerable, since we only need to know the first- and second-order statistics $m_X(t)$ and $K_X(t, s)$ (or $R_X(t, s)$). Furthermore, if $X(t)$ is WSS, then the mean vector

$$\mathbf{m}_X = m_X \left[\, 1 \ldots 1 \ldots 1\,\right]^T$$

and the autocovariance matrix

$$\mathbf{K}_X = (K_X(t_k - t_\ell))_{1 \le k, \ell \le N}$$

are unchanged under a time translation where $t_k \to t_k + h$ with h arbitrary. Thus $X(t)$ is SSS of order N for any N, so it is SSS. In other words, WSS implies SSS for Gaussian processes.

Example 4.7 The Ornstein–Uhlenbeck (OU) process $X(t)$ is a zero mean stationary Gaussian process with autocorrelation

$$R_X(\tau) = P_X \exp(-a|\tau|),$$

where $P_X = E[X^2(t)]$ denotes the average power of $X(t)$. It will be shown later that in an RC circuit in thermal equilibrium, the capacitor voltage is an OU process. Note that the OU process and the random telegraph wave have the same mean and autocorrelation, even though the two processes are extremely different: the OU process takes continuous values, whereas the telegraph wave takes only binary values ± 1. It will be shown later in Chap. 6 that the sample paths of the OU process are continuous almost surely, whereas the sample paths of the telegraph wave are almost surely discontinuous. In other words, two processes can have the same second order statistics, but yet behave very differently.

4.6.2 Markov Processes

Markov processes form another class for which the information needed to specify the finite joint distributions can be reduced drastically. $X(t)$ is a Markov process if for times $t > s > r_1 > r_2 > \ldots > r_K$, the conditional distribution of $X(t)$ given $X(s)$ is identical to the conditional distribution of $X(t)$ given $X(s)$, $X(r_k)$, $1 \leq k \leq K$. More colloquially, $X(t)$ is Markov if given the present value $X(s)$ of X, the future value $X(t)$ is conditionally independent of past values $X(r_k)$ with $1 \leq k \leq K$, i.e., knowledge of the previous values $X(r_k)$, $1 \leq k \leq K$ does not improve our knowledge of future value $X(t)$ once present value $X(s)$ is known. The concept of Markov process is similar to the concept of "state" in electrical and mechanical engineering systems. Specifically, the values of present capacitor voltages and inductor currents in an RLC circuit are all that is needed to analyze the circuit in the future. Further information about the past is not needed. Likewise, positions and velocities in a mechanical system are all that is needed to analyze the system in the future.

If we consider a process with absolutely continuous joint distributions, $X(t)$ is Markov if the conditional PDF

$$f_{X(t)|X(s)X(r_1)\cdots X(r_K)}(x_t, t|x_s, s; x_{r_1}, r_1; \cdots ; x_{r_K}, r_K) = f_{X(t)|X(s)}(x_t, t|x_s, s). \qquad (4.42)$$

If we use the compact notation

$$q(x_t, t; x_s, s) = f_{X(t)|X(s)}(x_t, t|x_s, s),$$

for the *transition density* (the conditional density) of the Markov process for $t \geq s$, then for $t_1 < \cdots < t_{N-1} < t_N$, the joint PDF of $X(t_1), \ldots, X(t_{N-1})$, and $X(t_N)$ can be expressed as

$$f_{X(t_1)\cdots X(t_{N-1})X(t_N)}(x_1, t_1; \cdots ; x_{N-1}, t_{N-1}; x_N; t_N)$$

$$= \prod_{k=2}^{N} q(x_k, t_k; x_{k-1}, t_{k-1}) f_{X(t_1)}(x_1, t_1). \qquad (4.43)$$

To prove (4.43), we proceed by induction. For $N = 2$, we have obviously

$$f_{X(t)X(s)}(x_t, t; x_s, s) = q(x_t, t; x_s, s) f_{X(s)}(x_s) \qquad (4.44)$$

for $t \geq s$. Next, assume that expression (4.43) holds for $N - 1$. We have

$$f_{X(t_N)X(t_{N-1})\cdots X(t_1)}(x_N, t_N; x_{N-1}, t_{N-1}; \ldots; x_1, t_1)$$

$$= f_{X(t_N)|X(t_{N-1})\cdots X(t_1)}(x_N, t_N | x_{N-1}, t_{N-1}; \cdots; x_1, t_1)$$

$$\times f_{X(t_{N-1})\cdots X(t_1)}(x_{N-1}, t_{N-1}; \cdots; x_1, t_1)$$

$$= q(x_N, t_N; x_{N-1}, t_{N-1}) f_{X(t_{N-1})\cdots X(t_1)}(x_{N-1}, t_{N-1}; \cdots; x_1, t_1), \qquad (4.45)$$

where to go from the second to the third line, we have used the Markov property (4.42). Since (4.43) was assumed to hold for $N - 1$, identity (4.45) implies it holds for N.

The expression (4.43) indicates that in order to specify completely the finite joint densities of a Markov process $X(t)$, we only need the transition density function $q(x_t, t; x_s, s)$ for $t \geq s$ and the PDF $f_{X(t_0)}(x_0, t_0)$ of $X(t_0)$ at some initial time t_0. By setting $t = t_1$ and $s = t_0$ inside the joint PDF (4.44) and marginalizing, we observe indeed that the density of X at a time $t_1 > t_0$ can be expressed in terms of the initial density $f_{X(t_0)}(x_0, t_0)$ as

$$f_{X(t_1)}(x_1, t_1) = \int q(x_1, t_1; x_0, t_0) f_{X(t_0)}(x_0, t_0) dx_0. \qquad (4.46)$$

Thus all quantities appearing in the joint PDF decomposition (4.43) depend only on the transition density and initial density. The transition density $q(x_t, t; x_s, s)$ needs obviously to be a PDF, i.e., it must be nonnegative and viewed as a function of x_t, its probability mass is one. But it must also satisfy a much stronger condition in order to be a valid Markov transition density. For $t \geq s \geq r$, by using (4.43), the joint conditional density of $X(t)$ and $X(s)$ given $X(r)$ can be expressed as

$$f_{X(t)X(s)|X(r)}(x_t, t; x_s, s | x_r, r) = q(x_t, t; x_s, s) q(x_s, s : x_r, r).$$

Then by using this expression to find the marginal density of $X(t)$ given $X(r)$, we find

$$q(x_t, t; x_r, r) = \int f_{X(t)X(s)|X(r)}(x_t, t; x_s, s | x_r, r) dx_s$$

$$= \int q(x_t, t; x_s, s) q(x_s, s; x_r, r) dx_s \qquad (4.47)$$

for $t > s > r$. The identity (4.47) is called the *Chapman–Kolmogorov* equation. It places an important constraint on eligible Markov transition densities. In fact, this constraint is the expression for Markov processes of Kolmogorov's consistency condition (4.1).

Whenever the transition density $q(x_t, t; x_s, s) = q(x_t, x_s, t - s)$ depends only on $t - s$, it is said to be *homogeneous*, in the sense that it is invariant under a change of time origin. Given a homogeneous transition density q, a PDF $\pi(x)$ is an *invariant density* if

$$\pi(x_1) = \int q(x_1, x_0, t) \pi(x_0) \qquad (4.48)$$

for all $t \geq 0$. In this case if the initial density $f_{X(t_0)}(x) = \pi(x)$, the density of $X(t)$ is $\pi(x)$ for all $t \geq t_0$. If the transition density of Markov process $X(t)$ is homogeneous, and if its initial density is invariant with respect to the transition density, the expression (4.43) for finite joint densities of $X(t)$ becomes

$$f_{X(t_1)\cdots X(t_{N-1})X(t_N)}(x_1, 0; \cdots ; x_{N-1}, t_{N-1} - t_1; x_N, t_N - t_1)$$

$$= \prod_{k=2}^{N} q(x_k, x_{k-1}, t_k - t_{k-1})\pi(x_1),$$

which is obviously unchanged under a time translation $t_k \to t_k + h$ with h arbitrary, so $X(t)$ is SSS. Thus a Markov process $X(t)$ is SSS if and only if its transition density is homogeneous and its initial density is q-invariant.

Example 4.4 (Continued) Although this was not mentioned earlier, the random telegraph wave $X(t)$ is a discrete-valued Markov process (a continuous-time Markov chain) taking values ± 1. In this case the specification of $X(t)$ given earlier in terms of its initial distribution

$$P(X(0) = 1) = P(X(0) = -1) = 1/2$$

and its conditional PMF

$$P(\epsilon_t, t; \epsilon_s, s) = P(X(t) = \epsilon_t | X(s) = \epsilon_s),$$

where $\epsilon_t, \epsilon_s \in \{1, -1\}$ and

$$P(\epsilon_t, t; \epsilon_s, s) = \frac{1}{2}[1 + \epsilon_t \epsilon_s \exp(-2a(t - s))]$$

is sufficient to compute all the joint PMFs of $X(t)$ at times $t_x < t_2 < \cdots < t_N$. Specifically, we have

$$P(X(t_1) = \epsilon_1, X(t_2) = \epsilon_2, \cdots , X(t_N) = \epsilon_N)$$

$$= \prod_{k=2}^{N} P(X(t_k) = \epsilon_k | X(t_{k-1}) = \epsilon_{k-1})P(X(t_1) = \epsilon_1)$$

$$= \frac{1}{2^N} \prod_{k=2}^{N}[1 + \epsilon_k \epsilon_{k-1} \exp(-2a(t_k - t_{k-1}))],$$

where $\epsilon_k = \pm 1$ for all $1 \leq k \leq N$. Furthermore, the discrete-space form of the Chapman–Kolmogorov identity (4.47) can be expressed as

$$\sum_{\epsilon_s \in \{1,-1\}} P(\epsilon_t, t; \epsilon_s, s)P(\epsilon_s, s; \epsilon_r, r)$$

$$= \frac{1}{4} \sum_{\epsilon_s \in \{1,-1\}} [1 + \epsilon_t \epsilon_s \exp(-2a(t - s))] \times [1 + \epsilon_s \epsilon_r \exp(-2a(s - r))]$$

$$= \frac{1}{2}[1 + \epsilon_t \epsilon_r \exp(-2a(t - r))] = P(\epsilon_t, t; \epsilon_r, r), \tag{4.49}$$

where $t > s > r$ and $\epsilon_t, \epsilon_r \in \{1, -1\}$. To derive (4.49), we have used the fact that

$$\epsilon_s^2 = 1 \quad \text{and} \quad \sum_{\epsilon_s \in \{1,-1\}} \epsilon_s = 0.$$

4.6.3 Independent Increments Processes

A random process $X(t)$ is said to have *independent increments* if for all integers $N \geq 1$, and all choices of nonoverlapping (but possibly touching) intervals $I_k = [s_k, t_k]$ with $1 \leq k \leq N$, the increments $Y_k = X(t_k) - X(s_k)$ are mutually independent for $1 \leq k \leq N$. Thus if

$$\mathbf{Y} = \begin{bmatrix} Y_1 \cdots Y_k \cdots Y_N \end{bmatrix}^T,$$

the joint CDF

$$F_{\mathbf{Y}}(\mathbf{y}) = \prod_{k=1}^{N} F_{Y_k}(y_k). \tag{4.50}$$

When the process $X(t)$ has absolutely continuous distributions, the joint PDF of the increments satisfies

$$f_{\mathbf{Y}}(\mathbf{y}) = \prod_{k=1}^{N} f_{Y_k}(y_k). \tag{4.51}$$

An important result is that if the initial value of an independent increments process $X(t)$ is known, say $X(t_0) = x_0$ is fixed, then $X(t)$ is Markov for $t \geq t_0$. To see why this is the case, let $t_0 < t_1 < t_2 < \cdots < t_N$ and consider intervals $I_k = [t_{k-1}, t_k]$ for $1 \leq k \leq N$. If

$$\mathbf{X} = \begin{bmatrix} X(t_1) \ldots X(t_k) \ldots X(t_N) \end{bmatrix}^T,$$

the vector $\mathbf{X}$ can be expressed in terms of increments vector $\mathbf{Y}$ as

$$\mathbf{X} = \mathbf{M}\mathbf{Y} + \mathbf{u}_1 x_0, \tag{4.52}$$

where

$$\mathbf{M} = \begin{bmatrix} 1 & 0 & & & \ldots & 0 \\ 1 & 1 & 0 & & & \\ 1 & 1 & 1 & 0 & & \\ \vdots & & \ddots & \ddots & & \vdots \\ 1 & \ldots & 1 & & 1 & 1 \end{bmatrix}$$

and

$$\mathbf{u} = \begin{bmatrix} 1 & 1 & \ldots & 1 \end{bmatrix}^T$$

is the vector of $\mathbb{R}^N$ with all entries equal to 1. The inverse transformation of (4.52) is given by

$$\mathbf{Y} = h(\mathbf{X}) = \mathbf{M}^{-1}\mathbf{X} - \mathbf{e}_1 x_0,$$

where

$$\mathbf{M}^{-1} = \begin{bmatrix} 1 & 0 & & & \ldots & 0 \\ -1 & 1 & 0 & & & \\ 0 & -1 & 1 & 0 & & \\ \vdots & & \ddots & \ddots & & \vdots \\ 0 & \ldots & 0 & & -1 & 1 \end{bmatrix}$$

is a triangular matrix with 1s on its diagonal, -1s on its first subdiagonal, and zeros everywhere else, and

$$\mathbf{e}_1 = \begin{bmatrix} 1 & 0 & \dots & 0 \end{bmatrix}^T$$

is a unit vector colinear with the first axis. Since the determinant of $\mathbf{M}$ is one, the Jacobian of the transformation (4.52) is one, and the PDF of vector $\mathbf{X}$ is given by

$$f_{\mathbf{X}}(\mathbf{x}) = f_{\mathbf{Y}}(\mathbf{M}^{-1}\mathbf{x} - \mathbf{e}_1 x_0)$$

$$= \prod_{k=1}^{N} f_{Y_k}(x_k - x_{k-1}). \tag{4.53}$$

The joint PDF (4.53) is in the Markov form (4.43) provided that we recognize that

$$q(x_k, t_k; x_{k-1}, t_{k-1}) = f_{Y_k}(x_k - x_{k-1}) \tag{4.54}$$

is the transition density of process $X(t)$, and the density of $X(t_1)$ is given by

$$f_{X(t_1)}(x_1) = f_{Y_1}(x_1 - x_0).$$

An interesting feature of the transition density (4.54) is that it is homogeneous in space, in the sense that if z is an arbitrary real translation vector

$$q(x_k + z, t_k; x_{k-1} + z, t_{k-1}) = q(x_k, t_k; x_{k-1}, t_{k-1}).$$

A process $X(t)$ is said to have stationary independent increments if in addition to property (4.50), the increment $X(t+h) - X(s+h)$ admits the same distribution as $X(t) - X(s)$ for any time translation h. In this case, the transition density of process $X(t)$ is homogeneous in both time and space, in the sense that

$$q(x_t, t; x_s, s) = q(x_t - x_s, t - s).$$

If we assume that $X(0) = 0$, $q(x, t)$ coincides with the PDF of $X(t)$ and the characteristic function of $X(t)$ is given by

$$\Phi_t(u) = E[\exp(iuX(t))] = \int_{-\infty}^{\infty} q(x, t) \exp(iux) dx. \tag{4.55}$$

By observing that

$$X(t + s) = X(t + s) - X(s) + X(s),$$

where increments $X(t + s) - X(s)$ and $X(s) - X(0) = X(s)$ are independent, and observing that since the increments of X are stationary, $X(t + s) - X(s)$ has the same distribution as $X(t)$, we find that the characteristic function of $X(t + s)$ satisfies

$$\Phi_{t+s}(u) = \Phi_t(u)\Phi_s(u) \tag{4.56}$$

for $t, s \geq 0$. Equivalently, the transition density $q(x, t + s)$ satisfies the convolution equation

$$q(x, t + s) = q(x, t) * q(x, s), \tag{4.57}$$

which is the form taken by the Chapman–Kolmogorov equation when $X(t)$ is a stationary independent increments process with zero initial value, i.e., with $X(0) = 0$.

Example 4.8 The Wiener process $W(t)$ is a stationary independent increments process such that $W(0) = 0$ and

$$W(t) \sim N(0, \sigma^2 t). \tag{4.58}$$

Since the transition density $q(x, t)$ coincides with the PDF of $W(t)$ we have

$$q(x, t) = \frac{1}{(2\pi\sigma^2)^{1/2}} \exp(-x^2/(2\sigma^2 t)), \tag{4.59}$$

and its Fourier transform

$$\Phi_t(u) = \exp(-\sigma^2 u^2 t/2)$$

obviously satisfies (4.56). By substituting the expression (4.59) for

$$f_{Y_k}(x_k - x_{k-1}) = q(x_k - x_{k-1}, t_k - t_{k-1})$$

inside expression (4.53) for the joint density of the $W(t_k)$s with $1 \le k \le N$, we conclude that in addition to being a Markov process, $W(t)$ is also Gaussian. According to (4.58), the mean and second moment of $W(t)$ are

$$m_W(t) = 0 \quad \text{and} \quad E[W^2(t)] = \sigma^2 t.$$

For $t \ge s$, by observing that

$$R_W(t, s) = E[W(t)W(s)] = E[(W(t) - W(s) + W(s))W(s)],$$

where increment $W(t) - W(s)$ is independent of $W(s)$, we find

$$R_W(t, s) = E[W(t) - W(s)]E[W(s)] + E[W^2(s)] = 0 + \sigma^2 s.$$

Since $R_W(t, s)$ needs to be symmetric in t and s, we conclude that

$$R_W(t, s) = \sigma^2 \min(t, s). \tag{4.60}$$

A key property of the Wiener process is that its sample paths are almost surely continuous, as will be proved in Chap. 6. The Poisson process is a stationary independent increments process which is the opposite of the Wiener process in the sense that it is a pure jumps process whose trajectories are piecewise constant between jumps and are thus discontinuous almost surely. This process can be characterized as follows.

Example 4.9 The Poisson process $N(t)$ is a stationary independent increments process such that $N(0) = 0$ and $N(t)$ is Poisson distributed with parameter λt, i.e.,

$$P(N(t) = n) = \frac{(\lambda t)^n}{n!} \exp(-\lambda t).$$

The characteristic function $\Phi_t(u) = E[\exp(juN(t))]$ is given by

$$\Phi_t(u) = \exp(\lambda t (\exp(ju) - 1)),$$

and satisfies (4.56). The PMF of $N(t)$ satisfies the discrete convolution equation

$$P(N(t) = n) * P(N(s) = n) = \sum_{m=0}^{n} P(N(t) = n - m)P(N(s) = m)$$

$$= \frac{\exp(-\lambda(t + s))}{n!}\left(\sum_{m=0}^{n} \binom{n}{m}(\lambda t)^{n-m}(\lambda s)^{m}\right)$$

$$= \frac{(\lambda(t + s))^n}{n!} \exp(-\lambda(t + s)) = P(N(t + s) = n),$$

which is the counterpart of (4.56) for discrete-valued distributions. Since $N(t + s) = N(t + s) - N(s) + N(s)$, where $N(t + s) - N(s)$ has the same distribution as $N(t)$, which takes values over positive integers, we conclude that $N(t + s) \geq N(s)$, for $t, s \geq 0$, so that all trajectories of the Poisson process are monotone nondecreasing. A typical sample path of $N(t)$ is depicted in Fig. 4.6, where the jump times, also called epochs, of the Poisson process are denoted by S_k with $k \geq 1$.

Finally, it is useful to observe that the random telegraph wave of Example 4.4 can be interpreted in terms of the Poisson process as

$$X(t) = X(0)(-1)^{N(t)}, \tag{4.61}$$

where $N(t)$ is independent of $X(0)$ and has rate $\lambda = a$. Indeed, expression (4.61) implies

$$X(t) = X(s)(-1)^{N(t)-N(s)}$$

for $t \geq s$, so that

$$P(X(t) = X(s)|X(s)) = P(N(t) - N(s) = \text{even})$$

$$= \sum_{m=0}^{\infty} \frac{(a(t - s))^{2m}}{(2m)!} \exp(-a(t - s)). \tag{4.62}$$

By using the power series expansion

$$\sum_{n=0}^{\infty} \frac{x^n}{n!} = \exp(-x)$$

Fig. 4.6 Sample path of a Poisson process

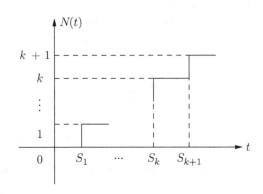

of the exponential, we deduce

$$\sum_{m=0}^{\infty} \frac{x^{2m}}{(2m)!} = \frac{1}{2}[\exp(x) + \exp(-x)],$$

which can be used to express (4.62) as

$$P(X(t) = X(s)|X(s)) = \frac{1}{2}[1 + \exp(-2a(t-s))].$$

This implies

$$P(X(t) = -X(s)|X(s)) = 1 - P(X(t) = X(s)|X(s)) = \frac{1}{2}[1 - \exp(-2a(t-s))],$$

which matches exactly the transition probabilities of the random telegraph wave.

4.7   Bibliographical Notes

Since the purpose of this textbook is introductory, several relevant topics were left out. Some applications, such as image processing or machine learning, rely on Markov fields models instead of Markov processes. A two-dimensional random field $X(t_1, t_2)$ is said to be a Markov field in the sense of Paul Lévy [5] if given its values on a closed contour C, the values of $X(t_1, t_2)$ in the interior of C are independent of its values outside C. Quite surprisingly, Markov fields do not reduce to Markov processes in one dimension, but to reciprocal processes. These processes were introduced by Sergei Bernstein [6] in 1932, while trying to give a stochastic derivation of Schrodinger's equation. Reciprocal processes were then analyzed systematically by Jamison [7]. The books by Grimmett [8] and Guyon [9] contain accessible presentations of random fields and processes over graphs. In continuous time, stationary independent increments processes such that $X(0) = 0$ whose sample paths are continuous in probability are called Lévy processes and have been the object of extensive studies [10]. The property (4.56) of their characteristic function implies that they admit infinitely divisible distributions, which are closely related to the stable laws discussed in the previous chapter. This observation was used by Lévy and Khinchin to characterize their distributions. Later, it was also shown that these processes can be decomposed as a sum of a Wiener process with drift and a jump process. Unfortunately, all these topics are too advanced for a meaningful discussion at this stage.

4.8   Problems

4.1 Let $p(t)$ be the square wave of period T defined by

$$p(t) = \begin{cases} 1 & \text{for } 0 \le t < T/2 \\ -1 & \text{for } T/2 \le t < T, \end{cases}$$

and $p(t) = p(t - nT)$ for n integer. Consider the random process

$$X(t) = Ap(t - D),$$

where A is a zero-mean Gaussian variable with variance σ^2, and D is a random delay independent of A and uniformly distributed over $[0, T)$. Find the mean and autocorrelation of $X(t)$.

4.2 Let $g(t)$ denote a periodic "sawtooth" signal of period T defined over the interval $[-T/2, T/2]$ as

$$g(t) = 1 - 4|t|/T.$$

Outside this interval $g(t)$ is extended periodically by observing that $g(t) = g(t - nT)$ with n integer. A random process $X(t)$ is then defined as

$$X(t) = g(t - D),$$

where D denotes a random delay.

a) Sketch some typical sample functions for the process $X(t)$ if D is uniformly distributed over the interval $[0, T]$.
b) Sketch all possible sample functions if D is distributed according to

$$P(D = mT/4) = 1/4$$

   for $m = 0, 1, 2, 3$.
c) Evaluate the mean and autocorrelation of $X(t)$ when D has the probability distribution of part b).

4.3 Consider the random processes

$$X(t) = A_c \cos(\omega_c t + \Theta)$$

and

$$Y(t) = A_s \sin(\omega_c t + \Theta),$$

where the amplitudes A_c and A_s, as well as the carrier frequency ω_c, are known, and Θ is a random phase uniformly distributed over $[0, 2\pi]$.

a) Is each of these two random processes wide-sense stationary?
b) Find the cross-correlation function $R_{XY}(t, s) = E[X(t)Y(s)]$ for the two random processes.
c) Are these two random processes jointly WSS? Note that in addition to requiring that $X(t)$ and $Y(t)$ are separately WSS, this requires that R_{XY} should be a function of $t - s$ only.
d) Are the two random processes uncorrelated?
e) Are they independent?

4.4 Consider a discrete-time random process $X(t)$ defined for $t \geq 0$ which takes only the values $\{0, 1\}$. The probability distribution of $X(0)$ is given by

$$P(X(0) = 0) = P(X(0) = 1) = \frac{1}{2},$$

and for $t \geq s$, the conditional distribution of $X(t)$ given $X(s)$ is specified by

$$P(X(t) = 0 \mid X(s) = 0) = P(X(t) = 1 \mid X(s) = 1) = \frac{1}{2}[1 + (1 - 2p)^{(t-s)}]$$

$$P(X(t) = 1 \mid X(s) = 0) = P(X(t) = 0 \mid X(s) = 1) = \frac{1}{2}[1 - (1 - 2p)^{(t-s)}],$$

where $0 < p < 1$.

a) Find the probability distribution of $X(t)$, i.e., evaluate $P(X(t) = 0)$ and $P(X(t) = 1)$.
b) Evaluate the mean $m_X(t)$ and autocorrelation $R_X(t, s)$ of $X(t)$. Is $X(t)$ wide-sense stationary?
c) Is $X(t)$ strict sense stationary? Please give a detailed justification of your answer.

4.5 Consider the pulse-amplitude modulated signal

$$Z(t) = \sum_{k=-\infty}^{\infty} X(k)p(t - kT - D),$$

where D is a random synchronization delay uniformly distributed over the interval $[0, T)$, and $p(t)$ is the rectangular pulse

$$p(t) = \begin{cases} 1 & 0 \leq t \leq T \\ 0 & \text{otherwise} . \end{cases}$$

The sequence $X(k)$ is generated by

$$X(k) = Y(k) - Y(k - 2),$$

where $\{Y(k)\}$ is a sequence of independent random variables taking the binary values ± 1 with equal probability.

a) Find the mean m_X and autocorrelation of the sequence $\{X(k)\}$.
b) Evaluate the mean and autocorrelation of $Z(t)$.

4.6 A pulse amplitude modulated partial response signal can be expressed as

$$Z(t) = \sum_{k=-\infty}^{\infty} X(k)p(t - kT - D),$$

where D denotes a random synchronization delay uniformly distributed over $[0, T]$. The sequence $X(k)$ is generated according to

$$X(k) = Y(k) + Y(k - 1),$$

where the $Y(k)$'s are independent binary digits taking the values $\{1, -1\}$ with equal probability. The pulse $p(t)$ has for Fourier transform

$$P(j\omega) = \begin{cases} AT & -\dfrac{\pi}{T} \leq \omega \leq \dfrac{\pi}{T} \\ 0 & \text{otherwise} . \end{cases}$$

Equivalently, in the time domain we have

$$p(t) = A \frac{\sin(\pi t/T)}{\pi t/T}.$$

Recall that the Fourier and inverse Fourier transform relations are given by

$$P(j\omega) = \int_{-\infty}^{\infty} p(t) \exp(-j\omega t) dt$$

$$p(t) = \frac{1}{2\pi} \int_{-\infty}^{\infty} P(j\omega) \exp(j\omega t) d\omega.$$

a) Find the mean and autocorrelation of the sequence $X(k)$.
b) Evaluate the deterministic autocorrelation

$$r_p(\tau) = p(\tau) * p(-\tau)$$

   of the pulse $p(t)$, where $*$ denotes the convolution of two functions.
c) Find the mean and autocorrelation of $Z(t)$.

4.7 A binary frequency-shift keying (FSK) signal switches between two oscillators according to the message bit being sent. The oscillator signals can be represented by

$$S_0(t) = A \cos(\omega_0 t + \Theta_0)$$

and

$$S_1(t) = A \cos(\omega_1 t + \Theta_1),$$

where the amplitude A and frequencies ω_0 and ω_1 are known, and the phase angles Θ_0 and Θ_1 are independent and uniformly distributed over $[0, 2\pi)$. A representation of the transmitted signal valid for all times is

$$S(t) = \frac{A}{2}(1 + Z(t)) \cos(\omega_0 t + \Theta_0) + \frac{A}{2}(1 - Z(t)) \cos(\omega_1 t + \Theta_1),$$

where

$$Z(t) = \sum_{k=-\infty}^{\infty} X(k) p(t - kT - D)$$

is a unit amplitude binary random wave. In this expression, the random variables $\{X(k)\}$ are independent and take the values ± 1 with equal probability, $p(t)$ is the rectangular pulse

$$p(t) = \begin{cases} 1 & 0 \leq t \leq T \\ 0 & \text{otherwise}, \end{cases}$$

and D is a random synchronization delay uniformly distributed over $[0, T)]$.

a) Evaluate the mean and autocorrelation of $Z(t)$.
b) Find the mean and autocorrelation of $S(t)$. Is $S(t)$ WSS?

4.8 Consider the quadrature phase-shift keyed (QPSK) signal

$$X(t) = A\cos(\omega_c t + \Theta),$$

where the amplitude A and carrier frequency ω_c are known, and Θ is a random phase which takes the values 0, $\pi/2$, π, and $3\pi/2$ with with equal probability, so that

$$P(\Theta = 0) = P(\Theta = \pi/2) = P(\Theta = \pi) = P(\Theta = 3\pi/2) = \frac{1}{4}.$$

a) Find the mean $m_X(t)$ and autocovariance $K_X(t, s)$ of $X(t)$. Determine whether the process is wide-sense stationary (WSS).
b) Find the probability density $f_{X(t)}(x)$ of $X(t)$ at a fixed time t. Is $X(t)$ strict-sense stationary?

4.9 Consider the random signal

$$X(t) = A\cos(\Omega t + \Theta),$$

where the amplitude A is known, the frequency Ω is random with density $f_\Omega(\omega)$, and the phase Θ is independent of Ω and uniformly distributed over the interval $[0, 2\pi)$.

a) Show that $X(t)$ is WSS with zero mean and autocorrelation

$$R_X(\tau) = E[X(\tau)X(0)] = \frac{A^2}{2}\operatorname{Re}\Phi_\Omega(\tau),$$

where

$$\Phi_\Omega(\tau) = E[\exp(j\Omega\tau)] = E[\cos(\Omega\tau)] + jE[\sin(\Omega\tau)]$$

is the characteristic function of Ω.
b) To illustrate the result obtained in part a), assume Ω has the Lorentzian probability density

$$f_\Omega(\omega) = \frac{1}{2\pi}\frac{2\lambda}{\omega^2 + \lambda^2}$$

for $-\infty < \omega < \infty$. Find and sketch $R_X(\tau)$ as a function of τ. To find the characteristic function of Ω, you may want to use the fact that

$$f(x) = \exp(-a|x|) \quad\leftrightarrow\quad F(u) = \frac{2a}{u^2 + a^2}$$

is a Fourier transform pair, as well as the duality property

$$F(x) \quad\leftrightarrow\quad 2\pi f(-u)$$

of the Fourier transform.

4.10 Let $X(t)$ and $Y(t)$ be two zero-mean jointly stationary Gaussian processes with auto- and cross-covariance functions $K_X(\tau)$, $K_Y(\tau)$ and $K_{XY}(\tau)$. Let

$$Z(t) = X(t)\cos(\omega_c t) + Y(t)\sin(\omega_c t),$$

where the carrier frequency ω_c is known.

a) Is $Z(t)$ a Gaussian process? Explain your reasoning briefly.
b) Find $m_Z(t)$ and $K_Z(t, s)$, the mean and autocovariance of $Z(t)$.
c) Give a set of conditions on $K_X(\tau)$, $K_Y(\tau)$, and $K_{XY}(\tau)$ which are both necessary and sufficient to ensure that $Z(t)$ is strict-sense stationary.

4.11 In quadrature phase shift keying modulation (QPSK), the in-phase and out-of-phase components of the carrier are each used to transmit separately one bit. Over the k-th signaling interval, the transmitted signal is given by

$$X(t) = I_k \cos(\omega_c t) + Q_k \sin(\omega_c t),$$

where I_k and Q_k take the values ± 1 with probability $1/2$. The two sequences I_k and Q_k are independent of each other, and symbols I_k and Q_k corresponding to different indices k are independent. Assuming that the signaling interval is T, and that there exists a random synchronization delay D which is independent of the I_ks and Q_ks, and is uniformly distributed over $[0, T]$, the transmitted signal admits therefore the representation

$$X(t) = X_c(t) \cos(\omega_c t) + X_s(t) \sin(\omega_c t)$$

with

$$X_c(t) = \sum_{k=-\infty}^{\infty} I_k p(t - kT - D)$$

$$X_s(t) = \sum_{k=-\infty}^{\infty} Q_k p(t - kT - D),$$

where $p(t)$ is the rectangular pulse sketched in Fig. 4.7.

a) Find the mean of $X_c(t)$ and $X_s(t)$.
b) Evaluate the auto- and cross-correlations

$$R_c(t, s) = E[X_c(t)X_c(s)]$$
$$R_s(t, s) = E[X_s(t)X_s(s)]$$
$$R_{cs}(t, s) = E[X_c(t)X_s(s)].$$

Are $X_c(t)$ and $X_s(t)$ jointly WSS?

Fig. 4.7 Rectangular
pulse $p(t)$

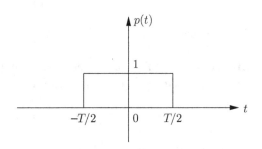

c) Find the mean and autocorrelation of $X(t)$. You may want to use the trigonometric identity

$$\cos(x - y) = \cos(x)\cos(y) + \sin(x)\sin(y).$$

Is $X(t)$ WSS?

d) Sketch the autocorrelation of $X(t)$. You may assume that the carrier frequency $f_c = \omega_c/2\pi$ is much larger than the signaling frequency $1/T$.

4.12 Consider a phase modulated (PM) signal

$$Z(t) = A\cos(\omega_c t + kX(t) + \Theta),$$

where the amplitude A, carrier frequency ω_c and modulation index k are known, $X(t)$ is a zero-mean WSS Gaussian process with autocorrelation function $R_X(\tau) = E[X(t + \tau)X(t)]$, and the phase Θ is uniformly distributed over $[0, 2\pi)$ and independent of $X(t)$.

a) Evaluate the mean of $Z(t)$.

b) Evaluate the autocorrelation $R_Z(t, s)$ of $Z(t)$. To do so, you are reminded that since the process $X(t)$ is Gaussian, the joint characteristic function of $X(t)$ and $X(s)$ is given by

$$\Phi_{X(t), X(s)}(u, v) = E[\exp(j(uX(t) + vX(s)))]$$

$$= \exp\left(-\frac{1}{2}[u\ v]\begin{bmatrix} R_X(0) & R_X(t - s) \\ R_X(t - s) & R_X(0) \end{bmatrix}\begin{bmatrix} u \\ v \end{bmatrix}\right).$$

Is $Z(t)$ WSS? If so, verify that $R_Z(\tau)$ with $\tau = t - s$ can be expressed as

$$R_Z(\tau) = \frac{R_C(\tau)}{2}\cos(\omega_c\tau),$$

where $R_C(\tau) = E[C(t + \tau)C^*(t)]$ denotes the autocorrelation of the complex envelope signal $C(t) = Ae^{jkX(t)}$.

c) Consider the wideband PM case with $k \gg 1$. Assume that

$$R_X(\tau) = P\exp(-(\tau/\tau_X)^2),$$

where the parameter τ_X denotes the coherence time of $X(t)$, i.e., it is the amount of time such that $R_X(\tau)/R_X(0) \le e^{-1}$ for $|\tau| > \tau_X$. In other words if $X(t)$ and $X(t+\tau)$ are separated by more than τ_X, they can be viewed as approximately uncorrelated. Obtain an approximation for the coherence time τ_C of $R_C(\tau)$ in terms of k, τ_X and the power $P = R_X(0)$ of signal $X(t)$.

4.13 For each of the following functions, determine whether it is an autocovariance function, and explain.

a) $K_X(t, s) = \exp(-|t - s|)$.
b) $K_X(t, s) = \exp(-(t + s))$.
c) $K_X(t, s) = \exp(t + s)$.
d) $K_X(t, s) = \exp(|t - s|)$.
e) $K_X(t, s) = \exp(-|t - s|)\cos(t - s)$.

f)

$$K_X(t, s) = \begin{cases} 1 & |t - s| \le T \\ 2 - (|t - s|/T) & T < |t - s| \le 2T \\ 0 & |t - s| > 2T. \end{cases}$$

4.14 A useful characterization of Markov processes can be stated as follows: given the present value of the process, its future and past values are conditionally independent. To prove this result, consider the times $t_1 > t_2 > \cdots > t_L > s > r_1 > r_2 > \cdots r_K$, where $X(s)$ represents the present value of the process, $X(t_1), X(t_2), \ldots X(t_L)$ correspond to future values, and $X(r_1), X(r_2), \ldots X(r_L)$ denote past values.

a) Prove that if $X(\cdot)$ is Markov, the joint density of the past and future given the present can be factored as

$$f_{X(t_1)X(t_2)\cdots X(t_L)X(r_1)\cdots X(r_K)|X(s)}(x_{t_1}, x_{t_2}, \cdots, x_{t_L}, x_{r_1}, \cdots x_{r_K} \mid x_s)$$
$$= f_{X(t_1)X(t_2)\cdots X(t_L)|X(s)}(x_{t_1}, x_{t_2} \cdots x_{t_L} \mid x_s)$$
$$\times f_{X(r_1)\cdots X(r_K)|X(s)}(x_{r_1} \cdots x_{r_K} \mid x_s).$$

b) Show that this implies that for arbitrary functions $g(\cdot)$ and $h(\cdot)$, we have

$$E[g(X(t_1), X(t_2), \cdots, X(t_L))h(X(r_1), X(r_2), \cdots, X(r_K)) \mid X(s)]$$
$$= E[g(X(t_1), X(t_2), \cdots X(t_L)) \mid X(s)]$$
$$\times E[h(X(r_1), X(r_2), \cdots X(r_K)) \mid X(s)].$$

In particular if $L = 1$, $K = 1$, $g(x) = x - m_X(t)$, $h(y) = y - m_X(r)$, where $m_X(t) = E[X(t)]$, we obtain

$$E[(X(t) - m_X(t))(X(r) - m_X(r)) \mid X(s)]$$
$$= E[X(t) - m_X(t) \mid X(s)]E[X(r) - m_X(r) \mid X(s)] \tag{4.63}$$

for $t > s > r$.

c) Given an arbitrary Gaussian process $X(t)$ with mean $m_X(t)$ and autocovariance $K_X(t, s)$, show that independently of whether s is smaller or larger than t,

$$E[X(t) - m_X(t) \mid X(s)] = K_X(t, s)K_X^{-1}(s, s)(X(s) - m_X(s)).$$

Combine this result with (4.63) to show that if $X(t)$ is both *Gaussian and Markov*, its autocovariance must satisfy

$$K_X(t, r) = K_X(t, s)K_X^{-1}(s, s)K_X(s, r) \tag{4.64}$$

for $t > s > r$. In fact, as shown in Problem 4.16, the converse is also true, i.e., if a Gaussian process $X(t)$ satisfies (4.64), it is Markov.

d) The Ornstein–Uhlenbeck process is a zero-mean Gaussian process with autocovariance

$$K_X(t, s) = \sigma^2 \exp(-a|t - s|).$$

Use the test (4.64) to determine whether it is a Markov process.

4.15 If $X(t)$ is a Gaussian process with mean $m_X(t)$ and autocovariance $K_X(t, s)$, it was shown in Problem 4.14 that

$$E[X(t)|X(s)] = m_X(t) + K_X(t, s)K_X^{-1}(s, s)[X(s) - m_x(s)]. \tag{4.65}$$

If $X(t)$ is a Markov process, its transition density $f_{X(t)|X(s)}(x_t, t|x_s, s) = q(x_t, t; x_s, s)$ satisfies the Chapman–Kolmogorov equation

$$q(x_t, t; x_r, r) = \int_{-\infty}^{\infty} q(x_t, t; x_s, s)q(x_s, s|x_r, r)dx_s$$

for $t > s > r$.

a) Use the Chapman Kolmogorov equation to prove that for a Markov process

$$E[E[X(t)|X(s)]|X(r)] = E[X(t)|X(r)] \tag{4.66}$$

for $t > s > r$.

b) Use (4.65) and (4.66) to show that the autocovariance function of a Gauss–Markov process must satisfy

$$K_X(t, r) = K_X(t, s)K_X^{-1}(s, s)K_X(s, r) \tag{4.67}$$

for $t > s > r$.

4.16 Let $X(t)$ be a Gaussian process with mean $m_X(t)$ and autocovariance $K_X(t, s)$. For $t > s > r$, show that if (4.67) holds, then $X(t)$ has the Markov property

$$f_{X(t)|X(s)X(r)}(x_t, t|x_s, s; x_r, r) = f_{X(t)|X(s)}(x_t, t|x_s, s).$$

4.17 Let $X(t)$ be a Gaussian process over the interval $[0, T]$. In Problems 4.14 and 4.15, it was proved that if $X(t)$ is Markov if and only if its autocovariance satisfies (4.67).

a) An alternative characterization which is often used is that $X(t)$ is Markov if and only if its autocovariance can be expressed as

$$K_X(t, s) = g(\max(t, s))h(\min(t, s)) \tag{4.68}$$

for some functions $g(\cdot)$ and $h(\cdot)$. Show that the properties (4.67) and (4.68) are equivalent. *Hint*: To show that (4.67) implies (4.68), you may want to try $g(t) = K_X(t, 0)$ and $h(t) = K_X^{-1}(t, 0)K_X(t, t)$.

b) The following table lists several zero-mean Gaussian processes, together with their autocovariance, and interval of definition. Use the test (4.68) to determine whether they are Markov, and if the response is affirmative, specify the functions $g(\cdot)$ and $h(\cdot)$. Note that these functions are only unique up to a constant, since multiplying $g(\cdot)$ by a constant c, and dividing $h(\cdot)$ by the same constant c preserves the structure (4.68).

Process	$K_X(t, s)$	Interval
Wiener	$\min(t, s)$	$[0, \infty)$
Brownian bridge	$\min(t, s) - \dfrac{ts}{T}$	$[0, T]$
Ornstein–Uhlenbeck	$\sigma^2 \exp(-a\lvert t - s\rvert)$	$[0, \infty)$
Slepian	$1 - \dfrac{\lvert t - s\rvert}{T}$	$[0, 2T]$

4.18   The expression (4.43) for the fine joint densities of a Markov process $X(t)$ relies on the *forward transition density*

$$f_{X(t)\mid X(s)}(x_t \mid x_s, s) = q(x_t, t; x_s, s),$$

where $t > s$. But the characterization of the Markov property derived in Problem 4.14, whereby $X(t)$ is Markov if and only if the future and past of the process are conditionally independent given the present indicates that the Markov property is time symmetric. Let

$$f_{X(s)\mid X(t)}(x_s, s \mid x_t, t) = q_B(x_s, s; x_t, t)$$

denote the *backward transition density* of $X(t)$ for $s < t$.

a) Express $q_B(x_s, s; x_t, t)$ in terms of the forward transition density $q(x_t, t; x_s, s)$ and the probability densities $f_{X(s)}(x_s, s)$ and $f_{X(t)}(x_t, t)$ of the Markov process at times s and t.
b) Show that for $t_1 < t_2 < \cdots < t_N$ the finite joint densities of the $X(t_k)$'s with $1 \leq k \leq N$ can be expressed as

$$f_{X(t_1)X(t_2)\cdots X(t_N)}(x_1, t_1; x_2, t_2; \cdots ; x_N, t_N)$$

$$= \prod_{k=1}^{N-1} q_B(x_k, t_k; X_{k+1}, t_{k+1}) f_{X(t_N)}(x_N, t_N), \tag{4.69}$$

so that the finite joint densities depend only on the backward transition density and the PDF of $X(t)$ at final time t_N. The expression (4.69) is the backwards counterpart of forward expression (4.43).
c) Suppose the forward transition density $q(x_t, x_s, t - s)$ is homogeneous. Is the backward transition density q_B necessarily homogeneous? If not, explain how the initial density $f_{X(t_0)}(x_0, t_0)$ should be selected to ensure that q_B is homogeneous.

4.19   Consider a Gauss–Markov process $X(t)$. It was shown in Problem 4.17 that its covariance function $K_X(t, s)$ can be expressed as $K_X(t, s) = g(\max(t, s))h(\min(t, s))$. Suppose that the process is sampled at times $t_1 < \ldots < t_k < \ldots < t_N$, and let $\mathbf{K}_X$ denote the $N \times N$ covariance matrix of

$$\mathbf{X} = \big[\, X(t_1) \ldots X(t_k) \ldots X(t_N) \,\big]^T.$$

We assume that $\mathbf{K}_X$ is positive definite, i.e., all its eigenvalues are strictly positive, or equivalently, its principal minors are strictly positive, so that $\mathbf{K}_X$ is invertible.

a) Verify that the entries $K_X(t_k, t_\ell)$ of $\mathbf{K}_X$ satisfy

$$K_X(t_{k+1}, t_\ell) = a_k K_X(t_k, t_\ell) \tag{4.70}$$

for $1 \leq \ell \leq k < N$ and express a_k in terms of the function $g(t)$.

b) Consider the lower triangular matrix

$$
\mathbf{A} = \begin{bmatrix}
1 & 0 & & \cdots & 0 \\
-a_1 & 1 & 0 & & 0 \\
0 & -a_2 & 1 & 0 & 0 \\
\vdots & \ddots & \ddots & \ddots & \\
0 & \cdots & 0 & -a_{N-1} & 1
\end{bmatrix}
$$

formed by 1s on its diagonal, $-a_k$'s on its first subdiagonal, and zeros everywhere else. Verify that

$$\mathbf{A}\mathbf{K}_X = \mathbf{U}, \tag{4.71}$$

where $\mathbf{U}$ is an *upper triangular* matrix whose diagonal elements are

$$q_1 = K_X(t_1, t_1)$$

$$q_k = K_X(t_k, t_k) - a_{k-1}K_X(t_{k-1}, t_k) \quad \text{for } 2 \le k \le N.$$

Prove that q_k is strictly positive for all k.

c) Use (4.71) and the symmetry to $\mathbf{K}_X$ to show that $\mathbf{K}_X^{-1}$ admits the UDL (upper triangular times diagonal times lower triangular) factorization

$$\mathbf{K}_X^{-1} = \mathbf{A}^T \mathbf{Q}^{-1} \mathbf{A}, \tag{4.72}$$

where

$$\mathbf{Q} = \text{diag}\,\{q_k, \ 1 \le k \le n\}.$$

Use this expression to prove that $\mathbf{M} = \mathbf{K}_X^{-1}$ is *tridiagonal*, i.e., its entries $m_{k\ell}$ with $1 \le k, \ell \le N$ satisfy

$$m_{k\ell} = 0$$

for $|k - \ell| > 1$. In other words only the diagonal, first subdiagonal, and first superdiagonal of $\mathbf{M}$ are nonzero.

d) If $X(t)$ is a Gaussian process with positive definite covariance function $K_X(t, s)$, show that $X(t)$ is Markov if and only if for all N, the covariance matrix $\mathbf{K}_X$ of any set of N ordered time samples is such that $\mathbf{K}_X^{-1}$ is tridiagonal.

4.20 Let $X(t)$ be a zero-mean discrete-time Gaussian process defined over $[1, N]$. Its autocovariance function is denoted as $K_X(t, s)$ with $1 \le t, s \le N$, and if

$$\mathbf{X} = \begin{bmatrix} X(1) \dots X(t) \dots X(N) \end{bmatrix},$$

the covariance matrix $\mathbf{K}_X$ is positive definite.

a) Show that

$$E[X(t+1)|X(t)] = a_t X(t),$$

and express a_t and the error variance $q_{t+1} = E[(X(t+1) - E[X(t+1)|X(t)])^2]$ in function of $K_X(t,s)$.

b) Prove that if $X(t)$ is Markov

$$E[(X(t+1) - E[X(t+1)|X(t)])X(r)] = 0$$

for $1 \leq r \leq t \leq N - 1$. To do so, you may want to use the Markov property and iterated conditioning.

c) By using the results of parts a) and b), show that if $X(t)$ is a Gauss–Markov process, the process

$$V(t) = X(t+1) - a_t X(t)$$

is Gaussian and white in the sense that

$$E[V(t)X(s)] = q_{t+1}\delta(t - s),$$

where

$$\delta(t - s) = \begin{cases} 1 & t = s \\ 0 & t \neq s \end{cases}$$

denotes the Kronecker delta function.

d) If $\mathbf{A}$ is defined as in Problem 4.19, and

$$\mathbf{Q} = \text{diag}\{q_t, \ 1 \leq t \leq N\}$$

with $q_1 = E[X^2(1)]$, verify that

$$\mathbf{A}\mathbf{K}_X = \mathbf{U},$$

where $\mathbf{U}$ is and upper triangular matrix, and deduce that $\mathbf{K}_X^{-1} = \mathbf{A}^T \mathbf{Q}^{-1} \mathbf{A}$.

e) Use the result of Problem 4.19 to conclude that a zero-mean Gaussian process $X(t)$ is Markov if and only if it admits a *state-space model*

$$X(t+1) = a_t X(t) + V(t)$$

where $V(t)$ is a zero-mean white Gaussian noise independent of the initial condition $X(0)$.

4.21 The r-th-order *joint moment* of the n random variables $X_1, \ X_2, \ \ldots, \ X_n$ is defined by

$$m(k_1, k_2, \ldots, k_n) = E[X_1^{k_1} X_2^{k_2} \cdots X_n^{k_n}],$$

where $r = k_1 + k_2 + \cdots + k_n$, and the nth-order *joint characteristic function* of these random variables is defined by

$$\Phi(u_1, u_2, \ldots, u_n) \overset{\Delta}{=} E[\exp(j(u_1 X_1 + u_2 X_2 + \cdots + u_n X_n))].$$

By expanding the exponential in an n-dimensional power series, the following identity can be obtained:

$$\frac{\partial^r \Phi(u_1, \ldots, u_n)}{\partial u_1^{k_1} \cdots \partial u_n^{k_n}} = j^r m(k_1, \ldots, k_n).$$

If we now use the expression

$$\Phi_X(u) = \exp(-\frac{1}{2}u^T K_X u)$$

with

$$u = \begin{bmatrix} u_1 \\ u_2 \\ \vdots \\ u_n \end{bmatrix}$$

for the characteristic function of jointly Gaussian random variables with zero-mean and covariance matrix K_X, it is easy to verify that for jointly Gaussian zero-mean random variables, all odd-order joint moments are zero, and the even-order joint moments are given by *Isserlis's formula* [11]

$$E[X_1 X_2 \cdots X_r] = \sum E[X_{j_1} X_{j_2}] \cdots E[X_{j_{r-1}} X_{j_r}], \tag{4.73}$$

where r is even, and the sum in (4.73) is taken over all possible ways of dividing r integers into $r/2$ combinations of pairs.

a) Use the identity (4.73) to show that the fourth-order joint moment of a zero-mean Gaussian process $X(t)$ with autocovariance $K_X(t, s)$ can be expressed as

$$E[X(t_1)X(t_2)X(t_3)X(t_4)]$$
$$= K_X(t_1, t_2)K_X(t_3, t_4) + K_X(t_1, t_3)K_X(t_2, t_4) + K_X(t_2, t_3)K_X(t_1, t_4).$$

b) Suppose now that $X(t)$ is a stationary zero-mean Gaussian process with autocovariance $K_X(\tau)$. This process is passed through a squarer with output $Y(t) = X^2(t)$, as shown in Fig. 4.8. Find the mean and autocorrelation of $Y(t)$.

4.22 *Price's theorem* [12] is a result which is often used to analyze the effect of passing zero-mean Gaussian signals through nonlinear memoryless devices. Let X_1 and X_2 be two jointly Gaussian random variables with zero-mean and covariance matrix

$$K = E[\begin{bmatrix} X_1 \\ X_2 \end{bmatrix} [X_1 \ X_2]] = \begin{bmatrix} \sigma_1^2 & k \\ k & \sigma_2^2 \end{bmatrix}.$$

Let $h(x_1, x_2)$ be an arbitrary function which either remains finite or grows only at a polynomial rate as $(x_1, x_2) \to \infty$. Show that

$$\frac{\partial^n}{\partial k^n} E[h(X_1, X_2)] = E[\frac{\partial^{2n} h(X_1, X_2)}{\partial X_1^n \partial X_2^n}]. \tag{4.74}$$

Fig. 4.8 Gaussian signal $X(t)$ passed through a squarer

$$X(t) \longrightarrow \boxed{(\cdot)^2} \longrightarrow Y(t) = X^2(t)$$

Hint: Observe that

$$\frac{\partial^n}{\partial k^n} f_{X_1 X_2}(x_1, x_2) = \frac{\partial^{2n}}{\partial x_1^n \partial x_2^n} f_{X_1 X_2}(x_1, x_2),$$

and repeatedly integrate

$$\int \int h(x_1, x_2) \frac{\partial^{2n}}{\partial x_1^n \partial x_2^n} f_{X_1 X_2}(x_1, x_2) dx_1 dx_2$$

by parts.

4.23 Consider a zero-mean stationary Gaussian process $X(t)$ with autocorrelation $R_X(t-s)$. Assume that $X(t)$ is passed through a time-invariant nonlinear memoryless function $g(\cdot)$, producing the output process $Y(t) = g(X(t))$.

a) For this part, assume that $g(\cdot)$ is one-to-one. Express the joint probability density of $Y_1 = Y(t_1)$, $Y_2 = Y(t_2)$, ..., $Y_N = Y(t_N)$ with $t_1 < t_2 < \cdots < t_N$ as a function of the joint density of $X_1 = X(t_1)$, $X_2 = X(t_2)$, ..., $X_N = X(t_N)$, and the derivative $\dot{g}$ of g. Use the expression you obtain to prove that $Y(t)$ is SSS (strict-sense stationary).

b) In the remainder of this problem, we apply Price's theorem to evaluate the autocorrelation function of the output process $Y(t)$. Let $X_1 = X(t_1)$ and $X_2 = X(t_2)$, so that

$$E[X_1 X_2] = k = R_X(t_1 - t_2).$$

By applying Price's theorem with $h(x_1, x_2) = g(x_1)g(x_2)$, show that if $Y_1 = Y(t_1)$ and $Y_2 = Y(t_2)$ are the outputs corresponding to X_1 and X_2, the autocorrelation

$$R_Y = E[Y_1 Y_2] = R_Y(t_1 - t_2)$$

satisfies

$$\frac{d^n R_Y}{dk^n} = E[\frac{d^n g}{dX_1^n}(X_1) \frac{d^n g}{dX_2^n}(X_2)]. \tag{4.75}$$

This formula can be used to evaluate R_Y by differentiating $g(\cdot)$ enough times to simplify the evaluation of the expectation on the right-hand-side of (4.74). Then R_Y can be evaluated by integration.

c) To illustrate the result of part b), consider the case where $g(\cdot)$ is the one-bit quantizer

$$g(x) = \begin{cases} +1 \ x > 0 \\ -1 \ x < 0 \end{cases} \tag{4.76}$$

shown in Fig. 4.9. Then, by setting $n = 1$ in (4.75), show that

$$\frac{d R_Y}{dk} = \frac{2}{\pi (R_X^2(0) - k^2)^{1/2}},$$

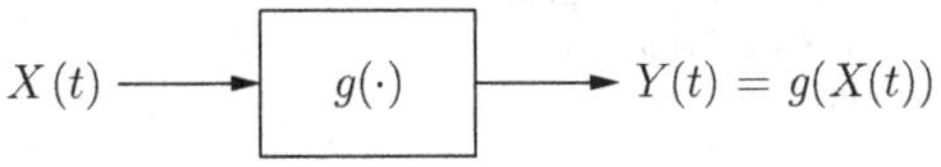

Fig. 4.9 Gaussian signal $X(t)$ passed through a one-bit quantizer $g(\cdot)$

$$X(t) \longrightarrow \boxed{g(\cdot)} \longrightarrow Y(t) = g(X(t))$$

which after integration with respect to k gives

$$R_Y = \frac{2}{\pi} \sin^{-1}(\frac{k}{R_X(0)}).$$

Consequently, if $t_2 - t_1 = \tau$, we get

$$R_Y(\tau) = \frac{2}{\pi} \sin^{-1}(\frac{R_X(\tau)}{R_X(0)}),$$

which gives a simple relation between the autocorrelations of the quantized process $Y(t)$ and of the input Gaussian process $X(t)$. Since $Y(t)$ takes only the values ± 1, its autocorrelation $R_Y(\tau)$ is easy to evaluate, and can be used to estimate $R_X(\tau)$ indirectly through the identity

$$R_X(\tau) = R_X(0) \sin(\pi R_Y(\tau)/2).$$

4.24 Another interesting consequence of Price's theorem is as follows. Let again $X(t)$ be a zero-mean stationary Gaussian process with autocorrelation $R_X(\tau)$. Let $Y(t) = g(X(t))$ be the process obtained by passing $X(t)$ through a time-invariant memoryless nonlinearity $g(x)$.

a) Prove that the cross-correlation between X and Y satisfies

$$R_{XY}(\tau) = K R_X(\tau) \quad \text{with} \quad K = E[\frac{dg}{dx}(X(t))]. \tag{4.77}$$

This result is known as *Bussgang's theorem* [13]. To prove it, let $X_1 = X(t_1)$ and $X_2 = X(t_2)$ with $t_2 - t_1 = \tau$, and apply Price's theorem with $h(X_1, X_2) = X_1 g(X_2)$ and $n = 1$. Then integrate with respect to $k = R_X(\tau)$. This result indicates that to estimate the autocorrelation R_X of X, we may instead estimate the cross-correlation R_{XY} of X and Y, which may be easier.

b) To illustrate this result consider again the case where $g(\cdot)$ is the one-bit quantizer given by (4.76). Show that in this case, the constant K appearing in (4.77) is given by

$$K = \left(\frac{2}{\pi R_X(0)}\right)^{1/2}.$$

4.25 The objective of this problem is to show that the autocorrelation function of a zero-mean stationary Gaussian process $X(t)$ can be obtained by passing $X(t)$ through a hard limiter, and computing the autocorrelation of of the resulting output $Y(t)$. This result is known as *Van Vleck's theorem*. It is derived in Problem 4.23 by using Price's theorem, but we consider here an alternate derivation.

a) Let X_1 and X_2 be two jointly Gaussian zero-mean random variables, with identical variance σ^2 and correlation coefficient ρ. Find the joint probability density of X_1 and $Z = X_2/X_1$, and use this result to show that Z admits a Cauchy probability density centered at ρ:

$$f_Z(z) = \frac{1}{\pi} \frac{\sqrt{1-\rho^2}}{(z-\rho)^2 + (1-\rho^2)}.$$

b) Use the result of part a) to show that

$$P[X_1 X_2 < 0] = P[Z < 0] = \frac{1}{\pi} \cos^{-1}(\rho). \tag{4.78}$$

c) Let $X(t)$ be a zero-mean stationary Gaussian process with autocorrelation function $R_X(\tau) = E[X(t + \tau)X(t)]$. Setting $X_1 = X(t + \tau)$ and $X_2 = X(t)$, use the identity (4.78) to evaluate the probability that the process $X(.)$ has an *odd* number of zeros in the interval $(t, t + \tau)$. What is the probability that $X(.)$ has an *even* number of zeros?

d) The process $X(t)$ is passed through a hard limiter $L(.)$ whose input–output relation is given by

$$y = L(x) = \begin{cases} 1 & x \geq 0 \\ -1 & x < 0. \end{cases}$$

Show that the autocorrelation of the output process $Y(t)$ satisfies

$$R_Y(\tau) = \frac{2}{\pi} \sin^{-1} \left(\frac{R_X(\tau)}{R_X(0)} \right),$$

so that

$$R_X(\tau) = R_X(0) \sin \left(\frac{\pi}{2} R_Y(\tau) \right).$$

Show that the last identity can be used to develop a simple scheme for evaluating the autocorrelation of $X(t)$ by using only the one-bit quantized signal $Y(t)$.

4.26 Let $X(t)$ be a zero-mean Gaussian WSS random process with autocorrelation

$$R_X(\tau) = E[X(t + \tau)X(t)].$$

This process is passed through a rectifier, yielding the output process

$$Y(t) = |X(t)|.$$

The objective of this problem is to evaluate the mean and autocorrelation of $Y(t)$, and to establish that $Y(t)$ is WSS.

a) Let X_1 and X_2 be two jointly Gaussian zero-mean random variables, with identical variance K_X and correlation coefficient ρ. Find the joint probability density of $U = (X_1)^2$ and $V = |X_2/X_1|$. To do so, observe that the transformation from (X_1, X_2) to (U, V) is 4 to 1, since the pairs (X_1, X_2), $(X_1, -X_2)$, $(-X_1, X_2)$, and $(-X_1, -X_2)$ all yield the same (U, V).

b) Use the result of part a) to evaluate the expectation

$$E[|X_1||X_2|] = E[UV].$$

To do so, you may want to use the following two integrals:

$$\int_0^\infty z \exp(-cz)dz = \frac{1}{c^2}, \quad c > 0$$

$$\int_0^z \frac{dw}{(w^2 + a^2)^2} = \frac{1}{2a^3} \arctan(z/a) + \frac{z}{2a^2(z^2 + a^2)}.$$

c) Let $X(t)$ be a zero mean WSS Gaussian process with autocorrelation $R_X(\tau)$. Evaluate the mean of $Y(t) = |X(t)|$.

d) Use the result of part b) with $X_1 = X(t + \tau)$ and $X_2 = X(t)$ to evaluate the autocorrelation

$$R_Y(t + \tau, t) = E[Y(t + \tau)Y(t)]$$

of the rectified process $Y(t)$. Show that the autocorrelation depends on τ only, so that the process is WSS, and express $R_Y(\tau)$ entirely in function of $R_X(\tau)$.

4.27 Consider a stationary independent increments process $X(t)$ with finite first- and second-order statistics.

a) Show that the mean $m_X(t)$ and variance $P_X(t) = E[(X(t) - m_X(t))^2]$ satisfy

$$m_X(t) = \mu t \ , \quad P_X(t) = \sigma^2 t.$$

To do so, you may want to use the property (4.56) of the characteristic function $\Phi_t(u)$ to show that

$$m_X(t + s) = m_X(t) + m_X(s)$$
$$P_X(t + s) = P_X(t) + P_X(s).$$

b) Show that the autocovariance $K_X(t, s)$ satisfies

$$K_X(t, s) = \sigma^2 \min(t, s).$$

c) What are the values of μ and σ^2 for the Wiener and Poisson processes $W(t)$ and $N(t)$, and for the binomial process Z_n of Example 4.3 (which is a discrete-time independent increments process)?

4.28 The Cauchy process $X(t)$ is a stationary independent increments process with $X(0) = 0$ whose PDF at time $t \geq 0$ is given by

$$f_{X(t)}(x) = \frac{1}{\pi} \frac{t}{x^2 + t^2}.$$

Verify that its characteristic function

$$\Phi_t(u) = E[\exp(iuX(t))]$$

obeys the relation (4.56). Even though this process has absolutely continuous distributions, its sample paths exhibit an infinite number of small jumps in a finite time interval.

References

1. A. N. Kolmogorov, *Foundations of the Theory of Probability*. New York, NY: Chelsea Pub. Co., 1956. Translation of a German 1933 monograph.
2. J. W. M. Bergmans, *Digital Baseband Transmission and Recording*. Dordrecht, The Netherlands: Kluwer Acad. Publishers, 1996.
3. R. A. Horn and C. R. Johnson, *Matrix Analysis*. Cambridge, United Kingdom: Cambridge University Press, 2005.

4. A. J. Laub, *Matrix Analysis for Scientists & Engineers*. Philadelphia, PA: Soc. Industrial and Applied Math., 2005.
5. P. Lévy, "A special problem of Brownian motion and a general theory of Gaussian random functions," in *Proc. 3rd Berkeley Symposium on Math. Statistics and Probability*, vol. 2, (Berkeley, CA), pp. 133–175, Univ. California Press, 1956.
6. S. Bernstein, "Sur les liaisons entre les variables aléatoires," in *Proc. Internat. Congress of Math.*, (Zurich, Switzerland), pp. 288–309, 1932.
7. B. Jamison, "Reciprocal processes," *Z. Wahrscheinlichkeitstheorie verw. Gebiete*, vol. 30, pp. 65–86, 1974.
8. G. Grimmett, *Probability on Graphs: Random processes on Graphs and Lattices*. Cambridge, United Kingdom: Cambridge University Press, 2010.
9. X. Guyon, *Random Fields on a Network: Modeling, Statistics and Applications*. New York: Springer Verlag, 1995.
10. K. Sato, *Lévy Processes and Infinitely Divisible Distributions*. Cambridge, United Kingdom: Cambridge University Press, 1999.
11. L. Isserlis, "On a formula for the product-moment coefficient of any order of a normal frequency distribution in any number of variables," *Biometrika*, vol. 12, pp. 134–139, Nov. 1918.
12. R. Price, "A useful theorem for nonlinear devices having Gaussian inputs," *IRE Trans. on Information Theory*, vol. 4, June 1958.
13. J. J. Bussgang, "Crosscorrelation functions of amplitude-distorted Gaussian signals," Tech. Rep. 216, Research Laboratory of Electronics, Massachusetts Institute of Technology, Mar. 1952.

Discrete-Time Finite Markov Chains

5

5.1 Introduction

In this chapter, we examine some of the properties of discrete-time finite Markov chains. These processes are discrete valued, and to keep the analysis as elementary as possible, we restrict our attention to homogeneous, i.e., time-invariant, and finite chains, i.e., chains where the process $X(t)$ can take only a finite number of values. Under these simplifying assumptions, Markov chains can be analyzed by using tools of linear algebra. Yet, the class of finite homogeneous Markov chains covers a wide range of engineering systems, since it is in essence identical to the class of synchronous stochastic automata.

In Sect. 5.2, it is shown that the initial distribution and one-step transition probability matrix are sufficient to specify all the finite joint probability distributions of the chain. The classification of the states between recurrent, i.e., states which are visited infinitely often, and transient states, i.e., states which are no longer visited after a certain time, is presented in Sect. 5.3. The convergence of irreducible and aperiodic Markov chains is analyzed in Sect. 5.4 by showing that the mapping of probability distributions induced by the one-step transition probability matrix is a contraction. This allows a relatively simple derivation of the Perron–Frobenius theorem characterizing the maximum eigenvalue and associated eigenvector of a stochastic matrix. The first-step analysis technique is used in Sect. 5.5 to perform several calculations involving the transient states of a Markov chain. Finally, digital modulation techniques which rely on a Markov chain encoder are analyzed in Sect. 5.6.

5.2 Transition Matrices and Probability Distribution

A discrete-time Markov process $X(t)$ which takes only discrete values is called a Markov chain. We restrict our attention to *finite* Markov chains, since they are easier to analyze, and since most engineering systems that can be described by Markov chains, such as buffers in computer networks, are finite (when packets arrive to a full buffer they are dropped). Without loss of generality the possible values, also called states, of the Markov chain are labeled from 1 to N, so that $X(t) \in \{1, 2, \ldots, N\}$ where N denotes the number of states. However, in situations where $X(t)$ represents a physical number, this convention will be adjusted as the situation warrants to allow other state labels. As explained in Chap. 4, a Markov process is completely described by its initial probability distribution, which is represented here by a row vector

© Springer Nature Switzerland AG 2020
B. C. Levy, *Random Processes with Applications to Circuits and Communications*,
https://doi.org/10.1007/978-3-030-22297-0_5

$$\boldsymbol{\pi}(0) = \left[\pi_1(0) \ldots \pi_i(0) \ldots \pi_N(0) \right],$$

where $\pi_i(0) = P(X(0) = i)$ for $1 \leq i \leq N$, and by its one-step transition probability matrix $\mathbf{P} = (p_{ij}, 1 \leq i, j \leq N)$, where

$$p_{ij} = P(X(t+1) = j | X(t) = i).$$

We assume here that the Markov chain is *homogeneous*, so that $\mathbf{P}$ is a constant matrix. The state i of $X(t)$ at time t specifies the row index of $\mathbf{P}$ and the state j at the next time $t + 1$ specifies the column index. The allowed transitions from one state to the next and their probabilities are described equivalently by a *state-transition diagram*. This diagram is a directed graph with N vertices representing the possible states $\{1, \ldots, i, \ldots, N\}$ and where an oriented edge exists between state i and state j whenever $p_{ij} > 0$. This edge is labeled by its probability p_{ij}. Thus, state transition diagrams play a role similar to the transition diagrams of finite-state machines, with the only difference being that transitions occur randomly. There is of course a one-to-one correspondence between one-step transition probability matrices and state-transition diagrams, as can be seen from the following example.

Example 5.1 Consider the state transition diagram shown in Fig. 5.1.

Its one-step transition matrix can be literally read off Fig. 5.1 and is given by

$$\mathbf{P} = \begin{bmatrix} 0 & 1/2 & 1/2 \\ 1/2 & 1/2 & 0 \\ 2/3 & 1/3 & 0 \end{bmatrix}.$$

In this respect it is worth noting that if no oriented edge exists between state i and state j, it means that $p_{ij} = 0$, i.e., zero entries of $\mathbf{P}$ do not give rise to any edge of the transition diagram. Also, observe that self-transitions are allowed. Specifically, if $p_{ii} > 0$, then an edge exists between state i and itself.

The one-step transition matrices $\mathbf{P}$ are called *stochastic matrices*. They have several interesting properties. First, all the entries p_{ij} of $\mathbf{P}$ are nonnegative. Second, by noting that for all i

$$\sum_{j=1}^{N} p_{ij} = P(X(t+1) \in \{1, 2, \ldots, N\} | X(t) = i) = 1 \tag{5.1}$$

Fig. 5.1 State-transition diagram of the Markov chain of Example 5.1

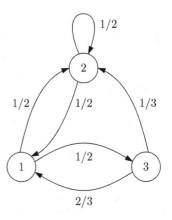

by the principle of total probability, we conclude that the sum of the elements in each row of $\mathbf{P}$ is one. If

$$\nu = \begin{bmatrix} 1 \ldots 1 \ldots 1 \end{bmatrix}^T$$

is the column vector with all elements equal to one, the property (5.1) can be rewritten as

$$\mathbf{P}\nu = \nu, \tag{5.2}$$

so that ν is a right eigenvector of $\mathbf{P}$ corresponding to the eigenvalue $\lambda = 1$. Note also that if $\mathbf{P}_1$ and $\mathbf{P}_2$ are two stochastic matrices, their product $\mathbf{P}_1\mathbf{P}_2$ is also a stochastic matrices, since all entries of the product are nonnegative and

$$\mathbf{P}_1\mathbf{P}_2\nu = \mathbf{P}_1\nu = \nu.$$

In this respect, it is also useful to note that although the eigenvalues of $\mathbf{P}$ may be complex, all eigenvalues λ have a magnitude less than or equal to 1, i.e., $|\lambda| \leq 1$. Indeed, let $\mathbf{x} \neq 0$ be an eigenvector of $\mathbf{P}$ corresponding to λ, so that $\mathbf{Px} = \lambda\mathbf{x}$. Let x_k be the component of $\mathbf{x}$ with largest magnitude, i.e.

$$|x_k| = \max_{1 \leq i \leq N} |x_i|.$$

We have

$$|\lambda||x_k| = |\sum_{j=1}^{N} p_{kj} x_j| \leq \sum_{j=1} p_{kj}|x_j| \leq |x_k|. \tag{5.3}$$

Since $|x_k| > 0$ (an eigenvector is necessarily nonzero), this implies $|\lambda| \leq 1$. Since $\lambda = 1$ is an eigenvalue of $\mathbf{P}$, $\mathbf{P}$ admits necessarily a left eigenvector π corresponding to $\lambda = 1$, i.e.,

$$\pi\mathbf{P} = \pi,$$

but this vector is not necessarily unique, since $\lambda = 1$ may have multiplicity greater than 1.

Since the transition distributions used to specify Markov processes in Chap. 4 involved transitions between two arbitrary times, our first task is to show how to compute the m-step transition probability matrices $\mathbf{F}(m) = (f_{ij}(m), 1 \leq i, j \leq N)$ for $m \geq 1$, where

$$f_{ij}(m) = P(X(t+m) = j|X(t) = i).$$

We have

$$P(X(t+m+1) = j, X(t+m) = \ell|X(t) = i)$$
$$= P(X(t+m+1) = j|X(t+m) = \ell, X(t) = i)P(X(t+m) = \ell|X(t) = i)$$
$$= P(X(t+m+1) = j|X(t+m) = \ell)P(X(t+m) = \ell|X(t) = i), \tag{5.4}$$

where to go from the second to the third line, we have used the Markov property of $X(t)$. By summing over ℓ on both sides of (5.4), we obtain the marginal conditional PMF

$$f_{ij}(m+1) = P(X(t+m+1) = j | X(t) = i)$$

$$= \sum_{\ell=1}^{N} P(X(t+m+1) = j | X(t+m) = \ell) P(X(t+m) = \ell | X(t) = i)$$

$$= \sum_{\ell=1}^{N} f_{i\ell}(m) p_{\ell j},$$

or equivalently, in matrix form

$$\mathbf{F}(m+1) = \mathbf{F}(m)\mathbf{P}. \tag{5.5}$$

Since $\mathbf{F}(1) = \mathbf{P}$, this implies that

$$\mathbf{F}(m) = \mathbf{P}^m, \tag{5.6}$$

so that the m-step transition matrix is obtained by m-fold multiplication of $\mathbf{P}$. Note also that the homogeneity property of $\mathbf{F}(m)$ is inherited from the homogeneity of $\mathbf{P}$. For $m, n \geq 1$, the expression (5.6) implies

$$\mathbf{F}(m+n) = \mathbf{F}(m)\mathbf{F}(n), \tag{5.7}$$

which is the expression of the Chapman–Kolmogorov equation for Markov chains.

Given the m-step transition probability matrix of the Markov chain, it is then possible to compute the probability distribution

$$\boldsymbol{\pi}(t) = \begin{bmatrix} \pi_1(t) \ \dots \ \pi_j(t) \ \dots \ \pi_N(t) \end{bmatrix}$$

with $\pi_j(t) = P(X(t) = j)$ for all times t. Observe indeed that the joint PMF of $X(0)$ and $X(t)$ can be expressed as

$$P(X(t) = j, \ X(0) = i) = P(X(t) = j | X(0) = i) P(X(0) = i) = \pi_i(0) f_{ij}(t),$$

and by marginalization, we find that the PMF of $X(t)$ is given by

$$\pi_j(t) = P(X(t) = j) = \sum_{i=1}^{N} \pi_i(0) f_{ij}(t),$$

for $1 \leq j \leq n$, or equivalently in vector form

$$\boldsymbol{\pi}(t) = \boldsymbol{\pi}(0)\mathbf{F}(t). \tag{5.8}$$

In light of the recursion (5.5) the probability distribution can also be computed recursively by using

$$\boldsymbol{\pi}(t+1) = \boldsymbol{\pi}(t)\mathbf{P}. \tag{5.9}$$

The recursion (5.9) illustrates an important property of stochastic matrices. Let $\mathcal{S}$ denotes the N-simplex formed by probability distributions over N states. Such distributions can be represented by row vectors $\boldsymbol{\pi}$ with entries $\pi_i \geq 0$ for $1 \leq i \leq N$ and such that

$$1 = \sum_{i=i}^{N} \pi_i = \boldsymbol{\pi}\boldsymbol{\nu}. \tag{5.10}$$

Fig. 5.2 Stochastic matrix
P viewed as a mapping of
simplex $\mathcal{S}$

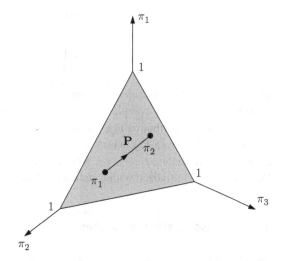

As shown in Fig. 5.2, $\mathcal{S}$ can be viewed as an $N-1$ dimensional polytope inside $\mathbb{R}^N$. This set is convex, since given two probability distributions π_1 and π_2 and a real number α such that $0 \le \alpha \le 1$, the linear combination

$$\alpha\pi_1 + (1-\alpha)\pi_2$$

is also a probability vector. $\mathcal{S}$ is clearly closed and bounded, so it is a compact set. Stochastic matrices have the interesting property that they map $\mathcal{S}$ into itself. Specifically, if π_1 is a row vector representing an element of $\mathcal{S}$, then

$$\pi_2 = \pi_1\mathbf{P}$$

also belongs to $\mathcal{S}$. Indeed, the nonnegativity of the entries of $\mathbf{P}$ ensures that the entries of π_2 are nonnegative, and since v is a right eigenvector of $\mathbf{P}$, we have

$$\pi_2 v = \pi_1 \mathbf{P} v = \pi_1 v = 1.$$

Accordingly, the propagation equation (5.9) can be viewed as an iteration inside $\mathcal{S}$. A beautiful theorem due to Brouwer [1, p. 46] ensures that any continuous mapping of a compact convex set into itself has necessarily a fixed point. Thus, given any stochastic matrix $\mathbf{P}$, there exists a stationary distribution π of $\mathcal{S}$ such that

$$\pi\mathbf{P} = \pi. \tag{5.11}$$

We already knew that there exists a left eigenvector of $\mathbf{P}$ corresponding to eigenvalue $\lambda = 1$, but we did not know that this eigenvector could be selected as a probability distribution with nonnegative entries. Note again that the multiplicity of the eigenvalue $\lambda = 1$ can be greater than one, so there is no guarantee that the fixed point (5.11) is unique. A probability distribution π which satisfies (5.11) is called a stationary distribution of the Markov chain, since selecting $\pi(0) = \pi$ ensures that the Markov chain probability distributions satisfies $\pi(t) = \pi$ for all t.

Having established how to compute the m-step transition probability matrix $F(m)$ and the probability distribution $\pi(t)$ at any time, we are now in a position to specify the finite joint probability distributions of the Markov chain for any K and any choice of times $t_1 < \ldots < t_k < \ldots < t_K$. By using the Markov property, we have

$$P(X(t_1) = i_1, \ldots, X(t_k) = i_k, \ldots, X(t_K) = i_K)$$

$$= \prod_{k=1}^{K-1} f_{i_k i_{k+1}}(t_{k+1} - t_k)\pi_{i_1}(t_1) . \tag{5.12}$$

This shows that the one-step transition probability matrix $\mathbf{P}$ and initial probability distribution $\boldsymbol{\pi}(0)$ are sufficient to specify completely all the finite joint distributions of the Markov chain. Finally, note that if the initial distribution is a stationary distribution $\boldsymbol{\pi}$ of the Markov chain, then for all t_1, π_{i_1} is independent of t_1 in (5.12), which implies that all the finite joint distributions of $X(t)$ are invariant under a time translation $t \rightarrow t + h$ with $h \in \mathbb{Z}$, so that $X(t)$ is SSS. In other words if a homogeneous Markov chain is initialized with a stationary distribution, it is SSS.

5.3   Classification of States

To move our analysis further, we need to investigate the detailed structure of Markov chains. A state j is accessible from state i, which is denoted as $i \rightarrow j$, if there exists an integer $m \geq 1$ such that $f_{ij}(m) > 0$. In other words there exists a chain of m successive transitions such that starting from $X(t) = i$, state j can be reached reached with nonzero probability after m transitions. The accessibility relation is transitive, i.e., $i \rightarrow j$ and $j \rightarrow k$ implies $i \rightarrow k$. Note indeed that if state k is accessible from j in n steps and j is accessible from i in m steps, the Chapman Kolmogorov equation (5.7) implies

$$f_{ik}(n + m) = \sum_{\ell=1}^{N} f_{i\ell}(m) f_{\ell k}(n) \geq f_{ij}(m) f_{jk}(n) > 0 ,$$

so k is accessible from i in $n+m$ steps. Two states i and j communicate, which is denotes as $i \leftrightarrow j$, if they are accessible from each other. The transitivity of the accessibility relation implies that if $i \leftrightarrow j$ and $j \leftrightarrow k$, then $i \leftrightarrow k$, so that the communication property defines an equivalence relation.

For a finite state Markov chain, a state i is said to be *recurrent* if it communicates with all states j that are accessible from i. In other words, if j can be reached from i, it is possible to return to i from j. A state is *transient* if it is not recurrent. This means that if i is transient, there exists at least one state j that can be reached from i, but such that $X(t)$ can never come back to i once it has reached j.

Based on the previous definitions, we conclude that the state space can be decomposed into one class of transient states and C classes of communicating recurrent states with $C \geq 1$, where states inside a recurrent class communicate with one another, but cannot access or be accessible from states in another recurrence class. This gives rise to a canonical decomposition of the stochastic matrix $\mathbf{P}$ of the form

$$\mathbf{P} = \begin{bmatrix} \mathbf{P}_1 & \mathbf{0} & & \ldots & \mathbf{0} \\ \mathbf{0} & \mathbf{P}_2 & \mathbf{0} & \ldots & \mathbf{0} \\ \vdots & \ddots & \ddots & \ddots & \vdots \\ \mathbf{0} & \ldots & \mathbf{0} & \mathbf{P}_C & \mathbf{0} \\ \mathbf{P}_{T1} & \mathbf{P}_{T2} & \ldots & \mathbf{P}_{TC} & \mathbf{P}_T \end{bmatrix} . \tag{5.13}$$

In this decomposition, $\mathbf{P}_1$, $\mathbf{P}_2$, $\ldots$, $\mathbf{P}_C$ are the one-step transition matrices of recurrence classes 1, 2, $\ldots$ C. $\mathbf{P}_T$ represents the transition probabilities between transient state, and $\mathbf{P}_{Tj}$ represents the transition probabilities between transient states and states in recurrence class j. To clarify this decomposition, consider the following example.

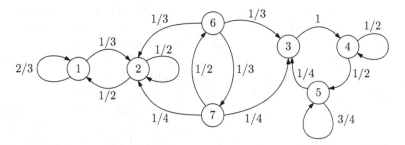

Fig. 5.3 State-transition diagram of the Markov chain of Example 5.2

Example 5.2 Consider the Markov chain with the state transition diagram shown in Fig. 5.3. By inspecting the diagram, it is easy to verify that there exists two classes of recurrent states: {1, 2} and {3, 4, 5}. On the other hand, the states {6, 7} are transient, since once $X(t)$ leaves these states to fall in one of the two recurrent classes, it is impossible to return.

For this chain the canonical decomposition (5.13) takes the form

$$\mathbf{P} = \begin{bmatrix} \mathbf{P}_1 & \mathbf{0} & \mathbf{0} \\ \mathbf{0} & \mathbf{P}_2 & \mathbf{0} \\ \mathbf{P}_{T1} & \mathbf{P}_{T2} & \mathbf{P}_T \end{bmatrix}$$

with

$$\mathbf{P}_1 = \begin{bmatrix} 2/3 & 1/3 \\ 1/2 & 1/2 \end{bmatrix}, \quad \mathbf{P}_2 = \begin{bmatrix} 0 & 1 & 0 \\ 0 & 1/2 & 1/2 \\ 1/4 & 0 & 3/4 \end{bmatrix}$$

and

$$\mathbf{P}_T = \begin{bmatrix} 0 & 1/3 \\ 1/2 & 0 \end{bmatrix}, \quad \mathbf{P}_{T1} = \begin{bmatrix} 0 & 1/3 \\ 0 & 1/4 \end{bmatrix}, \quad \mathbf{P}_{T2} = \begin{bmatrix} 1/3 & 0 & 0 \\ 1/4 & 0 & 0 \end{bmatrix}.$$

The form of the decomposition (5.13) indicates that the one-step transition matrices $\mathbf{P}_j$ corresponding to each of the recurrent classes are themselves stochastic matrices, since their entries are nonnegative and the sum of all entries in each row is one. Since a stochastic matrix has at least one eigenvalue equal to one, this indicates that for a Markov chain with C recurrent classes, the eigenvalue $\lambda = 1$ has at least multiplicity C. It is also useful to note that a finite Markov chain has at least one recurrent class. To see why all states cannot be transient, note that if i_1 is a transient state, there exists a state i_2 which is accessible from i_1 but such that i_1 is not accessible from i_2. If i_2 is recurrent, there is at least one recurrence class. On the other hand, if i_2 is transient, there exists a state i_3 accessible from i_2 but such that i_2 is not accessible from i_3. Note that i_3 cannot be i_1, since i_1 is not accessible from i_2. Thus, if all states were transient, we would be able to construct an infinite chain of transient states i_k, $k \geq 1$, where state i_{k+1} is accessible from i_k, and such that no state can be repeated. Since the Markov chain is finite, this is impossible. Therefore the decomposition (5.13) must include at least one class of recurrent states, and may also include a class of transient states. If a Markov chain does not have transient states, but includes several recurrent classes, this Markov chain can be decomposed in several unrelated Markov chains. We say that a Markov chain is *irreducible* if it does not have transient states and has a single class of recurrent states. Thus, in an irreducible Markov chain, all states communicate with each other.

Next, the *period* $d(i)$ of state i, with $1 \leq i \leq N$ is defined as the greatest common divisor (gcd) of all integers m such that $f_{ii}(m) > 0$. In plain words, $d(i)$ is the greatest common divisor of the lengths of all paths starting from state i and returning to i. Obviously if $p_{ii} > 0$, then $d(i) = 1$. A state i is said to be *aperiodic* if $d(i) = 1$. A key result is that all states in the same recurrence class have the same period. Consider two states i and j with periods $d(i)$ and $d(j)$ in the same class. Since i and j communicate with each other, there exists a path of length r from i to j and a path of length s from j to i, i.e., $f_{ij}(r) > 0$ and $f_{ji}(s) > 0$. Concatenating these two paths yields a path of length $r + s$ from i to itself, so $d(i)|r + s$, where $a|b$ denotes a divides b. For any closed path of length m from j to j, we can therefore construct a path of length $r + s + m$ from i to i since $f_{ii}(r + s + m) > f_{ij}(r)f_{jj}(m)f_{ji}(s) > 0$. This implies $d(i)|r + s + m$, and since $d(i)$ already divides $r + s$, it must divide m. But since $d(j)$ is the greatest common divisor of integers m such that $f_{jj}(m) > 0$, we conclude $d(i)|d(j)$. By reversing the roles of i and j, we have also $d(j)|d(i)$, so $d(j) = d(i)$. To illustrate the concept of period, we consider an example.

Example 5.3 Consider the Markov chain with the state transition diagram shown in Fig. 5.4.

It can be seen from the transition diagram that all four states communicate, so the chain is irreducible. The graph of Fig. 5.4 has also the interesting feature that the states can be partitioned in two groups $\{1, 2\}$ and $\{3, 4\}$, and all directed edges connect states in one group to states in the other group. This type of graph is called a bipartite graph. This means that starting from one group, $X(t)$ will oscillate periodically from one group to the other, so we would expect that the period of the chain is two. This can be established rigorously by noting that the one-step transition probability matrix has the form

$$\mathbf{P} = \begin{bmatrix} \mathbf{0} & \mathbf{P}_S \\ \mathbf{P}_S & \mathbf{0} \end{bmatrix},$$

where

$$\mathbf{P}_S = \begin{bmatrix} 1/2 & 1/2 \\ 1/2 & 1/2 \end{bmatrix} = \frac{1}{2} \begin{bmatrix} 1 \\ 1 \end{bmatrix} \begin{bmatrix} 1 & 1 \end{bmatrix}.$$

Observing that $\mathbf{P}_S^k = \mathbf{P}_S$ for all $k \geq 1$, we deduce that

$$\mathbf{F}(2n) = \begin{bmatrix} \mathbf{P}_S & \mathbf{0} \\ \mathbf{0} & \mathbf{P}_S \end{bmatrix}$$

Fig. 5.4 State-transition diagram of the Markov chain of Example 5.3

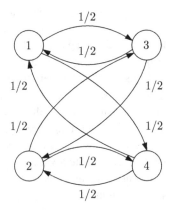

and $\mathbf{F}(2n + 1) = \mathbf{P}$. If we consider at an arbitrary state, say state 1, we see therefore that $f_{11}(2n) > 0$ and $f_{11}(2n + 1) = 0$. In other words there exists paths of even lengths connecting state 1 to itself, but there does not exist any path of odd length, which just reflects the bipartite property of the state transition diagram. Since the gcd of $2n$ with $n \in \mathbb{N}$ arbitrary is 2, the period of state 1 is $d(1) = 2$. Because of the symmetry existing between all states in the chain (the transition diagram or $\mathbf{P}$ are unchanged if $1 \leftrightarrow 2$ or $3 \leftrightarrow 4$, or if group 1 of states, i.e., states 1 and 2, is swapped with group 2 (states 3 and 4), we also deduce that $d(i) = 2$ for all states, which verifies that a recurrence class has the same period, here $d = 2$.

By using Laplace's expansion of the determinant of $\lambda\mathbf{I} - \mathbf{P}$, we obtain

$$\det(\lambda\mathbf{I} - \mathbf{P}) = \lambda^2(\lambda^2 - 1),$$

so that in addition to the eigenvalue $\lambda = 1$, $\mathbf{P}$ has also the eigenvalue $\lambda = -1$ of magnitude 1, which is the source of the Markov chain periodicity.

Finally, we say that a Markov chain is irreducible and aperiodic if the single recurrence class of the chain has period $d = 1$.

5.4   Convergence

In this section, we restrict our attention to irreducible and aperiodic Markov chains. This assumption is necessary to ensure that the invariant distribution of $\mathbf{P}$ is unique and that all eigenvalues of $\mathbf{P}$ other than $\lambda = 1$ have a magnitude strictly less than 1, which will be used to analyze the convergence of the probability distribution $\pi(t)$ from an arbitrary initial distribution $\pi(0)$. The analysis is rather intricate and relies on several preliminary results.

Lemma 1 *If $\mathbf{P}$ is the transition matrix of an irreducible aperiodic Markov chain, there exists a power, say M, such that all entries of $\mathbf{F}(M) = \mathbf{P}^M$ are strictly positive.*

A stochastic matrix $\mathbf{P}$ with all positive entries is said to be regular, and one such that $\mathbf{P}^M$ is regular for some M is called primitive. So we need to prove that the transition matrix of an irreducible aperiodic Markov chain is primitive. The proof is based on the following properties.

Property A   Given two arbitrary states i and j with $j \neq i$, there exists an integer m with $1 \leq m \leq N - 1$ such that $f_{ij}(m) > 0$. In other words, there exists a path of length m linking i and j. Let $\mathcal{R}(i, m)$ with $\mathcal{R}(i, 0) = \{i\}$ denote the states reachable from i in at most m steps. The sequence $\mathcal{R}(i, m)$ is clearly monotone nondecreasing, i.e., $\mathcal{R}(i, m) \subseteq \mathcal{R}(i, m + 1)$, and since the Markov chain is irreducible, all states ultimately belong to $\mathcal{R}(i, m)$ for m large enough. Also, if the set of reachable states stops growing as m increases from to $m + 1$, it will stop growing forever, since the states that can be reached in one step from $\mathcal{R}(i, m + 1)$ are the same as those that can be reached from $\mathcal{R}(i, m)$. This implies that the sequence $\mathcal{R}(i, m)$ is strictly increasing for up to $N - 1$ steps, and $\mathcal{R}(i, m) = \{1, 2, \ldots, N\}$ thereafter. We can also deduce from this result that for any state i, there exists $m \leq N$ such that $f_{ii}(m) > 0$. Indeed let j be any state reachable from i in one step. Then i must belong to $\mathcal{R}(j, N - 1)$, so there is a cycle of length at most N starting from i.

Property B   There is an m_0 such that $f_{ii}(m) > 0$ for all states i and all $m \geq m_0$. At the end of Property A, we noted that for any i, there exists $m \leq N$ such that $f_{ii}(m) > 0$. Since the chain is aperiodic, for any i, there exists m_1 and m_2 such that $f_{ii}(m_1) > 0$ and $f_{ii}(m_2) > 0$, and the gcd of m_1

and m_2 is 1 (see Prob. 5.18). In this case, since m_1 and m_2 are coprime, the Bezout identity [2, p. 7] ensures that for any integer $\ell \in \mathbb{N}$, there exists integers a and b in $\mathbb{Z}$ solving the linear diophantine equation

$$am_1 + bm_2 = \ell. \tag{5.14}$$

Integers a and b are not guaranteed to be both positive, but by Euclidean division, if a is negative, it can be selected such that $-a < m_2$, and if b is negative, it can be selected such that $-b < m_1$. One of them needs to be positive. This means that by adding m_2 to a when a is negative, or m_1 to b when b when b is negative, the equation

$$am_1 + bm_2 = m_1 m_2 + \ell \tag{5.15}$$

admits a solution with a and b positive. Thus by concatenating a cycles of length m_1 and b cycles of length m_2 starting and finishing in i, we have $f_{ii}(m_1 m_2 + \ell) > 0$.

By combining Properties A and B, we then deduce that for any i and j and $M = N - 1 + m_0$, we can ensure that $f_{ij}(M) > 0$. Indeed there exists $m \leq N - 1$ such that $f_{ij}(m) > 0$, so

$$f_{ij}(M) \geq f_{ii}(M - m) f_{ij}(m) > 0,$$

where we have used the fact that $f_{ii}(M - m) = f_{ii}(m_0 + N - 1 - m) > 0$.

The second important result needed to study the convergence of Markov chains is that the mapping on simplex $\mathcal{S}$ defined by a regular stochastic matrix is a strict contraction for the total variation metric. Let π and μ be two N-dimensional probability distributions of $\mathcal{S}$. The total variation metric is defined as

$$d(\pi, \mu) = \frac{1}{2} \sum_{j=1}^{N} |\pi_j - \mu_j|. \tag{5.16}$$

Then given a regular stochastic matrix $\mathbf{P}$, if

$$\epsilon_j = \min_{1 \leq i \leq N} p_{ij}$$

denotes the smallest element of column j, we have $\epsilon_j > 0$ and since

$$1 = \sum_{j=1}^{N} p_{ij} \geq \sum_{j=1}^{N} \epsilon_j$$

for any i, we deduce that the contraction coefficient

$$c(\mathbf{P}) \overset{\triangle}{=} 1 - \sum_{j=1}^{N} \epsilon_j$$

satisfies

$$0 \leq c(\mathbf{P}) < 1. \tag{5.17}$$

Whenever $c(\mathbf{P}) = 0$, by observing that $p_{ij} \geq \epsilon_j$, we deduce that $p_{ij} = \epsilon_j$ for all i. Otherwise if $p_{ij} > \epsilon_j$ for some i, we would have

$$\sum_{j=1}^{N} p_{ij} > \sum_{j=1}^{N} \epsilon_j = 1,$$

which is impossible. Thus if $c(\mathbf{P}) = 0$, we have

$$\mathbf{P} = \nu\epsilon, \tag{5.18}$$

where

$$\epsilon = \begin{bmatrix} \epsilon_1 \ldots \epsilon_j \ldots \epsilon_N \end{bmatrix}$$

and ν is the right eigenvector of $\mathbf{P}$ corresponding to $\lambda = 1$ with all entries equal to 1. On the other hand, if $c(\mathbf{P}) > 0$, we can decompose $\mathbf{P}$ as

$$\mathbf{P} = c(\mathbf{P})\mathbf{Q} + \nu\epsilon, \tag{5.19}$$

where $\mathbf{Q}$ is the stochastic matrix with entries

$$q_{ij} = \frac{p_{ij} - \epsilon_j}{c(\mathbf{P})} \tag{5.20}$$

for $1 \leq i, \ j \leq N$. Next, consider an arbitrary row probability distribution vector $\boldsymbol{\pi}$ of $\mathcal{S}$. We have

$$\boldsymbol{\pi}\mathbf{P} = c(\mathbf{P})\boldsymbol{\pi}\mathbf{Q} + \epsilon, \tag{5.21}$$

where we have used the normalization (5.10). For any two probability distribution vectors $\boldsymbol{\pi}$ and $\boldsymbol{\mu}$ of $\mathcal{S}$, we have therefore

$$d(\boldsymbol{\pi}\mathbf{P}, \boldsymbol{\mu}\mathbf{P}) = c(\mathbf{P})d(\boldsymbol{\pi}\mathbf{Q}, \boldsymbol{\mu}\mathbf{Q}), \tag{5.22}$$

where

$$d(\boldsymbol{\pi}\mathbf{Q}, \boldsymbol{\mu}\mathbf{Q}) = \frac{1}{2}\sum_{j=1}^{N} |\sum_{i=1}^{N}(\pi_i - \mu_i)q_{ij}|$$

$$\leq \frac{1}{2}\sum_{j=1}^{N}\sum_{i=1}^{N} |\pi_i - \mu_i| \ q_{ij} = \frac{1}{2}\sum_{i=1}^{N} |\pi_i - \mu_i| = d(\boldsymbol{\pi}, \boldsymbol{\mu}), \tag{5.23}$$

so that

$$d(\boldsymbol{\pi}\mathbf{P}, \boldsymbol{\mu}\mathbf{P}) \leq c(\mathbf{P})d(\boldsymbol{\pi}, \boldsymbol{\mu}). \tag{5.24}$$

Note that when $c(\mathbf{P}) = 0$, expression (5.18) implies

$$d(\boldsymbol{\pi}\mathbf{P}, \boldsymbol{\mu}\mathbf{P}) = 0,$$

so the contraction relation (5.24) still holds. We have therefore proved the following result.

Lemma 2 *If $\mathbf{P}$ is a regular stochastic matrix, the corresponding mapping on $\mathbf{S}$ is strictly contractive.*

Since $\mathcal{S}$ is closed, the Banach fixed-point theorem [1, p. 244] ensures that $\mathbf{P}$ has a unique fixed point over $\mathcal{S}$, i.e., there is a unique probability distribution $\boldsymbol{\pi}$ such that

$$\boldsymbol{\pi}\mathbf{P} = \boldsymbol{\pi}. \tag{5.25}$$

Since $\mathbf{P}$ is regular, all entries π_j of $\boldsymbol{\pi}$ must be strictly positive. Otherwise, we would have

$$0 = \pi_j = \sum_{i=1}^{N} \pi_i\, p_{ij}.$$

Since all $p_{ij} > 0$, this would imply that $\pi_i = 0$ for all i, which is impossible since the total probability mass must be one.

Next, we consider the case where $\mathbf{P}$ is primitive, which corresponds to the case of an irreducible aperiodic Markov chain. In this case, by Lemma 1, there exists an M such that P^M is regular, so there exists a unique probability distribution $\boldsymbol{\pi}$ satisfying

$$\boldsymbol{\pi}\mathbf{P}^M = \boldsymbol{\pi}. \tag{5.26}$$

Observing that $\mathbf{P}\mathbf{P}^M = \mathbf{P}^M\mathbf{P}$, we deduce that in this case we must have

$$(\boldsymbol{\pi}\mathbf{P})\mathbf{P}^M = \boldsymbol{\pi}\mathbf{P},$$

where $\boldsymbol{\pi}\mathbf{P}$ belongs to $\mathcal{S}$. But since the fixed point of $\mathbf{P}^M$ is unique, we must have (5.25), so that $\boldsymbol{\pi}$ is a fixed point of $\mathbf{P}$ with all positive entries. Note also that since any fixed point of $\mathbf{P}$ in $\mathcal{S}$ is a fixed point of $\mathbf{P}^M$, $\mathbf{P}$ has a unique fixed point. We are now in a position to prove the main result of this section.

Convergence Theorem The probability distribution $\boldsymbol{\pi}(t)$ of an irreducible aperiodic Markov chain with one-step transition probability matrix $\mathbf{P}$ and initial distribution $\boldsymbol{\pi}(0)$ converges to the unique fixed point $\boldsymbol{\pi}$ of $\mathbf{P}$ over $\mathcal{S}$ as t tends to infinity. Furthermore the t-step transition probability matrix $\mathbf{F}(t) = \mathbf{P}^t$ tends to the rank one matrix

$$\mathbf{F}(\infty) = \boldsymbol{v}\boldsymbol{\pi} \tag{5.27}$$

as t tends to infinity, where $\boldsymbol{v}$ is the right eigenvector with all 1 entries corresponding to $\lambda = 1$.

Proof Since $\mathbf{P}$ is primitive, there exists an integer M such that $\mathbf{P}^M$ is regular. Then, by Euclidean division of t by M, t can be expressed as $t = Ms + r$ with s and r integers such that with $0 \leq r \leq M - 1$, where $s \to \infty$ as $t \to \infty$. We have $\boldsymbol{\pi}(t) = \boldsymbol{\pi}(0)\mathbf{P}^t$, so if $\boldsymbol{\pi}_{ss}$ denotes the fixed point of $\mathbf{P}$

$$d(\boldsymbol{\pi}(t), \boldsymbol{\pi}_{ss}) \leq (c(\mathbf{P}^M))^s d(\boldsymbol{\pi}(r), \boldsymbol{\pi}_{ss}),$$

where $(c(\mathbf{P}^M)^s$ tends to zero as t tends to infinity. This implies

$$\lim_{t \to \infty} \boldsymbol{\pi}(t) = \boldsymbol{\pi}_{ss}. \tag{5.28}$$

Since the convergence result (5.28) does not depend on the choice of initial condition, by selecting $\boldsymbol{\pi}(0) = \mathbf{e}_i^T$, where $\mathbf{e}_i$ denotes the i-th unit vector of $\mathbb{R}^N$, i.e., the N dimensional vector whose entries are all zero except for a 1 in the i-th position, we find that $\mathbf{e}_i^T \mathbf{F}(t)$ tends to $\boldsymbol{\pi}_{ss}$ as t tends to infinity. Since this holds for all choices of i, this implies that the limit $\mathbf{F}(\infty)$ of $\mathbf{F}(t)$ has the structure (5.27).

From the asymptotic form (5.27) of $\mathbf{F}(t) = \mathbf{P}^t$, we can infer some information about the eigenvalues of $\mathbf{P}$. Recall that any matrix $\mathbf{P}$ admits a Jordan form [3, p. 82]

$$\mathbf{V}^{-1}\mathbf{P}\mathbf{V} = \mathrm{diag}\{\mathbf{J}_\ell, \ 1 \leq \ell \leq L\}, \tag{5.29}$$

with

$$\mathbf{J}_\ell = \lambda_\ell \mathbf{I}_{r_\ell} + \mathbf{Z}_{r_\ell}, \tag{5.30}$$

where $\mathbf{I}_{r_\ell}$ denotes the identity matrix of dimension r_ℓ and

$$\mathbf{Z}_{r_\ell} = \begin{bmatrix} 0 & 1 & 0 & & \dots & 0 \\ 0 & 0 & 1 & 0 & \dots & 0 \\ & 0 & 0 & 1 & & 0 \\ \vdots & & \ddots & \ddots & \ddots & \\ 0 & 0 & \dots & & 0 & 1 \\ 0 & 0 & \dots & & 0 & 0 \end{bmatrix}$$

is a shift matrix of dimension r_ℓ. The matrix $\mathbf{V}$ is formed by the eigenvectors and generalized eigenvectors of $\mathbf{P}$. To each Jordan block $\mathbf{J}_\ell$ corresponds a single eigenvector and $r_\ell - 1$ generalized eigenvectors. The eigenvalues λ_ℓ appearing in the decomposition (5.29) need not be distinct, i.e., an eigenvalue can have several corresponding Jordan blocks. The decomposition (5.29) implies that $\mathbf{P}^t$ can be expressed as

$$\mathbf{P}^t = \mathbf{V}\mathrm{diag}\{\mathbf{J}_\ell^t, \ 1 \leq \ell \leq L\}\mathbf{V}^{-1} \tag{5.31}$$

where because $\mathbf{I}_{r_\ell}$ and $\mathbf{Z}_{r_\ell}$ commute and $\mathbf{Z}_{r_\ell}$ is nilpotent, i.e., $(\mathbf{Z}_{r_\ell})^{r_{ell}} = 0$, we have

$$\mathbf{J}_\ell^t = \sum_{k=0}^{r_\ell - 1} \binom{t}{k} \lambda_\ell^{t-k} \mathbf{Z}_{r_\ell}^k. \tag{5.32}$$

By comparing the form (5.31) of $\mathbf{P}^t$ and its limit (5.27) and recalling that all eigenvalues of stochastic matrix $\mathbf{P}$ satisfy $|\lambda_\ell| \leq 1$, we deduce that in the Jordan decomposition (5.29), there is a single block of dimension 1 corresponding to eigenvalue $\lambda = 1$, i.e., $\lambda = 1$ has multiplicity one. In addition, since there is no oscillatory component in the limit, we deduce that $\mathbf{P}$ has no other eigenvalue such that $|\lambda| = 1$. This establishes the following result due to the German mathematicians Oskar Perron and Georg Frobenius.

Perron–Frobenius Theorem If $\mathbf{P}$ is a primitive stochastic matrix, the eigenvalue $\lambda = 1$ has multiplicity one and the corresponding left-eigenvector $\boldsymbol{\pi}_{ss}$ has positive elements. All other eigenvalues have a magnitude strictly less than one, and there is no other real eigenvector with all nonnegative elements.

The last observation is due to the fact that if $\boldsymbol{\pi}$ has nonnegative real entries, by multiplying

$$\boldsymbol{\pi}\mathbf{P} = \lambda\boldsymbol{\pi}$$

on the right by $\boldsymbol{v}$, we find

$$(\lambda - 1)\boldsymbol{\pi}\boldsymbol{v} = 0$$

so we must have either $\lambda = 1$ or

$$\boldsymbol{\pi}\boldsymbol{v} = \sum_{i=1}^{N} \pi_i = 0$$

But since all π_is are assumed nonnegative, this would imply $\boldsymbol{\pi} = 0$, which is impossible since an eigenvector must be nonzero.

Remark For an irreducible aperiodic Markov chain with one-step transition probability matrix $\mathbf{P}$, the form (5.31) of $\mathbf{F}(t) = \mathbf{P}^t$ allows us to infer some information not only about the limit $\mathbf{F}(\infty) = \boldsymbol{v}\boldsymbol{\pi}$, but also about the speed of convergence towards this limit. Assume that the eigenvalues λ_ℓ with $1 \leq \ell \leq L$ are ordered according to their magnitudes, so that

$$\lambda_1 = 1 > |\lambda_2| \geq |\lambda_3| \ldots \geq |\lambda_L|$$

and assume also that λ_2 has multiplicity one. Then for large t,

$$\mathbf{F}(t) - \boldsymbol{v}\boldsymbol{\pi} \approx \lambda_2^t \mathbf{v}_2 \mathbf{w}_2,$$

where $\mathbf{v}_2$ and $\mathbf{w}_2$ denote the right and left eigenvectors of $\mathbf{P}$ corresponding to λ_2, with the normalization $\mathbf{w}_2\mathbf{v}_2 = 1$. Then since

$$|\lambda_2|^t = \exp(-t/\tau),$$

where $\tau = 1/\ln(1/|\lambda_2|)$ denotes the time constant (also known as mixing time) of the Markov chain, we deduce that

$$\mathbf{F}(t) \approx \boldsymbol{v}\boldsymbol{\pi}$$

as long as $t > K\tau$ where K is a number between say 5 and 10. In this case, independently of the starting distribution, we have

$$\boldsymbol{\pi}(t) \approx \boldsymbol{\pi}.$$

This shows that the speed of convergence of a Markov chain is governed by its second largest eigenvalue magnitude.

Example 5.4 Consider a random walk over the triangulated pentagon shown in Fig. 5.5.

When two vertices, say i and j, are connected by an edge, the vertices i and j are neighbors, and if $\mathcal{N}(i)$ is the set of neighbors of node i and $d(i)$ denotes the degree of of node i, i.e., its number of neighbors, the one-step transition probability matrix of the Markov chain representing the random walk is given by

$$p_{ij} = p(X(t + 1) = j | X(t) = i) = \begin{cases} 1/d(i) \text{ if } j \in \mathcal{N}(i) \\ 0 \quad \text{otherwise}. \end{cases}$$

In other words, if X is located at node i at time t, it will move to one of the neighbors of i with equal probability at time $t + 1$. For the graph of Fig. 5.5, we have $d(1) = 4$ since vertex 1 has 4 neighbors, $d(3) = d(5) = 3$ and $d(2) = d(4) = 2$. The one-step transition matrix $\mathbf{P}$ is therefore given by

Fig. 5.5 Random walk over a triangulated pentagon

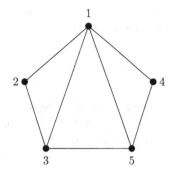

$$\mathbf{P} = \begin{bmatrix} 0 & 1/4 & 1/4 & 1/4 & 1/4 \\ 1/2 & 0 & 1/2 & 0 & 0 \\ 1/3 & 1/3 & 0 & 0 & 1/3 \\ 1/2 & 0 & 0 & 0 & 1/2 \\ 1/3 & 0 & 1/3 & 1/3 & 0 \end{bmatrix}. \tag{5.33}$$

The Markov chain is clearly irreducible, since all states communicate. To verify that the Markov chain is aperiodic, consider state 1. The cycles $1 - 3 - 1$ and $1 - 3 - 5 - 1$ have lengths 2 and 3 respectively, so $f_{11}(2) > 0$ and $f_{11}(3) > 0$. Since 2 and 3 are coprime, we deduce that $d = 1$, so the chain is aperiodic. To find the unique positive left eigenvector $\boldsymbol{\pi}$ satisfying (5.25), it is convenient to observe that the graph is invariant when vertices 2 and 4, and 3 and 5 are exchanged, so that the entries of

$$\boldsymbol{\pi} = \begin{bmatrix} \pi_1 & \pi_2 & \pi_3 & \pi_4 & \pi_5 \end{bmatrix}$$

must be such that $\pi_2 = \pi_4$ and $\pi_3 = \pi_5$. Then (5.25) yields

$$\pi_1 = \pi_2 + \frac{2}{3}\pi_3$$

$$\pi_2 = \frac{1}{4}\pi_1 + \frac{1}{3}\pi_3,$$

and since

$$1 = \sum_{i=1}^{5} \pi_i = \pi_1 + 2(\pi_2 + \pi_3),$$

by eliminating π_1, we obtain

$$1 = 3\pi_2 + \frac{8}{3}\pi_3$$

$$\frac{1}{2} = 3\pi_2 + \frac{1}{3}\pi_3,$$

so that

$$\pi_3 = \pi_5 = \frac{3}{14}, \quad \pi_2 = \pi_4 = \frac{1}{7} \text{ and } \pi_1 = \frac{2}{7}.$$

Remark It turns out that the stationary distribution for a random walk on a graph can be computed in closed form [4]. Let

$$d = \sum_{i=1}^{N} d(i)$$

denote the sum of the degrees of all graph vertices. We have $d = 2e$, where e is the number of graph edges, since when an edge connects nodes i and j, it is counted twice, as part of $d(i)$ and of $d(j)$. Then the entries π_i of the stationary distribution of the random walk are given by

$$\pi_i = \frac{d(i)}{d} . \tag{5.34}$$

To verify this expression, note that

$$\sum_{i=1}^{N} \pi_i p_{ij} = \sum_{i=1}^{N} \frac{d(i)}{d} \frac{1}{d(i)} 1_{\mathcal{N}(i)}(j)$$

$$= \frac{1}{d} \sum_{i=1}^{N} 1_{\mathcal{N}(i)}(j) = \frac{d(j)}{d} = \pi_j,$$

where

$$1_{\mathcal{N}(i)}(j) = \begin{cases} 1 & \text{if } j \in \mathcal{N}(i) \\ 0 & \text{otherwise} . \end{cases}$$

For the random walk of Example 5.4

$$d = d(1) + d(2) + d(3) + d(4) + d(5) = 14,$$

so $\pi_1 = d(1)/d = 4/14 = 2/7$, $\pi_2 = d(2)/2 = 2/14 = 1/7$, and $\pi_3 = d(3)/d = 3/14$.

5.5   First-Step Analysis

Let us turn our attention now to properties of transient states. By regrouping together all recurrent classes, the canonical decomposition (5.13) of the one-step transition probability matrix can be rewritten compactly as

$$\mathbf{P} = \begin{bmatrix} \mathbf{P}_R & \mathbf{0} \\ \mathbf{P}_{TR} & \mathbf{P}_T \end{bmatrix}, \tag{5.35}$$

so that $\mathbf{F}(t) = \mathbf{P}^t$ has the triangular structure

$$\mathbf{F}(t) = \begin{bmatrix} \mathbf{P}_R^t & \mathbf{0} \\ \mathbf{F}_{TR}(t) & \mathbf{P}_T^t \end{bmatrix} . \tag{5.36}$$

The triangular structure of $\mathbf{P}$ implies that the eigenvalues of $\mathbf{P}_T$ are eigenvalues of $\mathbf{P}$, so if λ is an eigenvalue of $\mathbf{P}_T$, we must have $|\lambda| \leq 1$. In fact, it turns out that all eigenvalues of $\mathbf{P}_T$ have a magnitude strictly less than 1. This is due to the fact that for each transient state i, there exists at least one path of finite length m_i and probability $q_i > 0$ leading to a recurrent state. By selecting $m = \max_{i \in T} m_i$, and $q = \min_{i \in T} q_i$, this means that if $X(t)$ starts in any transient state, it will have moved to a recurrent state with at least probability q after m steps. Equivalently, the probability of staying in the transient class after m steps is less than $1 - q < 1$, so after km steps it is less than $(1 - q)^{km}$, which tends to zero. If we consider the Jordan form representation of $\mathbf{P}_T$ and an expression similar to (5.31) and (5.32) for $\mathbf{P}_T^t$, since all entries of $\mathbf{P}_T^t$ tend to zero as t tends to infinity, we deduce that all eigenvalues of $\mathbf{P}_T$ must have magnitude strictly less than 1. This implies in turn that as t tends to infinity, the matrix power series

$$\sum_{k=0}^{t} \mathbf{P}_T^k$$

converges to the fundamental matrix $\mathbf{S} = (\mathbf{I} - \mathbf{P}_T)^{-1}$.

If a recurrent class is formed by a single state, this state is called an absorbing or trapping state. If all recurrent classes are formed by absorbing states, the Markov chain is called absorbing, and in this case $\mathbf{P}_R = \mathbf{I}_C$ in (5.34), where C denotes the number of absorbing/trapping states. In this case, it is easy to verify by induction that the matrix $\mathbf{F}_{TR}(t)$ in (5.36) can be expressed as

$$\mathbf{F}_{TR}(t) = \Big(\sum_{k=0}^{t-1} \mathbf{P}_T^k\Big)\mathbf{P}_{TR}, \tag{5.37}$$

which converges to $\mathbf{S}\mathbf{P}_{TR}$ as t tends to infinity.

5.5.1 Number of Visits to Transient States

A convenient technique for analyzing Markov chain problems is to use what is called a *first-step analysis*. If we try to evaluate the probability of an event or the expectation of a random variable conditioned on $X(0) = i$, the key idea is to condition on the value $X(1) = k$ of the Markov chain after one step, and then use the Markov property. To illustrate how this works, consider a Markov chain with transient states, and if i and j are transient states, suppose we are interested in computing the expected value $n_{ij} = E[N_j | X(0) = i]$ of the number N_j of visits to state j, starting from $X(0) = i$. By the iterated expectation principle, we have

$$n_{ij} = E_{X(1)}[E[N_j | X(1) = \ell, X(0) = i] | X(0) = i]], \tag{5.38}$$

where the Markov property implies

$$E[N_j | X(1) = \ell, X(0) = i] = \delta_{ij} + E[N_j | X(1) = \ell]$$
$$= \delta_{ij} + n_{\ell j}. \tag{5.39}$$

The Kronecker delta function δ_{ij} appearing in (5.39) just indicates that if $j = i$, the state i has been visited once at time $t = 0$. Combining (5.39) and (5.38) gives

$$n_{ij} = \delta_{ij} + \sum_{\ell \in T} p_{i\ell} n_{\ell j}$$

for all $i \in T$, which can be rewritten in matrix form as

$$\mathbf{N} = \mathbf{I} + \mathbf{P}_T \mathbf{N}, \tag{5.40}$$

so that

$$\mathbf{N} = (\mathbf{I} - \mathbf{P}_T)^{-1} = \mathbf{S}, \tag{5.41}$$

which provides an interpretation of fundamental matrix $\mathbf{S}$.

Remark 1 The expression (5.41) for the fundamental matrix could have also been derived directly by observing that the number of visits to state j can be expressed as

$$N_j = \sum_{t=0}^{\infty} 1_j(X(t)),$$

where the indicator function

$$1_j(X(t)) = \begin{cases} 1 & X(t) = j \\ 0 & X(t) \neq j. \end{cases}$$

Thus

$$E[N_j | X(0) = i] = \sum_{t=0}^{\infty} E[1_j(X(t)) | X(0) = i],$$

where

$$E[1_j(X(t)) | X(0) = i] = P(X(t) = j | X(0) = i) = f_{ij}(t).$$

This implies that when i and j are transient states,

$$n_{ij} = \sum_{t=0}^{\infty} f_{ij}(t),$$

or equivalently in matrix form

$$\mathbf{N} = \sum_{t=0}^{\infty} \mathbf{F}_T(t) = (\mathbf{I} - \mathbf{P}_T)^{-1}.$$

Remark 2 In some situations, instead of computing explicitly the fundamental matrix $\mathbf{S}$, it may be preferable to solve a linear system of N_T equations, where N_T denotes the number of transient states. For example, starting from transient state $X(0) = i$, suppose we seek to compute u_i, the expected number of steps until $X(t)$ leaves the transient class. Clearly

$$u_i = \sum_{j \in T} s_{ij},$$

so if

$$\mathbf{u} = \begin{bmatrix} u_1 \dots u_i \dots u_{N_T} \end{bmatrix}^T$$
$$\mathbf{v}_T = \begin{bmatrix} 1 \dots 1 \dots 1 \end{bmatrix},$$

we have

$$\mathbf{u} = \mathbf{S}\mathbf{v}_T, \tag{5.42}$$

so in light of (5.40), **u** satisfies the matrix equation

$$\mathbf{u} = \mathbf{P}_T\mathbf{u} + \boldsymbol{v}_T, \tag{5.43}$$

where in this format $\boldsymbol{v}_T$ is often called the reward vector. The reward here is that for every time step that $X(t)$ stays in the transient class, the transition counter is increased by 1. The solution of (5.43) is of course (5.42), so to compute **u**, we could compute the fundamental matrix **S** and multiply it by the reward vector $\boldsymbol{v}_T$. But in situations where the Markov chain admits symmetries, it is often easier to exploit these symmetries by solving system (5.43), as illustrated by the following example.

Example 5.4 (Continued) Consider a random walk over the triangulated pentagon of Fig. 5.5. Suppose that, starting in vertex i with $2 \leq i \leq 5$, we wish to evaluate the average number of steps u_i until vertex 1 is reached. To answer this question, we need to transform the one-step-transition matrix so that the walk is terminated as soon as vertex 1 is reached, i.e., we must transform state 1 into a trapping state. This is accomplished by replacing the first row of **P** in (5.33) by $\tilde{p}_{1j} = \delta_{1j}$, which yields the modified one-step transition matrix

$$\tilde{\mathbf{P}} = \begin{bmatrix} 1 & 0 & 0 & 0 & 0 \\ 1/2 & 0 & 1/2 & 0 & 0 \\ 1/3 & 1/3 & 0 & 0 & 1/3 \\ 1/2 & 0 & 0 & 0 & 1/2 \\ 1/3 & 0 & 1/3 & 1/3 & 0 \end{bmatrix}.$$

We have therefore

$$\mathbf{P}_T = \begin{bmatrix} 0 & 1/2 & 0 & 0 \\ 1/3 & 0 & 0 & 1/3 \\ 0 & 0 & 0 & 1/2 \\ 0 & 1/3 & 1/3 & 0 \end{bmatrix}, \quad \boldsymbol{v}_T = \begin{bmatrix} 1 \\ 1 \\ 1 \\ 1 \end{bmatrix},$$

and by taking into account the Markov chain symmetry $1 \leftrightarrow 2$ and $3 \leftrightarrow 5$, the solution **u** of (5.43) will be such that $u_2 = u_4$ and $u_3 = u_5$. This gives

$$u_2 = \frac{u_3}{2} + 1$$

$$u_3 = \frac{u_2}{3} + \frac{u_3}{3} + 1,$$

so

$$u_2 = 7/3 \quad \text{and} \quad u_3 = 8/3.$$

Note from the pentagonal graph that node 2 has only two neighbors, one of which is trapping node 1, whereas 3 has two neighbors, one of which is 1. This explains why u_3, the average number of steps until trapping if we start in node 3, is larger than the corresponding average number of steps u_2 if we start in node 2.

5.5.2 Absorption Probabilities

Consider an absorbing Markov chain with C trapping states. Given that $X(0) = i$, it is often of interest to compute the probability that $X(t)$ will be trapped by state j with $1 \leq j \leq C$ at t tends to infinity.

If T_j denotes the first time that state j is reached, the probability of absorption by j if $X(0) = i$ is given by

$$a_{ij} = P(T_j < \infty | X(0) = i) = \lim_{t \to \infty} P(T_j \le t | X(0) = i) . \tag{5.44}$$

Since j is absorbing

$$P(T_j \le t | X(0) = i) = P(X(t) = j | X(0) = i) = f_{ij}(t).$$

But we saw earlier that the limit of $\mathbf{F}_{TR}(t)$ as t tends to infinity is $\mathbf{SP}_{TR}$, so

$$a_{ij} = (\mathbf{SP}_{TR})_{ij} . \tag{5.45}$$

The expression (5.45) could have also been derived by performing a first-step analysis. By the principle of total probability

$$P(T_j < \infty | X(0) = i)$$

$$= \sum_{k=0}^{N} P(T_j < \infty) | X(1) = k, X(0) = i) P(X(1) = k | X(0) = i) , \tag{5.46}$$

where by the Markov property

$$P(T_j < \infty | X(1) = k, X(0) = i) = P(T_j < \infty | X(1) = k)$$

$$= \begin{cases} 1 & \text{if } k \in R, \ k = j \\ 0 & \text{if } k \in R, \ k \ne j \\ a_{kj} & k \in T . \end{cases} \tag{5.47}$$

In other words if $X(1) = k$ is an absorbing state other than j, then $X(t)$ will never be trapped by j, since it already got absorbed by k, but if $X(1) = j$, the probability of being absorbed by j is one. Substituting (5.47) inside (5.46) yields

$$a_{ij} = p_{ij} + \sum_{k=1}^{N_T} p_{ik} a_{kj}, \tag{5.48}$$

where the summation in (5.48) is performed over the transient states only. In matrix form, since i belongs to the transient class and j is an absorbing recurrent state, this gives

$$\mathbf{A} = \mathbf{P}_{TR} + \mathbf{P}_T \mathbf{A}, \tag{5.49}$$

whose solution is

$$\mathbf{A} = (\mathbf{I} - \mathbf{P}_T)^{-1} \mathbf{P}_{TR} = \mathbf{SP}_{TR}. \tag{5.50}$$

Example 5.5 Consider a random walk on a linear graph with N states shown in Fig. 5.6.

$X(t)$ represents the location of a drunk man at time t after exiting a bar after far too many drinks at time $t = 0$. Fortunately, the street is lined up with lampposts, and after clutching lamppost i at time t, the man moves to either lamppost $i - 1$ or $i + 1$ with probability $1/2$ each at time $t + 1$. The

Fig. 5.6 Random walk on a linear graph

man's house is at one end of the street, in front of lamppost 1, and the police station at the other end, in front of lamppost N. If the man reaches lamppost 1, his wife who is on the lookout for him will drag him inside the house and make sure he sleeps comfortably. If the man reaches the police station, the sergeant on guard in front of the police station will lock him up in jail for the night for public intoxication. We seek to compute v_i, the probability that the drunk will sleep in his bed, given that he started in position $X(0) = i$, with $2 \leq i \leq N - 1$.

Clearly, the Markov chain has two absorbing states: $\{1, N\}$, since once the drunk reaches one of these two states, he is removed from circulation for the rest of the night. The states $\{2, \ldots, N-1\}$ are transient, since states 1 and N can be accessed from any one of these states, but no return is possible once 1 or N are reached. By ordering the states as $1, N, 2, \ldots N - 1$, the one-step transition matrix $\mathbf{P}$ can be written as

$$\mathbf{P} = \begin{bmatrix} \mathbf{I}_2 & \mathbf{0} \\ \mathbf{P}_{TR} & \mathbf{P}_T \end{bmatrix},$$

where $\mathbf{I}_2$ denotes the identity matrix of dimension 2, and

$$\mathbf{P}_R = \begin{bmatrix} 1/2 & 0 \\ 0 & 0 \\ \vdots & \vdots \\ 0 & 0 \\ 0 & 1/2 \end{bmatrix}, \quad \mathbf{P}_T = \begin{bmatrix} 0 & 1/2 & 0 & & \cdots & 0 \\ 1/2 & 0 & 1/2 & 0 & & 0 \\ 0 & 1/2 & 0 & 1/2 & & 0 \\ \vdots & & \ddots & \ddots & \ddots & \vdots \\ 0 & & 0 & 1/2 & 0 & 1/2 \\ 0 & \cdots & & 0 & 1/2 & 0 \end{bmatrix}.$$

Then if

$$\mathbf{v} = \begin{bmatrix} v_2 & \ldots & v_i & \ldots & v_{N-1} \end{bmatrix}^T,$$

from (5.49) $\mathbf{v}$ satisfies the matrix equation

$$\mathbf{v} = \mathbf{P}_T \mathbf{v} + \mathbf{P}_{T1}, \tag{5.51}$$

where

$$\mathbf{P}_{T1} = \begin{bmatrix} 1/2 & 0 & \ldots & 0 \end{bmatrix}^T$$

denotes the first column of $\mathbf{P}_{TR}$. Thus v_i satisfies the second-order recursion

$$v_i - \frac{1}{2}(v_{i+1} + v_{i-1}) = 0 \tag{5.52}$$

for $3 \leq i \leq N - 2$ with the boundary conditions

$$v_2 = \frac{v_3}{2} + \frac{1}{2}$$

$$v_{N-1} = \frac{v_{N-2}}{2}. \tag{5.53}$$

Equation (5.52) is homogeneous with characteristic equation

$$a(z) = z^2 + 1 - 2z = (z - 1)^2 = 0.$$

The root $z = 1$ has multiplicity 2, so homogeneous solutions of (5.52) have the form

$$v_i = A + Bi,$$

where the constants A and B need to be selected so that boundary conditions (5.53) are satisfied. This gives

$$\begin{bmatrix} 1 & 1 \\ 1 & N \end{bmatrix} \begin{bmatrix} A \\ B \end{bmatrix} = \begin{bmatrix} 1 \\ 0 \end{bmatrix}. \tag{5.54}$$

The solution of (5.54) is given by

$$A = \frac{N}{N-1} \ , \quad B = -\frac{1}{N-1},$$

so that the solution

$$v_i = \frac{N-i}{N-1}$$

decreases linearly with i. Thus, the closer the bar location i is to home, the higher the chance that the drunkard will sleep in his own bed at night.

5.6   Markov Chain Modulation

In some applications, Markov chains are used to encode an input data stream which triggers Markov chain transitions [5, Chap. 4]. Then, if the Markov chain is in state $X(k) = i$ with $1 \le i \le N$ at the k-th encoding step, the signalling pulse $s_i(t)$ is transmitted during the k-th transmission period. If T denotes the transmission period and D denotes a random uniform synchronization delay uniformly distributed over $[0, T]$, the transmitted signal takes the form

$$Z(t) = \sum_{k=-\infty}^{\infty} s_{X(k)}(t - kT - D)). \tag{5.55}$$

The delay D is assumed independent of the Markov chain state sequence $X(k)$. The signal $Z(t)$ can be viewed as a generalization of the pulse amplitude modulated signals considered in Example 4.5. Typically, the purpose of Markov chain modulation is to shape the spectral content of $Z(t)$.

The Markov chain $X(k)$ is irreducible and aperiodic, and its one-step and m-step transition matrices are denoted respectively by $\mathbf{P}$ and $\mathbf{F}(m) = \mathbf{P}^m$. It is assumed that $X(k)$ has reached steady state and that its probability distribution $\pi(k) = \pi$, where the stationary distribution π satisfies

$$\pi \mathbf{P} = \pi.$$

The mean of $Z(t)$ is given by

$$E[Z(t)] = \sum_{k=-\infty}^{\infty} E[s_{X(k)}(t - kT - D)],$$

where

$$E[s_{X(k)}(t - kT - D)] = \frac{1}{T} \sum_{i=1}^{N} \pi_i \int_0^T s_i(t - kT - u)du$$

exchanging the summations with respect to k and i, and performing the change of variable $v = t - kT - u$, we obtain

$$m_Z(t) = \frac{1}{T} \sum_{i=1}^{N} \pi_i \left(\sum_{k=-\infty}^{\infty} \int_{t-(k+1)T}^{t-kT} s_i(v)dv \right)$$

$$= \frac{1}{T} \sum_{i=1}^{N} \pi_i \int_{-\infty}^{\infty} s_i(v)dv = \frac{1}{T} \sum_{i=1}^{N} \pi_i S_i(j0), \tag{5.56}$$

where if

$$S_i(j\omega) = \int_{-\infty}^{\infty} s_i(t) \exp(-j\omega t)dt$$

denotes the Fourier transform of $s_i(t)$, $S_i(j0)$ corresponds to its DC value. Thus, the mean m_Z is constant.

The autocorrelation of $Z(t)$ is given by

$$R_Z(t, s) = E[Z(t)Z(s)]$$

$$= \sum_{k=-\infty}^{\infty} \sum_{\ell=-\infty}^{\infty} E[s_{X(k)}(t - kT - D)s_{X(\ell)}(s - \ell T - D)] \tag{5.57}$$

with

$$E[s_{X(k)}(t - kT - D)s_{X(\ell)}(s - \ell T - D)]$$

$$= \frac{1}{T} \sum_{i=1}^{N} \sum_{j=1}^{N} q_{ij}(k - \ell) \int_0^T s_i(t - kT - u)s_j(s - \ell T - u)du$$

$$= \frac{1}{T} \sum_{i=1}^{N} \sum_{j=1}^{N} q_{ij}(k - \ell) \int_{t-(k+1)T}^T s_i(v)s_j(v - (t - s) + (k - \ell)T)dv, \tag{5.58}$$

where to go from the second to the third line, we have performed the change of variable $v = t - kT - u$. In the expressions above

$$q_{ij}(k - \ell) = P(X(k) = i, X(\ell) = j)$$

$$= \begin{cases} \pi_j f_{ji}(k - \ell) & k > \ell \\ \pi_i f_{ij}(\ell - k) & \ell > k \\ \pi_i \delta_{ij} & k = \ell \end{cases} \tag{5.59}$$

denotes the joint PMF of $X(k)$ and $X(\ell)$, where $f_{ij}(m)$ is the (i, j)-th entry of transition matrix $\mathbf{F}(m)$. By observing that

$$\sum_{k=-\infty}^{\infty} \int_{t-(k+1)T}^{t-kT} s_i(v)s_j(v-(t-s)+mT)dv = \int_{-\infty}^{\infty} s_i(v)s_j(v-(t-s)+mT)dv,$$

and denoting by

$$r_{ij}(\tau) = s_i(\tau) * s_j(-\tau) \tag{5.60}$$

the deterministic auto- and cross-correlation of waveforms $s_i(\tau)$ and $s_j(\tau)$ for $1 \le i, j \le N$, where $*$ is the convolution operation, we find that R_Z depends only on $\tau = t - s$ and

$$R_Z(\tau) = \frac{1}{T} \sum_{i=1}^{N} \sum_{j=1}^{N} \sum_{m=-\infty}^{\infty} q_{ij}(m)r_{ij}(\tau - mT). \tag{5.61}$$

This expression can be simplified further if we introduce the matrix sampling function

$$q_{ij}^{\mathrm{S}}(\tau) = \sum_{m=-\infty}^{\infty} q_{ij}(m)\delta(\tau - mT)$$

with $1 \le i, j \le N$. Then we find

$$R_Z(\tau) = \frac{1}{T} \sum_{i=1}^{N} \sum_{j=1}^{N} r_{ij}(\tau) * q_{ij}^{\mathrm{S}}(\tau), \tag{5.62}$$

where the effects of waveform selection and of Markov chain correlation shaping are captured separately by $r_{ij}(\tau)$ and $q_{ij}^{\mathrm{S}}(\tau)$.

5.7   Bibliographical Notes

The undergraduate textbook by Kemeny and Snell [6] gives a self-contained presentation of finite Markov chains at an elementary level. The textbooks by Brémaud [7], Ross [8], and Çinlar [9] provide accessible presentations of Markov chains, but with different emphases. The contraction approach used to prove the convergence of irreducible aperiodic Markov chains follows the presentation given in [10, 11]. In this respect, it is worth noting that while the total variation metric was used here to measure the discrepancy between two probability distributions, the mapping over the simplex of probability distributions induced by the one-step transition probability matrix is also contractive for other discrepancy measures, such as the relative entropy [12] or the Hilbert metric [13].

5.8   Problems

5.1 Consider a homogeneous Markov chain with 3 states. Its one-step transition probability matrix is

$$\mathbf{P} = \begin{bmatrix} 0.8 & 0.2 & 0 \\ 0.1 & 0.6 & 0.3 \\ 0 & 0.6 & 0.4 \end{bmatrix},$$

where we recall that the element p_{ij} in row i and column j of $\mathbf{P}$ is the one-step transition probability from state i to state j:

$$p_{ij} = P(X(t+1) = j | X(t) = i).$$

a) Draw the state transition diagram corresponding to $\mathbf{P}$. Make sure to label the transitions by their probability.
b) Find numerical values for
 i) $P(X(6) = 3 \mid X(4) = 1)$,
 ii) $P(X(12) = 2 \mid X(10) = 2)$,
 iii) $P(X(143) = 3 \mid X(52) = 1)$. For this question, you may assume that between the times $t_1 = 52$, and $t_2 = 143$, the Markov chain reaches a stationary steady state.
c) Suppose that

$$P(X(0) = i) = \begin{cases} 0, & i = 1 \\ \frac{1}{2}, & i = 2, 3. \end{cases}$$

 Let T be the time at which the process makes its first change of state, i.e., $X(t) = X(0)$ for $0 \le t \le T - 1$ and $X(T) \ne X(0)$. Find $E[T]$.
d) Assume an observer arrives at a random time instant t with the Markov chain in its stationary steady state. Find the probabilities of the following events.
 i) The state does not change at the first time instant after the observer arrives.
 ii) A $1 \to 2$ or a $2 \to 3$ transition takes place at the first time instant after the observer arrives.
  You may want to use the following two summation formulas

$$\sum_{k=0}^{\infty} p^k = \frac{1}{1-p}$$

$$\sum_{k=0}^{\infty} k p^k = \frac{p}{(1-p)^2},$$

 which are valid for $|p| < 1$.

5.2 The alternate mark inversion (AMI) line code [14] of digital communications encodes a sequence of 0's and 1's (bits) into a sequence of 0's, $+1$'s, and -1's as follows. If the input sequence contains a 0, the output sequence contains a 0 at the same place. If the input sequence contains a 1, then the output sequence will have either a -1 or a $+1$. The choice between -1 and $+1$ is performed in such a way that -1's and $+1$'s alternate in the output sequence. The first 1 is encoded as a $+1$. For instance, 011101 becomes $0, +1, -1, +1, 0, -1$. Find a finite state machine (FSM) with four states: $+1, -1,$

0_+, and 0_- for which the sequence of visited states, not counting the initial state 0_+, is exactly the encoded sequence (where 0_+ and 0_- are rewritten as 0) when it is fed by the input sequence.

a) Suppose that the input sequence is independent, identically distributed, where 0 and 1 have probability $1/2$. Verify that the FSM is then a homogeneous Markov chain. Draw its state transition diagram with the transitions labeled by their probability and find the one-step transition probability matrix $\mathbf{P}$ of the Markov chain.

b) Evaluate $\mathbf{P}^t$ for all $t \geq 0$ and find the steady-state probability distribution

$$\pi_{ss} = \lim_{t \to \infty} \pi(t),$$

where $\pi(t)$ is the row vector of dimension 4 formed by the probabilities of each of the four states at time t.

c) Assume that the Markov chain has reached steady state. Call $\{Y(t), t \geq 0\}$ the output sequence taking values in $\{0, +1, -1\}$. Find the mean and autocorrelation of $Y(t)$.

5.3 The Markov chain used for the generation of delay modulated (Miller-coded) signals in digital magnetic recording [5, 15] has for one-step transition probability matrix

$$\mathbf{P} = \begin{bmatrix} 0 & 1/2 & 0 & 1/2 \\ 0 & 0 & 1/2 & 1/2 \\ 1/2 & 1/2 & 0 & 0 \\ 1/2 & 0 & 1/2 & 0 \end{bmatrix}.$$

a) Write the state transition diagram for this Markov chain. Is the Markov chain irreducible?

b) Find the period of the Markov chain.

c) Verify that the Markov chain is invariant if the states 1 and 4 are exchanged while states 2 and 3 are exchanged. Use this observation to evaluate the steady-state probability distribution π_{ss} of the Markov chain.

5.4 Consider a random walk $X(t)$ over the graph shown in Fig. 5.7. The edges of the graph indicate which vertices are nearest neighbors of each other. For example, the neighbors of vertex 1 are $\{2, 3, 4, 5\}$, but the neighbors of 2 are only $\{1, 3\}$. If we assume that $X(t) = i$ and if i has n_i nearest neighbors, then

$$P(X(t+1) = j | X(t) = i) = \begin{cases} 1/n_i & j \in \mathcal{N}(i) \\ 0 & \text{otherwise}, \end{cases}$$

where $\mathcal{N}(i)$ denotes the set of nearest neighbors of i. In other words, if $X(t) = i$, $X(t+1)$ is any one of the nearest neighbors of i with probability $1/n_i$, but X cannot stay at vertex i, and it cannot jump to a vertex which is not a neighbor of i. For example if $X(t) = 2$, then $X(t+1)$ can be either 1 or 3 (the two neighbors of 1) with probability 1/2 each.

a) Explain why $X(t)$ is a Markov chain and find its one-step transition probability matrix $\mathbf{P}$.

b) Are there any transient states? Is the Markov chain irreducible? Is it periodic, and if so what is its period? Please justify your answers.

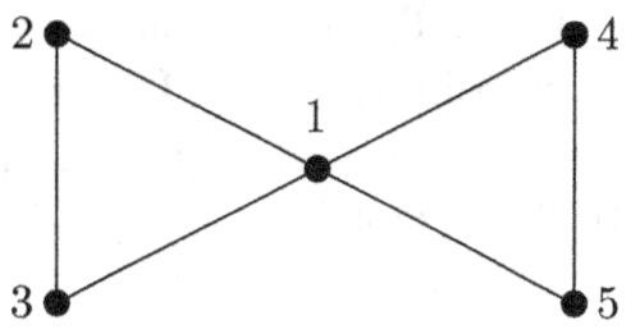

Fig. 5.7 Random walk
over a graph

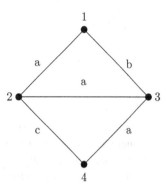

Fig. 5.8 Labeled graph

c) Find the steady-state probability distribution π_{ss} of **P**. In your analysis, you may want to use the fact that the Markov chain remains invariant if we exchange the states 2 and 3, and the states 4 and 5. The pair of states $(2, 3)$ can also be exchanged for the pair $(4, 5)$.

d) Let $u_i = E[N|X(0) = i]$ with $2 \leq i \leq 5$ denote the expected value of the number N of transitions needed for the state $X(t)$ to reach state 1, assuming that the initial state $X(0) = i$. Apply a first-step analysis to obtain equations for u_i with $2 \leq i \leq 5$, and evaluate u_2. To do so, you may want to use the symmetries of the Markov chain to deduce that

$$u_2 = u_3 = u_4 = u_5.$$

5.5 Consider the graph of Fig. 5.8 with four nodes and five edges. The symbol written next to an edge is called the label of the edge.

A particle executes a random walk on the graph. That is, if it is at node i at time t, it moves to one of the neighboring nodes at time $t + 1$ equiprobably. The $X(t)$ denotes the position of the particle at time t.

a) Verify that the evolution of $X(n)$ forms a Markov chain, and find the one-step transition probability matrix $\mathbf{P} = (p_{ij}; \ 1 \leq i, j \leq 4)$, where $p_{ij} = P(X(t + 1) = j \mid X(t) = i)$.

b) Find a probability distribution $\pi_i = P(X(0) = i)$ for the initial state $X(0)$ which ensures that $\{X(t), t \geq 0\}$ is SSS.

c) Let **U** be the orthogonal matrix

$$\mathbf{U} = \frac{1}{\sqrt{2}} \begin{bmatrix} 1 & 0 & 0 & 1 \\ 0 & 1 & 1 & 0 \\ 1 & 0 & 0 & -1 \\ 0 & 1 & -1 & 0 \end{bmatrix}.$$

Verify that

$$\mathbf{P} = \mathbf{U}^T \begin{bmatrix} \mathbf{P}_+ & 0 \\ 0 & \mathbf{P}_- \end{bmatrix} \mathbf{U},$$

where $\mathbf{P}_+$ and $\mathbf{P}_-$ are 2×2 matrices, and use this expression to evaluate the m step transition probability matrix $\mathbf{P}^m$. Note that to obtain a detailed expression, you will need to compute the eigenvalues and eigenvectors of $\mathbf{P}_+$.

d) Check that

$$\lim_{m \to \infty} \mathbf{P}^m = \mathbf{v}\boldsymbol{\pi}, \qquad (5.63)$$

where the vectors

$$\mathbf{v} = \begin{bmatrix} 1 \\ 1 \\ 1 \\ 1 \end{bmatrix} \qquad \boldsymbol{\pi} = \begin{bmatrix} \pi_1 & \pi_2 & \pi_3 & \pi_4 \end{bmatrix}$$

represent respectively to the right and left eigenvectors of $\mathbf{P}$ corresponding to the eigenvalue $\lambda = 1$. Conclude from the expression (5.63) that, independently of the initial state $X(0)$, the Markov chain $\{X(t)\}$ is asymptotically stationary.

e) Define a discrete-time stochastic process $\{Y(t); t \geq 0\}$ taking values in $\{a, b, c\}$ as follows. If $X(t) = i$ and $X(t+1) = j$, then $Y(t) =$ label of the edge (i, j). Suppose that the random walk on the graph has reached its stationary steady state, so that $P(Y(t) = a)$ does not depend on t. What is $P(Y(t) = a)$?

5.6 Consider the homogeneous Markov chain with the one-step transition diagram shown in Fig. 5.9.

a) Find the one-step transition probability matrix $\mathbf{P}$ of $X(t)$.
b) Are there any transient states? Is the Markov chain irreducible? Is it periodic, and if so what is its period? Please justify your answers.
c) Find the steady-state probability distribution π_{ss} of $\mathbf{P}$. In your analysis, you may wish to use the fact that the Markov chain remains invariant if we simultaneously exchange the states 1 and 3, and the states 2 and 4.
d) Let $u_i = E[N|X(0) = i]$ with $i = 2, 3, 4$ denote the expected value of the number N of transitions needed for the state $X(t)$ to reach state 1, assuming that the initial state $X(0) = i$. Apply a first-step analysis to obtain equations for u_i with $2 \leq i \leq 4$.

5.7 Consider the homogeneous Markov chain with the one-step transition diagram shown in Fig. 5.10.

a) Find the one-step transition probability matrix $\mathbf{P}$ of $X(t)$.
b) Are there any transient states? Is the Markov chain irreducible? Is it periodic, and if so what is its period? Please justify your answers.
c) Find the steady-state probability distribution π_{ss} of $\mathbf{P}$. In your analysis, you may wish to use the fact that the Markov chain remains invariant if we simultaneously exchange the states 1 and 3, and the states 2 and 4.

Fig. 5.9 State transition diagram of Problem 5.6

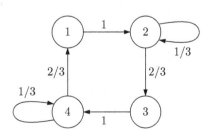

Fig. 5.10 State transition
diagram of Problem 5.7

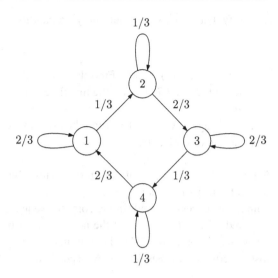

d) Let $u_i = E[N|X(0) = i]$ with $i = 1,\ 2,\ 3$ denote the expected value of the number N of transitions needed for the state $X(t)$ to reach state 4, assuming that the initial state $X(0) = i$. Apply a first-step analysis to obtain equations for u_i with $1 \leq i \leq 3$, and evaluate u_1.

5.8 Consider a cube whose sides have dimension 1, and label its eight vertices as 000, 100, 010, 001, 110, 101, 011, and 111, where indices ijk correspond to the three-dimensional x, y, z coordinates of the vertex. For example, the vertex 101 is the vertex with coordinates $x = 1$, $y = 0$ and $z = 1$. Each vertex has three nearest neighbors. For example, the three neighbors of 000 are 100, 010, and 001. For $t \geq 0$, consider a random walk $\{X(t),\ t \geq 0\}$ where if $X(t) = ijk$, then $X(t + 1)$ moves to any one of the three neighbors of ijk with probability 1/3. For example, if $X(t) = 000$, then

$$P(X(t + 1) = 100|X(t) = 000) = 1/3$$

$$P(X(t + 1) = 010|X(t) = 000) = 1/3$$

$$P(X(t + 1) = 001|X(t) = 000) = 1/3.$$

a) Verify that $X(t)$ is a Markov chain. Sketch its one-step transition probability diagram and find its one-step probability transition matrix **P**. To do so, you may want to partition the eight vertices into two groups: group A is formed by the vertices $\{000, 101, 110, 011\}$ such that $(i + j + k)$ mod $2 = 0$ and group B is formed by the vertices $\{100, 010, 001, 111\}$ such that $(i + j + k)$ mod $2 = 1$. Assuming now that the states are numbered so that the states of group A come before the states of group B, verify that **P** has the structure

$$\mathbf{P} = \begin{bmatrix} \mathbf{0} & \mathbf{P}_{AB} \\ \mathbf{P}_{BA} & \mathbf{0} \end{bmatrix}.$$

In other words, if the state is located in group A at time t, it must belong to group B at time $t + 1$, and vice versa.

b) What are the transient and recurrent states of the Markov chain (MC)? Is the MC irreducible? Find the period of the MC.

c) Verify that there exists a stationary probability distribution π_{ss} satisfying

$$\pi_{ss}\mathbf{P} = \pi_{ss}.$$

d) Assume that $X(0) = 000$. Find the expected number of transitions needed until the state reaches vertex 111, i.e., if T denotes the first time that the random walk visits vertex 111, find $E[T|X(0) = 000]$

e) Assume again $X(0) = 000$. Find the probability that the random walk will visit vertex 111 before returning to vertex 000.

5.9 Consider a sequence of coin tosses with a biased coin, where the probability of tails is p, and the probability of heads is $q = 1 - p$. We construct a Markov chain whose state $X(t)$ represents the number of successive heads in the coin toss sequence. Thus, if $X(t) = k$, i.e., if k heads in a row have occurred, $X(t + 1) = k + 1$ if the next coin toss produces heads, and $X(t) = 0$ if the next coin toss gives tails. In other words, if the coin toss produces tails, the heads count restarts at zero. The coin toss experiment stops as soon as N heads in a row have been obtained, with $N > 1$.

a) Draw the state transition diagram for the Markov chain we have just described and specify its one-step probability transition matrix $\mathbf{P}$.

b) Which states are transient, and which states are recurrent? Obtain the canonical representation

$$\mathbf{P} = \begin{bmatrix} \mathbf{P}_R & \mathbf{0} \\ \mathbf{P}_{TR} & \mathbf{P}_T \end{bmatrix}$$

of the one-step transition probability matrix.

c) Assume that the Markov chain starts in state k, so that the current heads count is k, with $1 \le k \le N - 1$. What is the probability that the heads count will reach N before it reaches 0?

d) Let u_k denote the expected number of coin tosses until the heads count reaches N, starting from state k. Use a first-step analysis to obtain an equation for all u_ks. If $\mathbf{u}$ denotes the vector

$$\mathbf{u} = \begin{bmatrix} u_0 \ u_1 \ u_2 \ \cdots \ u_{N-1} \end{bmatrix}^T,$$

show that $\mathbf{u}$ satisfies an equation of the form

$$\mathbf{u} = \mathbf{P}_T\mathbf{u} + \mathbf{r}$$

where $\mathbf{r}$ denotes a column reward vector. Find the vector $\mathbf{r}$.

e) Evaluate u_k for all $0 \le k \le N - 1$.

5.10 We employ a Markov chain to analyze the behavior of a customer line at the checkout counter of a department store. During each time period, there is a probability p that a new customer arrives at the back of the line, and there is a probability q that the customer at the head of the line leaves after paying the clerk. In the following, we denote by $p^c = 1 - p$ and $q^c = 1 - q$. The department store has also a policy of opening a new line as soon as the line contains N customers. In addition, if the customer line remains empty for T time periods after becoming empty, it is closed.

a) We are only interested in analyzing the behavior of the customer line until either a new line is opened, or the line is closed. To do so, we use a Markov chain where $X(t) = i$ if there are i customers in line at time t, with $0 \le i \le N$. We also set $X(t) = -i$ with $0 \le i \le T$ if the customer line is currently empty and has been empty for i time periods. Write the one-step transition diagram for this Markov chain, as well as the one-step transition probability matrix $\mathbf{P} = (p_{ij} ; -T \le i, j \le N)$, where $p_{ij} = P(X(t+1) = j | X(t) = i)$. What are the transient states and the recurrent states for the Markov chain? Obtain the canonical representation

$$\mathbf{P} = \begin{bmatrix} \mathbf{P}_R & \mathbf{0} \\ \mathbf{P}_{TR} & \mathbf{P}_T \end{bmatrix}$$

of $\mathbf{P}$.

b) Let v_i be the probability that starting in the state $X(0) = i$, $X(t)$ will hit N at some time t, so that a new line will be opened. By performing a first-step analysis, i.e., by conditioning with respect to $X(1) = j$, obtain a matrix equation for the vector

$$\mathbf{v}^T = \begin{bmatrix} v_{-(T-1)} \cdots v_{-1} \ v_0 \ v_1 \cdots v_{N-1} \end{bmatrix}.$$

c) Solve the equation of part b). To do so, you should observe that for $1 \le i \le N - 1$, the probability v_i satisfies the second order recurrence relation

$$(v_{i+1} - v_i) = \lambda(v_i - v_{i-1})$$

with

$$\lambda \overset{\triangle}{=} p_c q / (p q_c).$$

We assume in the following that $\lambda > 1$, which means that the customers are served at a faster rate than they arrive in line. Also, assume that $0 < p$ and $0 < q_c$, so that λ is finite. To check your result, examine the values you obtain for:

(i) $T \to \infty$, i.e., the line is never closed;
(ii) $T = 0$, i.e., the line is closed as soon as it is empty;
(iii) $N \to \infty$. i.e., a new line is never opened.

d) Assume that $X(0) = i$, so that i customers, where $0 < i < N$, are standing in line at the counter at time 0. Denote by u_i the average amount of time until either the line becomes empty or a new line needs to be opened after the length of the line reaches N. Note that for this problem the Markov chain of part a) needs to be modified, since we are no longer interested in what happens after the line becomes empty. Let $\tilde{P}$ be the one-step transition probability density for the modified chain. By using a first-step analysis, i.e., by conditioning with respect to $X(1) = j$, obtain a matrix equation

$$\mathbf{u} = \tilde{\mathbf{P}}_T \mathbf{u} + \mathbf{r}$$

for the vector

$$\mathbf{u}^T = \begin{bmatrix} u_1 \cdots u_i \cdots u_{N-1} \end{bmatrix},$$

where $\mathbf{r}$ denotes a reward vector of dimension $N - 1$. Solve this equation and find u_i.

5.11 Suppose that two sequences $\{X_i ; i \ge 1\}$ and $\{Y_i ; i \ge 1\}$ represent independent Bernoulli trials with *unknown* success probabilities p_1 and p_2, respectively. Thus X_i and Y_i take the binary values 0 and 1 with probabilities

$$P(X_i = 1) = 1 - P(X_i = 0) = p_1 \qquad P(Y_i = 1) = 1 - P(Y_i = 0) = p_2,$$

and all random variables are independent. To decide whether $p_1 > p_2$ or $p_2 > p_1$, we implement the following test. We define the random variables

$$U(n) = \sum_{i=1}^{n} X_i \qquad V(n) = \sum_{i=1}^{n} Y_i$$

and

$$X(n) = U(n) - V(n).$$

We choose some positive integer M and stop at N, the first value of n such that either $X(n) = M$ or $X(n) = -M$. If $X(N) = M$, we decide that $p_1 > p_2$, and if $X(N) = -M$, we decide that $p_2 > p_1$. To understand this test intuitively, note that by the strong law of large number $U(n)/n \to p_1$ and $V(n)/n \to p_2$ as $n \to \infty$. So, $X(n)/n \to p_1 - p_2$ as $n \to \infty$. Thus $X(n)$ should grow roughly as $(p_1 - p_2)n$ as n becomes large. Thus provided M is large enough, if $X(n)$ hits M first, it is reasonable to guess that $p_1 > p_2$. Similarly, if it hits $-M$ first, we should have $p_2 > p_1$. Of course, this is only approximately true, and we can use the theory of Markov chains to analyze the above test in detail.

a) Verify that $X(n)$ forms a Markov chain evolving among the $2M + 1$ states $\{-M, -(M - 1), \cdots, -1, 0, 1, \cdots, M - 1, M\}$. You can assume that after $X(n)$ has reached M or $-M$, it stays in that state, since the test is over. Write the one-step transition matrix $\mathbf{P} = (p_{ij} ; -M \leq i, j \leq M)$ for the Markov chain, where $p_{ij} = P(X(n + 1) = j | X(n) = i)$. What are the transient states and the recurrent states for the chain? Write $\mathbf{P}$ in the canonical form

$$\mathbf{P} = \begin{bmatrix} \mathbf{P}_R & \mathbf{0} \\ \mathbf{P}_{TR} & \mathbf{P}_T \end{bmatrix}.$$

b) Let v_i be the probability that starting in the state $X(0) = i$, $X(n)$ will hit $-M$, so that we will decide that $p_2 > p_1$. By performing a first-step analysis, i.e., by conditioning with respect to $X(1) = j$, where if we start from i, j can only take the values $\{i - 1, i, i + 1\}$, obtain a recursion linking v_{i-1}, v_i and v_{i+1}. Solve the 3 term recursion for v_i, and use your solution to show that when $p_1 \geq p_2$, the probability of making an error (that is, of deciding $p_2 > p_1$) is given by

$$P(E) = \frac{1}{1 + \lambda^M},$$

where

$$\lambda = \frac{p_1(1 - p_2)}{p_2(1 - p_1)}.$$

Verify that if $p_1 > p_2$, $\lambda > 1$, so that by selecting M large enough the probability of error $P(E)$ can be made arbitrarily small.

c) Let $u_i = E[N | X(0) = i]$ denote the expected number of sample pairs that need to be examined until a decision is reached, assuming we start in state $X(0) = i$. By using a first-step analysis, i.e., by conditioning with respect to $X(1) = j$, obtain a matrix equation

$$\mathbf{u} = \mathbf{P}_T \mathbf{u} + \mathbf{r}$$

for the vector

$$\mathbf{u} = \begin{bmatrix} u_{-(M - 1)} \cdots u_{-1}\, u_0\, u_1 \cdots u_{M-1} \end{bmatrix},$$

where $\mathbf{r}$ denotes a reward vector of dimension $2M - 1$. Use this equation to show that the expected duration of the test is given by

$$E[N] = u_0 = \frac{M(\lambda^M - 1)}{(p_1 - p_2)(\lambda^M + 1)}.$$

5.12 Consider the Miller code Markov chain of Problem 5.3. This Markov chain is used to generate a modulated signal of the form considered in Sect. 5.6 by using transmitting one of the four waveforms

$$s_1(t) = -s_4(t) = A \; 0 \le t \le T$$

$$s_2(t) = -s_3(t) = \begin{cases} A & 0 \le t < T/2 \\ -A & T/2 \le t < T \end{cases}$$

depending on the state $X(k)$ of the encoder at stage k.

a) If

$$\mathbf{V} = \begin{bmatrix} 1 & 1 & 1 & 0 \\ 1 & -1 & 0 & 1 \\ 1 & -1 & 0 & -1 \\ 1 & 1 & -1 & 0 \end{bmatrix},$$

verify that the one-step transition probability matrix $\mathbf{P}$ satisfies

$$\mathbf{PV} = \mathbf{VD}, \tag{5.64}$$

where the block diagonal $\mathbf{D}$ is given by

$$\mathbf{D} = \text{diag}\{1, 0, \mathbf{D}_c\},$$

with

$$\mathbf{D}_c = \frac{1}{2}\begin{bmatrix} -1 & 1 \\ -1 & -1 \end{bmatrix}.$$

What are the eigenvalues of $\mathbf{P}$? Use the identity (5.64) to obtain a general expression for $\mathbf{F}(m)$ for all $m \ge 1$, and verify that as m tends to infinity

$$\mathbf{F}(m) \to \boldsymbol{v}\boldsymbol{\pi},$$

where $\boldsymbol{v}$ is the vector with all entries equal to 1 and

$$\boldsymbol{\pi} = \begin{bmatrix} 1/4 & 1/4 & 1/4 & 1/4 \end{bmatrix}.$$

b) Compute the deterministic auto- and cross-correlations

$$r_{11}(\tau) = s_1(\tau) * s_1(-\tau)$$

$$r_{12}(\tau) = s_1(\tau) * s_2(-\tau)$$

$$r_{22}(\tau) = s_2(\tau) * s_2(-\tau),$$

of waveforms $s_1(\tau)$ and $s_2(\tau)$, and use them to specify the cross-correlations $r_{ij}(\tau)$ with $1 \le i, j \le 4$ specified by (5.60).

5.13 Let $X(t)$ be a finite homogeneous, irreducible, and aperiodic Markov chain with N states and one-step transition probability matrix $\mathbf{P}$. Its probability distribution at time t is $\boldsymbol{\pi}(t)$, and its unique invariant probability matrix is denoted as $\boldsymbol{\pi}$. Because of the time-symmetry of the Markov property, $X(t)$ admits also a backward in time description in terms of $\boldsymbol{\pi}(t)$ and a backward one-step transition matrix $\mathbf{P}^{\mathrm{B}}(t)$ whose entries are given by

$$p_{ij}^{\mathrm{B}}(t) = P(X(t) = j | X(t+1) = i)$$

for $1 \le i, j \le N$.

a) Use Bayes law to obtain an expression for $p_{ij}^{\mathrm{B}}(t)$ in terms of the forward transition probability p_{ji}, $\pi_j(t)$, and $\pi_i(t+1)$. Obtain a condition on the probability distribution $\boldsymbol{\pi}(t)$ which ensures that $\mathbf{P}^{\mathrm{B}}(t)$ is homogeneous.
b) The Markov chain is said to be *reversible* if its one-step forward and backward transition matrices are identical, so that $\mathbf{P} = \mathbf{P}^{\mathrm{B}}$. Obtain a reversibility condition in terms of the entries p_{ij} and π_i of $\mathbf{P}$ and $\boldsymbol{\pi}$ respectively.
c) Consider a random walk over a finite graph of the type examined in Example 5.4. For such a random walk, if $d(i)$ denotes the degree of node i, and $d = \sum_{i=1}^{N} d(i)$, the one-step transition probability matrix has for entries

$$p_{ij} = \begin{cases} 1/d(i) & j \in \mathcal{N}(i) \\ 0 & \text{otherwise}, \end{cases}$$

where $\mathcal{N}(i)$ denotes the set of neighbors of i (the nodes connected to i by an edge), and the invariant probability distribution is given by $\pi_i = d(i)/d$. Verify that the Markov chain is reversible.

5.14 Consider an urn which contains a total of N white and black balls. At time t, $X(t)$ balls are white, with $0 \le X(t) \le N$, and $N - X(t)$ are black. A ball is selected at random, and after inspection, is replaced by a ball of the opposite color. In other words $X(t) = X(t) + 1$ if the ball selected was black (and replaced by a white ball) and $X(t+1) = X(t) - 1$ if the ball selected was white (and replaced by a black ball).

a) Verify that $X(t)$ is a Markov chain. Find its state transition diagram, and obtain its one-step transition probability matrix $\mathbf{P}$.
b) What are the transient and recurrent states of the Markov chain (MC)? Is the MC irreducible? Find the period of the MC.
c) Verify that

$$\pi_i = \binom{N}{i} (1/2)^N \tag{5.65}$$

is a stationary distribution of the MC. Specifically, if

$$\boldsymbol{\pi} = \begin{bmatrix} \pi_0 \ \dots \ \pi_i \ \dots \ \pi_N \end{bmatrix},$$

verify that

$$\boldsymbol{\pi}\mathbf{P} = \boldsymbol{\pi}.$$

d) Even though, due to the periodicity of the chain, convergence of the state probability distribution $\pi(t)$ is not guaranteed from an arbitrary initial probability distribution $\pi(0)$, it is possible to show that several statistics of $X(t)$ converge as t tends to infinity. Consider processes

$$M_1(t) = \frac{X(t) - N/2}{(1 - 2/N)^t} \quad \text{and} \quad M_2(t) = \frac{(X(t) - N/2)^2 - N/4}{(1 - 4/N)^t} .$$

Verify that both $M_1(t)$ and $M_2(t)$ are *martingales*, i.e.

$$E[M_k(t+1)|M_k(s), \, 0 \le s \le t] = M_k(t)$$

for $k = 1, 2$. An important property of martingales is that they have a constant mean. Establish this fact, and use it to show that the mean and variance of $X(t)$ have limits as $t \to \infty$. What are these limits?

e) Perform a simulation of the MC for $N = 10$. To implement the simulation, the following scheme is proposed. Let $Y(t) =$ with $t \ge 1$ be a sequence of independent variables uniformly distributed over the set $\{1, 2, \ldots N\}$. Then set

$$X(t+1) = \begin{cases} X(t) - 1 \text{ if } Y(t+1) \le X(t) \\ X(t) + 1 \text{ if } Y(t+1) > X(t) . \end{cases}$$

Show that this algorithm models appropriately the random ball selection and replacement scheme. To generate the initial state $X(0)$, use a random variable uniformly distributed over $\{0, 1, \ldots, N\}$. Simulate the Markov chain for $1 \le t \le T$ with $T = 10^4$. Evaluate and plot the empirical probability distribution

$$\hat{\pi}_i = \frac{1}{T} \sum_{t=1}^{T} 1_i(X(t))$$

for $0 \le i \le N$, where

$$1_i(j) = \begin{cases} 1 & j = i \\ 0 & j \ne i , \end{cases}$$

and compare it to the stationary distribution π_i of part c).

5.15 Consider the random urn drawing scheme of Problem 5.14. Because the corresponding MC is periodic, convergence of the state probability distribution $\pi(t)$ to the invariant distribution (5.65) is not necessarily guaranteed as t tends to infinity. However, it is not difficult to analyze the asymptotic behavior of the chain by observing that if the number $X(t)$ of white balls is even at time t, it is odd at time $t + 1$. Likewise, if it is even at time t, it is odd at time $t + 1$.

a) By reordering the state $\{0, 1, \ldots, N\}$ so that the even number (of white balls) states $\{0, 2, 4, \ldots\}$ appear first, and the odd numbered states $\{1, 3, \ldots\}$ are considered next, verify that $\mathbf{P}$ has the structure

$$\mathbf{P} = \begin{bmatrix} \mathbf{0} & \mathbf{P}_{eo} \\ \mathbf{P}_{oe} & \mathbf{0,} \end{bmatrix} \tag{5.66}$$

where $\mathbf{P}_{eo}$ denotes the transition probability matrix from even to odd number (of white balls) states, and $\mathbf{P}_{oe}$ is the transition matrix from odd to even number states. Specify the matrices $\mathbf{P}_{eo}$ and $\mathbf{P}_{eo}$.

b) Use the structure (5.66) to verify that the m-step transition matrices have the structure

$$\mathbf{P}^{2m} = \begin{bmatrix} \mathbf{Q}_e^m & \mathbf{0} \\ \mathbf{0} & \mathbf{Q}_o^m \end{bmatrix}$$

for $n = 2m$ even and

$$\mathbf{P}^{2m+1} = \begin{bmatrix} \mathbf{0} & \mathbf{P}_{eo}\mathbf{Q}_o^m \\ \mathbf{P}_{oe}\mathbf{Q}_e^m & \mathbf{0} \end{bmatrix}$$

for $n = 2m + 1$ odd. Specify the matrices $\mathbf{Q}_e$ and $\mathbf{Q}_o$ and show that

$$\mathbf{Q}_e^m \mathbf{P}_{eo} = \mathbf{P}_{eo}\mathbf{Q}_o^m$$
$$\mathbf{Q}_o^m \mathbf{P}_{oe} = \mathbf{P}_{oe}\mathbf{Q}_e^m.$$

c) Show that $\mathbf{Q}_e$ and $\mathbf{Q}_o$ are transition matrices of irreducible aperiodic Markov chains, so they have unique steady-state distributions

$$\boldsymbol{\alpha} = \begin{bmatrix} \alpha_0 & \dots & \alpha_k & \dots & \alpha_{N_e} \end{bmatrix}$$
$$\boldsymbol{\beta} = \begin{bmatrix} \beta_0 & \dots & \beta_k & \dots & \alpha_{N_o} \end{bmatrix}$$

with

$$N_e = \lceil (N-1)/2 \rceil \quad \text{and} \quad N_o = \lfloor (N-1)/2 \rfloor,$$

which satisfy

$$\boldsymbol{\alpha}\mathbf{Q}_e = \boldsymbol{\alpha} \quad \text{and} \quad \boldsymbol{\beta}\mathbf{Q}_o = \boldsymbol{\beta}.$$

Verify that the entries of $\boldsymbol{\alpha}$ and $\boldsymbol{\beta}$ are given by

$$\alpha_k = \binom{N}{2k}(1/2)^{(N-1)} = 2\pi_{2k}$$

$$\beta_k = \binom{N}{2k+1}(1/2)^{(N-1)} = 2\pi_{2k+1}$$

with $k \geq 0$.

d) Let $\boldsymbol{v}_e$ and $\boldsymbol{v}_o$ be two column vectors of length N_e and N_o, respectively, with unit entries. Prove that

$$\lim_{m\to\infty} \mathbf{P}^{2m} = \begin{bmatrix} \boldsymbol{v}_e & \mathbf{0} \\ \mathbf{0} & \boldsymbol{v}_o \end{bmatrix}\begin{bmatrix} \boldsymbol{\alpha} & \mathbf{0} \\ \mathbf{0} & \boldsymbol{\beta} \end{bmatrix} \tag{5.67}$$

$$\lim_{m\to\infty} \mathbf{P}^{2m+1} = \begin{bmatrix} \mathbf{0} & \boldsymbol{v}_e \\ \boldsymbol{v}_o & \mathbf{0} \end{bmatrix}\begin{bmatrix} \boldsymbol{\alpha} & \mathbf{0} \\ \mathbf{0} & \boldsymbol{\beta} \end{bmatrix}. \tag{5.68}$$

Then consider an initial probability distribution which is partitioned based on whether the state indices are even or odd, so that

$$\boldsymbol{\pi}(0) = \begin{bmatrix} \boldsymbol{\pi}_e(0) & \boldsymbol{\pi}_o(0) \end{bmatrix}.$$

Then use the limits (5.67) and (5.68) to show that the state probability distribution $\pi(t)$ converges to the stationary distribution (5.65) as t tends to infinity as longs as the initial state distribution $\pi(0)$ assigns half of the total probability mass to even index states and half to odd index states, i.e., as long as

$$\pi_e(0)\mathbf{v}_e = \pi_o(0)\mathbf{v}_o = \frac{1}{2}.$$

The Problems 5.14 and 5.15 represent a variant of a model proposed in 1907 by Paul and Tatyana Ehrenfest to describe heat transfer between two containers. The white balls describe here gas molecules in the first container and the black balls are the molecules in the second container. This model was analyzed by Kac [16] and Klein [17], who showed that the relative entropy between the probability distribution $\pi(t)$ and the invariant distribution π decreases with time. In this context, the geometric rate of decrease derived in part d) of Problem 5.14 for the deviation of $E[X(t)]$ from its mean represents Newton's law of cooling.

5.16 In statistical mechanics, it is common to encounter random configurations X whose PMF takes the form

$$\pi_i = \frac{1}{Z(\beta)} \exp(-\beta E_i) \tag{5.69}$$

with $1 \le i \le N$, where N is so large that the normalizing constant $Z(\beta)$ (also called partition function) is hard to compute. This makes it extremely difficult to evaluate expectations of the form

$$E[g(X)] = \sum_{i=1}^{N} g(i)\pi_i.$$

To get around this difficulty, a class of random variable generation techniques called Markov chain Monte Carlo (MCMC) methods [18] have been developed which employ a Markov chain to generate random samples $X(t)$ with the desired target distribution π without ever having to compute $Z(\beta)$.

The Metropolis–Hastings algorithm [19,20] uses as starting point an irreducible aperiodic Markov chain with one-step transition matrix $\mathbf{Q}$ such that $q_{ji} > 0$ whenever $q_{ij} > 0$. In other words if the state transition diagram allows a transition from state i to state j, it also allows a transition in the reverse direction. Then given $\mathbf{Q}$, a modified one-step transition matrix $\mathbf{P}$ is constructed which has π as its invariant distribution. The entries p_{ij} of $\mathbf{P}$ are specified based on the acceptance function

$$a_{ij} = \frac{\pi_j q_{ji}}{\pi_i q_{ij}}.$$

Then

$$p_{ij} = \begin{cases} q_{ij} & \text{if } i \ne j, \, a_{ij} \ge 1 \\ q_{ij} a_{ij} & \text{if } i \ne j, \, a_{ij} < 1 \\ q_{ii} + \sum_{k:\, a_{ik}<1} q_{ik}(1 - a_{ik}) & \text{if } i = j. \end{cases}$$

Intuitively, this algorithm can be described as follows. If $X(t) = i$, select a candidate transition from i to j with probability q_{ij}. If the acceptance function $a_{ij} \ge 1$, the transition is accepted, and $X(t+1) = j$, but if $a_{ij} < 1$, a coin with probability of success a_{ij} is flipped. If the coin flip succeeds, the transition to j is accepted, so $X(t+1) = j$. Otherwise, the transition is rejected and $X(t+1) = i$. Note that because the acceptance function depends only on the ratio

$$\frac{\pi_j}{\pi_i} = \exp(\beta(E_i - E_j)),$$

the partition function $Z(\beta)$ never needs to be evaluated.

a) If $\mathbf{Q}$ is irreducible and aperiodic, show that $\mathbf{P}$ is irreducible and aperiodic.
b) Verify that

$$\pi_i p_{ij} = \pi_j p_{ji}, \tag{5.70}$$

and use this result to prove that

$$\pi \mathbf{P} = \pi$$

so that $\mathbf{P}$ has π as its steady-state distribution. Observe also that property (5.70) implies that the Markov chain is reversible.
c) One important feature of MCMC algorithms is that successive states $X(t)$ are not independent. In spite of this lack of independence, MCMC algorithms evaluate the expectation of a quantity of interest, say $g(X)$, by using the time-average approximation

$$E[g(X)] \approx \frac{1}{N} \sum_{k=1}^{N} g(X_k).$$

This type of averaging relies on an extension of the law of large numbers to ergodic stationary processes, which will be discussed in Chap. 9. Ergodic processes are typically characterized by a strong mixing property, whereby samples taken at different times become independent as their time separation increases. Assume that a Markov chain is in steady state at time t. Find a condition on the time separation m that ensures that states $X(t)$ and $X(t+m)$ are nearly independent, so that

$$P(X(t) = i, X(t+m) = j) \approx \pi_i \pi_j$$

for all states i and j.

The two-dimensional Ising model [21] describes interacting spins on a discrete square lattice, where the spin s_{uv} at position (u, v) takes the values ± 1. A spin configuration $x = \{s_{uv}, 1 \leq u, v \leq L\}$ can assume therefore 2^{L^2} values, where L denotes the lattice dimension. In Ising's model, the energy of configuration x is given by

$$E(x) = -J \sum_{<(u,v),(u',v')>} s_{uv} s_{u'v'} \tag{5.71}$$

where the summation is over nearest-neighbors sites, i.e., such that

$$(u', v') = (u \pm 1, v) \quad \text{or} \quad (u', v') = (u, v \pm 1).$$

The coupling constant $J > 0$ for ferromagnetic materials, and $J < 0$ in the antiferromagnetic case. We consider here the ferromagnetic case. In this case, we can always absorb the coupling constant J inside the parameter β appearing in (5.69), so without loss of generality, we can always set $J = 1$.

d) Write a MATLAB program to implement the Metropolis–Hastings algorithm to generate the steady-state distribution (5.69) for the Ising model (5.71). Assume $L = 20$. Use the checkerboard spin pattern where

$$s_{u'v'} = -s_{uv}$$

if (u', v') and (u, v) are nearest neighbors, as initial configuration $X(0)$. To specify the Markov chain transition matrix $\mathbf{Q}$, assume that a candidate site (u, v) is picked at random among the L^2 sites and the candidate transition consists of flipping the spin at this site, i.e., of deciding whether to allow $s_{uv}(t + 1) = -s_{uv}(t)$. For $\beta = 0.5$, verify that in steady state, the spins become aligned with their neighbors over large regions of the square lattice.

e) Let $L = 100$. Use the algorithm of d) to compute the expected energy per site $E[E(x)]/L^2$ without computing the partition function $Z(\beta)$. Do so for 20 values of β evenly spaced between 0.4 and 0.5. Verify that the energy per site undergoes a sharp decrease as β becomes larger than 0.44. This phenomenon is called a phase transition, where on one side of the transition $\beta < 0.44$, lattice spins are randomly oriented, but for $\beta > 0.44$ spins become aligned over large regions.

5.17 The *slice sampling method* [22] is a recently introduced sequential method for generating random variables X_n, $1 \leq n \leq N$ with a given density $f_X(x)$. This is achieved by generating at time n a pair (X_n, Y_n) of random variables with the joint density

$$f_{X,Y}(x, y) = \begin{cases} 1 & 0 \leq y \leq f_X(x) \\ 0 & \text{otherwise} . \end{cases} \tag{5.72}$$

a) Verify that the marginal density of $f_{X,Y}(x, y)$ is indeed $f_X(x)$.

In the slice sampling method, the next pair (X_{n+1}, Y_{n+1}) of random variables actually depends on (X_n, Y_n). The method works as follows. Given X_n, Y_{n+1} is a random variable uniformly distributed over interval $[0, f_X(X_n)]$. Then consider the set (called the Y_{n+1}-level slice)

$$A(Y_{n+1}) = \{x : f_X(x) \geq Y_{n+1}\}.$$

Its size is defined as

$$S(Y_{n+1}) = \int_{A(Y_{n+1})} dx,$$

and X_{n+1} is a random variable uniformly distributed over $A(Y_{n+1})$.

b) Verify that the sequence formed by the pairs (X_n, Y_n) is a Markov process with conditional density

$$f_{X_{n+1}, Y_{n+1} | X_n, Y_n}(x', y' | x, y) = \frac{1}{f_X(x)} I[0 \leq y' \leq f_X(x)] \frac{1}{S(y')} I[y' \leq f_X(x')] \tag{5.73}$$

where $I[\cdot]$ denotes the indicator function. For example

$$I[y' \leq f_X(x')] = \begin{cases} 1 \text{ for } y' \leq f_X(x') \\ 0 \quad \text{otherwise} . \end{cases}$$

c) Show that if the pair (X_n, Y_n) has the density $f_{X,Y}(x, y)$ given by (5.72), then the pair (X_{n+1}, Y_{n+1}) has the same density. To do so, it is recommended that you should start from the joint density of $(X_{n+1}, Y_{n+1}, X_n, Y_n)$ specified by the transition density (5.73) and density (5.72) for (X_n, Y_n) and verify that the marginal density of (X_{n+1}, Y_{n+1}) is (5.72).

Ultimately, in the slice method, the auxiliary random variables Y_n, $1 \leq n \leq N$ are discarded and only the X_n samples are retained. One advantage of this technique is that unlike the acceptance/rejection method, no random variable is discarded. The method falls in the general class of Markov Chain Monte Carlo (MCMC) techniques even though, strictly speaking, the pair (X_n, Y_n) takes continuous values and does not belong to a discrete set, so it is not truly a Markov chain. It has also the property, that you are not requested to prove, that if $f_X(x)$ is bounded and has a bounded support, then if the joint density of the initial pair (X_1, Y_1) is arbitrary, the density of (X_n, Y_n) converges asymptotically to the density (5.72).

d) To illustrate the slice method, consider the triangular density

$$f_X(x) = \begin{cases} 1 - |x| & -1 \leq x \leq 1 \\ 0 & \text{otherwise} . \end{cases} \tag{5.74}$$

shown in Fig. 5.11. Verify that the slice corresponding to Y_n is

$$A(Y_n) = \{x : -(1 - Y_n) \leq x \leq 1 - Y_n\},$$

and that its size is $S(Y_n) = 2(1 - Y_n)$. Write a MATLAB program generating $N = 10^6$ random variables X_n with density (5.74) by using the slice method. To initialize the algorithm, set $Y_1 = 0$ and let X_1 be uniformly distributed over $[-1, 1]$. Generate a histogram of random variables X_n, say with 40 bins, and verify it has the triangular shape of Fig. 5.11.

5.18 Consider an arbitrary state i of an irreducible aperiodic finite Markov chain. We seek to show that there exists two coprime excursion lengths m_1 and m_2 such that $f_{ii}(m_1) > 0$ and $f_{ii}(m_2) > 0$.

Fig. 5.11 Triangular
density $f_X(x)$

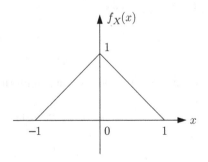

a) Prove first that there exists a finite number K of excursion lengths m_k, $1 \leq k \leq K$ such that $f_{ii}(m_k) > 0$ and such that the m_k's are coprime. *Hint:* Start with an arbitrary m_1 such that $f_{ii}(m_1) > 0$. If $m_1 = 1$, we are done. Otherwise, there must be an excursion length m_2 with $f_{ii}(m_2) > 0$ such that the greatest common divisor (gcd) of m_1 and m_2 satisfies $\gcd(m_1, m_2) < m_1$. If $\gcd(m_1, m_2) = 1$ we are done, otherwise there must be an excursion length m_3 such that $f_{ii}(m_3) > 0$ and $\gcd(m_1, m_2, m_3) < \gcd(m_1, m_2)$. If this process is continued, verify that only a finite number K of steps are needed to achieve $\gcd(m_k, 1 \leq k \leq K) = 1$.

b) Since the excursion lengths m_k, $1 \leq k \leq k$ of part a) are coprime, by Bezout's identity, there exists integers (possibly negative) a_k such that

$$\sum_{k=1}^{K} a_k m_k = 1.$$

Show that the a_ks with $2 \leq k \leq K$ can all be selected positive as long as a_1 is allowed to be negative, and let

$$m_2' = \sum_{k=2}^{K} a_k m_k.$$

Prove that $f_{ii}(m_2') > 0$ and m_1 and m_2' are coprime.

References

1. J.-P. Aubin and I. Ekeland, *Applied Nonlinear Analysis.* New York: John Wiley, 1984. Reprinted by Dover Publications, Mineola NY, 2006.
2. G. A. Jones and J. M. Jones, *Elementary Number Theory.* New York: Springer Verlag, 1998.
3. A. J. Laub, *Matrix Analysis for Scientists & Engineers.* Philadelphia, PA: Soc. Industrial and Applied Math., 2005.
4. L. Lovász, "Random walks on graphs: a survey," in *Combinatorics, Paul Erdös is Eighty* (D. Miklós, V. T. Sós, and T. Szönyi, eds.), vol. 2, pp. 1–46, Budapest, Hungary: Janos Bolyai Mathematical Society, 1993.
5. J. W. M. Bergmans, *Digital Baseband Transmission and Recording.* Dordrecht, The Netherlands: Kluwer Acad. Publishers, 1996.
6. J. G. Kemeny and J. L. Snell, *Finite Markov Chains, 3rd edition.* New York, NY: Springer Verlag, 1983.
7. P. Brémaud, *Markov Chains: Gibbs Fields, Monte Carlo Simulations, and Queues.* New York, NY: Springer Verlag, 1999.
8. S. Ross, *Stochastic Processes, 2nd edition.* New York, NY: J. Wiley, !996.
9. E. Çinlar, *Introduction to Stochastic Processes.* Englewood Cliffs, NJ: Prentice-Hall, 1975. Reprinted by Dover Publications, Mineola NY, 2013.
10. D. A. Levin, Y. Peres, and E. L. Wilmer, *Markov Chains and Mixing Times.* Providence, RI: American Mathematical Society, 2009.
11. W. Whitt, "Stochastic models I." Lecture notes for course IEOR 6711, Columbia University, 2013.
12. J. E. Cohen, Y. Iwasa, G. Rautu, M. B. Ruskai, E. Seneta, and G. Zbaganu, "Relative entropy under mappings by stochastic matrices," *Linear Algebra and its Appl.*, vol. 179, pp. 211–235, 1993.
13. G. D. Birkhoff, "Extensions of Jentzsch's theorem," *Trans. American Math. Soc.*, vol. 85, pp. 219–227, May 1957.
14. J. R. Barry, E. A. Lee, and D. G. Messerschmitt, *Digital Communications, 3rd edition.* New York: Springer Verlag, 2003.
15. M. K. Simon, S. M. Hinedi, and W. C. Lindsey, *Digital Communication Techniques: Signal Design and Detection.* Englewood Cliffs, NJ: Prentice-Hall, 1995.
16. M. Kac, "Random walk and the theory of Brownian motion," *Amer. Math. Monthly*, vol. 54, pp. 369–391, Sep. 1947. Reprinted in N. Wax, Ed., Selected Papers in Noise and Stochastic Processes, Dover Publ., Mineola, NY, 1954.
17. M. J. Klein, "Entropy and the Ehrenfest urn model," *Physica*, vol. 22, pp. 569–575, 1956.

18. P. Diaconis, "The Markov chain Monte Carlo revolution," *Bull. American Math. Soc.*, vol. 46, pp. 179–206, Apr. 2009.
19. N. Metropolis, A. W. Rosenbluth, M. N. Rosenbluth, A. H. Teller, and E. Teller, "Equations of state calculations by fast computing machines," *J. Chemical Phys.*, vol. 21, pp. pp. 1087–1092, 1953.
20. W. K. Hastings, "Monte Carlo sampling methods using Markov chains and their applications," *Biometrika*, vol. 57, pp. 97–109, Apr. 1970.
21. R. J. Baxter, *Exactly Solved Models in Statistical Mechanics*. San Diego, CA: Academic Press, 1982.
22. R. M. Neal, "Slice sampling," *Annals of Statistics*, vol. 31, pp. 705–767, 2003.

Wiener Process and White Gaussian Noise

6

6.1 Introduction

The Wiener process is a mathematical idealization of the physical Brownian motion describing the movement of suspended particles subjected to collisions by smaller atoms in a liquid. This phenomenon, first studied by the botanist Robert Brown in 1827, was investigated by Einstein [1] in one of his three famous 1906 papers, followed by Smoluchowski [2], Langevin [3], and other physicists. See [4] for an account of dynamical theories of Brownian motion from a mathematical viewpoint.

The Wiener process was first described and studied by MIT mathematician Norbert Wiener. This process, which is introduced in Sect. 6.2, has independent increments process and is also Gaussian and Markov. It has many fascinating properties since its sample paths are continuous almost surely, yet nowhere differentiable. In fact, its sample paths are fractal curves [5, Chap. 4]. This process is the simplest of a wider class of Markov processes called diffusion processes which share with it the property of having continuous but nowhere differentiable sample paths almost surely, and it plays a key role in their construction and characterization.

The construction of the Wiener process is discussed in Sect. 6.3. It turns out that the Wiener process can be constructed in two different ways. The first construction, as a limit of a scaled random walk, finds its origin in the work of Louis Bachelier [6] in his pioneering study of mathematical models of stock price fluctuations. The second method, due to Paul Lévy, which is particularly convenient for computer simulation purposes, expresses the Wiener process as the limit of a random series expressed in terms of the integrated Haar basis [5]. The sample path properties of the Wiener process are examined in Sect. 6.4. In spite of the fact that the Wiener process does not have a finite total variation, it is possible to define a generalization of the Stieltjes integral, called the Wiener integral, which as shown in Sect. 6.5 produces zero-mean Gaussian random variables whose correlation viewed as an inner product forms an isometry with the space of square-integrable functions over the interval of integration, say $[0, T]$. In this context, although the Wiener process is not mean-square differentiable in the sense of Chap. 4, its formal derivative called white Gaussian noise (WGN) yields the same inner-product expressions for integrated WGN as those obtained by using Wiener integrals. This provides a justification for the informal use of WGN used by engineers in linear systems models. This is illustrated in Sect. 6.6 by a description of the Nyquist–Johnson WGN model of thermal noise. This model is used to analyze a simple RC circuit, and in this context it is shown that in thermal equilibrium, the capacitor voltage forms an Ornstein–Uhlenbeck process.

© Springer Nature Switzerland AG 2020
B. C. Levy, *Random Processes with Applications to Circuits and Communications*,
https://doi.org/10.1007/978-3-030-22297-0_6

6.2   Definition and Basic Properties of the Wiener Process

The Wiener process $W(t)$ is defined for $t \geq 0$ and satisfies the following axioms:

i) $W(0) = 0$;
ii) $W(t)$ has stationary independent increments;
iii) $W(t)$ is $N(0, t)$ distributed.

Since $W(0) = 0$ is a known value and $W(t)$ has independent increments, it is a Markov process, as was shown at the end of Chap. 4. Because of the stationarity of the increments, $W(t) - W(s)$ has the same distribution as $W(t - s) \sim N(0, t - s)$ for $t \geq s$, so the transition density of $W(t)$ viewed as a Markov process is given by

$$q(w_t, t; w_s, s) = \frac{1}{(2\pi(t - s))^{1/2}} \exp\left(-\frac{(w_t - w_s)^2}{2(t - s)}\right). \tag{6.1}$$

The probability density

$$f_{W(t)}(w_t, t) = \frac{1}{(2\pi(t - s))^{1/2}} \exp\left(-\frac{w_t^2}{2t}\right) \tag{6.2}$$

of $W(t)$ satisfies the propagation equation

$$f_{W(t)}(w_t, t) = \int_{-\infty}^{\infty} q(w_t, t; w_s, s) f_{W(s)}(w_s, s) dw_s \tag{6.3}$$

for $t \geq s$. Also, by observing that $q(w_t, t; w_s, s)$ is the Green's function of the heat operator

$$L = \frac{\partial}{\partial t} - \frac{1}{2}\frac{\partial^2}{\partial w_t^2} \tag{6.4}$$

since it satisfies

$$Lq(w_t, t; w_s, s) = 0 \tag{6.5}$$

for $t > s$, with initial condition

$$\lim_{t \to s} q(w_t, t; w_s, s) = \delta(w_t - w_s),$$

we deduce that the probability density of $W(t)$ obeys the heat equation

$$L f_{W(t)}(w_t, t) = 0 \tag{6.6}$$

for $t > s$, with initial condition $f_{W(s)}(w_s, s)$.

In addition to being Markov, it was observed in Example 4.8 that $W(t)$ is a Gaussian process, since its finite joint densities are Gaussian, with zero mean and autocorrelation

$$R_W(t, s) = \min(t, s). \tag{6.7}$$

If we consider therefore the vector

$$\mathbf{W} = \left[\, W(t_1) \, \ldots \, W(t_k) \, \ldots \, W(t_N) \, \right]^T$$

formed by Wiener process samples at times $t_1 < \ldots < t_k < \ldots < t_N$, $\mathbf{W}$ is $N(\mathbf{0}, \mathbf{K}_W)$ distributed, where the covariance matrix

$$\mathbf{K}_W = \begin{bmatrix} t_1 \, t_1 & \ldots & t_1 \\ t_1 \, t_2 \, t_2 \ldots & t_2 \\ t_1 \, t_2 \, t_3 \ldots & t_3 \\ \vdots \, \vdots & \vdots \, \vdots \\ t_1 \, t_2 \, t_3 \ldots & t_N \end{bmatrix}$$

has a chevron-type pattern, with entries $K_{ij} = K_{ii} = K_{ji}$ for $j > i$. In this respect, it is also worth noting that as long as $t_1 > 0$, the covariance matrix $\mathbf{K}_W$ is positive definite, since given an arbitrary nonzero vector

$$\mathbf{a}^T = \left[\, a_1 \, \ldots \, a_k \, \ldots \, a_N \, \right]^T$$

of $\mathbb{R}^N$, we have

$$\mathbf{a}^T \mathbf{K}_W \mathbf{a} = \sum_{k=1}^{N} \sum_{\ell=1}^{N} a_k a_\ell \min(t_k, t_\ell)$$

$$= \sum_{k=1}^{N} \sum_{\ell=1}^{N} a_k a_\ell \int_0^{t_N} 1_{t_k}(s) 1_{t_\ell}(s) ds$$

$$= \int_0^{t_N} \left(\sum_{k=1}^{N} a_k 1_{t_k}(s) \right)^2 ds > 0 , \tag{6.8}$$

where

$$1_b(s) = u(s) - u(s - b)$$

denotes the indicator function of set $[0, b]$.

6.3   Constructions of the Wiener Process

6.3.1   Scaled Random Walk Model

The Wiener process can be constructed in two different ways, which separately illustrate several important features of this process. The first method, due to Bachelier [6] who was interested in describing fluctuations of stock market prices, models Brownian motion as a scaled random walk. Specifically, let X_k be a sequence of iid binary random variables such that

$$P[X_k = 1] = P[X_k = -1] = \frac{1}{2} .$$

Then

$$Z_n = \sum_{k=1}^{n} X_k$$

can be viewed as the position at time n of a particle undergoing a random walk on an infinite linear graph, starting from position $Z_0 = 0$ at time 0. The mean and variance of random variables X_k are $m_X = 0$ and $K_X = 1$. By the central limit theorem, we know that as n tends to infinity

$$\frac{Z_n}{n^{1/2}} \xrightarrow{d} \zeta, \tag{6.9}$$

where ζ is $N(0, 1)$ distributed. This suggests that the process

$$W_n(t) = \frac{Z_{\lfloor nt \rfloor}}{n^{1/2}} \tag{6.10}$$

will tend to the Wiener process $W(t)$ in distribution as n tends to infinity. Note indeed that

$$W_n(t) = \frac{Z_{\lfloor nt \rfloor}}{(\lfloor nt \rfloor)^{1/2}} \frac{(\lfloor nt \rfloor)^{1/2}}{n^{1/2}} \xrightarrow{d} t^{1/2} \zeta \tag{6.11}$$

as n tends to infinity, where $t^{1/2}\zeta$ is $N(0, t)$ distributed, since ζ is $N(0, 1)$ distributed. Also, observe that for $n' > m' \geq n > m$, the increments

$$Z_{n'} - Z_{m'} = \sum_{m'+1}^{n'} X_k \ \text{ and } \ Z_n - Z_m = \sum_{\ell=m+1}^{n} X_\ell$$

are nonoverlapping sums of iid random variables. So, the increments of the process $W_n(t)$ over disjoint intervals are independent, and thus $W(t)$ has independent increments. We also note that the increment $Z_n - Z_m$ involves a sum of $n - m$ independent binary valued random variables X_ℓ, so it has the same distribution as Z_{n-m}, and thus $W_n(t) - W_n(s)$ has the same distribution as $W_n(t-s)$, and $W(t) - W(s)$ has the same distribution as $W(t - s)$, i.e., the increments are stationary.

It is worth noting that the sample paths of approximating process $W_n(t)$ are piecewise constant over intervals $[k/n, (k + 1)/n)]$ with k integer, so they can be viewed as generated by a sample-and-hold device with hold time $\Delta t = 1/n$. At times $t = k/n$ with k integer, $W_n(t)$ jumps by $\pm \Delta x$, where $\Delta x = 1/\sqrt{n} = \sqrt{\Delta t}$. Since

$$\frac{\Delta x}{\Delta t} = \sqrt{n} \to \infty$$

as n tends to infinity, it is reasonable to expect that the sample paths of the limit process $W(t)$ will not be differentiable.

6.3.2 Paul Lévy's Construction

The random walk approximation of the Wiener process illustrates well the independent increments and Gaussian properties of the process, but the sample paths of $W_n(t)$ are discontinuous, even though as we shall see below, the sample paths of $W(t)$ are continuous almost surely. To illustrate the continuity of the sample paths of the Wiener process, Paul Lévy proposed a construction over a finite interval, say $[0, 1]$, where $W(t)$ is obtained as a uniform limit of continuous random functions. This is accomplished by generating recursively the samples of $W(t)$ at the dyadic points

$$\mathcal{D}_n = \{\frac{k}{2^n}, 0 \leq k \leq 2^n\} \tag{6.12}$$

for increasing values of n and performing a linear interpolation between dyadic points sample values. To do so, it is convenient to introduce the Haar orthonormal basis $\{\phi_0, \phi_{n,k}, n \geq 1, 1 \leq k \leq 2^{n-1}\}$ of the space of square-integrable functions $L^2[0, 1]$ as $\phi_0(t) = 1$ and for $\ell = 2k - 1$,

$$\phi_{n,k}(t) = \begin{cases} 2^{(n-1)/2} & \frac{\ell-1}{2^n} \leq t < \frac{\ell}{2^n} \\ -2^{(n-1)/2} & \frac{\ell}{2^n} \leq t < \frac{\ell+1}{2^n} \\ 0 & \text{otherwise.} \end{cases}$$

It is not difficult to verify that this basis is orthonormal and complete (this last property is due to the fact that Haar functions ϕ_0 and $\phi_{m,k}$ for $1 \leq m \leq n$ span piecewise constant functions over dyadic intervals $[\ell/n, \ell+1)/n)$, which are dense in $L^2[0, 1]$ as n tends to infinity). Thus any function $f \in L^2[0, 1]$ can be represented as

$$f(t) = < f, \phi_0 > \phi_0(t) + \sum_{n=1}^{\infty} \sum_{k=1}^{2^{n-1}} < f, \phi_{n,k} > \phi_{n,k}(t),$$

where

$$< f, g >= \int_0^1 f(t)g(t)dt$$

denotes the inner product of $L^2[0, 1]$. Furthermore, for arbitrary functions f and g of $L^2[0, 1]$, Parseval's identity implies

$$< f, g >=< f, \phi_0 >< g, \phi_0 > + \sum_{n=1}^{\infty} \sum_{k=1}^{2^{n-1}} < f, \phi_{n,k} >< g, \phi_{n,k} > . \tag{6.13}$$

The functions $\phi_{n,k}$ of the Haar basis are discontinuous, so we consider the basis of Schauder functions

$$\psi_0(t) = \int_0^t \phi_0(u)du =< I_t, \phi_0 >$$

$$\psi_{n,k}(t) = \int_0^t \phi_{n,k}(u)du =< I_t, \phi_{n,k} >$$

with $n \geq 1$ and $1 \leq k \leq 2^{n-1}$ obtained by integrating the Haar functions from 0 to t. These functions are no longer orthogonal, but they are continuous: $\psi_0(t) = t$ for $0 \leq t \leq 1$ and the functions $\psi_{n,k}(t)$ have a tent shape over interval $[(k-1)/2^{n-1}, (k+1)/2^{n-1}]$, as shown in Fig. 6.1.

Then the Wiener process $W(t)$ over $[0, 1]$ is approximated by the sequence

$$W_n(t) = X_0\psi_0(t) + \sum_{m=1}^{n} \sum_{k=1}^{2^{m-1}} X_{m,k}\psi_{m,k}(t), \tag{6.14}$$

Fig. 6.1 Schauder basis
function $\psi_{n,k}(t)$

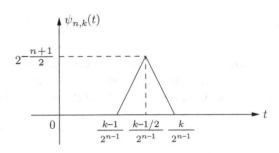

Fig. 6.2 Wiener process
approximation obtained by
dyadic sample points
interpolation

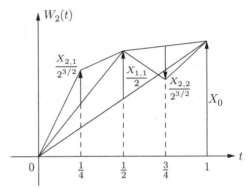

where the random variables $X_{m,k}$ are independent and $N(0, 1)$ distributed. To interpret this approximation, observe that $W_n(t)$ can be viewed as obtained by selecting the values of the Wiener process W at dyadic points $d \in \mathcal{D}_n$, and then interpolating linearly between the samples $W_n(d) = W(d)$. The selection of the samples at $\mathcal{D}_n$ is sequential in n, so that it is assumed that the samples at points of $\mathcal{D}_{n-1}$ have already been constructed at step $n - 1$. Then the samples $W(d)$ at points $\mathcal{D}_n \backslash \mathcal{D}_{n-1}$ (the points of $\mathcal{D}_n$ not already in $\mathcal{D}_{n-1}$) are chosen so that the joint distribution of the samples are those of the Wiener process, i.e., the joint distribution is Gaussian with zero mean (which is a consequence of the fact that the $X_{m,k}$'s are Gaussian with zero mean), with autocorrelation (6.7). This is accomplished by selecting

$$W(d) = \frac{1}{2}\Big(W(d - 2^{-n}) + W(d + 2^{-n})\Big) + 2^{-(n+1)/2}X_{n,k} \tag{6.15}$$

for $d = (2k - 1)/2^n \in \mathcal{D}_n \backslash \mathcal{D}_{n-1}$. The procedure is illustrated in Fig. 6.2 for $n = 1$ and 2. We start by selecting $W(1) = X_0 \sim N(0, 1)$. Then

$$W(1/2) = \frac{1}{2}W(1) + \frac{1}{2}X_{1,1}$$

is obtained by observing that $W(1)/2$ has exactly the desired correlation with $W(1)$, but its variance is only $1/4$ (instead of the desired $1/2$), which is adjusted by adding the component $X_{1,1}/2$, which does not change the correlation between $W(1/2)$ and $W(1)$, since $X_{1,1}$ is independent of X_0. Proceeding in a similar manner, the values of $W(d)$ at $\{1, 4, 3, 4\} = \mathcal{D}_2 \mathcal{D}_1$ are given by

$$W(1/4) = \frac{1}{2}W(1/2) + \frac{1}{2^{3/2}}X_{2,1}$$

$$W(3/4) = \frac{1}{2}(W(1/2) + W(1)) + \frac{1}{2^{3/2}}X_{2,2} .$$

The random interpolating curve $W_n(t)$ is continuous, but its slope changes at all points of $\mathcal{D}_n$, and since dyadic points are dense in $[0, 1]$, it is already reasonable to expect that the limit $W(t)$ will be continuous, but nowhere differentiable.

To verify that the sequence $W_n(t)$ converges to a Wiener process $W(t)$, note first that the autocorrelation of $W_n(t)$ is given by

$$R_{W_n}(t, s) = E[W_n(t)W_n(s)]$$

$$= \psi_0(t)\psi_0(s) + \sum_{m=1}^{n} \sum_{k=1}^{2^{m-1}} \psi_{m,k}(t)\psi_{m,k}(s), \qquad (6.16)$$

where to go from the first to the second line, we have used the independence of random variables $X_{m,k}$ in (6.14). Taking the limit gives

$$\lim_{n \to \infty} R_{W_n}(t, s) = \psi_0(t)\psi_0(s) + \sum_{m=1}^{\infty} \sum_{k=1}^{2^{m-1}} \psi_{m,k}(t)\psi_{m,k}(s)$$

$$= <I_t, \phi_0><I_s, \phi_0> + \sum_{m=1}^{\infty} \sum_{k=1}^{2^{m-1}} <I_t, \phi_{m,k}><I_s, \phi_{m,k}>$$

$$= <I_t, I_s> = \min(s, t) = R_W(t, s), \qquad (6.17)$$

where to go from the second to the third line we have used identity (6.13). This shows that the autocorrelation of $W_n(t)$ tends to the Wiener process autocorrelation as n tends to infinity. This implies that the finite joint densities of $W_n(t)$ converges in distribution to those of $W(t)$. Note indeed that for $0 < t_1 < \ldots < t_k < \ldots < t_N \le 1$, the joint characteristic function of samples $W_n(t_k)$ with $1 \le k \le N$ can be expressed as

$$\Phi_{\mathbf{W}_n}(\mathbf{u}) = \exp\left(-\mathbf{u}^T \mathbf{K}_{\mathbf{W}_n} \mathbf{u}/2\right),$$

where $\mathbf{K}_{\mathbf{W}_n}$ is the covariance matrix of sample vector

$$\mathbf{W}_n = \left[W_n(t_1) \ldots W_n(t_k) \ldots W_n(t_N) \right]^T$$

But as n tends to infinity, $\mathbf{K}_{\mathbf{W}_n}$ tends to the covariance matrix $\mathbf{K}_{\mathbf{W}}$ of

$$\mathbf{W} = \left[W(t_1) \ldots W(t_k) \ldots W(t_N) \right]^T, \qquad (6.18)$$

so $\Phi_{\mathbf{W}_n}(\mathbf{u})$ converges pointwise to the characteristic function $\Phi_{\mathbf{W}}(\mathbf{u})$ of Wiener process sample vector (6.18), so the finite joint distributions of W_n converge in distribution to those of W.

In addition, by using the Borel–Cantelli lemma, it can be shown [5, pp. 24–25] that the convergence of continuous random function $W_n(t)$ to $W(t)$ is uniform almost surely. This implies that $W(t)$ is continuous almost surely.

6.4   Sample Path Properties

6.4.1   Hölder Continuity

Although the construction we just considered indicates that the sample paths of the Wiener process are almost surely continuous, it turns out that the sample paths satisfy a stronger property. We start by noting that a function $f(t)$ is Hölder continuous with exponent γ if there exists a constant C such that for all t and s

$$|f(t) - f(s)| \leq C|t - s|^{\gamma} .$$

Since Hölder continuity characterizes the speed at which $f(t)$ tends towards $f(s)$ as t approaches s, it is stronger than pointwise or uniform continuity. To characterize the Hölder continuity of the Wiener process sample paths, we employ *Kolmogorov's continuity theorem* [7]: if a process $X(t)$ satisfies

$$E[|X(t) - X(s)|^{\alpha}] \leq D|t - s|^{1+\beta}, \tag{6.19}$$

then $X(t)$ is almost surely Hölder continuous with exponent $\gamma \in [0, \beta/\alpha)$. For the Wiener process, by using Isserlis's moment expansion formula (see Problem 4.21) for zero mean Gaussian random variables, we find

$$E[|W(t) - W(s)|^{2n}] \leq D(n)|t - s|^{n} , \tag{6.20}$$

where $D(n)$ is a constant depending on n. Thus $W(t)$ satisfies the continuity theorem conditions with $\alpha = 2n$ and $\beta = n - 1$. Letting n become large, we deduce that the Wiener process sample paths are almost surely Hölder continuous with exponent $\gamma < 1/2$.

6.4.2   Non-differentiability

In spite of their continuity, quite strikingly, the sample paths have the property of being nowhere differentiable with probability 1. Intuitively, this property should not be too surprising, since the construction of the Wiener process by dyadic sample points interpolation showed that the Wiener process constantly changes direction. To prove the nondifferentiability property, we follow the argument given in [8]. Suppose that $W(t)$ is differentiable at point $t_0 \in (0, 1)$. Then there exists $\epsilon > 0$ and $C > 0$ such that

$$|W_t - W_{t_0}| \leq C|t - t_0| \tag{6.21}$$

for $t_0 - \epsilon < t < t_0 + \epsilon$. But we can find an integer n_0 such that for all $n \geq n_0$, four points $(k + i)/4^n$ with $i = 0, 1, 2, 3$ belong to $(t_0 - \epsilon, t_0 + \epsilon)$. By the triangle inequality, this implies

$$|W((k + i + 1)/4^n) - W((k + i)/4^n)| \leq 4C/4^n \text{ for } i = 0, 1, 2 . \tag{6.22}$$

Let $B_{n,k}$ denote the event (6.22). We show that the probability that $B_{n,k}$ occurs infinitely often is zero. If $B_n = \cup_{0 \leq k \leq 4^n - 3} B_{n,k}$, by the Borel–Cantelli lemma, it is enough to prove that

$$\sum_{n=n_0}^{\infty} P(B_n) < \infty . \tag{6.23}$$

By observing that the three increments appearing in (6.22) are independent and $N(0, 1/4^n)$ distributed, we deduce that

$$P(B_{n,k}) \leq D/2^{3n},$$

where $D = (4C/(2\pi)^{1/2})^3$, so that

$$P(B_n) \leq 4^n \times D/2^{3n} = D/2^n,$$

which is summable in n so that (6.23) holds.

6.4.3 Quadratic and Total Variation

An important property of the Wiener process is that it has finite quadratic variation. Specifically, consider an arbitrary interval $[a, b]$ and consider a uniform partition $t_0 = a < \ldots < t_k < \ldots < t_{N_n} = b$ of this interval into N_n subintervals with mesh size $h_n = t_{k+1} - t_k = (b - a)/N_n$. The quadratic variation of W with respect to this partition is defined as

$$Q_n = \sum_{k=1}^{N_n} (W(t_k) - W(t_{k-1}))^2 . \tag{6.24}$$

Then Q_n converges in the mean-square sense to $b - a$ as h_n tends to zero. To see why this is the case, note first that the increments

$$Y_k = (W(t_k) - W(t_{k-1}))^2$$

are iid with mean

$$m_Y = E[Y_k] = t_k - t_{k-1} = h_n$$

and variance

$$E[(Y_k - m_Y)^2] = E[Y_k^2] - m_Y^2 = 2h_n^2 , \tag{6.25}$$

where we have used Isserlis's formula (see Problem 4.21) for evaluating the fourth order moments of Gaussian random variables. Then

$$E[(Q_n - (b - a))^2] = E\Big[\Big(\sum_{k=1}^{N_n}(Y_k - m_Y)\Big)^2\Big]$$

$$= \sum_{k=1}^{N_n}\sum_{\ell=1}^{N_n} E[(Y_k - m_Y)(Y_\ell - m_Y)]$$

$$= \sum_{k=1}^{N_n} E[(Y_k - m_Y)^2] = 2N_n h_n^2 = 2(b - a)h_n \to 0 \tag{6.26}$$

where we have used the independence of the Y_k's to go from the second to the third line. This shows that Q_n converges in the mean-square to $b - a$.

It can also be shown that Q_n tends to $b - a$ almost surely if the mesh size h_n tends to zero sufficiently rapidly. Let $D_n = Q_n - (b - a)$ represent the deviation of Q_n from its limit. The Chebyshev inequality implies

$$P[|D_n| > \epsilon] \leq \frac{E[D_n^2]}{\epsilon^2} = \frac{2(b-a)}{\epsilon^2} h_n .$$

Hence if the mesh size satisfies

$$\sum_{n=1}^{\infty} h_n < \infty , \tag{6.27}$$

we have

$$\sum_{n=1}^{\infty} P[|D_n| > \epsilon] \leq \frac{2(b-a)}{\epsilon^2} \sum_{n=1}^{\infty} h_n , \tag{6.28}$$

so that by the Borel Cantelli lemma, Q_n converges to $b - a$ almost surely. Note in particular that the dyadic subdivision with $h_n = 2^{-n}$ satisfies (6.27), since in this case

$$\sum_{n=1}^{\infty} 2^{-n} = 1 .$$

The fact that the limit of the quadratic variation is finite for a sequence of uniform partitions whose mesh size h_n satisfies (6.27) implies that for such a sequence, the total variation

$$T_n = \sum_{k=1}^{N_n} |W(t_k) - W(t_{k-1})|$$

will tend to infinity almost surely. Note indeed that

$$Q_n \leq \max_{1 \leq k \leq N_n} |W(t_k) - W(t_{k-1})| T_n. \tag{6.29}$$

But since $W(t)$ is continuous over $[a, b]$, it is uniformly continuous over this interval, so

$$\lim_{n \to \infty} \max_{1 \leq k \leq N_n} |W(t_k) - W(t_{k-1})| = 0$$

as h_n tends to zero. Since the limit of Q_n on the left-hand side of (6.29) is $b - a > 0$, this implies that the total variation T_n must tend to infinity.

6.5   Wiener Integrals and White Gaussian Noise

Consider a deterministic (nonrandom) function $h \in L^2[0, T]$. We seek to define the integral

$$I(h) = \int_0^T h(t) dW(t) , \tag{6.30}$$

where $W(t)$ is a Wiener process. Since W has infinite total variation, the integral (6.30) cannot be defined as a Stieltjes integral. The Wiener integral (6.30) is constructed by observing that the function h can be approximated by a sequence of piecewise constant functions h_n converging to h in $L^2[0, T]$. Then, if $h_n(t) = c_k$ for $t_k \leq t < t_{k+1}$, with $1 \leq k \leq N - 1$, $t_1 = 0$, and $t_N = T$, $I(h_n)$ can be defined as

$$I(h_n) = \sum_{k=1}^{N} c_k(W(t_{k+1}) - W(t_k)),$$

which is clearly zero-mean and Gaussian with variance

$$E[I^2(h_n)] = \sum_{k=1}^{N} c_k^2(t_{k+1} - t_k) = \int_0^T h_n^2(t)dt = ||h_n||^2.$$

By observing that

$$E[(I(h_n) - I(h_m))^2] = \int_0^T (h_n(t) - h_m(t))^2 dt$$

tends to zero as n and m tend to infinity, we deduce that $I(h_n)$ forms a Cauchy sequence in the space of random variables with finite second-order moments, so

$$I(h_n) \overset{m.s.}{\to} I(h),$$

where the random variable $I(h)$ is Gaussian with zero mean and variance

$$E[I^2(h)] = \int_0^T h^2(t)dt = ||h||^2.$$

This last property is due to the fact that if $I(h_n)$ converges to $I(h)$ in the mean-square, it converges in distribution. But the characteristic function

$$\Phi_{I(h_n)}(u) = \exp(-u^2||h_n||^2/2)$$

converges pointwise to

$$\Phi_{I(h)}(u) = \exp(-u^2||h||^2/2),$$

which indicates that $I(h)$ is $N(0, ||h||^2)$ distributed. The random variable $I(h)$ is called the *Wiener integral* of function $h \in L^2[0, T]$. An interesting consequence of this construction is that we consider the correlation $E[I(h_1)I(h_2)]$ of random variables $I(h_1)$ and $I(h_2)$, by observing that

$$I(h_1)I(h_2) = \frac{1}{2}[(I(h_1) + I(h_2))^2 - I^2(h_1) - I^2(h_2)]$$

we deduce that

$$E[I(h_1)I(h_2)] = \int_0^T h_1(t)h_2(t)dt.$$

This indicates that if we consider the inner product $E[X_1 X_2]$ in the space of random variables with finite second-order moments, the inner product of the Wiener integrals $I(h_1)$ and $I(h_2)$ forms an isometry with the inner product

$$<h_1, h_2> = \int_0^T h_1(t)h_2(t)dt$$

of functions of $L^2[0, T]$.

We are now in a position to introduce the concept of white Gaussian noise (WGN), the formal derivative of the Wiener process $W(t)$. Note that $W(t)$ is not mean-square differentiable in the sense of Chap. 4, since

$$\frac{\partial}{\partial s} R_W(t, s) = u(t - s) \tag{6.31}$$

is discontinuous at $t = s$, so the second cross derivative of $R_W(t, s)$ with respect to t and s does not exist as an ordinary function. Formally, if we define white Gaussian noise $N(t)$ as $N(t) = dW/dt$, its autocorrelation would be

$$R_N(t, s) = \frac{\partial^2}{\partial t \partial s} R_W(t, s) = \delta(t - s) \tag{6.32}$$

which is a Dirac delta function, thereby indicating that white Gaussian noise is not an ordinary random process. Nevertheless, suppose that we consider a white noise integral of the form

$$I = \int_0^T h(t) N(t) dt \tag{6.33}$$

and apply ordinary rules of mean-square calculus based on the generalized autocorrelation function (6.32) to evaluate its variance. We obtain

$$E[I^2] = \int_0^T \int_0^T h(t) R_N(t, s) h(s) dt ds$$
$$= \int_0^T h^2(t) dt , \tag{6.34}$$

which is consistent with the variance of the Wiener integral $I(h)$. This indicates that as long as we realize that we are actually evaluating Wiener integrals, there is really no harm in writing $I(h)$ in the WGN form (6.33). This notation is particularly convenient when analyzing linear differential systems of the form

$$\frac{dX}{dt} = a(t) X(t) + b(t) U(t) , \tag{6.35}$$

for $t \geq 0$, with initial condition $X(0)$. When the input $U(t)$ is deterministic, the solution of this equation can be expressed as

$$X(t) = \phi(t, 0) X(0) + \int_0^t h(t, s) U(s) ds , \tag{6.36}$$

where the transition function $\phi(t, s)$ solves

$$\frac{d\phi}{dt} = a(t) \phi(t, s)$$

for $t \geq s$ with initial condition $\phi(s, s) = 1$, and

$$h(t, s) = \phi(t, s) b(s)$$

is the time-varying impulse response of the system, i.e., the response at time t to an impulse at time $s \leq t$. When the input $U(t)$ is a WGN, the solution (6.36) remains valid formally, as long as we realize it really corresponds to

$$X(t) = \phi(t, 0)X(0) + \int_0^t h(t, s)dW(s) \, .$$

For this reason, in this case the differential equation (6.35) is often written in the form of a Langevin equation

$$dX(t) = a(t)X(t)dt + b(t)dW(t),$$

whose notation indicates clearly that its solution needs to be expressed as a Wiener integral.

At this point, it is worth pointing out that the use of Wiener integrals to solve stochastic differential equations is restricted to the linear case. In the nonlinear case, a more general form of stochastic integral developed by Ito [9], where the integrand h is random, is required. This type of stochastic calculus will be described later in Chap. 10. Finally, to address concerns of physically inclined readers who might object to the fact that WGN has infinite power, it will be shown later in Chap. 8 that WGN is really a mathematical idealization of bandlimited white Gaussian noise, which is mean-square differentiable and has finite power. However, calculations involving bandlimited noise are often unwieldy, whereas white noise calculations are elementary. Yet, it will be shown that as long as the physical systems excited by bandlimited noise have a bandwidth which is less than the noise bandwidth, white noise calculations yield exactly the same results as bandlimited noise. Thus both in a mathematical sense and in a physical sense, WGN should be viewed as a convenient computational device used to simplify noise calculations.

6.6 Thermal Noise

In two seminal papers Nyquist [10] and Johnson [11] proposed in 1928 a model of voltage fluctuations across the terminals of a resistor in thermal equilibrium at temperature T. The thermal fluctuations induce a small voltage imbalance across the resistor terminals, so that the noisy resistor can be modeled by the series combination of an ideal resistor with a white Gaussian noise (WGN) voltage source $N(t)$ as shown in Fig. 6.3. The white noise $N(t)$ has zero mean and intensity q, so that

$$E[N(t)N(s)] = q\delta(t - s) \, ,$$

where the intensity q will be determined below by using physical considerations, and where $\delta(t - s)$ denotes the Dirac delta function. Assume that this noisy resistor is connected to a capacitor with capacitance C, and denote by $V(t)$ the voltage across the capacitor terminals. The current flowing through the capacitor is denoted as $I(t)$, as shown in Fig. 6.3. Then by applying the Kirchoff voltage law, we find that

$$N(t) = V(t) + RI(t), \tag{6.37}$$

where the constituent relation of a capacitor implies

$$I(t) = C\frac{dV}{dt} \, ,$$

so that the RC circuit satisfies the first-order differential equation

$$RC\frac{dV}{dt} + V(t) = N(t) \, , \tag{6.38}$$

Fig. 6.3 Noisy resistor
connected to a capacitor

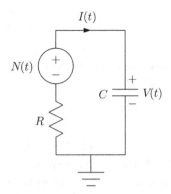

which if we recall that $N(t)$ can be expressed formally as $N(t) = q^{1/2} dW/dt$, can be written equivalently as the Langevin equation

$$dV(t) = -\frac{1}{RC}V(t)dt + \frac{q^{1/2}}{RC}dW(t), \tag{6.39}$$

where $W(t)$ is a standard Wiener process.

By Laplace transforming equation (6.38) we obtain

$$RC(s\hat{V}(s) - V(0)) + \hat{V}(s) = \hat{N}(s), \tag{6.40}$$

where $\hat{V}(s)$ and $\hat{N}(s)$ denote the one-sided Laplace transforms of $V(t)$ and $N(t)$, respectively, and $V(0)$ denotes the initial voltage. It is assumed that $V(0)$ is $N(0, P(0))$ distributed and is independent of $N(t)$ for $t \geq 0$. By solving (6.40), we obtain

$$\hat{V}(s) = \frac{1}{s + 1/(RC)}(V(0) + \frac{1}{RC}\hat{N}(s)). \tag{6.41}$$

Then, by noting that

$$H(s) = \frac{1/(RC)}{s + 1/(RC)}$$

is the Laplace transform of

$$h(t) = \frac{1}{RC}\exp(-t/(RC))u(t),$$

where $u(t)$ denotes the unit step function

$$u(t) = \begin{cases} 1 & t \geq 0 \\ 0 & t < 0, \end{cases}$$

we find that the solution of (6.38) is given by

$$V(t) = \exp(-t/(RC))V(0) + h(t) * N(t)$$

$$= \exp(-t/(RC))V(0) + \frac{1}{RC}\int_0^t \exp(-(t-u)/(RC))N(u)du. \tag{6.42}$$

In this expression, the two terms are Gaussian with zero mean (recall that a white noise integral is just a formal notation for a Wiener integral, which is Gaussian with zero mean). The two terms are also independent since $V(0)$ has been assumed independent of $N(t)$ for $t \geq 0$. Note also that if instead of solving the RC circuit from $t = 0$, we solve it for $t \geq s$, we have the equivalent solution

$$V(t) = \exp(-(t-s)/(RC))V(s) + \frac{1}{RC}\int_s^t \exp(-(t-u)/(RC))N(u)du. \tag{6.43}$$

Accordingly, $V(t)$ is $N(0, P(t))$ distributed, where the variance $P(t)$ is obtained by summing the variances of each term in (6.42). We find

$$P(t) = \exp(-2t/(RC))P(0) + \frac{q}{(RC)^2}\int_0^t \exp(-2(t-u)/(RC))du$$

$$= \exp(-2t/(RC))P(0) + \frac{q}{2RC}\exp(-2(t-u)/(RC)) \;|_{u=0}^{u=t}$$

$$= \frac{q}{2RC} + (P(0) - \frac{q}{2RC})\exp(-2t/(RC)). \tag{6.44}$$

Let $\tau = RC$ denote the time constant of the circuit. The expression (6.44) shows that when t exceeds about 5τ, the variance

$$P(t) \approx P_{ss} = \frac{q}{2RC}. \tag{6.45}$$

Furthermore, if the initial variance $P(0)$ is equal to the steady-state variance P_{ss}, $P(t)$ is constant and equal to the steady-state variance for all times.

Consider now expression (6.43) for $V(t)$, where $t \geq s$. In this expression $V(s)$ is independent of the second term, since $V(s)$ depends only on $V(0)$ and $N(u)$ over $0 \leq u \leq s$. Accordingly, the autocorrelation

$$R_V(t,s) = E[V(t)V(s)]$$

$$= \exp(-(t-s)/(RC))E[V^2(s)]$$

$$+ \frac{1}{RC}E[\int_s^t \exp(-(t-u)/(RC))N(u)du\,V(s)]$$

$$= \exp(-(t-s)/(RC))P(s). \tag{6.46}$$

Since the autocorrelation is symmetric in t and s,

$$R_V(t,s) = \exp(-|t-s|/(RC))P(\min(t,s)) \tag{6.47}$$

holds for all t, s. In this expression

$$P(\min(t,s)) \approx P_{ss}$$

as long as $\min(t, s) > 5\tau$, and in this case

$$R_V(t-s) = \exp(-|t-s|/(RC))P_{ss} \tag{6.48}$$

depends only on $t - s$. Thus if the initial capacitor variance $P(0)$ is arbitrary, $V(t)$ becomes WSS within a few RC circuit time constants. Since $V(t)$ is Gaussian, this means that it is SSS. In this case, we say that the circuit has reached thermal equilibrium. Of course, if the circuit is initially in thermal equilibrium, so that $P(0) = P_{\text{ss}}$, then it stays in equilibrium forever. By observing that the autocorrelation function (6.48) has the same form as the Ornstein–Uhlenbeck process of Example 4.6, we conclude that the capacitor voltage $V(t)$ in an RC circuit in thermal equilibrium is an Ornstein–Uhlenbeck (OU) process. In this respect, it is worth noting that in addition to being Gaussian and stationary with zero-mean and autocorrelation (6.48), $V(t)$ is also Markov, since expression (6.43) indicates that given the present value $V(s)$ of the voltage, knowledge of past value $V(r)$ with $r < s$ would be of no use to predict future values $V(t)$ for $t \geq s$. This is just due to the fact that $V(t)$ is the state of the RC circuit, and that knowledge of the initial state, here $V(s)$, and future white noise inputs is all what is needed to evaluate $V(t)$.

To find the white noise intensity q, we observe that in thermal equilibrium the average energy stored in capacitor C is

$$E = E[\frac{CV^2(t)}{2}] = CP_{\text{ss}}/2 = \frac{q}{4R} \, . \tag{6.49}$$

But the *equipartition theorem* of statistical mechanics [12] states that in a closed system in thermal equilibrium at temperature T, the energy stored per degree of freedom is $kT/2$ where $k = 1.38 \times 10^{-23}$ denotes Boltzmann's constant, and T is measured with the Kelvin temperature scale. This result is independent of the type of physical system considered. Equating E in (6.49) to $kT/2$, we find

$$q = 2kRT \, , \tag{6.50}$$

which is the main result of Nyquist and Johnson's twin papers (the theoretical analysis is due to Nyquist, and the experimental result to Johnson). Another interesting aspect of expression (6.50) is that the intensity q of white noise fluctuations is linearly proportional to the resistance R characterizing the dissipation of energy in the circuit. This result is called the *fluctuation-dissipation theorem* and it holds for all systems in thermal equilibrium subjected to a linear dissipation mechanism. For example, it is shown in Problem 6.3 that the noise intensity of a Brownian motion in a viscous fluid is proportional to the friction coefficient of the fluid. This result was first proved by Callen and Welton [13], and its extensions are discussed in [14].

Note that the statistical mechanics considerations employed above to determine the value of q depend in a crucial way on the assumption that the system is closed. For open systems, such as active circuits, where an external energy source or energy sink delivers or withdraws energy from the system, the equipartition theorem of energy is no longer valid. Finally, by substituting $q = 2kRT$, we find that the steady-state value of the capacitor variance is given by

$$P_{\text{ss}} = \frac{kT}{C} \, , \tag{6.51}$$

which is often a useful figure of merit for sizing capacitors to ensure a desired thermal noise rms (root-mean-square) value.

6.7   Bibliographical Notes

Bachelier's random walk model of stock market fluctuations is at the origin of Eugene Fama's efficient market hypothesis [15], which was later popularized by a number of authors, such as Malkiel [16]. Since Brownian motion is at the center of studies of Markov diffusions and stochastic calculus, it has been the focus of extensive examination. Most works in this area are highly technical, at least for engineering readers, but accessible presentations are given in [17, Chap 7.] and [5]. We have focused here on topics useful for Wiener process simulation or noise calculations, but the Wiener process admits also interesting level crossings properties. Although the construction of Wiener integrals has been superseded by Ito integration, it remains popular among engineers who find it convenient to use white noise in their calculations. A description of integrals of deterministic functions with respect to orthogonal increments processes, which includes the Wiener integral as a special case, is given in [18, Chap. 9].

6.8   Problems

6.1 Consider the random processes

  i) $X_1(t) = -W(t)$
 ii) $X_2(t) = W(t + s) - W(s)$
iii) $X_3(t) = cW(t/c^2)$
 iv) $X_4(t) = tW(1/t)$
  v) $X_5(t) = W(T) - W(T - t)$

where $W(t)$ is a standard Wiener process. Which of these processes, if any, are Wiener processes? Justify your answers.

6.2 A Wiener process $X(t)$ that is known to go through 0 at time $t = T$, so that $X(T) = 0$, is called a *Brownian bridge*.

a) Find the conditional probability density functions

$$f_{X(t)|X(T)}(x, t \mid X(T) = 0)$$

and

$$f_{X(t),X(s)|X(T)}(x_t, t; x_s, s \mid X(T) = 0)$$

from the original joint density functions of the Wiener process for $0 < t < s < T$.
b) Find the conditional autocovariance function

$$\text{Cov}\,(X(t), X(s) \mid X(T) = 0)\,.$$

c) Is the Brownian bridge Gaussian? Is it Markov?
d) Compare the Brownian bridge to the process $W(t) - tW(T)/T$.

6.3 It is shown in Problem 4.17 of that a zero-mean Gaussian process $X(t)$ has the Markov property if and only if its autocovariance function has the structure

$$K_X(t, s) = g(\max(t, s))h(\min(t, s)) \tag{6.52}$$

for some functions $g(\cdot)$ and $h(\cdot)$.

a) Consider the time-function $\tau(t) = h(t)/g(t)$. Verify that since $K_X(t, t)$ is positive and

$$K_X^2(t, s) \leq K_X(t, t)K_X(s, s),$$

$\tau(t)$ must be positive, and an increasing function of t.

b) Use the structure (6.52) to show that a Gauss–Markov process $X(t)$ can be represented as

$$X(t) = g(t)W(\tau(t)),$$ (6.53)

where $W(t)$ is the Wiener process. In other words, all Gauss–Markov processes can be constructed from a Wiener process by a simple time change $\tau(t)$ and a scaling $g(t)$.

c) The Ornstein–Uhlenbeck process $U(t)$ is a zero-mean Gauss–Markov process with covariance

$$K_U(t, s) = P_U \exp(-a|t - s|).$$

Obtain a representation of the form (6.53) for this process.

6.4 Whereas Brownian motion can be used as a conceptual model of stock price fluctuations, it cannot model the long-term evolution of stock prices since it can take negative values. A long-term model consistent with the Brownian motion representation of short-term fluctuations is given by the *geometric Brownian motion* process

$$X(t) = \exp(\mu t + \sigma W(t))X(0),$$

where the constant μ represents the effect of productivity gains and inflation, $W(t)$ is a standard Wiener process and σ is a volatility parameter characterizing the size of fluctuations. The initial price $X(0)$ is assumed known, and without loss of generality we can set $X(0) = 1$.

a) Show that $X(t)$ is a Markov process. To do so, you may want to observe that the transformation from $W(t)$ to $X(t)$ is one-to-one, since

$$W(t) = \frac{1}{\sigma}(\ln(X(t)) - \mu t).$$

b) For $t \geq s$, compute the transition density $q(x_t, t; x_s, s)$ of $X(t)$. As a first step you may wish to compute the conditional CDF

$$P(X(t) \leq x_t | X(s) = x_s)$$

by using the transformation existing between $X(t)$ and $W(t)$.

c) Compute the conditional mean $E[X(t)|X(s)]$ and the conditional variance

$$E[(X(t) - E[X(t)|X(s)])^2 | X(s)]$$

of $X(t)$ given $X(s)$ for $t \geq s$. To do so, you are reminded that the generating function of a $N(m_Y, K_Y)$ random variable Y is

$$E[\exp(sY)] = \exp(m_Y s + K_Y s^2/2).$$

Fig. 6.4 Sample path
(black) and its reflection
(grey)

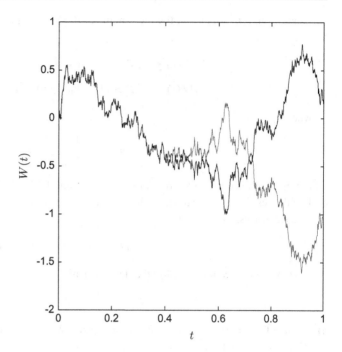

6.5 The sample paths of the Wiener process $W(t)$ have an interesting property, which can be described as follows. If there exists a sample path $W(\cdot, \omega_1)$ which crosses the level a for the first time at $\tau < t$, there exists a second sample path $W(\cdot, \omega_2)$, which coincides with $W(\cdot, \omega_1)$ up to the crossing time τ, and which is the reflection of $W(\cdot, \omega_1)$ with respect to the level a after the time τ. Specifically,

$$W(s, \omega_2) = \begin{cases} W(s, \omega_1) & 0 \leq s \leq \tau \\ 2a - W(s, \omega_1) & \tau \leq t \leq t. \end{cases}$$

The original path $W(\cdot, \omega_1)$ and the "flipped" path $W(\cdot, \omega_2)$ are plotted in Fig. 6.4.

a) The property of the Wiener process we have just described is usually called the *reflection principle*. It can be expressed in quantitative terms as follows. Let T be the first time that $W(\cdot)$ crosses the level a. Note that T is a random variable. Then, according to the reflection principle, for any x (which may be either greater or smaller than a), we must have

$$P[W(t) \leq x \mid T \leq t] = P[W(t) \geq 2a - x \mid T \leq t] \,,$$

or equivalently, for $\tau \leq t$,

$$P[W(t) \leq x \mid T = \tau] = P[W(t) \geq 2a - x \mid T = \tau] \,. \tag{6.54}$$

Prove the identity (6.54). To do so, you may want to use the fact that the events $A = \{W(t) \leq x \mid T = \tau\}$ and $B = \{W(t) \geq 2a - x \mid T = \tau\}$ can be represented as

$$A = \{W(t) - W(\tau) \leq x - a \mid W(\tau) = a\}$$
$$B = \{W(t) - W(\tau) \geq a - x \mid W(\tau) = a\} \,.$$

In addition, remember that the increments of the Wiener process are stationary and independent.
b) If $x \leq a$, show that

$$P[W(t) \leq x, T \leq t] = 1 - F_{W(t)}(2a - x)$$

$$P[W(t) \leq x, T > t] = F_{W(t)}(x) + F_{W(t)}(2a - x) - 1,$$

where

$$F_{W(t)}(w) = \frac{1}{\sqrt{2\pi t}} \int_{-\infty}^{w} \exp(-\frac{u^2}{2t}) du$$

represents the cumulative probability distribution of $W(t)$.
c) Use the result of part b) to show that the probability distribution for the first level crossing T of the Wiener process is given by

$$P[T \leq t] = 2(1 - F_{W(t)}(a)).$$

Hint: You may want to employ the decomposition

$$P[T \leq t] = P[T \leq t, W(t) \leq a] + P[T \leq t, W(t) \geq a].$$

d) Deduce from part c) that the trajectories of the Wiener process cross a fixed level a in finite time with probability 1, i.e.,

$$P[T < \infty] = 1.$$

6.6 A *Brownian motion* corresponds to the movement of a suspended particle in a viscous fluid, subjected to thermal collisions of the fluid molecules. Typically, the mass M of the suspended particle is much larger than the mass m of the fluid molecules with which it collides. We consider only the projection of the particle's motion along one space direction. Then, if $X(t)$ denotes the position of the particle at time t, the motion obeys Newton's law

$$M\frac{d^2}{dt^2}X(t) = -f\frac{d}{dt}X(t) + F(t), \tag{6.55}$$

where $-f dX/dt$ represents a friction force due to the viscosity of the fluid, and $F(t)$ is a fluctuating force modeling the random collisions to which the suspended particle is subjected. Typically, $F(t)$ can be modeled by a white Gaussian noise with zero mean and autocorrelation

$$R_F(t, s) = q\, \delta(t - s),$$

where q will be determined below based on physical considerations. The friction coefficient f is proportional to the viscosity coefficient η of the fluid: if the suspended particle is spherical with radius r, $f = 6\pi\eta r$.

a) Let $V(t) = dX(t)/dt$ be the particle velocity. We assume that the initial position of the particle is $X(0) = 0$, and that the particle is initially at rest, i.e., its initial velocity $V(0) = 0$. By rewriting (6.55) in terms of the velocity $V(t)$ as

$$\frac{d}{dt}V(t) = -aV(t) + \frac{F(t)}{M}$$

with $a = f/M$, and employing standard techniques for solving differential equations (such as the Laplace transform, or the superposition of homogeneous and particular solutions), show that $V(t)$ can be expressed as

$$V(t) = \frac{1}{M} \int_0^t \exp(-a(t - u))F(u)\, du \,. \tag{6.56}$$

Use this solution to evaluate the mean $m_V(t)$ and autocorrelation $K_V(t, s)$ of $V(t)$.

b) Prove that as $t \to \infty$, $V(t)$ becomes an Ornstein–Uhlenbeck process, i.e., it is asymptotically stationary and Gaussian, with zero-mean and autocorrelation

$$K_V(t, s) = P \exp(-a|t - s|) \,,$$

where

$$P = \frac{q}{2aM^2} = \frac{q}{2fM} \,.$$

c) The intensity q of the fluctuations $F(t)$ can be evaluated as follows. According to statistical mechanics, the average energy associated with each degree of freedom for a system in thermal equilibrium at the absolute temperature T is $kT/2$, where $k = 1.38 \times 10^{-23}$ J/° is Boltzmann's constant. For the suspended particle we consider, the average energy is

$$E[\frac{MV^2(t)}{2}] = \frac{MP}{2} \,,$$

and a single degree of freedom is involved, since we look at a the motion of the particle along only one space direction. Obtain q. You will find that q is proportional to f. This relationship is again a manifestation of the *fluctuation dissipation theorem*, according to which, when energy is dissipated and transformed into heat in a system in thermal equilibrium, the energy dissipation gives rise to random fluctuations whose intensity (here q) is proportional to the constant (here f) determining the rate of conversion of energy into heat.

d) By integrating the expression (6.56) for the velocity, show that the particle's position can be written as

$$X(t) = \int_0^t V(r)dr$$

$$= \frac{1}{f} \int_0^t [1 - \exp(-a(t - u))]F(u)du \,. \tag{6.57}$$

Compute the mean and autocovariance of $X(t)$. Use these quantities to show that after a long time $t >> a^{-1} = M/f$, $X(t)$ is a zero-mean Gaussian process with autocovariance

$$K_X(t, s) = D \min(t, s) \,,$$

i.e., $X(t)$ is a Wiener process! As a consequence the Wiener process is often called "Brownian motion." What is the value of the diffusion constant D? The expression you have just derived was first obtained by Albert Einstein in his famous 1905 paper [1] on Brownian motion.

6.7 Consider a Wiener process $W(t)$ with intensity q, i.e., $W(0) = 0$, $W(t)$ has independent increments, and $W(t) \sim N(0, qt)$. $W(t)$ can sometimes be used to model random phase variations in signals, where the signal phase $\Theta(t)$ with $t \geq 0$ is modeled by

$$\exp(j\Theta(t)) = \exp(j(\Theta(0) + W(t))) \, .$$

The initial phase $\Theta(0)$ is assumed to be independent of $W(t)$. In other words

$$\Theta(t) = \Theta(0) + W(t) \quad \mathrm{mod}\ 2\pi \, , \tag{6.58}$$

where the modulo 2π operation has the effect of ensuring that $-\pi < \Theta(t) \leq \pi$. The process $\Theta(t)$ can be viewed as obtained by wrapping sample paths of the Wiener process around a cylinder whose circular cross-section has radius 1.

a) We assume first that $\Theta(0) = 0$. By observing that

$$P(-\pi < \Theta(t) \leq \theta) = \sum_{n=-\infty}^{\infty} P(-\pi + n2\pi < W(t) \leq \theta + n2\pi)$$

deduce that the CDF of $\Theta(t)$ satisfies

$$F_{\Theta(t)}(\theta) = \begin{cases} 0 & \theta < -\pi \\ \sum_{n=-\infty}^{\infty} [F_{W(t)}(\theta + n2\pi) - F_{W(t)}((2n-1)\pi)] & -\pi < \theta \leq \pi \\ 1 & \theta > \pi \end{cases}$$

Conclude therefore that the PDF of $\Theta(t)$ can be expressed as

$$f_{\Theta(t)}(\theta) = \sum_{n=-\infty}^{\infty} f_{W(t)}(\theta + n2\pi) \tag{6.59}$$

$$= \frac{1}{(2\pi qt)^{1/2}} \sum_{n=-\infty}^{\infty} \exp\left(-\frac{(\theta + n2\pi)^2}{2qt}\right)$$

for $-\pi < \theta \leq \pi$ and $f_{\Theta(t)}(\theta) = 0$ otherwise.

b) Since the density $f_{\Theta(t)}(\theta)$ is nonzero only over interval $(-\pi, \pi]$, it can be extended periodically into a function $f_{\Theta(t)}^p(\theta)$ which admits a Fourier series of the form

$$f_{\Theta(t)}^p(\theta) = \sum_{k=-\infty}^{\infty} G_k(t) \exp(ik\theta) \, ,$$

where the Fourier coefficients

$$G_k(t) = \frac{1}{2\pi} \int_{-\pi}^{\pi} f_{\Theta(t)}(\theta) e^{-ik\theta} d\theta \, .$$

Prove that

$$G_k(t) = \frac{1}{2\pi} \Phi_{W(t)}(-k), \tag{6.60}$$

where

$$\Phi_{W(t)}(u) = E[\exp(iuW(t))] = \exp(-qtu^2/2)$$

denotes the characteristic function of $W(t)$.

c) Use the observation (6.60) to prove that $\Theta(t)$ converges in distribution to a uniform distribution over $(-\pi, \pi]$ as t tends to infinity.

d) Consider now the general case where $\Theta(0)$ is not necessarily zero. Show that $\Theta(t)$ a Markov process. In your analysis, you might want to observe that for $t \geq s$

$$\Theta(t) = W(t) - W(s) + \Theta(s) \quad \mathrm{mod}\ 2\pi\ ,$$

where increment $W(t) - W(s)$ is independent of $W(r)$ and thus all $\Theta(r)$ for all $r \leq s$.

e) Find the transition density $q(\theta_t, t; \theta_s, t)$ of $\Theta(t)$. Is it homogeneous, i.e., does it depend only on $t - s$?

f) Show that the uniform density

$$f(\theta) = \begin{cases} \frac{1}{2\pi} & -\pi < \theta \leq \pi \\ 0 & \text{otherwise} \end{cases}$$

is an invariant density for the $\Theta(t)$ process, i.e., if $\Theta(0)$ is uniformly distributed, show that $\Theta(t)$ is uniformly distributed for all $t \geq 0$.

g) How should the distribution of $\Theta(0)$ be selected to ensure that $\Theta(t)$ is SSS?

6.8 In a switched capacitor circuit, a switch S is used to connect a noisy resistor R at temperature T to a capacitor C during sampling phases of duration $T_s/2$. The sampling phases alternate with holding phases of duration $T_s/2$, during which the switch S is placed in the "off" position. The binary variable $b(t) = 1$ whenever the switch S is on, and $b(t) = 0$ when S is off, as indicated in Fig. 6.5. The sampling circuit is shown in Fig. 6.6. In this circuit, $U(t)$ is the voltage source representing the thermal fluctuations of the resistor. Thus, $U(t)$ is a zero-mean white Gaussian noise of intensity $2kRT$, so that

$$E[U(t)U(s)] = 2kRT\delta(t - s)\ .$$

Here $k = 1.37 \times 10^{-23}$ J/° denotes Boltzmann's constant. Assume that the initial value $V_C(0) = v_0$ of the capacitor voltage is known.

a) Obtain the differential equation satisfied by the capacitor voltage $V_C(t)$ during the sampling phases $nT_s \leq t < (n + 1/2)T_s$ with n integer. Express the solution $V_C(t)$ of this differential equation in terms of the voltage $V_C(nT_s)$ at the beginning of the sampling phase, and the noise $U(\cdot)$ during the interval $[nT_s, t)$.

b) Obtain differential equations for the mean $m(t) = E[V_C(t)]$ and variance $P(t) = E[(V_C(t) - m(t))^2]$ of the capacitor voltage during the sampling phases. Express the solutions of these equations over the interval $[nT_s, (n + 1/2)T_s)$ in terms of the mean $m(nT_s)$ and variance $P(nT_s)$ at the beginning of the nth sampling phase.

Fig. 6.5 Switch activation signal

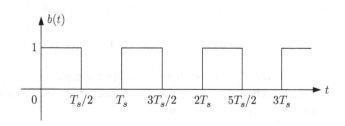

Fig. 6.6 Switched
capacitor sampling circuit

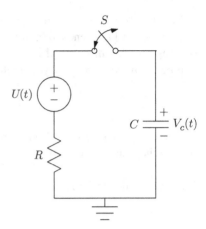

c) During the holding phases $(n + 1/2)T_s \leq t < (n + 1)T_s$ with n integer, we have $V_C(t) = V_C((n + 1/2)T_s)$. Use the solution of the differential equation obtained in part a) to find a discrete-time state-space model of the form

$$V_C((n + 1)T_s) = a_d V_C(nT_s) + U_d(n) \tag{6.61}$$

for the capacitor voltage at the beginning of each successive sampling phase. The driving noise $U_d(n)$ depends on the thermal noise $U(\cdot)$ over the interval $[nT_s, (n + 1/2)T_s)$. Verify that $U_d(n)$ is a zero-mean discrete-time white Gaussian noise, i.e.

$$E[U_d(n)U_d(m)] = q_d\delta(n - m),$$

and evaluate its intensity q_d in terms of the intensity of $U(\cdot)$ and the circuit parameters R and C.

d) Consider the discrete-time stochastic process formed by the sampled values $\{V_C(nT_s); n \geq 0\}$ of the capacitor voltage at the beginning of each sampling phase. Use the state-space model (6.61) to show that this process is asymptotically WSS. Find the steady-state values of its mean and variance.

e) Consider now the continuous-time process $\{V_C(t); t \geq 0\}$. Is it asymptotically WSS? Please justify your answer, and explain any difference with the result of part d). Is $V_C(t)$ asymptotically first-order stationary? If yes, please specify its asymptotic mean and variance.

f) Indicate whether the discrete-time sampled process $\{V_C(nT_s); n \geq 0\}$ and/or the continuous-time process $\{V_C(t); t \geq 0\}$ are Markov. Explain your answer.

6.9 Consider the ideal op-amp circuit shown in Fig. 6.7. The two resistors have temperature T and are affected by thermal fluctuations represented by the independent white Gaussian noise voltage sources $N_1(t)$ and $N_2(t)$ with intensity $2kRT$, so that

$$E[N_i(t)N_i(s)] = 2kRT\delta(t - s)$$

for $i = 1, 2$. The switch S is open for $t < 0$ and closed for $t \geq 0$.

a) Find the differential equation satisfied by the capacitor voltage $V(t)$ for $t < 0$. Note that since the switch has been open for a long time, steady state has been reached, so that $V(t)$ is a WSS process. Find the mean, variance, and autocorrelation of $V(t)$ for $t < 0$.

Fig. 6.7 Switched
capacitor circuit in thermal
equilibrium

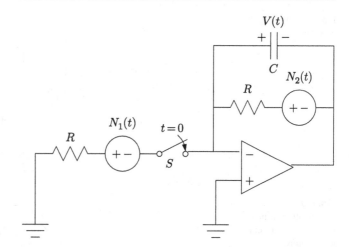

b) Find the differential equation satisfied by the capacitor voltage $V(t)$ for $t \geq 0$. Find the mean, variance, and autocorrelation of $V(t)$ for $t \geq 0$.

c) After the switch has been closed, the capacitor voltage $V(t)$ reaches a steady state as $t \to \infty$. Find the average energy

$$E = \frac{1}{2} E[C V^2(t)]$$

stored in the capacitor for large t. Does this result satisfy the equipartition theorem of energy of statistical physics? If not, why?

6.10 Consider the op-amp circuit shown below, where the resistors R_s and R_F are at the temperature T and the voltage sources $N_s(t)$ and $N_F(t)$ represent thermal fluctuations. Thus $N_s(t)$ and $N_F(t)$ are two independent WGNs with zero mean and autocorrelation

$$R_{N_s}(t - s) = E[N_s(t)N_s(s)] = 2kR_sT\delta(t - s)$$

$$R_{N_F}(t - s) = E[N_F(t)N_F(s)] = 2kR_FT\delta(t - s) \,. \tag{6.62}$$

The op-amp is ideal. For non-EEs, this means that the currents $I_+(t)$ and $I_-(t)$ into the plus and minus input terminals of the op-amp are zero, and the voltages $V_+(t)$ and $V_-(t)$ at the plus and minus input terminals of the op-amp are equal, i.e.,

$$V_+(t) = V_-(t) = 0 \,.$$

Note that in addition to the two input terminals and the output terminal of the op-amp, which carry information bearing signals, there exists two-power supply terminals (Fig. 6.8). We assume that at time $t = 0$, the voltages $V_s(0)$ and $V_F(0)$ across the source and feedback capacitors are zero.

a) To analyze the op-amp circuit, it is convenient to use the method of superposition where we examine separately the effect of each source on the circuit, setting the other source to zero. So we start by setting $N_F(t) = 0$. In this case obtain differential equations describing the evolution of $V_s(t)$ and $V_F(t)$. Solve these equations and find the mean, variance, and autocorrelation of the two capacitor voltages. Are $V_s(t)$ and $V_F(t)$ independent? If not, evaluate their cross-correlation

Fig. 6.8 Op-amp circuit in
thermal equilibrium

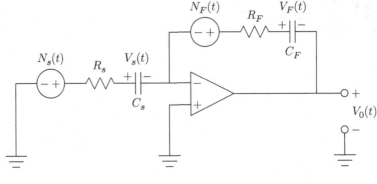

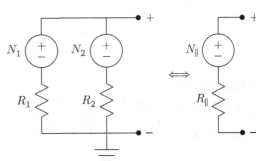

Fig. 6.9 Parallel
combination of noisy
resistors at different
temperatures

$$R_{V_s V_F}(t, s) = E[V_s(t) V_F(s)],$$

b) Repeat part a) with $N_s(t) = 0$.

c) Combining now the results of parts a) and b), as $t \to \infty$ find the average energies

$$E_s = E[C_s V_s^2(t)/2] \quad \text{and} \quad E_F = E[C_F V_F^2(t)/2]$$

stored in the source and feedback capacitors. Does your result satisfy the equipartition theorem of
energy? If not, why not?

d) Use the results of parts a) and b) to express the output voltage $V_0(t)$ in function of $N_s(t)$ and $N_F(t)$.
Evaluate the mean and autocorrelation of $V_0(t)$.

6.11   Consider two noisy resistors in parallel as shown in Fig. 6.9. The first resistor R_1 has temperature
T_1, while R_2 has temperature T_2, so that the independent thermal WGN voltages V_1 and V_2 have
intensity $2kR_1 T_1$ and $2kR_2 T_2$, respectively.

a) Assume first that the two resistors have the same temperature $T_1 = T_2 = T$. By using Thevenin's
theorem show that the parallel combination of the two noisy resistors can be replaced by a noisy
resistor with resistance

$$R_\parallel = \frac{R_1 R_2}{R_1 + R_2}$$

at the temperature T.

b) For the case when $T_1 \neq T_2$, show that the parallel combination of the two resistors can be replaced
by a noisy resistor with resistance $R_\parallel$ and effective temperature

Fig. 6.10 Noisy switched
RC circuit

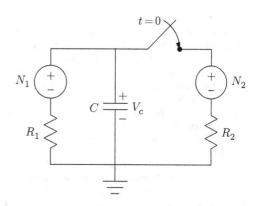

$$T_{eff} = aT_1 + (1 - a)T_2$$

with $0 \leq a \leq 1$. Express a in terms of R_1 and R_2.

c) The following general principle can be used to determine the effective temperature of the Thevenin resistance R_{Th} replacing a one-port network formed by n resistors R_i at temperature T_i, with $1 \leq i \leq n$. Assume that the resistors are noiseless, and place a voltage source V_s across the terminals of the network that will be replaced by R_{Th}. Let a_i be the fraction of the power delivered by V_s which is dissipated in resistor R_i, with $1 \leq i \leq n$, so that we have

$$\sum_{i=1}^{n} a_i = 1 .$$

Then, the effective temperature of this network is given by

$$T_{eff} = \sum_{i=1}^{n} a_i T_i .$$

Use this procedure to verify the expression obtained for T_{eff} in part b).

d) Assume that for $t < 0$ the resistor R_1 with temperature T_1 is placed in series with the capacitor C. Prior to $t = 0$, thermal equilibrium has been reached. Find the mean and variance of the capacitor voltage $V_C(t)$ just prior to $t = 0$. At time $t = 0$, the switch S is closed as shown in Fig. 6.10, so that the capacitor C is now connected to the resistors R_1 and R_2 at temperature T_1 and T_2 in parallel. Use the result of part b) to evaluate the mean, variance, and autocorrelation of the capacitor voltage $V_C(t)$ for $t \geq 0$.

References

1. A. Einstein, "Investigations on the theory of Brownian movement," *Annalen der Physik*, vol. 17, no. 8, pp. 549–560, 1905. Translated and republished in 1956 by Dover Publications, Mineola NY.
2. M. von Smoluchowski, "Zur kinetischen theorie der Brownschen molekularbeweguung und der suspensionen," *Annalen der Physik*, vol. 17, pp. 756–780, 1906.
3. P. Langevin, "Sur la théorie du mouvement Brownien," *Comptes Rendus Académie des Sciences*, vol. 146, pp. 530–532, 1908.
4. E. Nelson, *Dynamical Theories of Brownian Motion*. Princeton, NJ: Princeton University Press, 1967.

5. P. Morters and Y. Peres, *Brownian Motion*. Cambridge, United Kingdom: Cambridge University Press, 2010.

6. L. Bachelier, *Théorie de la Spéculation*. Paris, France: Gauthier-Villars, 1900.

7. D. Revuz and M. Yor, *Continuous Martingales and Brownian Motion*. Berlin, Germany: Springer Verlag, 1991.

8. A. Dvoretsky, P. Erdös, and S. Kakutani, "Non increase everywhere of the Brownian motion process," in *Proc. Fourth Berkeley Symposium*, vol. 2, pp. 103–116, 1961.

9. B. Oksendal, *Stochastic Differential Equations: An Introduction with Applications, 6th edition*. Berlin, Germany: Springer, 2003.

10. H. Nyquist, "Thermal agitation of electric charge in conductors," *Physical Review*, vol. 32, pp. 110–113, July 1928.

11. J. B. Johnson, "Thermal agitation of electricity in conductors," *Physical Review*, vol. 32, pp. 97–109, July 1928.

12. W. Greiner, L. Neise, and H. Stöcker, *Thermodynamics and Statistical Mechanics*. New York: Springer-Verlag, 2001.

13. H. B. Callen and T. A. Welton, "Irreversibility and generalized noise," *Physical Review*, vol. 83, July 1951.

14. R. Kubo, "The fluctuation-dissipation theorem," *Rep. Prog. Phys.*, vol. 29, pp. 255–284, 1966.

15. E. F. Fama, "Random walks in stock market prices," *Financial Analysts Journal*, pp. 55–59, Sep.-Oct. 1965.

16. B. B. Malkiel, *A Random Walk Down Wall Street: The Time Tested Strategy for Successful Investing, (11th edition)*. New York: W. W. Norton, 2016.

17. S. Karlin and H. M. Taylor, *A Second Course in Stochastic Processes*. New York, NY: Academic Press, 1981.

18. J. L. Doob, *Stochastic Processes*. New York, NY: J. Wiley & Sons, 1953.

Poisson Process and Shot Noise

7.1 Introduction

While the Wiener process has continuous sample paths and continuously changes by small increments, the Poisson process is integer valued with discontinuous sample paths and changes infrequently by unit increments. So, it is well adapted to model discrete random phenomena, such as photon counts in optics, the number of tasks processed by a computer system, or random spikes in a neural pathway. Like the Wiener process, the Poisson process has independent increments and can be constructed in several equivalent ways, which are described in Sect. 7.2. The times at which the Poisson process undergoes jumps are called epochs, and the interarrival times (the time between two epochs) are exponentially distributed. In this context, as explained in Sect. 7.3, the memoryless property of exponential distributions has an interesting consequence concerning the distribution of the residual time until the occurrence of the next epoch after an arbitrary reference time. As shown in Sect. 7.5, Poisson processes have also the the interesting feature that they remain Poisson after either merging, or splitting randomly into two lower rate processes. Finally, Sect. 7.6 examines the statistical properties of shot noise, which is a form of noise occurring in electronic devices, such as vacuum tubes or optical detectors, when electrons or photons arriving at random instants (modeled typically by the epochs of a Poisson process) trigger an electronic circuit response.

7.2 Poisson Process Properties

A random process $N(t)$ defined for $t \geq 0$ is said to be a counting process if it is nonnegative, integer valued, and monotone nondecreasing in the sense that $N(t) \geq N(s)$ if $t \geq s$. For $t > s$, the increase $N(t) - N(s)$ measures the number of events that have occurred over interval $(s, t]$. Many electrical and computer engineering devices and systems can be modeled in terms of counting processes. For example, the number of photons hitting a detector, or the number of electrons hitting the plates of a vacuum tube are integer valued and increase over time. Likewise, the number of packets arriving at a a computer router, the number of memory accesses performed by a computer processor are counting phenomena.

The Poisson process is probably the most important representative of the class of counting processes. It admits several equivalent definitions, the first of which takes the following form.

© Springer Nature Switzerland AG 2020
B. C. Levy, *Random Processes with Applications to Circuits and Communications*,
https://doi.org/10.1007/978-3-030-22297-0_7

Definition 1

i) $N(0) = 0$;

ii) $N(t)$ has stationary independent increments;

iii) $N(t)$ is Poisson distributed with parameter λt, so that

$$P(N(t) = k) = \frac{(\lambda t)^k}{k!} \exp(-\lambda t) \, . \tag{7.1}$$

Thus the Poisson and Wiener processes are both stationary independent increments processes and are zero at time $t = 0$. Of course, their one-time distributions are very different, since $N(t)$ is Poisson distributed, whereas $W(t)$ is Gaussian. Because of the independence of the increments, we deduce that $N(t)$ is a Markov process which for $t \geq s$ and $k \geq \ell$ admits the conditional PMF

$$
\begin{aligned}
P(N(t) = k | N(s) = \ell) &= P(N(t) - N(s) = k - \ell | N(s) - N(0) = \ell) \\
&= P(N(t) - N(s) = k - \ell) = P(N(t - s) = k - \ell) \\
&= \frac{(\lambda (t - s))^{k-\ell}}{(k - \ell)!} \exp(-\lambda(t - s)) \, ,
\end{aligned}
\tag{7.2}
$$

where the first equality of the second line uses the independence of the increments, and the second equality their stationarity. Since the conditional distribution (7.2) depends only on $k - \ell$ and $t - s$, it is homogeneous in both space and time, as was the case for the Wiener process.

Definition 2 The second definition retains axioms i) and ii) of Definition 1, but replaces axiom iii) by

iii') For small h

$$
\begin{aligned}
P(N(h) = 0) &= 1 - \lambda h + o(h) \\
P(N(h) = 1) &= \lambda h + o(h) \\
P(N(h) > 1) &= o(h),
\end{aligned}
\tag{7.3}
$$

where the "little o" notation $o(h)$ is used to represent any function $f(h)$ such that

$$\lim_{h \to 0} \frac{f(h)}{h} = 0 \, ,$$

so that $o(h)$ is any function of h that goes to zero faster than h, such as $h^{4/3}$.

Definition 2 provides an intuitive characterization of the behavior of the Poisson process over small intervals. To prove the equivalence of Definitions 1 and 2, we follow the approach of [1, p. 61–62]. Clearly Definition 1 implies Definition 2. To show the converse, let $P_k(t) = P(N(t) = k)$. We prove by induction that $P_k(t)$ is given by (7.1). For $k = 0$, we have

$$
\begin{aligned}
P_0(t + h) &= P(N(t + h) = 0) = P(N(t + h) - N(t) = 0, \, N(t) = 0) \\
&= P(N(t + h) - N(t) = 0)P_0(t) = (1 - \lambda h + o(h))P_0(t) \, ,
\end{aligned}
\tag{7.4}
$$

where the first equality of the second line uses the independence of the increments of $N(t)$ and the second equality their stationarity, so that

$$\frac{P_0(t+h) - P_0(t)}{h} = -\lambda P_0(t) + \frac{o(h)}{h} .$$

Letting h tend to zero implies that $P_0(t)$ obeys the differential equation

$$\frac{d}{dt} P_0(t) = -\lambda P_0(t) , \tag{7.5}$$

which in light of the initial condition $P_0(0) = 1$ implies $P_0(t) = \exp(-\lambda t)$. Then assume that $P_k(t)$ satisfies (7.1). We have

$$
\begin{aligned}
P_{k+1}(t+h) &= P(N(t+h) - N(t) = 1, N(t) = k) \\
&\quad + P(N(t+h) - N(t) = 0, N(t) = k+1) \\
&= P(N(t+h) - N(t) = 1)P_k(t) + P(N(t+h) - N(t) = 0)P_{k+1}(t) \\
&= (\lambda h + o(h))P_k(t) + (1 - \lambda h + o(h))P_{k+1}(t) ,
\end{aligned}
\tag{7.6}
$$

where we have used again the independence and stationarity of the increments of $N(t)$. This implies that $P_{k+1}(t)$ satisfies the differential equation

$$\frac{d}{dt} P_{k+1}(t) = -\lambda P_{k+1}(t) + \lambda P_k(t) . \tag{7.7}$$

Performing the change of variable $f(t) = \exp(\lambda t)P_{k+1}(t)$, we find

$$\frac{df}{dt} = \lambda \exp(\lambda t) P_k(t) ,$$

which together with the initial condition $f(0) = P_{k+1}(0) = 0$ implies

$$f(t) = \lambda \int_0^t \frac{(\lambda u)^k}{k!} du = \frac{(\lambda t)^{k+1}}{(k+1)!}$$

thus proving that $P_{k+1}(t)$ satisfies (7.1).

Definition 3 The third definition of the Poisson process is more useful for simulation purposes. Let X_k, with $k \geq 1$ denote a sequence of independent exponential random variables with parameter λ, and let

$$S_k = \sum_{\ell=1}^{k} X_\ell . \tag{7.8}$$

Then

$$N(t) = \sum_{k=1}^{\infty} u(t - S_k) \tag{7.9}$$

for $t \geq 0$, where $u(t)$ denotes the unit step function. A sample path of $N(t)$ is depicted in Fig. 4.6. The definition (7.9) highlights the fact that sample paths are monotone nondecreasing and integer valued. The exponential random variables

$$X_k = S_k - S_{k-1} \tag{7.10}$$

are called the interarrival times of the process and the times S_k at which $N(t)$ increases are called the epochs of the Poisson process. When the interarrival times X_k are independent and identically distributed, but not necessarily exponential, processes of the form (7.8) and (7.9) are called *renewal processes*, since the can be used to count the number of parts that need to be replaced due to wear in a system of interest in a fixed time t. In this context, the exponential random variable model for interarrival times is usually most appropriate for elementary components, such as simple electronic devices.

Note that since the characteristic function of an exponential random variable X with parameter λ is

$$\Phi_X(u) = \frac{\lambda}{\lambda - iu}$$

and the interarrival times X_k are independent, the characteristic function of the k-th epoch S_k is

$$\Phi_{S_k}(u) = E[\exp(iuS_k)] = \left(\Phi_X(u)\right)^k = \left(\frac{\lambda}{\lambda - iu}\right)^k$$

which after inverse Fourier transformation gives the PDF

$$f_{S_k}(x) = \frac{\lambda(\lambda x)^{k-1}}{(k-1)!} \exp(-\lambda x) u(x) \tag{7.11}$$

of an Erlang distribution with parameters k (the shape parameter) and λ. This observation can be used to show that Definition 3 implies axiom iii) of Definition 1 (and thus axiom iii)' of Definition 2). Note indeed that

$$P(N(t) = k) = P(N(t) \geq k) - P(N(t) \geq k+1)$$

$$= P(S_k \leq t) - P(S_{k+1} \leq t) = \int_0^t (f_{S_k}(x) - f_{S_{k+1}}(x))dx$$

$$= \int_0^t \left(\frac{\lambda(\lambda x)^{k-1}}{(k-1)!} - \frac{\lambda(\lambda x)^k}{k!}\right) \exp(-\lambda x)dx$$

$$= \int_0^t \frac{d}{dx}\left(\frac{(\lambda x)^k}{k!} \exp(-\lambda x)\right)dx = \frac{(\lambda t)^k}{k!} \exp(-\lambda t), \tag{7.12}$$

as desired.

We observe also that since $X_1 > 0$ almost surely, Definition 3 implies $N(0) = 0$ with probability 1. To show that Definition 3 implies axiom ii) (the independence and stationarity of the increments of $N(t)$) we need first to establish several preliminary results.

Joint PDF of the Poisson Epochs Given the first $n+1$ interarrival times X_k, $1 \leq k \leq n+1$ with joint PDF

$$f_\mathbf{X}(\mathbf{x}) = \lambda^{n+1} \exp(-\lambda \sum_{k=1}^{n+1} x_k) \prod_{k=1}^{n+1} u(x_k), \tag{7.13}$$

the joint PDF of the first $n + 1$ epochs S_k, $1 \leq k \leq n + 1$ can be evaluated by using the transformation (7.8) and inverse transformation (7.10). The Jacobian of the transformation is $J = 1$, so the joint density of the epochs is given by

$$f_{\mathbf{S}}(\mathbf{s}) = \lambda^{n+1} \exp(-\lambda s_{n+1}) u(s_1) \prod_{k=2}^{n+1} u(s_k - s_{k-1}) . \tag{7.14}$$

Conditional PDF of the Epochs Suppose now that $N(t) = n$. We seek to compute the conditional PDF of the epochs S_k with $1 \leq k \leq n$ given $N(t) = n$. For small h, this density can be approximated as

$$
\begin{aligned}
f_{\mathbf{S}|N(t)}(\mathbf{s}|N(t) = n) &\approx \frac{P(s_k < S_k \leq s_k + h, 1 \leq k \leq n, N(t) = n)}{h^n P(N(t) = n)} \\
&= \frac{P(s_k < S_k \leq s_k + h, 1 \leq k \leq n, S_{n+1} > t)}{h^n P(N(t) = n)} \\
&= \lambda^n \int_t^\infty \lambda \exp(-\lambda s_{n+1}) ds_{n+1} u(s_1) u(t - s_n) \frac{\prod_{k=1}^n u(s_k - s_{k-1})}{P(N(t) = n)} \\
&= \frac{n!}{t^n} u(s_1) u(t - s_n) \prod_{k=2}^n u(s_k - s_{k-1}) . \tag{7.15}
\end{aligned}
$$

This last result is quite interesting, since it can be interpreted in terms of the order statistics of n independent random variables U_k, $1 \leq k \leq n$ uniformly distributed over $[0, t]$. Specifically, the joint PDF of the uniform random variables Uk is given by

$$f_{\mathbf{u}}(\mathbf{u}) = \prod_{k=1}^n f_{U_k}(u_k) = \frac{1}{t^n}$$

with $1 \leq u_k \leq t$ for all t. Then is we rank order the U_ks and select

$$
\begin{aligned}
S_1 &= \min_{1 \leq \ell \leq n} U_\ell \\
S_k &= \text{k-th smallest } U_\ell \qquad\qquad (7.16) \\
S_n &= \max_{1 \leq \ell \leq n} U_\ell ,
\end{aligned}
$$

the random variables S_k with $1 \leq k \leq n$ will have the joint PDF (7.15).

Poisson Process Simulation The combination of the Poisson distribution (7.1) for $N(t)$ and the characterization (7.15) and (7.16) of the epochs of the Poisson process yields a simple technique to simulate the sample paths of the Poisson process over an interval $[0, t]$:

Step 1: Generate a sample n of $N(t)$ with the Poisson PMF (7.1).

Step 2: Given the value n of $N(t)$ generated in Step 1, generate n independent random variables U_k, $1 \leq k \leq n$ uniformly distributed over $[0, t]$. Order these samples to obtain epochs S_k, $1 \leq k \leq n$. Then the sample path of the Poisson process is given by

$$N(s) = \sum_{k=1}^{n} u(s - S_k)$$

for $0 \le s \le t$.

Definition 3 Implies Axiom ii) The conditional PDF (7.15) of the epochs of the Poisson process can then be used to show that for $n = 1$ times t_k, $0 \le k \le n$ such that $t_0 = 0$ and $t_k < t_{k+1}$, the increments $N(t_k) - N(t_{k-1})$ are independent and Poisson distributed with parameter $\lambda(t_k - t_{k-1})$, so that they are stationary. Indeed if $m = \sum_{k=1}^{n} m_k$, we have

$$P(E) \overset{\triangle}{=} P(N(t_k) - N(t_{k-1}) = m_k, \ 1 \le k \le n)$$

$$= P(N(t_k) - N(t_{k-1}) = m_k, \ 1 \le k \le n | N(t_n) = m) P(N(t_n) = m). \tag{7.17}$$

But given $N(t_n) = m$, the unordered epochs U_ℓ, $1 \le \ell \le m$ are independent uniformly distributed over $[0, t]$. The probability that any one of these unordered epochs belongs to $(t_{k-1} t_k]$ is $(t_k - t_{k-1})/t_n$. Accordingly, given $N(t_n) = m$, the probability of a configuration with m_1 epochs in $(0, t_1]$, m_2 epochs in $(t_1, t_2]$, ..., and m_n epochs in $(t_{n-1}, t_n]$ is

$$P(E | N(t_n) = m) = \binom{m}{m_1, \dots, m_n} \prod_{k=1}^{n} \left(\frac{t_k - t_{k-1}}{t_n} \right)^{m_k}, \tag{7.18}$$

where we have used the fact that the number of partitions of m objects into n subsets of m_k objects each with $1 \le k \le n$ is the multinomial coefficient

$$\binom{m}{m_1, \dots, m_n} = \frac{m!}{\prod_{k=1}^{n} m_k!}. \tag{7.19}$$

Multiplying (7.18) by $P(N_{t_n} = m)$ yields after simplification

$$P(E) = \prod_{k=1}^{n} P(N(t_k) - N(t_{k-1}) = m_k), \tag{7.20}$$

where

$$P(N(t_k) - N(t_{k-1}) = m_k) = \frac{(\lambda(t_k - t_{k-1}))^{m_k}}{m_k!} \exp(-\lambda(t_k - t_{k-1})). \tag{7.21}$$

The factored form (7.20) indicates that the increments are independent, and since the increment PMF (7.21) depends only on $t_k - t_{k-1}$, the increments are stationary. Thus Definition 3 implies Definitions 1 and 2.

To show the equivalence between the three definitions, all what is left is showing that Definition 1 implies Definition 3. To do so, we compute the conditional distribution of interarrival time X_n given that the prior interarrival times $X_k = x_k$ for $1 \le k \le n - 1$, or equivalently, given that the first $n - 1$ epochs $S_k = s_k$ with $s_k = \sum_{\ell=1}^{k} x_\ell$. We have

$$P(X_n > x | X_k = x_k, \ 1 \le k \le n - 1) = P(X_n > x | S_k = s_k, \ 1 \le k \le n - 1)$$

$$= P(X_n > x | N(t), \ 0 \le t \le s_{n-1})$$

$$= P(N(s_{n-1} + x) - N(s_{n-1}) = 0 | N(t), \, 0 \le t \le s_{n-1})$$

$$= P(N(s_{n-1} + x) - N(s_{n-1}) = 0) = P(N(x) = 0) = \exp(-\lambda x) \,, \tag{7.22}$$

where to go from the first to the second line, we have used the fact that knowledge of the first $n - 1$ epochs of $N(t)$ is equivalent to knowing $N(t)$ for $0 \le t \le s_{n-1}$, The last line uses the independence and then the stationarity of the increments of $N(t)$. The identity (7.22) shows that the interarrival times are independent and exponentially distributed with parameter λ, so that Definition 1 implies Definition 3.

7.3 Residual Waiting Time and Elapsed Time

Consider an arbitrary time t. Then $S_{N(t)+1}$ denotes the first epoch after t and $S_{N(t)}$ the last epoch just before t, as depicted in Fig. 7.1. We are interested in the distributions of the residual waiting time $R(t) = S_{N(t)+1} - t$ and of the elapsed time $E(t) = t - S_{N(t)}$. Letting $N(t) = n$, the conditional distribution of $R(t)$ given $E(t)$ satisfies

$$P(R(t) > w | E(t) = e) = P(X_{n+1} > w + e | X_{n+1} > e)$$

$$= \exp(-\lambda w) \,, \tag{7.23}$$

where we have used the memoryless property of the exponential distribution to go from the first to the second line. This shows that $R(t)$ is exponential with parameter λ and independent of $E(t)$. Note also that the distribution of $R(t)$ is independent of time t. Conversely, by performing the same reasoning backwards in time, we find

$$P(E(t) > e | R(t) = w) = \exp(-\lambda e), \tag{7.24}$$

so that the elapsed time $E(t)$ is exponential with parameter λ and independent of $R(t)$.

Bus Waiting Paradox This leads us to the following paradox. We have

$$X_{N(t)+1} = S_{N(t)+1} - S_{N(t)} = R(t) + E(t) \,, \tag{7.25}$$

so, on one hand, the Definition 3 of Poisson processes seems to suggest that the interarrival time $X_{N(t)}$ between epochs $S_{N(t)+1}$ and $S_{N(t)}$ is exponentially distributed with parameter λ, and thus with

Fig. 7.1 Elapsed and residual times with respect to reference time t

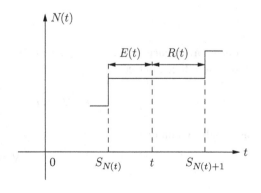

mean $m_X = 1/\lambda$. On the other hand, $S_{N(t)+1} - S_{N(t)}$ is the sum $R(t) + E(t)$ of two independent exponential random variables with parameter λ, so it is Erlang with parameters 2 and λ, with mean $2/\lambda$. So how can these two apparently contradictory models be reconciled? The answer is that when a time t is selected at random, it has a higher probability of falling inside a long interarrival period rather than a short one, so $X_{N(t)+1}$ is not exponentially but Erlang distributed. In every day life, this means that if buses arrive at a bus stop according to a Poisson process with rate λ, so interarrival times are exponential with parameter λ and mean $1/\lambda$, when a rider arrives at the bus stop at an arbitrary time t, the rider will need to wait for an exponentially distributed with parameter λ residual time $R(t)$ until the next bus arrives, independently of the value $E(t)$ of the elapsed time. In contrast, if the periods between successive buses are fixed and equal to T, a rider arriving at random at the bus stop will only need to wait on average $T/2$ until the next bus. Furthermore, if the elapsed time since the previous bus is $E(t)$, the residual time is $R(t) = T - E(t)$, so that residual and elapsed times are not independent in the fixed bus schedule case.

7.4   Poisson Process Asymptotics

In this section we examine the deviations of $N(t)$ with respect to its mean slope λt. We start by evaluating the mean, variance, and autocovariance of $N(t)$. Since the mean and variance of a Poisson random variable with parameter μ are both equal to μ, we have

$$m_N(t) = E[N(t)] = \lambda t \ \text{ and } \ P_N(t) = E[(N(t) - m_N(t))^2] = \lambda t \ . \tag{7.26}$$

Let $t \geq s$. The independence of the increments of $N(t)$ implies

$$R_N(t, s) = E[N(t)N(s)] = E[(N(t) - N(s) + N(s))N(s)]$$
$$= E[N(t) - N(s)]E[N(s)] + E[N^2(s)]$$
$$= \lambda^2(t - s)s + \lambda^2 s^2 + \lambda s = \lambda^2 ts + \lambda s \ .$$

Since $R_N(t, s)$ is symmetric in t and s, we have

$$R_N(t, s) = \lambda^2 ts + \lambda \min(t, s) \ , \tag{7.27}$$

and

$$K_N(t, s) = E[(N(t) - m_N(t))(N(s) - m_N(s))] = \lambda \min(t, s) \ . \tag{7.28}$$

Slope Convergence Then $N(t)/t$ converges to λ both in the mean-square and almost sure senses as t tends to infinity. Indeed

$$E[(N(t)/t - \lambda)^2] = \frac{P_N(t)}{t^2} = \frac{\lambda}{t} \to 0$$

as t tends to infinity, so

$$\frac{N(t)}{t} \overset{m.s.}{\to} \lambda \ .$$

To prove almost sure convergence, observe first that $N(t)$ tends to infinity almost surely as t tends to infinity (verify this). Then by the SLLN

$$\frac{S_{N(t)}}{N(t)} \overset{a.s}{\to} m_X = \frac{1}{\lambda} \, ,$$

where $m_X = 1/\lambda$ denotes the mean of the interarrival times X_k of the Poisson process. By construction of $N(t)$, we have the inequality

$$S_{N(t)} \leq t < S_{N(t)+1} \, ,$$

so by dividing by $N(t)$ we find

$$\frac{S_{N(t)}}{N(t)} \leq \frac{t}{N(t)} \leq \frac{S_{N(t)+1}}{N(t)+1} \frac{N(t)+1}{N(t)} \, .$$

Thus $t/N(t)$ is bracketed by two functions which tend to $1/\lambda$ as t to infinity. Accordingly, the pinching theorem implies

$$\frac{N(t)}{t} \overset{a.s.}{\to} \lambda \, . \tag{7.29}$$

as t tends to infinity.

Gaussian Approximation Although $N(t)$ is integer valued, it is interesting to note its deviations from the slope λt are approximately Gaussian, when properly scaled. Specifically, we have

$$Z(t) \overset{\triangle}{=} \frac{N(t) - \lambda t}{(\lambda t)^{1/2}} \overset{d.}{\to} N(0, 1) \tag{7.30}$$

when either t or λ tend to infinity. This can be shown by noting that $N(t)$ has for characteristic function

$$\Phi_{N(t)}(u) = \exp(\lambda t (\exp(iu) - 1)) \tag{7.31}$$

so the characteristic function of $Z(t)$ is

$$\Phi_{Z(t)}(u) = E[\exp(i(u/(\lambda t)^{1/2})(N(t) - \lambda t))]$$
$$= \Phi_{N(t)}(\frac{u}{(\lambda t)^{1/2}}) \exp(-iu(\lambda t)^{1/2}) \, . \tag{7.32}$$

But for large t or λ, we have

$$\exp(iu/(\lambda t)^{1/2}) \approx 1 + \frac{iu}{(\lambda t)^{1/2}} - \frac{u^2}{2\lambda t} + O(1/(\lambda t)^{3/2}) \, ,$$

where the "big O" notation $O(z)$ indicates a term proportional to z. Substituting this expression inside (7.31) and (7.32) we find that as either t or λ tend to infinity

$$\Phi_{Z(t)}(u) \to \exp(-u^2/2)$$

which is the characteristic function of a $N(0, 1)$ distributed random variable, thus proving (7.30).

Remark The asymptotics results described above also hold, with appropriate modifications, for the class of renewal processes, for which the iid interarrival times have an arbitrary distribution.

7.5   Merging and Splitting of Poisson Processes

Poisson processes have the interesting feature that they remain Poisson under merging and splitting operations. Specifically, if $N_1(t)$ and $N_2(t)$ be two independent Poisson processes with rates λ_1 and λ_2, respectively, let

$$N(t) = N_1(t) + N_2(t) \, .$$

Then $N(t)$ is a Poisson process with rate $\lambda = \lambda_1 + \lambda_2$. To prove this result, note first that $N(0) = N_1(0) + N_2(0) = 0$. Also observe that for any K and any $0 \leq t_0 < t_1 < \ldots < t_K$, the increments

$$N(t_k) - N(t_{k-1}) = N_1(t_k)) - N_1(t_{k-1}) + N_2(t_k) - N_2(t_{k-1})$$

are independent since the components $N_i(t_k) - N_i(t_{k-1})$ are independent for $1 \leq k \leq K$ and $i = 1, 2$ and since the increments $N_1(t_k) - N_1(t_{k-1})$ and $N_2(t_\ell) - N_2(t_{\ell-1})$ for $1 \leq k, \ell \leq K$ are independent because the processes $N_1(\cdot)$ and $N_2(\cdot)$ are independent. In addition, since $N_i(t) - N_i(s)$ with $i = 1, 2$ is Poisson distributed with characteristic function

$$\Phi_{N_i(t)-N_i(s)}(u) = \exp\left(\lambda_i(t-s)(\exp(iu) - 1)\right),$$

the independence of the increments $N_1(t) - N_1(s)$ and $N_2(t) - N_2(s)$ implies that the characteristic function of

$$N(t) - N(s) = N_1(t) - N_1(s) + N_2(t) - N_2(s)$$

is

$$\begin{aligned} \Phi_{N(t)-N(s)}(u) &= \Phi_{N_1(t)-N_1(s)}(u)\Phi_{N_2(t)-N_2(s)}(u) \\ &= \exp\left((\lambda_1 + \lambda_2)(t-s)(\exp(iu) - 1)\right), \end{aligned} \tag{7.33}$$

so that the increments are stationary and Poisson distributed with parameter $\lambda = \lambda_1 + \lambda_2$.

If we consider now the epochs S_k, $k \geq 1$ of the merged process $N(t) = N_1(t) + N_2(t)$, they correspond to epochs of either $N_1(t)$ or $N_2(t)$. If an epoch of $N(t)$ occurs over interval $(t, t+h]$ with h small, the probability it belongs to $N_1(t)$ is

$$\begin{aligned} &P(N_1(t+h) - N_1(t) = 1 | N(t+h) - N(t) = 1) \\ &= \frac{P(N_1(t+h) - N_1(t) = 1, N_2(t+h) - N_2(t) = 0)}{P(N(t+h) - N(t) = 1)} \\ &\approx \frac{\lambda_1 h(1 - \lambda_2 h)}{\lambda h} = \frac{\lambda_1}{\lambda} + o(h) \, . \end{aligned} \tag{7.34}$$

By symmetry, the probability that an epoch of $N(t)$ belongs to $N_2(t)$ is λ_2/λ.

Example 7.1 A service station sells both diesel and gasoline to customers. Diesel customers arrive at a rate λ and gasoline customers arrive at a rate μ. If n customers arrive in a certain period of time, the probability that k of them were diesel customers is

$$P = \binom{n}{k}\left(\frac{\lambda}{\lambda + \mu}\right)^k\left(\frac{\mu}{\lambda + \mu}\right)^{n-k} \, .$$

Conversely, consider a Poisson process $N(t)$ with rate λ, and split $N(t)$ into two processes $N_1(t)$ and $N_2(t)$ by allocating each epoch of $N(t)$ randomly to $N_1(t)$ with probability p, or to $N_2(t)$ with probability $1 - p$. Formally, let Y_k, $k \geq 1$ denote the sequence of independent Bernoulli random variables where $Y_k = 1$ if epoch S_k of $N(t)$ is allocated to $N_1(t)$ and $Y_k = 0$ otherwise. We have

$$P(Y_k = 1) = p \; , \quad P(Y_k = 0) = 1 - p \, ,$$

for $k \geq 1$ and

$$N_1(t) = \sum_{k=1}^{\infty} Y_k u(t - S_k) = \sum_{k=1}^{N(t)} Y_k \tag{7.35}$$

$$N_2(t) = \sum_{k=1}^{\infty} (1 - Y_k) u(t - S_k) = N(t) - N_1(t) \, . \tag{7.36}$$

Processes of the form (7.35) where the random variables Y_ks are iid but not necessarily Bernoulli are called *compound Poisson processes*. Clearly $N(0) = 0$ implies $N_1(0) = N_2(0) = 0$. Over disjoint intervals, the increments

$$N_1(t) - N_1(s) = \sum_{N(s)+1}^{N(t)} Y_k$$

are independent since they are sums with $N(t) - N(s)$ elements of nonoverlapping independent random variables. Thus the processes $N_1(t)$ and $N_2(t)$ have independent increments. To evaluate the joint PMF of increments $N_1(t) - N_1(s)$ and $N_2(t) - N_2(s)$, we note that

$$P(N_1(t) - N_1(s) = k, N_2(t) - N_2(s) = \ell)$$
$$= P(N_1(t) - N_1(s) = k, N(t) - N(s) = k + \ell)$$
$$= P(N_1(t) - N_1(s) = k | N(t) - N(s) = k + \ell) P(N(t) - N(s) = k + \ell), \tag{7.37}$$

where given $N(t) - N(s) = k + \ell$, since the increment $N_1(t) - N_1(s)$ is the sum of $k + \ell$ Bernoulli random variables Y_j, it admits the binomial distribution distribution

$$P(N_1(t) - N_1(s) = k | N(t) - N(s) = k + \ell) = \binom{k + \ell}{k} p^k (1 - p)^\ell \, . \tag{7.38}$$

Substituting (7.38) inside (7.37) and using the Poisson distribution of $N(t) - N(s)$ gives

$$P(N_1(t) - N_1(s) = k, N_2(t) - N_2(s) = \ell)$$
$$= \binom{k + \ell}{k} p^k (1 - p)^\ell \frac{(\lambda(t - s))^{k+\ell}}{(k + \ell)!} \exp(-\lambda(t - s))$$
$$= \frac{(\lambda p(t - s))^k}{k!} \exp(-\lambda p(t - s))$$
$$\times \frac{(\lambda(1 - p)(t - s))^\ell}{\ell!} \exp(-\lambda(1 - p)(t - s)) \, . \tag{7.39}$$

The joint PMF has a factored form and depends only on $t - s$, so the increments $N_1(t) - N_1(s)$ and $N_2(t) - N_2(s)$ are stationary and independent. Furthermore we recognize

$$P(N_1(t) - N_1(s) = k) = \frac{(\lambda p(t - s))^k}{k!} \exp(-\lambda p(t - s)) \tag{7.40}$$

$$P(N_2(t) - N_2(s) = \ell) = \frac{(\lambda(1 - p)(t - s))^\ell}{\ell!} \exp(-\lambda(1 - p)(t - s)), \tag{7.41}$$

so $N_1(t)$ and $N_2(t)$ are independent Poisson processes with rates λp and $\lambda(1 - p)$, respectively. This establishes that the Bernoulli switching scheme splits Poisson process $N(t)$ into two independent Poisson processes $N_1(t)$ and $N_2(t)$.

7.6 Shot Noise

Shot noise occurs whenever a system responds to discrete random excitations such as electrons hitting the plates of a vacuum tube, or photons hitting a photodetector. Shot noise was first investigated in 1918 by the German physicist Walter Schottky [2] who was studying fluctuations in vacuum tubes. Assume that the system we consider is causal and linear time-invariant (LTI) with impulse response $h(t)$. Then, assuming that the system is initially at rest and the interarrival times between excitations is exponential with parameter λ, shot noise can be modeled as

$$Y(t) = \sum_{k=1}^{N(t)} h(t - S_k) = \int_0^t h(t - u)dN(u), \tag{7.42}$$

where $N(t)$ is a Poisson process with parameter λ and epochs S_k. The integral appearing in (7.42) should be interpreted as a pathwise Stieltjes integral. The evaluation of the first- and second-order statistics of this integral is described below. From expression (7.42) we see that $Y(t)$ can be viewed as the output

$$Y(t) = h(t) * X(t)$$

of a LTI system with impulse response $h(t)$ when it is excited by the random impulse train

$$X(t) = \sum_{k=1}^{\infty} \delta(t - S_k). \tag{7.43}$$

In other words, every random event triggers a response, say some electrical current, $h(t - S_k)$, and the shot noise $Y(t)$ is the superposition of the effect of the responses triggered by events prior to t. Typically, $h(t)$ decays exponentially with time, so it is the response $h(t - S_{N(t)})$ to the event $S_{N(t)}$ closest to t which dominates in $Y(t)$, but earlier events have also a small effect, as illustrated in Fig. 7.2 for the case where the system impulse response $h(t) = \exp(-t/2)$. The epochs S_k over [0, 10] are generated by using the independence and uniform distribution of the unordered epochs U_k.

To analyze the statistical properties of shot noise, both expressions in (7.42) can be employed. In this respect, it is convenient to introduce the deviation

$$D(t) = N(t) - \lambda t \tag{7.44}$$

Fig. 7.2 Shot noise
sample path

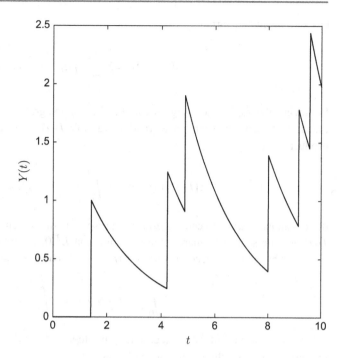

of $N(t)$ with respect to in terms of its mean λt. The process $D(t)$ still has independent increments,
but the increments are also *orthogonal* in the sense that

$$E[(D(t_1) - D(s_1))(D(t_2) - D(s_2))] = 0 \tag{7.45}$$

for two disjoint intervals $(s_1, t_1]$ and $(s_2, t_2]$. Furthermore, over an interval $(s, t]$, we have

$$
\begin{aligned}
E[(D(t) - D(s))^2] &= E[(N(t) - N(s) - \lambda(t - s))^2] \\
&= E[(N(t - s) - \lambda(t - s)^2] = \lambda(t - s),
\end{aligned}
\tag{7.46}
$$

so the variance of the increments is proportional to the length of interval $(s, t]$. Except for the fact
that $D(t) - D(s)$ is not Gaussian, the properties (7.45) and (7.46) are similar to those satisfied by the
increments of a Wiener process and if g is a square integrable function of $L^2[0, T]$, the procedure
used in Sect. 6.5 to construct Wiener integrals can be used to define integrals of the form

$$I(g) = \int_0^T g(u) dD(u). \tag{7.47}$$

Specifically, the function g can be approximated by a sequence of piecewise constant functions g_n
converging to g in $L_2[0, T]$. If $g_n(t) = c_i$ over $[t_{i-1}, t_i]$ for $1 \le i \le I$, with $t_0 = 0$, $t_I = T$, the
random variable

$$I(g_n) = \sum_{i=1}^{I} c_i (D(t_i) - D(t_{i-1}))$$

has zero mean and variance

$$E[I^2(g_n)] = \lambda \sum_{i=1}^{I} c_i^2 (t_i - t_{i-1}) = \lambda ||g_n||^2 .$$

Following the same reasoning as for the Wiener integral, we deduce that $I(g_n)$ converges in the mean square sense to a zero-mean random variable $I(g)$ with variance $\lambda ||g||^2$. Furthermore, the inner product

$$E[I(g_1)I(g_2)] = \lambda \int_0^T g_1(t)g_2(t)dt = \lambda < g_1, g_2 >$$

defines an isometry between integrals with respect to the increments of orthogonal increments process $D(t)$ and the space of square integrable function $L^2[0, T]$. Returning now to the integral appearing in (7.42), by using the decomposition $N(t) = \lambda t + D(t)$, we deduce that for $g \in L^2[0, T]$

$$Y = \int_0^T g(u)dN(u) = \lambda \int_0^T g(u)du + I(g) \tag{7.48}$$

is the sum of a deterministic integral and an integral of the form (7.47) with respect to increments of process $D(t)$. Accordingly the mean of Y is

$$m_Y = \lambda \int_0^T g(u)du$$

and its variance is

$$K_Y = \lambda ||g||^2 = \lambda \int_0^T g^2(u)du .$$

Furthermore any two random variables Y_1 and Y_2 of the form (7.48) and corresponding to g_1 and g_2 in $L^2[0, T]$, respectively, have for covariance

$$K_{Y_1 Y_2} = E[(Y_1 - m_{Y_1})(Y_2 - m_{Y_2})] = \lambda < g_1, g_2 > .$$

We can now apply directly the expressions derived above to the evaluation of the mean, variance, and autocorrelation of the shot noise process $Y(t)$ defined in (7.42). Its mean is

$$m_Y(t) = \lambda \int_0^t h(t - u)du = \lambda \int_0^t h(v)dv , \tag{7.49}$$

where the change of variable $v = t - u$ has been used to obtain the second equality. As t tends to infinity, $m_Y(t)$ tends to

$$m_Y = \lambda \int_0^\infty h(v)dv = \lambda H(i0), \tag{7.50}$$

where

$$H(j\omega) = \int_0^\infty h(u) \exp(-j\omega u)du$$

denotes the frequency response of the filter $h(t)$. The variance of $Y(t)$ is given by

$$K_Y(t) = \lambda \int_0^t h^2(t-u)du = \lambda \int_0^t h^2(v)dv .$$ (7.51)

As t tends to infinity, it tends to

$$K_Y = \lambda \int_0^\infty h^2(v)dv = \lambda E_h,$$ (7.52)

where E_h denotes the energy of the impulse response $h(t)$. The formulas (7.50) and (7.52) are usually called Campbell's theorem, since they were first derived in 1909 by the British physicist Norman Campbell [3] who was investigating thermionic noise in vacuum tubes. The autocorrelation of $Y(t)$ is given by

$$R_Y(t, s) = m_Y(t)m_Y(s) + \lambda \int_0^{\min(t,s)} h(t-u)h(s-u)du$$
$$= m_Y(t)m_Y(s) + \lambda \int_0^{\min(t,s)} h(|\tau|+v)h(v)dv$$ (7.53)

with $\tau = t - s$, where we performed the change of variable $v = \min(t, s) - u$ to go from the first to the second line. Then as t and s tend jointly to infinity, we find that the autocorrelation $R_Y(t, s)$ tends to

$$R_Y(\tau) = m_Y^2 + \lambda r_h(\tau),$$ (7.54)

where

$$r_h(\tau) = h(\tau) * h(-\tau)$$

denotes the deterministic autocorrelation of $h(\tau)$. This shows that $Y(t)$ is asymptotically WSS. The transient period is due to the fact that we assumed that the system was at rest at $t = 0$ and that the train of Poisson impulses started only at $t = 0$. If the start of the Poisson process is shifted to $-\infty$, the shot noise is WSS with mean (7.50) and autocorrelation (7.54).

It is also possible to characterize the probability distribution of $Y(t)$ by using the first expression in (7.42). If we condition with respect to $N(t) = n$, then

$$Y(t) = \sum_{k=1}^n h(t - S_k) = \sum_{k=1}^n h(t - U_k),$$ (7.55)

where U_k with $1 \le kn$ denotes the unordered epochs of the Poisson process. Since these epochs are independent and uniformly distributed over $[0, t]$, conditioned on $N(t) = n$, the characteristic function of $Y(t)$ can be expressed as

$$\Phi_{Y(t)|N(t)}(u|n) = E[\exp(juY(t))|N(t) = n]$$
$$= \prod_{k=1}^n E[\exp(juh(t - U_k))]$$
$$= \left(\frac{1}{t} \int_0^t \exp(juh(w))dw\right)^n .$$ (7.56)

Then the characteristic function of $Y(t)$ is

$$\Phi_{Y(t)}(u) = \sum_{n=0}^{\infty} \Phi_{Y(t)|N(t)}(u|n) P(N(t) = n)$$

$$= \exp(-\lambda t) \sum_{n=0}^{\infty} \frac{1}{n!} \left(\lambda \int_0^t \exp(juh(w)) dw \right)^n$$

$$= \exp \left(\lambda \int_0^t (\exp(juh(w)) - 1) dw \right). \tag{7.57}$$

Computing the first and second derivatives of $\Phi_{Y(t)}(u)$ with respect to u and setting $u = 0$ gives

$$\frac{1}{j} \frac{d}{du} \Phi_{Y(t)}(u) \mid_{u=0} = \lambda \int_0^t h(w) dw = m_Y(t)$$

$$\frac{1}{j^2} \frac{d^2}{du^2} \Phi_{Y(t)}(u) \mid_{u=0} = (m_Y(t))^2 + \lambda \int_0^t h^2(w) dw,$$

which is consistent with the expressions (7.49) and (7.51) for the first- and second-order moments of $Y(t)$.

If we assume that the impulse response $h(t)$ is such that the integral

$$\int_0^{\infty} (\exp(juh(w)) - 1) dw$$

exists for all u, as t tends to infinity

$$\Phi_{Y(t)}(u) = \exp \left(\lambda \int_0^{\infty} (\exp(juh(w)) - 1) dw \right) \tag{7.58}$$

does not depend on t, so that $Y(t)$ is asymptotically stationary. The asymptotic characteristic function $\Phi_{Y(t)}(u)$ given by (7.58) can be used to obtain a Gaussian approximation of $Y(t)$ in the high rate (large λ) regime. Consider the random variable

$$Z(t) \triangleq \frac{Y(t) - \lambda H(j0)}{(\lambda E_h)^{1/2}}. \tag{7.59}$$

Its characteristic function is

$$\Phi_{Z(t)}(u) = \Phi_{Y(t)}(u/(\lambda E_h)^{1/2}) \exp(-ju\lambda^{1/2} H(j0)/E_h^{1/2}), \tag{7.60}$$

and by using the Taylor series approximation

$$\int_0^{\infty} (\exp(juh(w)/(\lambda E_h)^{1/2}) - 1) dw = \frac{juH(j0)}{(\lambda E_h)^{1/2}} - \frac{u^2}{2\lambda} + O(\lambda^{-3/2})$$

we find that

$$\lim_{\lambda \to \infty} \Phi_{Z(t)}(u) = \exp(-u^2/2),$$

which shows that $Z(t)$ converges in distribution to a $N(0, 1)$ random variable as λ tends to infinity.

This result can be interpreted as follows: suppose $H(j0) = 0$ so that the system DC response is zero. Suppose also that as λ increases and the shocks to which the system is subjected become more frequent, the energy E_h of the response to each shock scales like $1/\lambda$. In other words, we are trading off a very large energy released by infrequent shocks against much smaller energies in response to frequent shocks (the average released energy per unit time remains constant). Then as λ becomes large, the shot noise process $Y(t)$ becomes Gaussian and thus indistinguishable from thermal noise.

7.7 Bibliographical Notes

Good discussions of Poisson processes can be found in [1, 4]. Renewal processes form a natural extension of Poisson processes, and in addition to the aforementioned references, the books by Karlin and Taylor [5] and Çinlar [6] contain concise expositions of renewal processes. In this respect, renewal processes provide a convenient tool to analyze successive visits to a Markov chain state, as explained in [6]. The approach used in Sect. 7.6 to define stochastic integrals with respect to the increments of orthogonal increments process $D(t) = N(t) - \lambda t$ is described in detail in [7, Chap. 9]. Beyond Campbell and Schottky's early studies of shot noise, Rice [8] evaluated its autocorrelation function and the characteristic function of its probability distribution, and Gilbert and Pollak [9] analyzed its CDF. The papers by Picinbono et al. [10] and Papoulis [11] examine the validity or lack thereof of Gaussian approximations of shot noise. Finally, Lowen and Teich [12] used a shot noise process with a power law impulse response to construct a model of $1/f$ noise. It should also be mentioned that other processes, such as compound Poisson processes, doubly stochastic Poisson processes, or self-exciting Poisson processes with a wide variety of applications to quantum electronics [13] or medical imaging [14] can be constructed by extending various features of Poisson processes.

7.8 Problems

7.1 Let $N(t)$ be a Poisson process with rate λ. We showed that $N(t)$ is a Markov process with forward conditional PMF (7.2). Since the Markov property is preserved by reversing the time direction, for $s < t$ and $0 \le k \le \ell$, find the backward conditional PMF

$$P(N(s) = k|N(t) = \ell) \,.$$

7.2 In Example 4.4, the random telegraph wave process $X(t)$ with $t \ge 0$ was studied from the perspective of Markov processes. However, it can also be viewed as a jump process whose values alternate between 1 and -1, with the following properties:

(i) $X(0)$ takes the values 1 and -1 with equal probability.
(ii) The number of jumps in any time interval is independent of $X(0)$.
(iii) $X(t) = X(0)$ if the number of jumps between 0 and t is even; $X(t) = -X(0)$ if the number of jumps is odd.
(iv) If k is a positive integer, the probability $P(k; t, T)$ that $X(t)$ jumps k times over the interval $(t, t + T]$ is

$$P(k; t, T) = \frac{(\lambda T)^k}{k!} \exp(-\lambda T) \,,$$

where $\lambda > 0$ is a constant.

(v) If K_1, K_2, ..., K_N denote the number of jumps over a set of disjoint time intervals $(s_1, t_1]$, $(s_2, t_2]$, ..., $(s_N, t_N]$, the random variables K_i with $1 \leq i \leq N$ are independent.

A typical sample path of the random telegraph wave is depicted in Fig. 4.2.

a) Show that the above axioms are equivalent to defining $X(t)$ as

$$X(t) = X(0) (-1)^{N(t)},$$

where $N(t)$ is a Poisson counting process with rate λ.

b) Show that $X(t)$ is a Markov process, and use the characterization of part a) to compute the PMF $P[X(t) = x_t]$ with $x_t = \pm 1$, and the conditional PMF $P(X(t) = x_t | X(s) = x_s)$ with x_t, $x_s = \pm 1$. Verify that they coincide with the PMFs used to study $X(t)$ in Example 4.4.

7.3 Consider a Poisson process with epochs S_k, $k \geq 1$.

a) Conditioned on $N(t) = n$, show that the conditional PDF $f_{S_k|N(t)}(s|n)$ of the k-th epoch S_k, with $k \leq n$ is given by

$$f_{S_k|N(t)}(s|n) = \frac{n!}{(k-1)!(n-k)!} \frac{1}{t} (s/t)^{k-1} (1 - s/t)^{n-k} .$$

b) Prove that

$$E[S_k|N(t) = n] = \frac{kt}{n+1} .$$

Hint: First show that

$$E[S_1|N(t) = n] = \frac{t}{n+1} ,$$

and then use integration by parts to express $E[S_{k+1}|N(t) = n]$ in terms of $E[S_k|N(t) = n]$.

7.4 A bus arrives at its terminal and leaves T minutes later. The bus is initially empty, but riders arrive according to a Poisson process $N(t)$ with rate λ/min and take a seat on the bus until the bus leaves. If the epochs of the Poisson process are S_k, $k \geq 1$, the average waiting time experienced by the riders before the bus leaves is

$$W = \frac{1}{N(T)} \sum_{k=1}^{N(T)} (T - S_k) .$$

Find the mean of W. *Hint:* Condition with respect to $N(T) = n$.

7.5 Consider a Poisson process with epochs S_k, $k \geq 1$. Use the joint PDF of the first $n + 1$ epochs obtained in (7.14) to evaluate the conditional density

$$f_{S_1,...,S_n|S_{n+1}}(s_1, ..., s_n|s_{n+1})$$

of the first n epochs given that the $n + 1$-th epoch $S_{n+1} = s_{n+1}$.

7.6 Suppose buses arrive at a bus stop according to a Poisson process $N(t)$ with rate λ. The interarrival times of the process are X_k, $k \geq 1$ and its epochs are $S_n = \sum_{k=1}^{n} X_k$, $n \geq 1$. We saw in

Fig. 7.3 Sample path of
the residual interarrival
time process $R(t)$

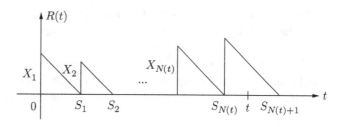

Sect. 7.3 that if a rider arrives at time t, the rider's waiting time $R(t) = S_{N(t)+1} - t$ is exponentially distributed with parameter λ, so the mean waiting time is $m_X = 1/\lambda$. We consider here an alternative verification of this result. First, consider the sample path of the residual process $R(t)$ depicted in Fig. 7.3. Suppose that the arrival time U of the rider is uniformly distributed over $[0, T]$, so its PDF is

$$f_U(u) = \begin{cases} 1/T & 0 \le u \le T \\ 0 & \text{otherwise} . \end{cases}$$

Then the expected waiting time is

$$\bar{R}(T) = E_U[R(U)] = \frac{1}{T} \int_0^T R(u)du ,$$

where the expectation $E_U[.]$ is taken only with respect to U, not with respect to the statistics of the Poisson process, so $\bar{R}(T)$ is a random variable depending on the Poisson process sample path between 0 and T.

a) By observing that the integral of $R(u)$ over $[0, T]$ represents the area under the sample path of Fig. 7.3, verify that

$$\bar{R}(T) \approx \frac{1}{T} \sum_{k=1}^{N(T)} \frac{X_k^2}{2} \tag{7.61}$$

for large T.

b) By rewriting expression (7.61) as

$$\bar{R}(T) = \frac{N(T)}{T} \frac{1}{N(T)} \sum_{k=1}^{N(T)} \frac{X_k^2}{2} , \tag{7.62}$$

use the SLLN and the almost sure convergence of the slope $N(T)/T$ to show that the expected waiting time $\bar{R}(T)$ converges almost surely as T tends to infinity. Is the value you obtain consistent with the exponential model of the residual interarrival time $R(t)$?

7.7 Buses and riders arrive at a bus stop according to two independent Poisson processes with rates λ and μ, respectively. To ensure that each bus is sufficiently full μ will usually be much larger than λ. However, it should not be too large if we want to avoid situations where the number of riders waiting for a bus exceeds the bus capacity.

a) Let C denote the bus capacity. if M denote the number of riders waiting at a stop when a bus arrives, what is the probability that $P(M > C)$?

b) Assuming μ is fixed, how should the rate of bus arrival λ be selected to ensure that $E[M] \leq C/4$, so that on average not more than 25% of the bus capacity gets filled at a single stop.

7.8 Buses and riders arrive at a bus stop according to two independent Poisson processes with rates λ and μ respectively. Usually, riders arrive at a much faster rate than buses, so that $\mu > \lambda$. Suppose M denotes the number of riders waiting in line for the next bus at any given time.

a) What is the PMF $P(M = k)$?
b) What is the conditional PDF of the waiting time $R(t)$ for the next bus, given $M = k$?
c) What is the conditional PDF of the elapsed time $E(t)$ since the previous bus arrived, given $M = k$? Can you recognize this distribution?
d) What is the conditional expectation $E[E(t)|M = k]$?

7.9 After landing at a European airport, travelers arrive to the passport control station according to a Poisson process with rate λ. With probability p, a traveler holds a EU passport, and this traveler goes through with only minimal control, without leaving any record. With probability $1 - p$, the traveler arrives from from a non-EU country, and his/her passport is checked and the entry is recorded. Let $N(t)$ be the total number of arriving passengers over interval $[0, t]$, $S(t)$ is the number of passengers who hold EU passports and $C(t)$ the number of non-EU passengers. Given that $C(t) = k$, what is the conditional mean and variance of the total number of passengers $N(t)$ who passed through passport control during $[0, t]$?

7.10 You have been invited with several friends to a wedding that will take place at 5 pm exactly. You arrived early, and while waiting you observe that guests arrive according to a Poisson process at a rate of 1 per minute. Your friends suggest betting on the last guest to arrive before the wedding. Your friends are all English and business majors who have never heard of probability distributions, but after taking an engineering random processes course, Poisson processes hold no secret for you. You decide that the optimum strategy is to wait until s minutes before 5 pm, and bet on the first guest through the door after that time. So you will lose if either no guest arrives after that time, or more than one arrive before 5pm. How should you select s to maximize your probability of winning, and for this choice, what is your winning probability?

7.11 A detector is capable of detecting both α and β particles. Alpha and beta particles hit the detector according to two independent Poisson processes with rates λ_A and λ_B, respectively. After a particle is recorded by the detector, there is a reset period of duration T during which arriving particles are ignored. Such particles are said to be erased.

a) If a particle hits the detector, what is the probability it is an α particle?
b) Find the probability that among the first 10 recorded particles, 7 are alpha particles and 3 are beta particles.
c) If a particle hits the detector, what is the probability it will be erased?
d) Is the arrival process formed by the recorded particles a Poisson process? If so, what is its rate? If not, explain why it cannot be a Poisson process.
e) Find the probability distribution (either PDF or CDF) of the interval of time between two successive recorded particles.

Fig. 7.4 Sample path of the compound process $Z(t)$

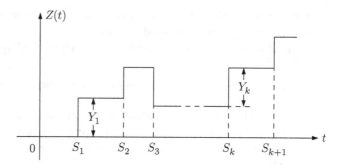

7.12 Let $N(t)$ be a Poisson process with rate λ. Let $\{Y_k, \ 1 \le k < \infty\}$ be a sequence of independent identically distributed random variables with probability density $f_Y(y)$. Then, consider the *compound Poisson process*:

$$Z(t) = \begin{cases} 0 & N(t) = 0 \\ \displaystyle\sum_{k=1}^{N(t)} Y_k & N(t) > 0. \end{cases}$$

The sample functions of this process are random staircases with exponentially distributed runs, and $f_Y(y)$ distributed rises, as shown in Fig. 7.4. In this figure, the epochs of Poisson process $N(t)$, i.e., the times at which $N(t)$ increases, are denoted as S_k. It is assumed that the first and second moments of $f_Y(y)$ are finite and denoted by $m_1 = E[Y]$ and $m_2 = E[Y^2]$.

a) Is $Z(t)$ an independent increments process? Is it Markov?

b) If $\Phi_Y(u) = E[\exp(juY)]$ denotes the characteristic function corresponding to the density $f_Y(y)$. show that the characteristic function $\Phi_{Z(t)-Z(s)}(u) = E[\exp(ju(Z(t) - Z(s)))]$ of the increments $Z(t) - Z(s)$ for $t > s$ can be expressed as

$$\Phi_{Z(t)-Z(s)}(u) = \exp[\lambda(t - s)(\Phi_Y(u) - 1)].$$

To do so, you may want to first evaluate $E[\exp(iu(Z(t) - Z(s)) \mid N(t) - N(s)]$, and then use iterated conditioning to compute $\Phi_{Z(t)-Z(s)}(u)$. Are the increments of $Z(t)$ stationary?

c) Use the results of parts a) and b) to evaluate the mean $m_Z(t)$, variance $K_Z(t, t)$ and autocovariance $K_Z(t, s)$ of $Z(t)$. Express your results in terms of λ, m_1 and m_2.

d) When the rate λ and length $t - s$ of interval $[s, t]$ are such that $\lambda(t - s) \gg 1$, it is reasonable to expect that the increment $N(t) - N(s)$ of the Poisson process over $[s, t]$ is large, so the increment

$$Z(t) - Z(s) = \sum_{N(s)+1}^{N(t)} Y_k$$

is the sum of a large but random number of independent identically distributed random variables Y_k. A modified version of the central limit theorem should therefore be applicable to this sum. By using the expression for $\Phi_{Z(t)-Z(s)}(u)$ obtained in part b), show that the normalized increments

$$\frac{(Z(t) - Z(s)) - \lambda(t - s)m_1}{(\lambda(t - s))^{1/2}}$$

converge in distribution to a zero-mean Gaussian random variable as $\lambda(t - s)$ tends to infinity. What is the variance of this random variable? To prove this result, you may want to observe that for small v, the characteristic function $\Phi_Y(v)$ admits the Taylor series approximation

$$\Phi_Y(v) \approx 1 + jvm_1 - \frac{v^2}{2}m_2 .$$

e) Suppose we are interested in generating samples of the compound Poisson process $Z(t)$ at times $t_0 = 0 < t_1 < \ldots < t_N$, where successive time increments are such that $\lambda(t_i - t_{i-1}) \gg 1$ for all i between 1 and N, can you suggest a simple simulation technique based on the approximation of part d)?

7.13 Let $N(t)$ be a Poisson process with rate λ. Let $\{Y_k, k \geq 1\}$ be a sequence of independent identically distributed exponential random variables with parameter μ, so that each Y_k admits the density

$$f_Y(y) = \mu \exp(-\mu y)u(y),$$

where $u(y)$ denotes the unit step function. The random variables Y_k are independent of the process $N(t)$. Let $\{S_k,\ k \geq 1\}$ denote the epochs of $N(t)$. Then, consider the generalized shot-noise process

$$Z(t) = \begin{cases} 0 & N(t) = 0 \\ \displaystyle\sum_{k=1}^{N(t)} Y_k \exp(-a(t - S_k)) & N(t) > 0 . \end{cases}$$

The main difference between this process and the version of the shot-noise process examined in Sect. 7.6 is that the amplitudes Y_k of the excitations are random instead of having unit value.

a) Is $Z(t)$ a Markov process? To answer this question, you may want to use the decomposition

$$Z(t) = \exp(-a(t - s))Z(s) + \sum_{N(s)+1}^{N(t)} Y_k \exp(-a(t - S_k))$$

  for $t \geq s$.
b) Find the characteristic function $\Phi_Y(u) = E[\exp(juY)]$ of the random excitations Y_k.
c) Given that $N(t) = n$, it was shown in Sect. 7.2 that the unordered epochs U_k with $1 \leq k \leq n$ are independent and uniformly distributed over the interval $[0, t)$, so that their density takes the form

$$f_U(u) = \begin{cases} 1/t & 0 \leq u < t \\ 0 & \text{otherwise} . \end{cases}$$

Then the epochs S_k are obtained by rank ordering the random variables U_k, so that S_1 is the smallest of the U_k's, S_2 is the second smallest, and S_n is the largest. By observing that given that $N(t) = n$, $Z(t)$ can be expressed as

$$Z(t) = \sum_{k=1}^{n} Y_k \exp(-a(t - U_k)) ,$$

evaluate the conditional characteristic function

$$\Phi_{Z(t)|N(t)}(u|n) = E[\exp(juZ(t)) \mid N(t) = n].$$

d) Use the result of part c) to find the characteristic function

$$\Phi_{Z(t)}(u) = E[\exp(juZ(t)]$$

of the process $Z(t)$.

e) Use the results of parts a) and d) to evaluate the mean $m_Z(t)$, variance $K_Z(t, t)$, and autocovariance $K_Z(t, s)$ of $Z(t)$. Is $Z(t)$ asymptotically WSS? Explain your response.

f) Show that as $t \to \infty$, we have

$$\lim \Phi_{Z(t)}(u) = \left(\frac{\mu}{\mu - ju}\right)^b.$$

Express b in terms of λ and a. But $(\mu/(\mu - ju))^b$ is the characteristic function of the gamma density function

$$f_Z(z) = \frac{\mu(\mu z)^{b-1}}{\Gamma(b)} \exp(-\mu z).$$

7.14 Let $N(t)$ be a Poisson process with rate λ and epochs $\{S_k, k \geq 1\}$ and let $\{Y_k, k \geq 1\}$ denote a sequence of independent $N(0, K_Y)$ distributed random variables. Consider the generalized shot noise process

$$Z(t) = \begin{cases} 0 & N(t) = 0 \\ \sum_{k=1}^{N(t)} Y_k h(t - S_k) & N(t) \geq 1 \end{cases}$$

obtained by passing a random impulsive train

$$X(t) = \sum_{k=1}^{\infty} Y_k \delta(t - S_k)$$

through a LTI system with impulse response $h(t)$. The energy of $h(t)$ is denoted as

$$E_h = \int_0^{\infty} h^2(w)dw.$$

a) Assume that $N(t) = n$. By rewriting $Z(t)$ in terms of the unordered epochs U_k, $1 \leq k \leq n$ of $N(t)$ as

$$Z(t) = \sum_{k=1}^{n} Y_k h(t - U_k)$$

compute the conditional characteristic function

$$\Phi_{Z(t)|N(t)}(u|n) = E[\exp(juZ(t))|N(t) = n].$$

Recall that the characteristic function of the Y_ks is

$$\Phi_Y(u) = \exp(-u^2 K_Y/2)\,.$$

b) Obtain the characteristic function $\Phi_{Z(t)}(u)$.
c) Consider the high rate case where λ tends to infinity, but at the same time $K_Y = \bar{K}_Y/\lambda$, so that the standard deviation of the shocks varies like $\lambda^{-1/2}$ as the shock rate increases. Show that as λ tends to infinity, $Z(t)$ is Gaussian distributed. What are its mean and variance?

References

1. S. Ross, *Stochastic Processes, 2nd edition*. New York, NY: J. Wiley, !996.
2. W. Schottky, "Über spontane stromschwankungen in verschiedenen elektrizitätsleitern," *Annalen der Physik*, vol. 57, pp. 541–567, 1918.
3. N. Campbell, "Discontinuous phenomena," *Proc. Cambridge Phil. Soc.*, vol. 15, pp. 117–136, 1909.
4. R. G. Gallager, *Stochastic Processes: Theory for Applications*. Cambridge, United Kingdom: Cambridge University Press, 2014.
5. S. Karlin and H. M. Taylor, *A First Course in Stochastic Processes, 2nd edition*. New York, NY: Academic Press, 1975.
6. E. Çinlar, *Introduction to Stochastic Processes*. Englewood Cliffs, NJ: Prentice-Hall, 1975. Reprinted by Dover Publications, Mineola NY, 2013.
7. J. L. Doob, *Stochastic Processes*. New York, NY: J. Wiley & Sons, 1953.
8. S. O. Rice, "Mathematical analysis of random noise, parts I and II," *Bell System Tech J.*, vol. 23, pp. 282–332, July 1944.
9. E. N. Gilbert and H. O. Pollak, "Amplitude distribution of shot noise," *Bell System Tech. J.*, vol. 39, pp. 333–350, 1960.
10. B. Picinbono, C. Bendjaballah, and J. Pouget, "Photoelectron shot noise," *J. Math. Phys.*, vol. 11, pp. 2166–2176, July 1970.
11. A. Papoulis, "High density shot noise and Gaussianity," *J. Applied Probability*, vol. 8, pp. 118–127, 1971.
12. S. B. Lowen and M. C. Teich, "Power-law shot noise," *IEEE Trans. Informat. Theory*, vol. 36, pp. 1302–1318, Nov. 1990.
13. M. C. Teich and B. E. A. Saleh, "Branching processes in quantum electronics," *IEEE J. Selected Topics in Quantum Electronics*, vol. 6, pp. 1450–1457, Nov/Dec 2000.
14. D. L. Snyder and M. I. Miller, *Random Point Processes in Time and Space, second edition*. New York: Springer Verlag, 1991.

Processing and Frequency Analysis of Random Signals

8

8.1 Introduction

In this chapter, we discuss the effect of basic signal processing operations performed on random signals, such as linear filtering and modulation. Most of the attention is focused on WSS random signals, since such signals can be analyzed in the Fourier domain by using their power spectral density (PSD), which as explained in Sect. 8.3 provides a picture of the frequency composition of the signal of interest. The effect of linear filtering on the mean and autocorrelation of random signals is described in Sect. 8.2. The PSD and its properties are analyzed in Sect. 8.3. In this context, it is explained that WGN is undistinguishable from bandlimited WGN when passed through filters with a finite bandwidth, which explains why WGN is so convenient for performing noise computations. The Nyquist–Johnson model of noisy resistors is extended to RLC circuits in thermal equilibrium in Sect. 8.4. Techniques for evaluating the PSDs of PAM and Markov-chain modulated signals are described in Sect. 8.5. Then in Sect. 8.6, it is shown that a bandlimited WSS signal with bandwidth B can be reconstructed in the mean-square sense from its uniform samples provided the sampling frequency is greater than twice the signal bandwidth. This result extends to the WSS case the classical reconstruction result of Nyquist and Shannon for deterministic signals. Finally, Sect. 8.7 presents Rice's model of bandpass WSS signals, which represents a bandpass WSS signal with occupied bandwidth $2B$ centered about ω_c as a quadrature modulated signal with carrier frequency ω_c and WSS in-phase and quadrature components of bandwidth B. This model is used routinely to analyze modulated bandpass communications or radar receivers by examining an equivalent baseband problem, thereby removing the the effect of modulation from consideration.

8.2 Random Signals Through Linear Systems

Consider a CT random signal with mean $m_X(t) = E[X(t)]$, autocorrelation $R_X(t, s) = E[X(t)X(s)]$, and autocovariance

$$K_X(t, s) = R_X(t, s) - m_X(t)m_X(s) .$$

This signal is passed through a CT linear system with time varying impulse response $h(t, s)$, where $h(t, s)$ denotes the system response at time t to an impulse applied at time s. The corresponding output can therefore be expressed as

© Springer Nature Switzerland AG 2020
B. C. Levy, *Random Processes with Applications to Circuits and Communications*,
https://doi.org/10.1007/978-3-030-22297-0_8

$$Y(t) = \int_{-\infty}^{\infty} h(t, u)X(u)du \ . \tag{8.1}$$

We are interested in evaluating the first- and second-order joint statistics of $Y(t)$ and $X(t)$. By taking expectations on both sides of (8.1), we find that the mean of $Y(t)$ is given by

$$m_Y(t) = E[Y(t)] = \int_{-\infty}^{\infty} h(t, u)m_X(u)du \ , \tag{8.2}$$

and the cross-correlation between the output $Y(t)$ and input $X(s)$ is given by

$$R_{YX}(t, s) = E[Y(t)X(s)] = E\left[\int_{-\infty}^{\infty} h(t, u)X(u)du \ X(s) \right]$$

$$= \int_{-\infty}^{\infty} h(t, u)E[X(u)X(s)]du = \int_{-\infty}^{\infty} h(t, u)R_X(u, s)du \ . \tag{8.3}$$

Finally, the autocorrelation of $Y(t)$ is given by

$$R_Y(t, s) = E[Y(t)Y(s)] = E\left[Y(t) \int_{-\infty}^{\infty} h(s, v)X(v)dv \right]$$

$$= \int_{-\infty}^{\infty} h(s, v)R_{YX}(t, v)dv$$

$$= \int_{-\infty}^{\infty} \int_{-\infty}^{\infty} h(t, u)R_X(u, v)h(s, v)dudv \ , \tag{8.4}$$

where we have used (8.3) to go from the second to the third line.

WSS Linear Time-Invariant Case Up to this point no assumption has been made except the linearity of the filter h through which the random signal $X(t)$ is passed. Suppose now that $X(t)$ is WSS, so that its mean $m_X(t) = m_X$ is constant, and its autocorrelation $R_X(t - s)$ depends only on $\tau = t - s$. Assume also that in addition to being linear, the filter h is time invariant, so that its impulse response $h(t - s)$ depends only on the time elapsed between the time s at which the impulse is applied, and the observation time t. Then, as depicted in Fig. 8.1, the output

$$Y(t) = h(t) * X(t) = \int_{-\infty}^{\infty} h(t - u)X(u)du \tag{8.5}$$

is now the convolution of $h(t)$ and $X(t)$.

Then the mean of $Y(t)$ given by (8.2) becomes

$$m_Y = m_X \int_{-\infty}^{\infty} h(t - u)du = m_X \int_{-\infty}^{\infty} h(v)dv, \tag{8.6}$$

Fig. 8.1 WSS random signal $X(t)$ through LTI filter $h(t)$

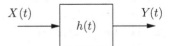

where to go from the first to the second equality, we performed the change of variable $v = t - u$. Thus $Y(t)$ has a *constant mean*, and if

$$H(j\omega) = \int_{-\infty}^{\infty} h(t)\exp(-j\omega t)dt$$

denotes the Fourier transform of $h(t)$, we can also write m_Y as

$$m_Y = H(j0)m_X,$$

where $H(j0)$ denotes the DC value, i.e., the response at $\omega = 0$ of filter $H(j\omega)$. Likewise, the expression (8.3) for the cross-correlation between output process $Y(t)$ and input $X(t)$ becomes

$$R_{YX}(t, s) = \int_{-\infty}^{\infty} h(t - u)R_X(u - s)du$$

$$= \int_{-\infty}^{\infty} h(v)R_X(\tau - v)dv,$$

where $\tau = t - s$ and to go from the first to the second line we performed the change of variable $v = t - u$. This shows that the cross-correlation of $Y(t)$ and $X(s)$ depends only on τ and

$$R_{YX}(\tau) = h(\tau) * R_X(\tau) \tag{8.7}$$

is just the convolution of filter $h(\tau)$ with autocorrelation function $R_X(\tau)$. Similarly, by using the second line of (8.4) we find that the autocorrelation of $Y(t)$ satisfies

$$R_Y(t, s) = \int_{-\infty}^{\infty} R_{YX}(t - v)h(s - v)dv$$

$$= \int_{-\infty}^{\infty} R_{YX}(\tau - u)h(-u)du,$$

where $\tau = t - s$ and to go from the first to the second line we have performed the change of variable $u = v - s$. Since the autocorrelation depends on τ only, $Y(t)$ is WSS with

$$R_Y(\tau) = R_{YX}(\tau) * h(-\tau) = h(\tau) * R_X(\tau) * h(-\tau). \tag{8.8}$$

From identities (8.6)–(8.8), we conclude that when a WSS signal $X(t)$ is passed through a LTI filter, the output $Y(t)$ is *jointly WSS* with the input $X(t)$, i.e., $Y(t)$ itself is WSS and the cross-correlation of $Y(t)$ and $X(s)$ depends on $\tau = t - s$ only.

Example 8.1 In our analysis of the effect of thermal noise on a RC circuit in Sect. 6.6, we used a circuit viewpoint. However, it is possible to derive the same results by using a systems viewpoint. Specifically the RC circuit shown in Fig. 6.3 can be rewritten in the impedance domain as Fig. 8.2, where $\hat{N}(s)$ and $\hat{V}(s)$ denote the Laplace transforms of input voltage source $N(t)$ and capacitor voltage $V(t)$, respectively.

Fig. 8.2 RC circuit in the
impedance domain

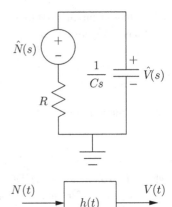

Fig. 8.3 Systems model
of the noisy RC circuit of
Fig. 6.3

Then, by using the voltage divider rule, the transfer function $H(s)$ relating output $\hat{V}(s)$ to input $\hat{N}(s)$ is given by

$$H(s) = \frac{1/(Cs)}{R + 1/(Cs)} = \frac{1/(RC)}{s + 1/(RC)},$$

and the corresponding impulse response is given by

$$h(t) = \frac{1}{RC} \exp(-t/(RC))u(t) \, .$$

In steady state, the circuit of Fig. 6.3 admits therefore the equivalent input–output model shown in Fig. 8.3. If the input $N(t)$ is a WGN with autocorrelation

$$R_N(\tau) = 2kRT\delta(\tau) \, ,$$

we deduce that the cross-correlation between $V(t)$ and $N(s)$ is given by

$$R_{VN}(\tau) = h(\tau) * R_N(\tau) = \frac{2kT}{C} \exp(-\tau/(RC))u(\tau) \, ,$$

where $\tau = t - s$. Then the autocorrelation of $V(t)$ is given by

$$R_V(\tau) = R_{VN}(\tau) * h(-\tau) = \int_{-\infty}^{\infty} h(-v)R_{VN}(\tau - v)du.$$

For $\tau > 0$, this gives

$$R_V(\tau) = \frac{2kT}{RC^2} \int_{-\infty}^{0} \exp(v/(RC)) \exp(-(\tau - v)/(RC))dv$$

$$= \frac{2kT}{RC^2} \exp(-\tau/(RC)) \int_{-\infty}^{0} \exp(2v/(RC))dv$$

$$= \frac{kT}{C} \exp(-\tau/(RC)) \, ,$$

and by symmetry of the autocorrelation function we deduce that

$$R_V(\tau) = \frac{kT}{C} \exp(-|\tau|/(RC)) \,,$$

which is precisely the steady-state autocorrelation of the Ornstein–Uhlenbeck process obtained in Sect. 6.6.

8.3 Power Spectral Density

The power spectral density (PSD) of a WSS random signal $X(t)$ is defined as the Fourier transform of its autocorrelation function. In the continuous-time (CT) case, the PSD of $X(t)$ is therefore given by

$$S_X(\omega) = \int_{-\infty}^{\infty} R_X(\tau) \exp(-j\omega\tau)d\tau \,, \tag{8.9}$$

and conversely, the autocorrelation function $R_X(\tau)$ can be recovered from the PSD by using the inverse Fourier transform

$$R_X(\tau) = \frac{1}{2\pi} \int_{-\infty}^{\infty} S_X(\omega) \exp(j\omega\tau)d\omega \,. \tag{8.10}$$

Since $R_X(\tau)$ is real and even, so is $S_X(\omega)$. In addition, it will be shown below that the nonnegativity property (4.27) of the autocorrelation function $R_X(\tau)$ ensures that $S_X(\omega) \geq 0$ for all ω. Similarly, in the discrete-time (DT) case, the PSD is defined as

$$S_X(e^{j\omega}) = \sum_{m=-\infty}^{\infty} R_X(m)e^{-j\omega m} \,, \tag{8.11}$$

and the autocorrelation function $R_X(m)$ with $m \in \mathbb{Z}$ can be recovered by using the inverse discrete-time Fourier transform

$$R_X(m) = \frac{1}{2\pi} \int_{-\pi}^{\pi} S_X(e^{j\omega})e^{j\omega m}d\omega \,. \tag{8.12}$$

In this case $S_X(e^{j\omega})$ is again real and even, and like all DT Fourier transforms, it is periodic with period 2π. By setting $\tau = 0$ in (8.10), we find that the total power of CT WSS signal $X(t)$ can be expressed in terms of the PSD as

$$P_X = E[X^2(t)] = R_X(0) = \frac{1}{2\pi} \int_{-\infty}^{\infty} S_X(\omega)d\omega \,. \tag{8.13}$$

As we shall see later this expression can be interpreted as providing a decomposition of the total power P_X into elementary components $S_X(\omega)\Delta/2\pi$ representing the power of $X(t)$ in frequency band $[\omega - \Delta/2, \omega + \Delta/2]$. The corresponding DT expression is given by

$$P_X = E[X^2(t)] = R_X(0) = \frac{1}{2\pi} \int_{-\pi}^{\pi} S_X(e^{j\omega})d\omega \tag{8.14}$$

and admits a similar interpretation.

Remark Suppose that the mean m_X of $X(t)$ is nonzero. Then in the CT case

$$R_X(\tau) = m_X^2 + K_X(\tau),$$

where we assume that $K_X(\tau)$ is summable, so that its Fourier transform $\Sigma_X(\omega)$ exists. In this case

$$S_X(\omega) = 2\pi m_X^2 \delta(\omega) + \Sigma_X(\omega),$$

so that the nonzero mean m_X manifests itself by the presence of a spectral line in $S_X(\omega)$ at $\omega = 0$. Similarly, suppose that in the DT case $R_X(m) = m_X^2 + K_X(m)$, where the covariance function $K_X(m)$ has DTFT $\Sigma_X(e^{j\omega})$. Then the PSD of $X(t)$ can be written as

$$S_X(e^{j\omega}) = 2\pi m_X^2 \Big(\sum_{k=-\infty}^{\infty} \delta(\omega - 2\pi k)\Big) + \Sigma_X(e^{j\omega}),$$

so that in this case, due to the periodicity of the PSD, the spectral line at $\omega = 0$ is repeated at all integer multiples of 2π.

If $Y(t)$ and $X(t)$ are two jointly WSS random processes with cross-correlation $R_{YX}(\tau) = E[Y(t + \tau)X(t)]$ in the CT case or $R_{YX}(m) = E[Y(n + m)X(n)]$ in the DT case, their cross-spectral density is defined by

$$S_{YX}(\omega) = \int_{-\infty}^{\infty} R_{YX}(\tau) \exp(-j\omega\tau)d\tau \tag{8.15}$$

in the CT case, and

$$S_{YX}(e^{j\omega}) = \sum_{m=-\infty}^{\infty} R_{YX}(m)e^{-j\omega m} \tag{8.16}$$

in the DT case. By observing that

$$R_{XY}(\tau) = E[X(t + \tau)Y(t)] = E[Y(t)X(t + \tau)] = R_{YX}(-\tau) \tag{8.17}$$

and using the fact that if $F(j\omega)$ is the CT Fourier transform of $f(\tau)$, the Fourier transform of $f(-\tau)$ is $F^*(j\omega) = F(-j\omega)$, we deduce that in the CT case

$$S_{XY}(\omega) = S_{YX}^*(\omega) = S_{YX}(-\omega). \tag{8.18}$$

The corresponding DT symmetry property is

$$S_{XY}(e^{j\omega}) = S_{YX}^*(e^{j\omega}) = S_{YX}(e^{-j\omega}). \tag{8.19}$$

WSS Random Signals Through LTI Filters Assume that CT WSS signal $X(t)$ is passed through a LTI filter with impulse response $h(t)$ and frequency response $H(j\omega)$, as shown in Fig. 8.1. We found in the previous section that the output $Y(t)$ is jointly WSS with input $X(t)$, with cross-correlation and auto-correlation (8.7) and (8.8). Since the Fourier transform converts the convolution into a product, we deduce that the cross-spectral density of $Y(t)$ and $X(t)$ and the PSD of $Y(t)$ are given by

$$S_{YX}(\omega) = H(j\omega)S_X(\omega) \tag{8.20}$$

$$S_Y(\omega) = |H(j\omega)|^2 S_X(\omega) . \tag{8.21}$$

Also, the symmetry property (8.10) implies

$$S_{XY}(\omega) = H^*(j\omega)S_X(\omega) . \tag{8.22}$$

If we assume that DT WSS random signal is passed through a LTI filter with impulse response $h(n)$ and frequency response $H(e^{j\omega})$, the corresponding expressions for the DT cross-spectral densities of $X(t)$ and $Y(t)$ and PSD of $Y(t)$ are given by

$$S_{YX}(e^{j\omega}) = H(e^{j\omega})S_X(e^{j\omega})$$

$$S_{XY}(e^{j\omega}) = H^*(e^{j\omega})S_X(e^{j\omega})$$

$$S_Y(e^{\omega}) = |H(e^{j\omega})|^2 S_X(e^{\omega}) .$$

Nonnegativity of the PSD We are now in a position to establish the nonnegativity of the PSD. Without loss of generality, we consider the CT case only. Assume that WSS signal $X(t)$ with PSD $S_X(\omega)$ is passed through a narrowband bandpass filter whose frequency response

$$H(j\omega) = \begin{cases} 1 & |\omega \pm \omega_0| \le \Delta/2 \\ 0 & \text{otherwise} \end{cases} \tag{8.23}$$

extracts the frequency content of $X(t)$ in the band $[\omega_0 - \Delta/2, \omega_0 + \Delta/2]$ and its mirror negative frequency image. The center frequency ω_0 is arbitrary, and the bandwidth Δ of the filter is sufficiently narrow to ensure that $S_X(\omega) \approx S_X(\omega_0)$ over the entire passband of the filter. The filter impulse response

$$h(t) = 2\Delta \text{sinc}(\Delta t/2) \cos(\omega_0 t) ,$$

where

$$\text{sinc}(t) \triangleq \frac{\sin(t)}{t}$$

denotes the sinc function, is noncausal but causality is not required for our argument. Then if $Y(t)$ is the WSS random signal at the output of $h(t)$, its PSD is given by

$$S_Y(\omega) = |H(j\omega)|^2 S_X(\omega) \approx \begin{cases} S_X(\omega_0) & |\omega \pm \omega_0| \le \Delta/2 \\ 0 & \text{otherwise} . \end{cases}$$

By using expression (8.21), we find that the power of signal $Y(t)$ is given by

$$P_Y = \frac{1}{2\pi} \int_{-\pi}^{\pi} S_Y(\omega) d\omega \approx \frac{\Delta}{\pi} S_X(\omega_0), \tag{8.24}$$

and since $P_Y = E[Y^2(t)] \ge 0$, this implies $S_X(\omega_0) \ge 0$ for all choices of center frequency ω_0. Since $Y(t)$ was obtained by extracting the frequency content of $X(t)$ in band $[\omega_0 - \Delta/2, \omega_0 + \Delta/2]$ and its mirror image, we deduce that as indicated earlier, (8.13) can be viewed as a decomposition of the total power of $X(t)$ as a sum of the powers of all its frequency band components. To verify this interpretation, we consider several examples.

Example 8.2 In Example 4.1, the sinusoidal signal

$$X(t) = A\cos(\omega_c t + \Theta)$$

with known amplitude A and frequency ω_c and uniformly distributed phase Θ was found to be WSS with zero mean and autocorrelation

$$R_X(\tau) = \frac{A^2}{2}\cos(\omega_c \tau). \tag{8.25}$$

By Fourier transforming (8.25) we find

$$S_X(\omega) = A^2\pi(\delta(\omega - \omega_c) + \delta(\omega + \omega_c)). \tag{8.26}$$

Thus, all the power of signal $X(t)$ is concentrated at frequencies $\pm\omega_c$, as expected since the signal is sinusoidal with frequency ω_c.

Example 8.3 The Ornstein–Uhlenbeck process is a zero-mean Gaussian signal with autocorrelation function

$$R_X(\tau) = P_X \exp(-a|\tau|) \tag{8.27}$$

with $a > 0$, where $P_X = E[X^2(t)]$ denotes the power of $X(t)$. By observing that

$$P_X \exp(-a|\tau|) = P_X[\exp(-a\tau)u(\tau) + \exp(a\tau)u(-\tau)],$$

and using the FT pair

$$\exp(-a\tau)u(\tau) \longleftrightarrow \frac{1}{j\omega + a}$$

we find

$$S_X(\omega) = P_X[\frac{1}{j\omega + a} + \frac{1}{-j\omega + a}] = \frac{2aP_X}{\omega^2 + a^2}, \tag{8.28}$$

which is sometimes called a Lorentz PSD and is plotted in Fig. 8.4 for $P_X = 1$ and $a = 1$. Observing that

$$S_X(a) = \frac{1}{2}S_X(0),$$

we deduce that the parameter a corresponds to the half-power bandwidth of $X(t)$, i.e., the frequency at which the PSD has half as much power per rad/s as at its maximum, the zero frequency. For the RC circuit driven by thermal noise considered in Sect. 6.6, we saw that $a = 1/RC$, where RC is the circuit time constant, which illustrates the inverse relation existing between a circuit's bandwidth and its time constant.

Example 8.4 As was seen in last chapter, a CT WGN $N(t)$ with intensity q can be viewed as $N(t) = q^{1/2}dW/dt$, where $W(t)$ is a standard Wiener process. This process does not exist as an ordinary

Fig. 8.4 Lorentz PSD
with $P_X = 1$ and $a = 1$

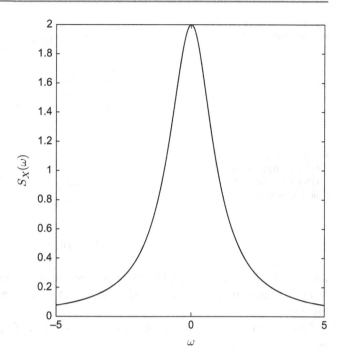

random process, but can be viewed as a generalized zero-mean Gaussian process with autocorrelation

$$R_N(\tau) = q\delta(\tau),$$

and PSD

$$S_N(\omega) = q \tag{8.29}$$

for all ω. Thus, in the same way as white light contains all visible frequencies, WGN has the same power density q at all frequencies. Of course, because $P_N = R_N(0)$ is infinite, WGN is only an idealization for *bandlimited WGN*, whose PSD

$$S_{N_B}(\omega) = \begin{cases} q & |\omega| \le B \\ 0 & \text{otherwise} \end{cases} \tag{8.30}$$

is identical to the WGN PSD, except that it is cut-off at $\pm B$, where B denotes the bandwidth of the WGN. The corresponding autocorrelation

$$R_{N_B}(\tau) = P_{N_B}\text{sinc}(B\tau), \tag{8.31}$$

where the signal power $P_{N_B} = qB/\pi$ is finite but large when B is large. The PSDs of WGN and bandlimited WGN are plotted in parts a) and b) of Fig. 8.5, respectively.

Physical Justification of WGN We are now in a position to justify the use of WGN instead of bandlimited WGN in the analysis of LTI systems. Consider a LTI system whose frequency response $H(j\omega)$ has a finite bandwidth W, so that

Fig. 8.5 PSD of (a) WGN and (b) bandlimited WGN

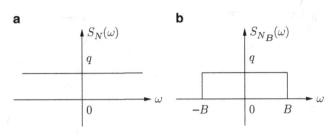

Fig. 8.6 Effect of (a) WGN and (b) bandlimited WGN passed through a bandlimited filter

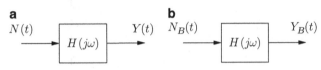

$$|H(j\omega)| = 0$$

for $|\omega| > W$. Let $Y(t)$ and $Y_B(t)$ be the respective outputs of this system when it is excited by a WGN $N(t)$ with intensity q or a bandlimited WGN $N_B(t)$ with intensity q and bandwidth B, as shown in Fig. 8.6.

Then the cross-spectral densities of $Y(t)$ and $N(t)$, and of $Y_B(t)$ and $N_B(t)$ are given respectively by

$$S_{YN}(\omega) = q H(j\omega)$$

$$S_{Y_B N_B}(\omega) = \begin{cases} q H(j\omega) & |\omega| \leq B \\ 0 & \text{otherwise} , \end{cases}$$

and the PSDs of $Y(t)$ and $Y_B(t)$ are given respectively by

$$S_Y(\omega) = q |H(j\omega)|^2$$

$$S_{Y_B}(\omega) = \begin{cases} q |H(j\omega)|^2 & |\omega| \leq B \\ 0 & \text{otherwise} . \end{cases}$$

From these expressions we deduce that

$$S_{YN}(\omega) = S_{Y_B N_B}(\omega) \ , \ \ S_Y(\omega) = S_{Y_B}(\omega)$$

or equivalently

$$R_{YN}(\tau) = R_{Y_B N_B}(\tau) \ , \ \ R_Y(\tau) = R_{Y_B}(\tau)$$

if $W < B$, i.e., if the bandwidth of filter $H(j\omega)$ is smaller than the bandwidth of the bandlimited WGN. In other words, by evaluating the autocorrelation of the filter output, or its cross-correlation with the input, it is impossible to determine whether the filter input is WGN or bandlimited WGN. This means that when computing first- and second-order statistics of random signals obtained by passing bandlimited WGN through bandlimited filters with a narrower bandwidth than the noise, it is preferable to assume that the input is WGN instead of bandlimited WGN, since WGN computations are usually much simpler to perform.

Remark In fact, the Nyquist–Johnson model of WGN described in Sect. 6.6 is only a low-frequency approximation of the model described by Nyquist [1] which is actually bandlimited with bandwidth $B/(2\pi) = kT/h$ (Hz), where h denotes Planck's constant. At room temperature, this bandwidth is corresponds to about 6THz, which is well above the bandwidth W of modern circuits. This explains why there is no harm in ignoring the cutoff frequency of the Nyquist–Johnson model.

PSD of Shot Noise It was shown in Chap. 7 that if $N(t)$ is a Poisson process with rate λ, the shot noise process

$$X(t) = \int_{-\infty}^{\infty} h(t - u) dN(u)$$

is WSS with mean

$$m_X = \lambda \int_{-\infty}^{\infty} h(u) du$$

and autocorrelation

$$R_X(\tau) = m_X^2 + \lambda r_h(\tau), \tag{8.32}$$

where $r_h(\tau) = h(\tau) * h(-\tau)$ denotes the deterministic autocorrelation of $h(\tau)$. By Fourier transforming (8.32), we obtain

$$S_X(\omega) = 2\pi m_X^2 \delta(\omega) + \lambda |H(j\omega)|^2, \tag{8.33}$$

where $H(j\omega)$ denotes the Fourier transform of $h(\tau)$. When $h(\tau)$ is approximated in terms of its dominant pole as $h(\tau) = c \exp(-a\tau) u(\tau)$, this gives

$$S_X(\omega) = 2\pi m_X^2 \delta(\omega) + \frac{\lambda c^2}{\omega^2 + a^2},$$

which, except for the presence of a spectral line at $\omega = 0$, is identical to the PSD of an Ornstein–Uhlenbeck (OU) process, even though the underlying physical models of shot noise and the OU process are quite different.

Wiener–Khinchin Theorem We saw earlier that the power of the signal $Y(t)$ obtained by extracting frequency band $[\omega_0 - \Delta/2, \omega_0 + \Delta/2]$ and the matching negative frequency band from $X(t)$ is $S_X(\omega_0)\Delta/\pi$, which suggests that the PSD $S_X(\omega)$ is really a measure of how much power $X(t)$ contains in each frequency band. Observe that since a WSS signal $X(t)$ has a constant second-order moment $E[X^2(t)]$, it is necessarily a *finite power* signal, with *infinite energy*. Indeed, the energy

$$E = \int_{-\infty}^{\infty} |f(t)|^2 dt$$

of a signal is finite only if $|f(t)|$ decays at a rate greater than $|t|^{-1/2}$. In this case, if

$$\check{F}(j\omega) = \int_{-\infty}^{\infty} f(t) \exp(-j\omega_t) dt$$

denotes the CT Fourier transform of $f(t)$, by using Parseval's identity, the energy can also be expressed as

$$E = \frac{1}{2\pi} \int_{-\infty}^{\infty} |F(j\omega)|^2 d\omega,$$

so that $|F(j\omega)|^2$ can also be viewed as representing the *energy density* of signal $f(t)$. Consider the truncated random signal

$$X_T(t) = \begin{cases} X(t) & |t| \le T/2 \\ 0 & \text{otherwise} \end{cases} \tag{8.34}$$

of length T obtained by chopping the tails of $X(t)$ with $|t| > T/2$. The resulting signal is a finite energy signal, and if

$$\hat{X}_T(j\omega) = \int_{-T/2}^{T/2} X(t) \exp(-j\omega t) dt \tag{8.35}$$

denotes its CT Fourier transform, the ratio $|\hat{X}_T(j\omega)|^2/T$ of the energy density divided by the length T of the truncated signal can be viewed as a measure of the signal power density at frequency ω. Then the *Wiener–Khinchin theorem* consists of showing that

$$S_X(\omega) = \lim_{T\to\infty} E[\frac{|\hat{X}_T(j\omega)|^2}{T}] . \tag{8.36}$$

In other words, the power spectral density can be interpreted as the ensemble average of the power density $|\hat{X}_T(j\omega)|^2/T$ as the observation interval T goes to infinity.

To prove (8.36), we start by noting that

$$\frac{1}{T}|\hat{X}_T(j\omega)|^2 = \frac{1}{T} \int_{-T/2}^{T/2} \int_{-T/2}^{T/2} X(t)X(s) \exp(-j\omega(t-s)) dt ds .$$

Taking expectations yields

$$E[\frac{1}{T}|\hat{X}_T(j\omega)|^2] = \frac{1}{T} \int_{-T/2}^{T/2} \int_{-T/2}^{T/2} R_X(t-s) \exp(-j\omega(t-s)) dt ds . \tag{8.37}$$

Since the integrand on the right-hand side of (8.37) depends only on $\tau = t - s$, we can perform the change of variable

$$\begin{bmatrix} \tau \\ \sigma \end{bmatrix} = \begin{bmatrix} 1 & -1 \\ 1 & 1 \end{bmatrix} \begin{bmatrix} t \\ s \end{bmatrix} . \tag{8.38}$$

The Jacobian of the inverse transformation

$$\begin{bmatrix} t \\ s \end{bmatrix} = \frac{1}{2} \begin{bmatrix} 1 & 1 \\ -1 & 1 \end{bmatrix} \begin{bmatrix} \tau \\ \sigma \end{bmatrix}$$

is $J = 1/2$, and under transformation (8.38), the square domain $[-T/2, T/2]^2$ for (t, s) is transformed into the rotated square domain for (τ, σ) shown in Fig. 8.7. Accordingly, the double integral in (8.37) becomes

$$\frac{1}{2T} \int_{-T}^{T} R_X(\tau) \exp(-j\omega\tau) [\int_{-(T-|\tau|)}^{T-|\tau|} d\sigma] d\tau$$

Fig. 8.7 Transformed domain of integration

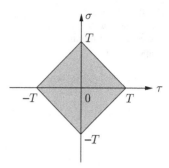

$$= \int_{-T}^{T} (1 - |\tau|/T) R_X(\tau) \exp(-j\omega\tau) d\tau \,.$$

In this integral $|\tau|/T$ tends to zero for a fixed τ as T tends to infinity, and the integral limits tend to infinity, so that

$$\lim_{T\to\infty} E[\frac{\hat{X}_T(j\omega)|^2}{T}] = \int_{-\infty}^{\infty} R_X(\tau) \exp(-j\omega\tau) d\tau = S_X(j\omega) \,, \qquad (8.39)$$

which proves the theorem.

The Wiener Khinchin theorem provides the justification for the averaged periodogram spectral estimation method [2, 3] which, given N approximately independent segments $X^i(t)$, $1 \le i \le N$ of length T of a WSS random process, estimates the power spectral density $S_X(\omega)$ by using the approximation

$$S_X(\omega) \approx \frac{1}{N} \sum_{i=1}^{N} \frac{|\hat{X}_T^i(\omega)|^2}{T} \,.$$

Since the segments are assumed independent, the law of large numbers ensures that for large N the average converges to

$$E[\frac{|\hat{X}_T(\omega)|^2}{T}] \,,$$

which in turn converges to $S_X(\omega)$ as T tends to infinity.

8.4 Nyquist–Johnson Model of RLC Circuits

In Sect. 6.6, it was shown that a noisy resistor of resistance R in thermal equilibrium at temperature T is functionally equivalent to a noiseless resistor R and a WGN voltage source with intensity $q = 2kRT$. But engineers typically encounter large RLC networks which can be replaced by a Thevenin equivalent circuit formed by the equivalent impedance of the circuit in series with a voltage source. Thus, consider an RLC circuit in thermal equilibrium at temperature T shown in Fig. 8.8, and let $Z_{Th}(j\omega)$ denote its equivalent impedance. Due to the presence of thermal fluctuations, each resistor in the RLC network could be replaced by a noiseless resistor in series with its matching $2kRT$ WGN

Fig. 8.8 Thevenin equivalent of a RLC circuit in thermal equilibrium at temperature T

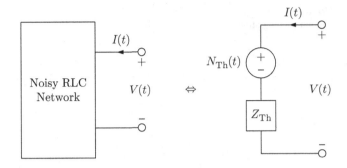

source, so the Thevenin equivalent circuit of the noisy RLC network appearing on the right of Fig. 8.8 must include a voltage source $N_{\mathrm{Th}}(t)$ representing the combined effect of all individual thermal noise sources attached to each resistor in the RLC network.

In his landmark paper [1], Nyquist showed that the voltage source $N_{\mathrm{Th}}(t)$ is a zero-mean WSS Gaussian process with PSD

$$S_{N_{\mathrm{Th}}}(\omega) = 2k\Re\{Z_{\mathrm{Th}}(j\omega)\}T \,, \tag{8.40}$$

or equivalently, with autocorrelation

$$R_{N_{\mathrm{Th}}}(\tau) = 2kz_{\mathrm{Th}}^{e}(\tau)T \tag{8.41}$$

where, if $z_{Th}(\tau)$ denotes the inverse Fourier transform of impedance $Z_{Th}(j\omega)$,

$$z_{\mathrm{Th}}^{e}(\tau) = \frac{1}{2}(z_{\mathrm{Th}}(\tau) + z_{\mathrm{Th}}(-\tau)) \tag{8.42}$$

represents the even part of $z_{\mathrm{Th}}(\tau)$. The argument given by Nyquist uses reciprocity when two networks in thermal equilibrium are connected to each other (see also [4, p. 427]). We will use here a direct argument based on analyzing the network obtained by replacing all resistors in the RLC network on the left of Fig. 8.8 by resistors in series with WGN sources. While this derivation is not as elegant as the one of [1], it has the advantage of providing a good illustration of noise computations in RLC networks, using the approach described in [5].

Suppose that M resistors R_i with $1 \leq i \leq M$ appear in the RLC network appearing on the left of Fig. 8.8. Then the RLC network can be rewritten as a $M + 1$-port as shown in Fig. 8.9, where in addition to the V–I port of interest, each resistor R_i and its matching thermal WGN source $N_i(t)$ is connected by using one network port. The remaining network is now purely reactive and consists exclusively of capacitors and inductors. The PSD of the WGN noise source $N_i(t)$ representing the fluctuations of resistor R_i is given by

$$S_{N_i}(\omega) = 2kR_iT$$

for $1 \leq i \leq M$. Then if

$$\mathbf{V}_s = \begin{bmatrix} V_1 \ V_2 \ \dots \ V_M \end{bmatrix}^T$$

$$\mathbf{I}_s = \begin{bmatrix} I_1 \ I_2 \ \dots \ I_M \end{bmatrix}^T$$

Fig. 8.9 $M + 1$ port representation of an RLC circuit with M noisy resistors

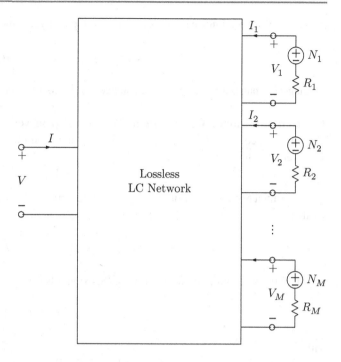

denote the vectors regrouping the voltages and currents respectively at the M ports connecting the noisy resistors, and if V and I represent the voltage and current at the port of interest as shown in Fig. 8.9, the $M + 1$-port admits the representation

$$\begin{bmatrix} V \\ \mathbf{V}_s \end{bmatrix} = j\mathbf{X}(\omega) \begin{bmatrix} I \\ \mathbf{I}_s \end{bmatrix}, \tag{8.43}$$

where we have used the fact that after removal of all resistors, the network is purely reactive. The reactance matrix $\mathbf{X}(\omega)$ can be partitioned as

$$\mathbf{X}(\omega) = \begin{bmatrix} X_{oo} & \mathbf{X}_{os} \\ \mathbf{X}_{so} & \mathbf{X}_s \end{bmatrix},$$

where the network reciprocity implies

$$\mathbf{X}^T(\omega) = \mathbf{X}(\omega), \tag{8.44}$$

which implies in particular

$$\mathbf{X}_{so}(\omega) = \mathbf{X}_{os}^T(\omega) \text{ and } \mathbf{X}_s(\omega) = \mathbf{X}_s^T(\omega).$$

In the following we will also denote by

$$\mathbf{N}_s(t) = \begin{bmatrix} N_1(t) & N_2(t) & \dots & N_M(t) \end{bmatrix}^T$$

the vector of thermal noise sources attached to the resistors, and by

$$\mathbf{R}_s = \mathrm{diag}\,\{R_i,\ 1 \le i \le M\}$$

the diagonal matrix formed by the resistances of the M resistors.

Thevenin Impedance Computation The first step in verifying model (8.40) is to evaluate the Thevenin impedance of the RLC network. This is accomplished by setting the noise source vector $\mathbf{V}_s$ to zero, so that

$$\mathbf{V}_s = -\mathbf{R}_s\mathbf{I}_s \tag{8.45}$$

and by using a test current source to inject some current I in the I–V port. Then by substituting (8.45) inside (8.43), we find

$$(\mathbf{R}_s + j\mathbf{X}_s)\mathbf{I}_s = -j\mathbf{X}_{so}I,$$

which after substitution in the first row of (8.43) yields

$$V = Z_{\mathrm{Th}}(j\omega)I$$

with

$$Z_{\mathrm{Th}} = jX_{oo} + \mathbf{X}_{os}(\mathbf{R}_s + j\mathbf{X}_s)^{-1}\mathbf{X}_{so} . \tag{8.46}$$

Using the symmetry $\mathbf{X}_{so} = \mathbf{X}_{os}^T$, we find after some simple algebraic manipulations

$$\begin{aligned}
\Re\{Z_{\mathrm{Th}}\} &= \frac{1}{2}(Z_{\mathrm{Th}} + Z_{\mathrm{Th}}^*)\\
&= \mathbf{X}_{os}(\mathbf{R}_s + j\mathbf{X}_s)^{-1}\mathbf{R}_s(\mathbf{R}_s - j\mathbf{X}_s^T)^{-1}\mathbf{X}_{os}^T .
\end{aligned} \tag{8.47}$$

Transfer Function from $\mathbf{N}_s$ to V Next, with the I–V terminal left open, so that $I = 0$ in (8.43), we compute the transfer function $\mathbf{H}_s$ used to express the output voltage in terms of the voltage sources $\mathbf{N}_s$ as

$$V = \mathbf{H}_s\mathbf{N}_s . \tag{8.48}$$

By Kirchoff's voltage law at the WGN source terminals on the right of Fig. 8.9, we have

$$\mathbf{V}_s = \mathbf{N}_s - \mathbf{R}_s\mathbf{I}_s,$$

which after substitution inside (8.43) while taking $I = 0$ into account gives

$$\mathbf{I}_s = (\mathbf{R}_s + j\mathbf{X}_s)^{-1}\mathbf{N}_s . \tag{8.49}$$

Substituting this identity in the first line of (8.43) gives (8.48) with

$$\mathbf{H}_s = \mathbf{X}_{os}(\mathbf{R}_s + j\mathbf{X}_s)^{-1} . \tag{8.50}$$

Fig. 8.10 Thevenin equivalent of RC circuit in thermal equilibrium at temperature T

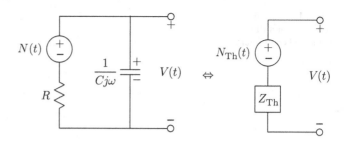

PSD of the Thevenin Voltage Source Since the equivalent voltage source representing the effect of the thermal fluctuations at the $I-V$ terminals is obtained by passing the M-dimensional WGN source $N_s(t)$ through the $1 \times M$ filter with frequency response $\mathbf{H}_s$, it is necessarily zero-mean Gaussian and WSS with PSD

$$S_{N_{\text{Th}}}(\omega) = \mathbf{H}_s \times 2k\mathbf{R}_s T \times \mathbf{H}_s^H \tag{8.51}$$

where $\mathbf{H}_s^H = (\mathbf{H}_s^*)^T$ denotes the Hermitian transpose (the conjugate transpose) of $\mathbf{H}_s$. Substituting (8.50) inside (8.51) and comparing with (8.47) give

$$S_{N_{\text{Th}}}(\omega) = 2kT\mathbf{X}_{os}(\mathbf{R}_s + j\mathbf{X}_s)^{-1}\mathbf{R}_s(\mathbf{R}_s - j\mathbf{X}_s^T)^{-1}\mathbf{X}_{os}^T$$
$$= 2k\Re\{Z_{\text{Th}}(j\omega)\}T,$$

which proves Nyquist's theorem.

Example 8.1 (Continued) Consider again the RC circuit in thermal equilibrium considered in Sect. 6.6. This circuit can be replaced by its Thevenin equivalent as shown in Fig. 8.10. The Thevenin impedance is obtained by setting the thermal WGN source $N(t)$ to zero, so that

$$Z_{\text{Th}}(j\omega) = R \parallel \frac{1}{Cj\omega} = \frac{R}{1 + RCj\omega} .$$

By multiplying the numerator and denominator of Z_{Th} by the complex conjugate of the denominator, we obtain

$$Z_{\text{Th}} = \frac{R(1 - RCj\omega)}{1 + (RC\omega)^2} ,$$

so that the PSD of the zero-mean WSS Thevenin noise source $N_{\text{Th}}(t)$ is

$$S_{N_{\text{Th}}}(\omega) = 2k\Re\{Z_{\text{Th}}(j\omega)\}T = \frac{2kRT}{1 + (RC\omega)^2} .$$

By using the approach of Example 8.3 to compute the inverse Fourier transform of $S_{N_{\text{Th}}}(\omega)$, we find that $N_{\text{Th}}(t)$ has the Ornstein–Uhlenbeck autocorrelation function

$$R_{N_{\text{Th}}}(\tau) = \frac{kT}{C} \exp(-|\tau|/RC) ,$$

as expected.

Fig. 8.11 Norton
equivalent of a RLC circuit
in thermal equilibrium

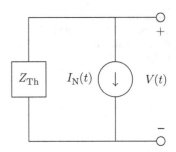

Norton Equivalent Circuit In some situations, it may be preferable to replace a noisy RLC circuit in thermal equilibrium at temperature T by its Norton equivalent shown in Fig. 8.11. Since the Norton current $I_N(t)$ representing the thermal fluctuations is obtained by passing the Thevenin voltage $N_{Th}(t)$ through a LTI filter whose frequency response is the Thevenin admittance

$$Y_{Th}(j\omega) = \frac{1}{Z_{Th}(j\omega)},$$

we deduce that $I_N(t)$ is zero mean Gaussian and WSS with PSD

$$S_{I_N}(\omega) = |Y_{Th}(j\omega)|^2 S_{N_{Th}}(\omega) = 2k\Re\{Y_{Th}(j\omega)\}T. \tag{8.52}$$

8.5   Power Spectral Densities of Modulated Signals

8.5.1   Pulse Amplitude Modulated Signals

In Example 4.5, we found that the autocorrelation function of a PAM signal

$$Z(t) = \sum_{k=-\infty}^{\infty} X_k p(t - kT - D) \tag{8.53}$$

is given by (4.22)–(4.24). By noting that the Fourier transform of $R_S(\tau)$ is

$$S_S(\omega) = \sum_{m=-\infty}^{\infty} R_X(m) \exp(-j\omega mT) = S_X(e^{j\omega T}), \tag{8.54}$$

where $S_X(e^{j\omega_d})$ denotes the PSD of discrete-time signal X_k, and observing that if $P(j\omega)$ represents the Fourier transform of $p(\tau)$, then the Fourier transform of the deterministic autocorrelation $r_p(\tau) = p(\tau) * p(-\tau)$ is $|P(j\omega)|^2$, we find that the PSD of WSS signal $Z(t)$ is

$$S_Z(\omega) = S_X(e^{j\omega T})\frac{|P(j\omega)|^2}{T}. \tag{8.55}$$

Since the signaling period of PAM signal (8.51) is T, its baud frequency, i.e., the frequency at which symbols X_k are transmitted, is $\omega_b = 2\pi/T$. In the PSD expression (8.53), the component $S_X(e^{j\omega T})$, which reflects the spectral information of the symbol stream X_k, $k \in \mathbb{Z}$ is periodic with period ω_b. Since it is real and symmetric, this means that $S_X(e^{j\omega T})$ can be recovered from $S_Z(\omega)$ as long as $P(j\omega)$ is nonzero for $|\omega| \leq \omega_b/2$. A minimum bandwidth pulse is given by

$$P(j\omega) = \begin{cases} T & |\omega| \leq \omega_b/2 \\ 0 & |\omega| > \omega_b/2, \end{cases} \tag{8.56}$$

which corresponds to the sinc function

$$p(t) = \operatorname{sinc}(\pi t/T) \,.$$

This pulse decays only like $1/t$, so it creates intersymbol interference (ISI) in the sense that the effect of the transmission of symbol X_k in the k-th interval $kT \leq t \leq (k+1)T$ leaks into adjacent transmission intervals. On the other hand, since

$$p(kT) = \begin{cases} 1 & k = 0 \\ 0 & k \neq 0 \end{cases}$$

we deduce that the effect of intersymbol interference disappears if the synchronization delay D is estimated exactly and signal $Z(t)$ is sampled precisely at time $kT + D$. However this is difficult to achieve exactly. An opposite choice for the pulse $p(t)$ consists of selecting the rectangular time-limited pulse

$$p(t) = \begin{cases} A & |t| \leq T/2 \\ 0 & \text{otherwise.} \end{cases}$$

This choice avoids ISI entirely, since this case $Z(kT) = AX_k$ independently of the delay D (which is assumed uniform over $[-T/2, T/2]$), but at the expense of a significant bandwidth expansion since in this case

$$P(j\omega) = AT \operatorname{sinc}(\omega T/2) \,. \tag{8.57}$$

The magnitude of $|P(j\omega)|^2$ is plotted as a function of ω/ω_b in Fig. 8.12. The plot indicates that the bandwidth is approximately $2\omega_b$, which is four times larger than the minimum bandwidth pulse (8.56). The large increase in bandwidth is due to the time discontinuities of pulse $p(t)$ at $t = \pm T/2$.

Finally, it is worth noting that when the mean m_X of sequence X_k is nonzero, so that

$$R_X(m) = m_X^2 + K_X(m),$$

where $K_X(m) = E[(X_{k+m} - m_X)(X_k - m_X)]$ denotes the covariance function of sequence X_k, the periodic spectrum $S_X(e^{j\omega T})$ can be decomposed into continuous and discrete components as

$$S_X(e^{j\omega T}) = S_c(e^{\omega T}) + S_d(e^{j\omega T}),$$

where if $\Sigma_X(e^{j\omega_d})$ denotes the discrete-time Fourier transform of $K_X(m)$, we have

$$S_c(e^{j\omega T}) = \Sigma_X(e^{j\omega T})$$

$$S_d(e^{j\omega T}) = m_X^2 \omega_b \sum_{k=-\infty}^{\infty} \delta(\omega - k\omega_b) \,.$$

Fig. 8.12 Plot of the squared Fourier transform magnitude $|P(j\omega)|^2$ for a rectangular pulse of duration T

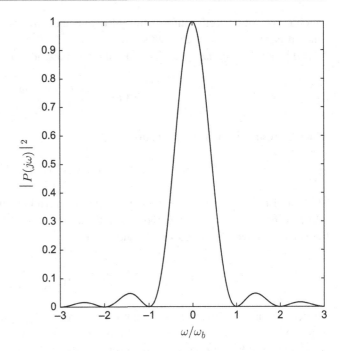

Since the bandwidth of $P(j\omega)|^2$ may extend well beyond $\omega_b/2$, this indicates that when m_X is nonzero, tones located at integer multiples of the baud frequency ω_b may be present in the PSD $S_Z(\omega)$ of signal $Z(t)$.

8.5.2 Markov Chain Modulated Signals

Let us now turn to the class of Markov chain modulated signals

$$Z(t) = \sum_{k=-\infty}^{\infty} s_{X(k)}(t - kT - D) \tag{8.58}$$

considered in Sect. 5.6, where we assume that the Markov chain $X(k)$ is irreducible, aperiodic, and in steady state. Then $Z(t)$ is WSS and its autocorrelation is given by (5.60)–(5.62). If $S_i(j\omega)$ denotes the Fourier transform of $s_i(\tau)$, the Fourier transform of $r_{ij}(\tau) = s_i(\tau) * s_j(-\tau)$ is $S_i(j\omega)S_j^*(j\omega)$. Next, if

$$\mathbf{Q}(m) = (q_{ij}(m),\ 1 \le i,\ j \le N)$$

denotes the $N \times N$ matrix function whose entries $q_{ij}(m)$ are given by (5.59), we have

$$\mathbf{Q}(m) = \begin{cases} \mathbf{D}(\boldsymbol{\pi}) & m = 0 \\ \mathbf{F}^T(m)\mathbf{D}(\boldsymbol{\pi}) & m > 0 \\ \mathbf{D}(\boldsymbol{\pi})\mathbf{F}(-m) & m < 0, \end{cases} \tag{8.59}$$

with $\mathbf{F}(m) = \mathbf{P}^m$ and

$$\mathbf{D}(\boldsymbol{\pi}) = \mathrm{diag}\left\{\pi_i, \ 1 \le i \le N\right\}, \tag{8.60}$$

where $\boldsymbol{\pi}$ denotes the steady-state PMF of the Markov chain. Since

$$\lim_{m \to \infty} \mathbf{P}^m = \boldsymbol{v}\boldsymbol{\pi},$$

we deduce that

$$\lim_{|m| \to \infty} \mathbf{Q}(m) = \boldsymbol{\pi}^T \boldsymbol{\pi}$$

is nonzero, so that the matrix function $\mathbf{Q}(m)$ is not summable. Accordingly, it is convenient to decompose the one-step transition matrix $\mathbf{P}$ as

$$\mathbf{P} = \boldsymbol{v}\boldsymbol{\pi} + \mathbf{P}_c, \tag{8.61}$$

where $\mathbf{P}_c$ is the matrix obtained by removing the Jordan block of dimension 1 corresponding to $\lambda = 1$ in the Jordan form decomposition of $\mathbf{P}$. Then $\mathbf{F}_c(m) = \mathbf{P}_c^m$ converges to zero as m tends to infinity, since all the remaining eigenvalues of $\mathbf{P}$ have a magnitude strictly less than one, and from (5.31), we have

$$\mathbf{F}(m) = \boldsymbol{v}\boldsymbol{\pi} + \mathbf{F}_c(m). \tag{8.62}$$

Thus the matrix function $\mathbf{Q}(m)$ introduced in (8.59) can be decomposed as

$$\mathbf{Q}(m) = \boldsymbol{\pi}^T \boldsymbol{\pi} + \mathbf{Q}_c(m), \tag{8.63}$$

where

$$\mathbf{Q}_c(m) = \begin{cases} \mathbf{D}(\boldsymbol{\pi}) - \boldsymbol{\pi}^T \boldsymbol{\pi} & m = 0 \\ \mathbf{F}_c^T(m)\mathbf{D}(\boldsymbol{\pi}) & m > 0 \\ \mathbf{D}(\boldsymbol{\pi})\mathbf{F}_c(-m) & m < 0, \end{cases} \tag{8.64}$$

is summable. The Fourier transform of

$$\mathbf{Q}^S(\tau) = \sum_{m=-\infty}^{\infty} \mathbf{Q}(m)\delta(\tau - mT) \tag{8.65}$$

can therefore be decomposed as

$$\boldsymbol{\Phi}^S(e^{j\omega T}) = \boldsymbol{\Phi}_c(e^{j\omega T}) + \boldsymbol{\Phi}_d(e^{j\omega T}), \tag{8.66}$$

where

$$\boldsymbol{\Phi}_c(e^{j\omega T}) = \sum_{m=0}^{\infty} \mathbf{Q}_c(m)e^{-j\omega mT} = \sum_{m=0}^{\infty}(\mathbf{P}_c^T e^{-j\omega T})^m \mathbf{D}(\boldsymbol{\pi})$$

$$+\mathbf{D}(\boldsymbol{\pi}) \sum_{m=-\infty}^{0}(\mathbf{P}_c e^{j\omega T})^{-m} - \mathbf{D}(\boldsymbol{\pi}) - \boldsymbol{\pi}^T \boldsymbol{\pi}. \tag{8.67}$$

and

$$\boldsymbol{\Phi}_d(e^{j\omega T}) = \boldsymbol{\pi}^T \boldsymbol{\pi} \omega_b \sum_{k=-\infty}^{\infty} \delta(\omega - k\omega_b) , \tag{8.68}$$

where $\omega_b = 2\pi/T$ denotes the baud frequency. Since

$$\sum_{m=0}^{\infty} (\mathbf{P}_c e^{-j\omega T})^m = (\mathbf{I}_N - \mathbf{P}_c e^{-j\omega T})^{-1},$$

we find

$$\boldsymbol{\Phi}_c(e^{j\omega T}) = (\mathbf{I}_N - \mathbf{P}_c^T e^{-j\omega T})^{-1} \mathbf{D}(\boldsymbol{\pi}) + \mathbf{D}(\boldsymbol{\pi})(\mathbf{I}_N - \mathbf{P}_c e^{j\omega T})^{-1}$$
$$-\mathbf{D}(\boldsymbol{\pi}) - \boldsymbol{\pi}^T \boldsymbol{\pi} . \tag{8.69}$$

Then by combining these identities, and denoting by

$$\mathbf{S}(j\omega) = \begin{bmatrix} S_1(j\omega) & S_2(j\omega) & \dots & S_N(j\omega) \end{bmatrix}^T$$

the column vector whose entries are the Fourier transforms of the state signals $s_i(t)$, we find that by Fourier transforming the autocorrelation (5.62), the PSD of $Z(t)$ can be decomposed in continuous and discrete components as

$$S_Z(\omega) = S_c(\omega) + S_d(\omega), \tag{8.70}$$

where

$$S_c(\omega) = \frac{2}{T} \Re\{\mathbf{S}^H(j\omega)\mathbf{D}(\boldsymbol{\pi})(\mathbf{I}_N - \mathbf{P}_c e^{-j\omega T})^{-1}\mathbf{S}(j\omega)\}$$
$$-\frac{1}{T}[\sum_{i=1}^{N} \pi_i |S_i(j\omega)|^2 + |\sum_{i=1}^{N} \pi_i S_i(j\omega)|^2] \tag{8.71}$$

and

$$S_d(\omega) = \frac{\omega_b}{T} |\sum_{i=1}^{N} \pi_i S_i(j\omega)|^2 \sum_{k=-\infty}^{\infty} \delta(\omega - k\omega_b) . \tag{8.72}$$

Remark The matrix $(\mathbf{I}_N - \mathbf{P}_c)^{-1}$ obtained by setting $\omega_d = 0$ in $(\mathbf{I}_N - \mathbf{P}_c e^{-j\omega_d})^{-1}$ is called the *fundamental matrix* of the Markov chain. As explained in [6, Chap. 4], it plays a key role in the analysis of several problems, such as the first passage time to a state starting from another state.

Example 8.5 The formulas (8.70)–(8.72) can be used even in situations where the Markov chain modulation is trivial. Consider for example the case of a frequency-shift keying (FSK) modulation scheme where independent equally likely bits $X(k)$ are transmitted. Depending on whether $X(k) = 0$ or 1, we transmit one or two cycles of a sinusoidal carrier. Specifically, if $X(k) = 0$ the pulse

$$s_0(t) = \begin{cases} A \sin(2\pi t/T) & \text{for } 0 \le t \le T \\ 0 & \text{otherwise} , \end{cases}$$

is transmitted, and if $X(k) = 1$, the pulse

$$s_1(t) = \begin{cases} A\sin(4\pi t/T) & \text{for } 0 \le t \le T \\ 0 & \text{otherwise} \end{cases}$$

is sent. The resulting modulated signal $Z(t)$ can be viewed as produced by a Markov chain modulation with one-step transition probability matrix

$$\mathbf{P} = \begin{bmatrix} 1/2 & 1/2 \\ 1/2 & 1/2 \end{bmatrix} = \begin{bmatrix} 1 \\ 1 \end{bmatrix} \pi$$

where

$$\pi = \begin{bmatrix} 1/2 & 1/2 \end{bmatrix}$$

just reflects the fact that the two bit values 0 and 1 are equally likely. In this case $\mathbf{P}_c = \mathbf{0}$ and the continuous component $S_c(\omega)$ of the PSD $S_Z(\omega)$ reduces to

$$S_c(\omega) = \frac{1}{2T}[|S_0(j\omega)|^2 + |S_1(j\omega)|^2] - \frac{1}{4T}|S_0(j\omega) + S_1(j\omega)|^2$$

$$= \frac{1}{4T}|S_0(j\omega) - S_1(j\omega)|^2 .$$

The discrete spectrum is given by

$$S_d(\omega) = \frac{\omega_b}{4T}|S_0(j\omega) + S_1(j\omega)|^2 \sum_{k=-\infty}^{\infty} \delta(\omega - k\omega_b) .$$

8.6 Sampling of Bandlimited WSS Signals

A WSS random signal $X(t)$ with PSD $S_X(\omega)$ is said to be bandlimited with bandwidth B if

$$S_X(\omega) = 0$$

for $|\omega| > B$ and B is the largest frequency contained in the PSD of $X(t)$. When a signal $X(t)$ is bandlimited and nonrandom, Nyquist and Shannon [7] showed that the signal can be reconstructed exactly from its samples $X(nT)$ as long as the sampling frequency $\omega_s = 2\pi/T > 2B$, i.e., as long as the sample frequency exceeds twice the bandwidth of the signal. It turns out that the same result holds for WSS random signals. Specifically, let $H(j\omega)$ denote the ideal low-pass filter with frequency response

$$H(j\omega) = \begin{cases} T & |\omega| \le \omega_s/2 \\ 0 & |\omega| > \omega_s/2 \end{cases} \tag{8.73}$$

and impulse response

$$h(t) = \mathrm{sinc}(\pi t/T) = \frac{\sin(\pi t/T)}{\pi t/T} , \tag{8.74}$$

where $\omega_s = 2\pi/T$ denotes the sampling frequency. Then if $\omega_s > 2B$ and

$$\hat{X}_N(t) = \sum_{n=-N}^{N} X(nT)h(t-nT) \tag{8.75}$$

denotes the reconstructed CT random signal obtained from the $2N+1$ samples $X(nT)$ with $-N \le n \le N$, we have

$$\hat{X}_N(t) \overset{m.s.}{\to} X(t) \tag{8.76}$$

as N tends to infinity. In other words the reconstructed signal $\hat{X}_N(t)$ converges in the mean-square to $X(t)$ as the number $2N+1$ of interpolation samples tends to infinity. To prove this result, we need to show that

$$M_E(N) = E[(X(t) - \hat{X}_N(t))^2] \tag{8.77}$$

tends to zero as N tends to infinity. By substituting (8.10) inside (8.77), we find

$$M_E(N) = R_X(0) - 2 \sum_{n=-N}^{N} R_X(t-nT)h(t-nT)$$

$$+ \sum_{n=-N}^{N} \sum_{m=-N}^{N} R_X((n-m)T)h(t-nT)h(t-mT)$$

$$= \frac{1}{2\pi} \int_{-\infty}^{\infty} |\exp(-j\omega t) - \sum_{n=-N}^{N} \exp(-j\omega nT)h(t-nT)|^2 S_X(\omega)d\omega , \tag{8.78}$$

where the second equality uses the inverse Fourier transform expression (8.10) for the autocorrelation function. Since the PSD $S_X(\omega) = 0$ for $|\omega| > B$, and $\pi/T > B$, we can rewrite

$$M_E(N) = \frac{1}{2\pi} \int_{-\pi/T}^{\pi/T} |\exp(j\omega t) - \sum_{n=-N}^{N} \exp(j\omega nT)h(t-nT)|^2 S_X(\omega)d\omega .$$

With t viewed as a fixed parameter, consider the Fourier series representation of the function $\exp(j\omega t)$ over interval $-\pi/T \le \omega \le \pi/T$. The n-th Fourier coefficient is given by

$$F_n(t) = \frac{T}{2\pi} \int_{-\pi/T}^{\pi/T} \exp(j\omega(t-nT))d\omega = h(t-nT) .$$

Thus for each ω such that $-\pi/T \le \omega \le \pi/T$, $\exp(j\omega t)$ admits the Fourier series representation

$$\exp(j\omega t) = \sum_{n=-\infty}^{\infty} h(t-nT)\exp(j\omega nT) .$$

This shows that $M_E(N)$ tends to zero as N tends to infinity.

8.7 Rice's Model of Bandpass WSS Random Signals

WSS communication signals and radar signals are often bandpass signals in the sense that their PSD is nonzero over a band $[\omega_c - B, \omega_c + B]$ and the matching negative frequency band $[-(\omega_c + B), -(\omega_c - B)]$ as shown in Fig. 8.13, where typically $\omega_c \gg B$. For signals of this type, even though the highest signal frequency is $\omega_c + B$, the occupied bandwidth of the signal, that is to say the range of positive frequencies over which $S_X(\omega)$ is nonzero is only $2B$, so instead of viewing the signal as a baseband signal with bandwidth $\omega_c + B$, it is more fruitful to view it as obtained by quadrature modulation of two baseband signals of bandwidth B with carrier frequency ω_c.

Hilbert Transform A Hilbert transform filter has for frequency response

$$H(j\omega) = -j\,\mathrm{sgn}(\omega) = \begin{cases} e^{-j\pi/2} & \omega > 0 \\ e^{j\pi/2} & \omega < 0. \end{cases} \tag{8.79}$$

Since the filter magnitude $|H(j\omega)| = 1$, it is an all pass filter, and as the expression (8.79) indicates, the phase has a $-\pi$ discontinuity at $\omega = 0$. The Hilbert transform impulse response

$$h(t) = \frac{1}{\pi t}$$

is singular at $t = 0$, which indicates that the filter is difficult to implement in general. However, when it is applied to bandlimited signals, this filter is quite easy to implement in digital signal processing (DSP) form, as explained in [8, Chap. 12].

Then consider the effect of passing a WSS process $X(t)$ with autocorrelation $R_X(\tau)$ and PSD $S_X(\omega)$ through a Hilbert transform filter as shown in Fig. 8.14.

If $\hat{X}(t)$ denotes the filter output, we know that $X(t)$ and $\hat{X}(t)$ are jointly WSS. The PSD of $\hat{X}(t)$ and its cross-spectral density with $X(t)$ are given respectively by

$$S_{\hat{X}}(\omega) = |H(j\omega)|^2 S_X(\omega) = S_X(\omega) \tag{8.80}$$

$$S_{\hat{X}X}(\omega) = H(j\omega) S_X(\omega), \tag{8.81}$$

so that the Hilbert transform does not change the PSD of a signal. Thus the autocorrelation of $\hat{X}(t)$ and its cross-correlation with $X(t)$ are

$$R_{\hat{X}}(\tau) = R_X(\tau) \tag{8.82}$$

$$R_{\hat{X}X}(\tau) = \hat{R}_X(\tau), \tag{8.83}$$

Fig. 8.13 Power spectral density of a bandpass signal

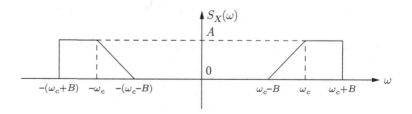

Fig. 8.14 Hilbert
transform of a WSS
random process

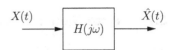

Fig. 8.15 Specification of
the quadrature components
$X_c(t)$ and $X_s(t)$

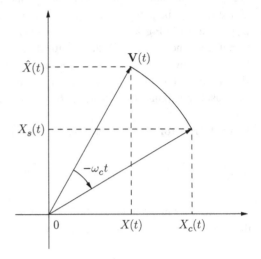

which indicates that the cross-correlation of $\hat{X}$ and X is the Hilbert transform of the autocorrelation
of X. Note also that since $H^*(j\omega) = -H(j\omega)$, we have

$$R_{X\hat{X}}(\tau) = -R_{\hat{X}X}(\tau) = -\hat{R}_X(\tau) . \tag{8.84}$$

In-Phase and Quadrature Components Suppose $X(t)$ is a WSS signal with the bandpass PSD
shown in Fig. 8.13. Let

$$\mathbf{Q}(\theta) = \begin{bmatrix} \cos(\theta) & -\sin(\theta) \\ \sin(\theta) & \cos(\theta) \end{bmatrix}$$

denote the 2×2 matrix representing a rotation by an angle θ in $\mathbb{R}^2$, and observe that

$$\mathbf{Q}(\theta_1 + \theta_2) = \mathbf{Q}(\theta_2)\mathbf{Q}(\theta_1)$$

and $\mathbf{Q}^{-1}(\theta) = \mathbf{Q}(-\theta) = \mathbf{Q}^T(\theta)$. Then the quadrature components of $X(t)$ with respect to center
frequency ω_c are defined by

$$\begin{bmatrix} X_c(t) \\ X_s(t) \end{bmatrix} = \mathbf{Q}(-\omega_c t) \begin{bmatrix} X(t) \\ \hat{X}(t) \end{bmatrix} ,$$

so they can be viewed as obtained by applying a rotation with angle $-\omega_c t \mod 2\pi$ to the vector

$$\mathbf{V}(t) = \begin{bmatrix} X(t) \\ \hat{X}(t) \end{bmatrix}$$

as shown in Fig. 8.15.

The inverse transformation is of course the opposite rotation by an angle $\omega_c t \mod 2\pi$:

$$\begin{bmatrix} X(t) \\ \hat{X}(t) \end{bmatrix} = \mathbf{Q}(\omega_c t) \begin{bmatrix} X_c(t) \\ X_s(t) \end{bmatrix}, \tag{8.85}$$

so that $X(t)$ can be written in terms of its in-phase and quadrature components as

$$X(t) = X_c(t)\cos(\omega_x t) - X_s(t)\sin(\omega_c t) \tag{8.86}$$

$$= M(t)\cos(\omega_c t + \Theta(t)),$$

where

$$M(t) = (X_c^2(t) + X_s^2(t))^{1/2}$$

$$\Theta(t) = \arctan(X_s(t)/X_c(t))$$

represent the magnitude and phase of the complex envelope

$$E(t) = X_c(t) + jX_s(t) = M(t)\exp(j\Theta(t)).$$

Suppose now that $X(t)$ is a bandpass process whose PSD $S_X(\omega)$ is nonzero only in the bands $[\omega_c - B, \omega_c + B]$ and $[-(\omega_c + B), -\omega_c + B]$. Then $X_c(t)$ and $X_s(t)$ are jointly WSS baseband processes with bandwidth B. To prove this result and compute the auto- and cross-correlation of these processes and the matching PSDs, note that

$$E\Big[\begin{bmatrix} X_c(t+\tau) \\ X_s(t+\tau) \end{bmatrix} \big[X_c(t)\ X_s(t) \big] \Big]$$

$$= \mathbf{Q}(-\omega_c(t+\tau))E\Big[\begin{bmatrix} X(t+\tau) \\ \hat{X}(t+\tau) \end{bmatrix} \big[X(t)\ \hat{X}(t) \big] \Big]\mathbf{Q}^T(-\omega_c t), \tag{8.87}$$

where

$$E\Big[\begin{bmatrix} X(t+\tau) \\ \hat{X}(t+\tau) \end{bmatrix} \big[X(t)\ \hat{X}(t) \big] \Big] = \begin{bmatrix} R_X(\tau) & -\hat{R}_X(\tau) \\ \hat{R}_X(\tau) & R_X(\tau) \end{bmatrix}$$

$$= R_X(\tau)\mathbf{I}_2 + \hat{R}_X(\tau)\mathbf{Q}(\pi/2) \tag{8.88}$$

Combining (8.87) and (8.88) and using the properties of rotation matrices, we find

$$E\Big[\begin{bmatrix} X_c(t+\tau) \\ X_s(t+\tau) \end{bmatrix} \big[X_c(t)\ X_s(t) \big] \Big] = R_X(\tau)\mathbf{Q}(-\omega_c\tau) + \hat{R}_X(\tau)\mathbf{Q}(\pi/2 - \omega_c\tau) \tag{8.89}$$

which depends only on τ, so that $X_c(t)$ and $X_s(t)$ are jointly WSS. The diagonal entries of (8.89) are identical and given by

$$R_{X_c}(\tau) = R_{X_s}(\tau) = R_X(\tau)\cos(\omega_c\tau) + \hat{R}_X(\tau)\sin(\omega_c\tau), \tag{8.90}$$

and the off-diagonal terms are given by

$$R_{X_cX_s}(\tau) = -R_{X_sX_c}(\tau) = R_X(\tau)\sin(\omega_c\tau) - \hat{R}_X(\tau)\cos(\omega_c\tau) \,. \tag{8.91}$$

Then by Fourier transforming (8.90) and using the definition (8.79) of the Hilbert transform, we find

$$S_{X_c}(\omega) = S_{X_s}(\omega)$$

$$= \frac{1}{2}\Big[(1 - \mathrm{sgn}\,(\omega - \omega_c))S_X(\omega - \omega_c) + (1 + \mathrm{sgn}\,(\omega - \omega_c))S_X(\omega + \omega_c)\Big]$$

$$= \begin{cases} S_X(\omega - \omega_c) + S_X(\omega + \omega_c) & |\omega| \le \omega_c \\ 0 & |\omega| > \omega_c \,. \end{cases} \tag{8.92}$$

Likewise, by Fourier transforming (8.91), we obtain

$$S_{X_cX_s}(\omega) = -S_{X_sX_c}(\omega)$$

$$= \frac{1}{2j}\Big[(1 - \mathrm{sgn}\,(\omega - \omega_c))S_X(\omega - \omega_c) - (1 + \mathrm{sgn}\,(\omega + \omega_c))S_X(\omega + \omega_c)\Big]$$

$$= \begin{cases} j(-S_X(\omega - \omega_c) + S_X(\omega + \omega_c)) & |\omega| \le \omega_c \\ 0 & |\omega| > \omega_c \,. \end{cases} \tag{8.93}$$

The PSD expressions (8.92) and (8.93) combined with the assumption that $S_X(\omega)$ is nonzero only in bands $[\omega_c - B, \omega_c + B]$ and $[-(\omega_c + B), -\omega_c + B]$ imply that the bandwidth of the in-phase and quadrature components X_c and X_s is B, i.e.,

$$S_{X_c}(\omega) = S_{X_s}(\omega) = S_{X_cX_s}(\omega) = 0$$

for $|\omega| > B$.

Example 8.6 To fix ideas, consider the bandpass WSS signal $X(t)$ with the PSD depicted in Fig. 8.13. Then according to expression (8.92), the PSD of the quadrature components $X_c(t)$ and $X_s(t)$ is obtained by shifting the negative frequency part of the spectrum $S_X(\omega)$ to the right by ω_c, while shifting the positive frequency part to the left by $-\omega_c$. Adding the two components yields the PSD shown in part a) of Fig. 8.16. Likewise, consider now the cross spectral density of $X_c(t)$ and $X_s(\omega)$. It is purely imaginary, and its imaginary part is obtained by flipping the negative part of the spectrum of $S_X(\omega)$ and shifting it to the right by ω_c and combining it with the positive part of the spectrum shifted to the left by $-\omega_c$. The resulting cross-spectrum is shown in part b) of Fig. 8.16.

Remark It is useful to note that if the PSD $S_X(\omega)$ is locally symmetric about $\pm\omega_c$ in the sense that

$$S_X(\omega_c + v) = S_X(\omega_c - v) = S_X(-\omega_c + v) = S_X(-\omega_c - v)$$

for $|v| < B$, then the relation (8.93) implies

$$S_{X_cX_s}(\omega) = 0 \,,$$

Fig. 8.16 (a) Power spectral density and (b) cross-spectral density of the quadrature components of $X(t)$

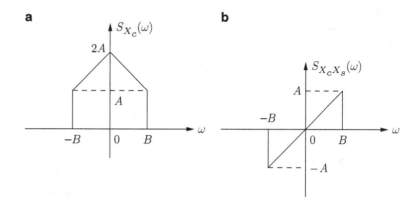

so that the in-phase and quadrature components $X_c(t)$ and $X_s(t)$ are uncorrelated.

8.8 Bibliographical Notes

The material presented in this chapter forms the core of classical engineering applications of random processes and similar material can be found in [4, 9–11]. The extension of the sampling theorem to bandlimited WSS random processes is due to Balakrishnan [12] and Parzen [13] (see also [14]). The computation of autocorrelation and spectra of Markov chains modulated signal was first considered by Titsworth and Welch [15], but the presentation given here is closer to the analysis described in [16]. Finally, the representation of bandpass processes in terms of bandlimited quadrature components was first described by S. O. Rice in his pioneering work on random processes and their applications to the analysis of communications receivers [17–20].

8.9 Problems

8.1 Figure 8.17 describes a noise cross-correlation method for measuring part of the frequency response of a linear time-invariant (LTI) system. In this figure:

i) All signals have a discrete time index, so that t takes only the integer values $0, \pm 1, \pm 2, \cdots$. The input $V(t)$ of the system is a zero-mean white Gaussian noise with intensity q, i.e.,

$$E[V(t)V(s)] = q\delta(t - s).$$

ii) The first LTI filter has an unknown real-valued impulse response $h_1(t)$, and unknown frequency response $H_1(e^{j\omega})$. Its output $X(t)$ can be expressed as

$$X(t) = h_1(t) * V(t) = \sum_{u=-\infty}^{\infty} h_1(t - u)V(u).$$

iii) The second LTI filter is a known passband reference filter with impulse response

$$h_2(t) = \frac{2}{q\pi t} \sin(\frac{\Delta t}{2}) \cos(\omega_0 t)$$

Fig. 8.17 Frequency
response measurement
system

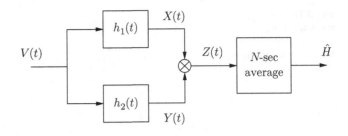

and frequency response

$$H_2(e^{j\omega}) = \begin{cases} q^{-1} & |\omega - \omega_0| \le \Delta/2 \text{ or } |\omega + \omega_0| \le \Delta/2 \\ 0 & \text{elsewhere over } [-\pi, \pi], \end{cases}$$

where $0 < \omega_0 - \Delta/2 < \omega_0 + \Delta/2 < \pi$. Its output is given by

$$Y(t) = h_2(t) * V(t) = \sum_{u=-\infty}^{\infty} h_2(t-u)V(u).$$

iv) $Z(t) = X(t)Y(t)$, and

$$\hat{H} = \frac{1}{N} \sum_{t=-(N-1)/2}^{(N-1)/2} Z(t),$$

where N is an odd integer.

a) Find the mean $m_Z(t)$ and autocovariance function $K_Z(t, s)$ of the process $Z(t)$. Express your mean function answer as an explicit function of the unknown frequency response $H_1(e^{j\omega})$. You may leave your covariance function answer in terms of $h_1(t)$ and $h_2(t)$, or $H_1(e^{j\omega})$ and $H_2(e^{j\omega})$. Is $Z(t)$ WSS?

b) Find the mean and variance of $\hat{H}$ in terms of $m_Z(t)$ and $K_Z(t, s)$.

c) Combine the results of parts a) and b) to obtain a simple approximation for $E[\hat{H}]$ which applies when

$$H_1(e^{j\omega}) \approx H_1(e^{j\omega_0})$$

over the passband of H_2.

d) Under the assumption that $|H_1(e^{j\omega})| \le M$ for all ω, where M is a known constant, show that the variance of $\hat{H}$ tends to zero as $N \to \infty$.

In this problem, you may want to use the following results.

1) If $X_k, k = 1, 2, 3, 4$ are jointly Gaussian zero-mean random variables, we have

$$E[X_1 X_2 X_3 X_4] = E[X_1 X_2]E[X_3 X_4] + E[X_1 X_3]E[X_2 X_4] + E[X_1 X_4]E[X_2 X_3].$$

Fig. 8.18 Power spectral density of $X(t)$

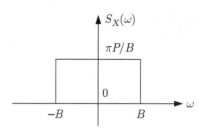

This identity is a special case of a general formula due to Isserlis for computing the higher order moments of Gaussian random variables from the first and second moments. In the above formula, the Gaussian variables X_1, X_2, X_3 and X_4 need not be distinct. For example, if $X_1 = X_2 = X_3 = X_4 = X$, we obtain

$$E[X^4] = 3(E[X^2])^2 \,,$$

which is the standard expression for the moment of order 4 of a zero mean Gaussian random variable X.

2) For an arbitrary function $f(\cdot)$ and N an odd integer,

$$\sum_{t=-(N-1)/2}^{(N-1)/2} \sum_{s=-(N-1)/2}^{(N-1)/2} f(t-s) = \sum_{\tau=-(N-1)}^{(N-1)} (1 - \frac{|\tau|}{N}) f(\tau) \,.$$

3) Two arbitrary discrete-time functions $x_1(t)$ and $x_2(t)$ with discrete-time Fourier transforms $X_1(e^{j\omega})$ and $X_2(e^{j\omega})$ satisfy Parseval's identity

$$\sum_{t=-\infty}^{\infty} x_1(t) x_2^*(t) = \frac{1}{2\pi} \int_{-\pi}^{\pi} X_1(e^{j\omega}) X_2^*(e^{j\omega}) d\omega \,.$$

8.2 Let $X(t)$ be a zero-mean WSS random signal with the power spectral density

$$S_X(\omega) = \begin{cases} \dfrac{\pi P}{B} & \text{for } 0 \le |\omega| \le B \\ 0 & \text{otherwise} \end{cases} \tag{8.94}$$

which is sketched in Fig. 8.18.

Since $X(t)$ is bandlimited with bandwidth B rad/s, it can be reconstructed with zero mean-square error from its samples $X_k = X(kT_s)$ as long as the sampling frequency $\omega_s = 2\pi/T_s$ is larger than the Nyquist frequency $\omega_N = 2B$. However, the digitization of the signal $X(t)$ is not complete at this stage, since the samples X_k take continuous values. For digital storage, the samples X_k need to be quantized, i.e., to be represented by a finite sequence of ones and zeros. The quantization of X_k is performed by an analog-to-digital converter (ADC), but depending on the choice of ADC architecture, the required circuitry may be quite complex. For a flash ADC, $2^M - 1$ comparators are needed to implement an M-bit quantizer, and for a pipeline ADC, M comparators are required, but at the expense of a greater latency. To overcome this difficulty, and taking advantage of the increased speed of modern analog circuits, *oversampled delta–sigma modulator* ADC architectures have been developed. In this type of architecture, the signal $X(t)$ is sampled much faster than the Nyquist rate, say with a frequency $\omega_s = N\omega_N$, where the integer N, which is called the oversampling ratio, can be fairly large. Then,

Fig. 8.19 Delta–sigma
ADC architecture

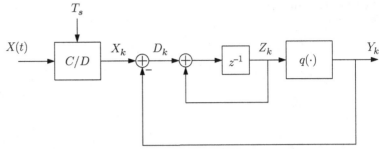

Fig. 8.20 Input–output
characteristic of one-bit
quantizer $q(\cdot)$

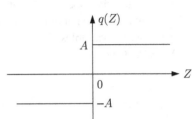

Fig. 8.21 Linearized
model of the delta–sigma
converter loop

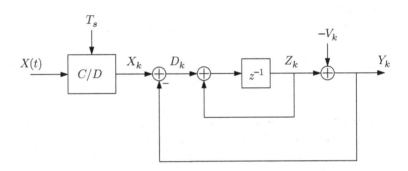

as shown in Fig. 8.19, the samples X_k are passed through a negative feedback loop which contains a digital integrator and a one-bit quantizer. The output Y_k of the delta–sigma modulator (DSM) takes therefore the discrete values $\pm A$, where $\Delta = 2A$ denotes the quantizer step-size. The input–output characteristic $Y = q(Z)$ of the quantizer is depicted in Fig. 8.20.

Since the quantizer contained in sigma–delta modulator loop is a nonlinear device, the analysis of sigma–delta modulators is fairly complex. However, a simple model which allows a fairly good understanding of the behavior of delta–sigma modulator loops consists of modeling the quantization error

$$V_k = Z_k - Y_k = Z_k - q(Z_k)$$

as a zero-mean white noise sequence uncorrelated with X_k with variance $\Delta^2/12 = A^2/3$, i.e.

$$E[V_k V_l] = \frac{A^2}{3}\delta(k - l).$$

The resulting linear model of a delta–sigma modulator is shown in Fig. 8.21. The remainder of this problem focuses on the analysis of the linearized loop and the reconstruction of X_k from the binary samples Y_k. We assume that the oversampling ratio is exactly N, i.e., $\omega_s = 2NB$.

Fig. 8.22 Samples
reconstruction system

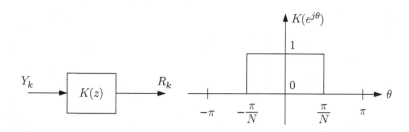

a) Given that the power spectral density of the continuous WSS signal $X(t)$ is given by (8.94), evaluate and sketch the power spectral density $S_s(e^{j\theta})$ of the sampled signal $X_k = X(kT_s)$ at the input of the DSM. For this question, you may want to use the fact that if $f(t)$ is a deterministic signal with continuous-time Fourier transform $F(j\omega)$, the discrete-time Fourier transform of the sampled signal $f_k = f(kT_s)$ is

$$F_s(e^{j\theta}) = \frac{1}{T_s} \sum_{m=-\infty}^{\infty} F(j\frac{(\theta - m2\pi)}{T_s}) .$$

b) By analyzing the linearized DSM loop, show that its response to the input signals X_k and V_k takes the form

$$Y_k = g_k * X_k + h_k * V_k , \qquad (8.95)$$

where $*$ denotes the discrete-time convolution, and g_k and h_k denote the impulse responses of the SDM loop to the X_k and V_k inputs. Equivalently, in the z–transform domain, we have

$$Y(z) = G(z)X(z) + H(z)V(z) .$$

Find $G(z)$ and $H(z)$.
c) Use the filters $G(z)$ and $H(z)$ obtained in part b) to evaluate the power spectral density $S_Y(e^{j\theta})$ of the binary samples Y_k.
d) Given the binary samples Y_k, it is still necessary to approximate the input signal samples X_k. This is usually accomplished by passing the samples Y_k through an ideal low-pass filter $K(e^{j\theta})$ with cutoff frequency π/N as shown in Fig. 8.22, producing an output R_k.

Show that R_k can be represented as

$$R_k = X_{k-1} + E_k,$$

where E_k can be viewed as the quantization error of the delta–sigma modulator ADC. Evaluate and sketch the power spectral density $S_E(e^{j\theta})$ of the error process E_k. Evaluate the mean and variance of E_k. To evaluate the variance, you may need to use the trigonometric identity

$$2\sin^2(\theta/2) = 1 - \cos(\theta) .$$

8.3 The series RLC circuit shown in Fig. 8.23 is in thermal equilibrium. The thermal voltage fluctuations are modeled by a voltage source $N(t)$ in series with R, where $N(t)$ is a zero-mean white Gaussian noise with intensity $2kRT$, i.e.

$$E[N(t)N(s)] = 2kRT\delta(t - s) .$$

T denotes the absolute temperature of the circuit and k is Boltzmann's constant.

Fig. 8.23 RLC circuit at
temperature T

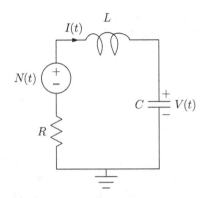

a) Find the transfer functions $H_C(s)$ and $H_L(s)$ from $N(t)$ to the capacitor voltage $V(t)$ and inductor current $I(t)$, respectively.
b) Find the power spectral densities $S_V(\omega)$ and $S_I(\omega)$ of $V(t)$ and $I(t)$, respectively. You may find useful to express your answers in terms of the resonance frequency $\omega_0 = (LC)^{-1/2}$ and damping ratio

$$\zeta = \frac{RC}{2}\omega_0 = \frac{R}{2}\sqrt{\frac{C}{L}}\,.$$

Verify that the spectral densities $S_V(\omega)$ and $S_I(\omega)$ obey Nyquist's theorem, i.e., they satisfy

$$S_V(\omega) = 2k\Re\{Z_C(j\omega)\}T$$

$$S_I(\omega) = 2k\Re\{Y_L(j\omega)\}T\,,$$

where $Z_C(j\omega)$ denotes the Thevenin impedance seen across the capacitor terminals, and $Y_L(j\omega)$ denotes the admittance across two terminals formed by opening the inductor branch.
c) Compute the autocorrelation functions $K_V(\tau)$ and $K_I(\tau)$ of the capacitor voltage and inductor current, respectively. Find the variances of $V(t)$ and $I(t)$ and verify that the average energies stored in the capacitor and inductor satisfy the theorem of equipartition of energy.

8.4 When an airplane flying over the ocean sends a radar pulse $s(t)$, it receives a multitude of echoes, which are Doppler-shifted copies of its transmitted pulse reflected from numerous points on the surface of the water. Those points reflect the pulse with random amplitudes and delays which for the case of a windswept surface can be considered statistically independent. The echoes constitute a random signal known as *clutter* which takes the form

$$X(t) = \sum_{k=-\infty}^{\infty} A_k s(t - S_k)\,,$$

where signal $s(t)$ represents the transmitted radar pulse, S_k the time at which the k-th pulse reaches the receiver, and A_k its amplitude. We study the statistics of this process assuming it lasts from $-\infty$ to ∞, and that the arrival times S_k constitute a Poisson point process with rate λ. The amplitudes A_k are assumed to be independent, identically distributed with density $f_A(a)$, and independent of the Poisson epochs S_k.

a) The process $X(t)$ can be viewed as a generalized shot noise process

$$X(t) = \int_{-\infty}^{\infty} s(t-u)dZ(u),$$ (8.96)

where (8.96) is interpreted as pathwise Stieltjes integral and $Z(t)$ is a compound Poisson process with independent increments

$$Z(t) - Z(s) = \sum_{k=N(s)+1}^{N(t)} A_k .$$

Verify that the mean and variance of increment $Z(t) - Z(s)$ are

$$E[Z(t) - Z(s)] = \lambda m_A(t-s)$$
$$E[(Z(t) - Z(s) - \lambda m_A(t-s))^2] = \lambda E[A^2](t-s) ,$$

where m_A and $E[A^2]$ are the mean and second moment of random variables A_k. Then by introducing the centered process $D(t) = Z(t) - \lambda m_A t$ and decomposing the integral (8.96) as

$$X(t) = \lambda m_A \int_{-\infty}^{\infty} s(t-u)du + \int_{\infty}^{\infty} s(t-u)dD(u)$$

extend the shot noise analysis of Sect. 7.6 to evaluate the mean $m_X(t)$ and autocorrelation $R_X(t,s)$ of $X(t)$. Express your answers in terms of λ, $s(t)$, m_A, and $E[A^2]$.

b) Evaluate the power spectral density $S_X(\omega)$ of $X(\cdot)$.

c) Compute and sketch $S_X(\omega)$ when $s(t)$ is the modulated square pulse

$$s(t) = p(t)\cos(\omega_c t)$$

with

$$p(t) = \begin{cases} P^{1/2} & 0 \le |t| \le T/2 \\ 0 & \text{otherwise} . \end{cases}$$

Assume that the length T of the pulse is much larger than the period $T_c = 2\pi/\omega_c$ of the carrier. Also, do not hesitate to make approximations, provided they are justified.

8.5 Consider the quadrature amplitude modulated process

$$Z(t) = X(t)\cos(1000t) - Y(t)\sin(1000t),$$

where $X(t)$ and $Y(t)$ are two zero-mean jointly WSS random processes whose auto- and cross-correlation functions satisfy

$$R_X(\tau) = R_Y(\tau) \quad , \quad R_{XY}(\tau) = -R_{XY}(-\tau)$$

and admit the power spectral densities

$$S_X(\omega) = S_Y(\omega) = \begin{cases} 10 & |\omega| < 100 \\ 0 & |\omega| > 100 \end{cases}$$

$$S_{XY}(\omega) = \begin{cases} j\omega/10 & |\omega| < 100 \\ 0 & |\omega| > 100 \end{cases}$$

a) Find the mean and autocorrelation of $Z(t)$ and verify it is a WSS process.
b) Evaluate and sketch the power spectral density $S_Z(\omega)$ of $Z(t)$.
c) Suppose that instead of expressing $Z(t)$ as a quadrature modulated signal with center frequency $\omega_c = 1000$ rad/s, we wish to express it in terms of the center frequency $\omega_c' = 900$ rad/s, i.e.,

$$Z(t) = U(t)\cos(900t) - V(t)\sin(900t) .$$

Find and sketch the power spectral densities $S_U(\omega)$, $S_V(\omega)$ and $S_{UV}(\omega)$ of the in-phase and quadrature components $U(t)$ and $V(t)$.

For this problem you may need the continuous-time Fourier transform pairs

$$\cos(\omega_0 t) \longleftrightarrow \pi(\delta(\omega - \omega_0) + \delta(\omega + \omega_0))$$

$$\sin(\omega_0 t) \longleftrightarrow -j\pi(\delta(\omega - \omega_0) - \delta(\omega + \omega_0)) .$$

8.6 Consider random process

$$X(t) = A\cos(\Omega_0 t + \Theta),$$

where the amplitude A is known, the phase Θ is uniformly distributed over $[0, 2\pi)$, and frequency Ω_0 is random and uniformly distributed over $[\omega_L, \omega_H]$ where $0 < \omega_L < \omega_H$. Thus

$$f_{\Omega_0}(\omega) = \begin{cases} \dfrac{1}{\omega_H - \omega_L} & \omega_L \leq \omega \leq \omega_H \\ 0 & \text{otherwise} . \end{cases}$$

The frequency Ω_0 is independent of phase Θ.

a) Evaluate the mean and autocorrelation of $X(t)$. To do so, you may want to use iterated expectation, i.e.,

$$E[X(t + \tau)X(t)] = E[E[X(t + \tau)X(t)|\Omega_0]] .$$

Is $X(t)$ WSS?
b) Evaluate the power spectral density $S_X(\omega)$. To do so, you are reminded that the Fourier transform of

$$f(\tau) = \text{sinc}\,(Wt) = \frac{\sin(Wt)}{Wt}$$

Fig. 8.24 Power spectral density of $X(t)$

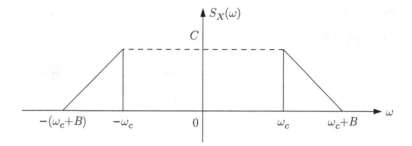

is

$$F(j\omega) = \begin{cases} \pi/W & |\omega| \le W \\ 0 & \text{otherwise}. \end{cases}$$

c) Verify that $X(t)$ is a bandpass process. Then consider Rice's representation

$$X(t) = X_c(t)\cos(\omega_c t) - X_s(t)\sin(\omega_c t),$$

where $\omega_c = (\omega_L + \omega_H)/2$. Evaluate and sketch the PSDs of $X_c(t)$ and $X_s(t)$, as well as their cross-spectral density $S_{X_c X_s}(\omega)$.

8.7 The power spectral density of a zero-mean WSS random signal $X(t)$ is sketched in Fig. 8.24. Assume that $X(t)$ is real valued.

a) Draw a complete block diagram of a system used to generate the in-phase and quadrature component $X_c(t)$ and $X_s(t)$ in the representation

$$X(t) = X_c(t)\cos(\omega_c t) - X_s(t)\sin(\omega_c t).$$

Your block diagram should employ the Hilbert transform $\check{X}(t)$ of $X(t)$. Evaluate and sketch the power spectral densities $S_{X_c}(\omega)$, $S_{X_s}(\omega)$, and cross-density $S_{X_c X_s}(\omega)$ of $X_c(t)$ and $X_s(t)$.

b) Although $X(t)$ is a bandpass process, its total bandwidth is only B (we are counting only positive frequencies). This means that the minimum sampling frequency required to sample $X(t)$ and its Hilbert transform $\check{X}(t)$ is only $\omega_s = 2B$. Express $X(t)$ in terms of the samples $X(kT_s)$ and $\check{X}(kT_s)$ with k integer, and $T_s = 2\pi/\omega_s = \pi/B$.

8.8 Spread-spectrum communication systems were developed to allow a transmitter and receiver to communicate covertly without being detected by a potential eavesdropper monitoring the communications channel. The system shown in Fig. 8.25 is a simple baseband model for a spread-spectrum modulation scheme that accomplishes this objective.

In this system, $M(t)$ is the message waveform that needs to be transmitted. We assume that $M(t)$ is a zero-mean first-order Gauss–Markov stationary process with autocorrelation

$$R_M(\tau) = P_M \exp(-2\lambda_M|\tau|).$$

The spread-spectrum transmitter multiplies $M(t)$ by $X(t)$, which is a random telegraph wave statistically independent of $M(t)$. It is shown in Problem 7.2 that $X(t)$ can be expressed as

Fig. 8.25 Baseband model of a spread-spectrum system

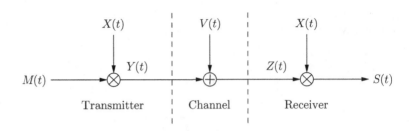

$$X(t) = X(0)(-1)^{N(t)} ,$$

where $N(t)$ is a Poisson process with rate λ_X, and $X(0)$ is independent of $N(t)$ and takes the values 1 and -1 with probability 1/2. Thus, $X(t)$ takes values 1 or -1, and as shown in Problem 7.2, it has zero-mean and autocorrelation

$$R_X(\tau) = \exp(-2\lambda_X|\tau|) .$$

The transmitted signal $Y(t)$ is corrupted in the communication channel by an additive white Gaussian noise process $V(t)$ which is independent of $M(t)$ and $X(t)$ and such that

$$m_V = 0 \quad , \quad R_V(\tau) = q\delta(\tau) .$$

The received waveform $Z(t) = Y(t) + V(t)$ is multiplied by the *same* random telegraph wave that was used in the transmitter. We want to show that the spectrum of $Y(t)$ can be hidden in the white noise $V(t)$ without increasing the noise contained in the signal $S(t)$ at the receiver output.

a) Find the autocorrelation function $R_Y(\tau)$ and its associated power spectrum $S_Y(\omega)$. Make a labeled sketch of $S_Y(\omega)$.
b) Show that λ_X can be chosen so that the spectrum of $Z(t)$ satisfies

$$\max_{\omega} |S_Z(\omega) - q| \le 10^{-3}q .$$

c) Show that the signal $S(t)$ at the receiver output can be expressed as

$$S(t) = M(t) + W(t) ,$$

where the noise $W(t)$ depends only on $X(t)$ and $V(t)$. Find the autocorrelation $R_W(\tau)$ and power spectral density $S_W(\omega)$ of $W(t)$. For this question, you will need to use the fact that for any continuous function $f(\tau)$, we have $f(\tau)\delta(\tau) = f(0)\delta(\tau)$.

To solve this problem, you may need the FT pair

$$\exp(-a|\tau|) \quad \leftrightarrow \quad \frac{2a}{\omega^2 + a^2} \quad \text{for} \quad a > 0 .$$

8.9 Let $X(t) = A\cos(\omega_c t + \Theta)$, where the amplitude A and phase Θ are two random variables. Consider the received signal

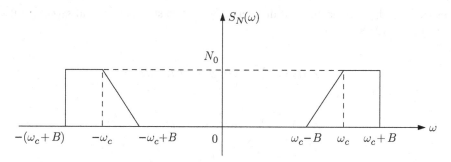

Fig. 8.26 Power spectral density of $N(t)$

$$Z(t) = X(t) + N(t) \,,$$

where the noise $N(t)$ is a zero-mean WSS process independent of $X(t)$, with the power spectral density shown in Fig. 8.26.

a) Obtain the in-phase/quadrature representation

$$N(t) = U(t) \cos(\omega_c t + \Theta) - V(t) \sin(\omega_c t + \Theta)$$

of the noise $N(t)$, i.e., evaluate and sketch the power spectral densities $S_U(\omega)$, $S_V(\omega)$ and cross-spectral density $S_{UV}(\omega)$ of $U(t)$ and $V(t)$.

b) Consider now the envelope and phase representation

$$Z(t) = E(t) \cos(\omega_c t + \Psi(t))$$

of the received signal $Z(t)$. Show that when the signal to noise ratio $E[A^2]/E[N^2(t)]$ is large, the following approximations are satisfied:

$$E(t) \approx A + U(t) \qquad \Psi(t) = \Theta + \frac{1}{A} V(t) \,. \tag{8.97}$$

To derive approximations (8.97), you may want to use a phasor representation of the signal $X(t)$ and noise $N(t)$, and employ geometric considerations. The approximations (8.97) are very convenient to analyze amplitude and/or angle modulation schemes.

8.10 Consider Example 8.5. The power spectral density of binary FSK modulated signal $Z(t)$ can also be computed by observing that $Z(t)$ can be written as the sum of two PAM signals

$$Z(t) = Z_c(t) + Z_d(t) \,,$$

where if

$$a(t) = \frac{1}{2}(s_0(t) + s_1(t)) \,, \quad d(t) = \frac{1}{2}(s_1(t) - s_0(t))$$

denote respectively the average and half difference of the two signals $s_1(t)$ and $s_0(t)$, and if symbol $Y(k) = 2(X(k) - 1/2)$, we have

$$Z_c(t) = \sum_{k=-\infty}^{\infty} Y(k)d(t - kT - D)$$

$$Z_d(t) = \sum_{k=-\infty}^{\infty} a(t - kT - D).$$

a) Verify that the iid symbol sequence $Y(k)$ takes values ± 1 with equal probability and compute its mean m_Y and autocorrelation $R_Y(m) = E[Y(k + m)Y(k)]$.
b) Compute the means of signals $Z_c(t)$, $Z_d(t)$ and $Z(t)$.
c) Compute the cross-correlation $R_{cd}(\tau) = E[Z_c(t + \tau)Z_d(t)]$ and autocorrelations $R_c(\tau) = E[Z_c(t + \tau)Z_c(t)]$, $R_d(\tau) = E[Z_d(t + \tau)Z_d(t)]$ of signals $Z_c(t)$ and $Z_d(t)$.
d) Use expression (8.55) to compute the PSDs of signal $Z_c(t)$ and $Z_d(t)$, and verify that your results match those of Example 8.5.

8.11 Consider a sequence $\{X(k), k \in \mathbb{Z}\}$ of independent identically distributed random symbols taking the values 0 and 1 with probability $1/2$. To transmit this sequence, we implement a binary pulse position modulation scheme where to transmit a zero, we send a pulse $p(t)$, and to transmit a one, we send the pulse $p(t - T/2)$, which is the pulse $p(t)$ delayed by half a baud period (T denotes the baud period). For simplicity, we assume that $p(t)$ is just a rectangular pulse of duration $T/2$, i.e.

$$p(t) = \begin{cases} 1 & 0 < t < T/2 \\ 0 & \text{otherwise}. \end{cases}$$

We assume that a random synchronization delay D is present, where D is uniformly distributed over the interval $[0, T)$, so that its probability density is given by

$$f_D(d) = \begin{cases} 1/T & 0 \le d < T \\ 0 & \text{otherwise}. \end{cases}$$

a) Show that the transmitted signal can be written as

$$Z(t) = Z_0(t) + Z_1(t),$$

where

$$Z_0(t) = \sum_{k=-\infty}^{\infty} U(k)p(t - kT - D)$$

$$Z_1(t) = \sum_{k=-\infty}^{\infty} V(k)p(t - kT - T/2 - D),$$

where $U(k)$ and $V(k)$ depend linearly on $X(k)$.
b) Find the mean and autocorrelation of the signals $U(k)$ and $V(k)$ and the cross-correlation between $U(k)$ and $V(k)$.

c) Evaluate the means of the signals $Z_0(t)$ and $Z_1(t)$. Evaluate their autocorrelations $R_{Z_0}(t, s)$ and $R_{Z_1}(t, s)$ as well as the cross-correlation

$$R_{Z_0 Z_1}(t, s) = E[Z_0(t) Z_1(s)] \, .$$

Are $Z_0(t)$ and $Z_1(t)$ jointly WSS?

d) Use the results of part b) to obtain the mean $m_Z(t)$ and autocorrelation $R_Z(t, s)$ of $Z(t)$. Is $Z(t)$ WSS?

e) Find the power spectral density $S_Z(\omega)$ of $Z(t)$. To do so, you may want to use the fact that the Fourier transform of the pulse $p(t)$ is given by

$$P(j\omega) = \frac{T}{2} \frac{\sin(\omega T/4)}{\omega T/4} \exp(-jT/4) \, .$$

8.12 For ultra wideband (UWB) wireless communications, it has been proposed to employ pulse-position modulated signals of the form

$$Z(t) = \sum_{k=-\infty}^{\infty} p(t - kT - Q_k - D) \, , \tag{8.98}$$

where $p(t)$ is a known signalling pulse, T denotes the baud period, and where the transmitted information is encoded by the pulse positions Q_k which take values in the discrete set $\{0, \Delta, \ldots, (M-1)\Delta\}$ with $\Delta = T/M$, where M is an integer. For example for $M = 4$, to encode the bit pairs $(0, 0)$, $(0, 1)$, $(1, 1)$, or $(1, 0)$ we transmit respectively the waveforms

$$p_0(t) = p(t) \qquad\qquad p_1(t) = p(t - T/4)$$
$$p_2(t) = p(t - T/2) \qquad\qquad p_3(t) = p(t - 3T/4) \, .$$

The random positions Q_k are i.i.d., with PMF

$$P[Q_k = i\Delta] = 1/M$$

for $i = 0, 1, \ldots, (M - 1)$. In expression (8.98) for the modulated signal, D is a random synchronization delay independent of the positions Q_k and uniformly distributed over $[0, T]$, i.e., its probability density is given by

$$f_D(d) = \begin{cases} 1/T & 0 \leq d < T \\ 0 & \text{otherwise} \, . \end{cases}$$

a) Determine the mean function of the signal $Z(t)$ and verify that it is constant. Express your answer in terms of the pulse integral

$$I = \int_{-\infty}^{\infty} p(t) dt \, .$$

b) Show that the autocorrelation function of $Z(t)$ is a function of $\tau = t - s$, so that the process is WSS, and that it can be decomposed as

$$R_Z(\tau) = R_c(\tau) + R_d(\tau) , \tag{8.99}$$

where

$$R_c(\tau) = \frac{1}{T}[r_p(\tau) - r_p(\tau) * r_q(\tau)] \tag{8.100}$$

and

$$R_d(\tau) = \frac{1}{MT} \sum_{k=-\infty}^{\infty} r_p(\tau - k\Delta) . \tag{8.101}$$

In the above expressions $*$ denotes the convolution of two functions,

$$r_p(\tau) = p(\tau) * p(-\tau)$$

is the deterministic autocorrelation of the pulse $p(\tau)$, and

$$r_q(\tau) = \frac{1}{M} \sum_{k=-(M-1)}^{M-1} (1 - |k|/M)\delta(\tau - k\Delta) . \tag{8.102}$$

To derive the expressions (8.99)–(8.101) for the autocorrelation $R_Z(\tau)$, you should first verify that

$$R_Z(t+\tau, t) = E\left[\sum_{k=-\infty}^{\infty} E[p(t+\tau-kT-Q_k-D)p(t-kT-Q_k-d) \mid D] \right]$$

$$+ E\left[\sum_{r \neq 0} \sum_{k=-\infty}^{\infty} E[p(t+\tau-kT-Q_k-D) \mid D] \right.$$

$$\left. \times E[p(t-(k+r)T-Q_{k+r}-D) \mid D] \right].$$

c) The significance of the decomposition (8.99) is that $R_d(\tau)$ is a periodic function with period $\Delta = T/M$, so that its Fourier transform will consist of discrete impulsive spectral lines located at multiples of $M\omega_b$, where $\omega_b = 2\pi/T$ denotes the baud/symbol frequency of the pulse position modulation scheme. By using the fact that the Fourier transform of the sampling function

$$g(t) = \sum_{k=-\infty}^{\infty} \delta(t - k\Delta)$$

is given by

$$G(j\omega) = M\omega_b \sum_{n=-\infty}^{\infty} \delta(\omega - nM\omega_b)$$

verify that the power spectral density of $Z(t)$ can be decomposed as

$$S_Z(\omega) = S_c(\omega) + S_d(\omega) \,, \tag{8.103}$$

where if $P(j\omega)$ denotes the Fourier transform of the pulse $p(t)$ and if

$$\Phi_Q(\omega) = E[\exp(i\omega Q)]$$

denotes the characteristic function of position Q_k, we have

$$S_c(\omega) = \frac{1}{T}|P(j\omega)|^2[1 - |\Phi_Q(\omega)|^2] \tag{8.104}$$

and

$$S_d(\omega) = \frac{1}{MT}|P(j\omega)|^2 G(j\omega) \,. \tag{8.105}$$

d) For UWB wireless communications, $p(t)$ is an "ultrawideband" pulse, which means that its bandwidth must represent a large proportion, say more than 25%, of its center frequency. We consider here a modulated Gaussian pulse of the form

$$p(t) = A\exp(-2(t/T)^2)\cos(4\pi t/T) \,. \tag{8.106}$$

To evaluate its Fourier transform, you may want to use the Fourier transform pair:

$$\exp(-\frac{t^2}{2\sigma^2}) \qquad \longleftrightarrow \qquad \sigma\sqrt{2\pi}\exp(-\sigma^2\omega^2/2) \,. \tag{8.107}$$

The bandwidth of the pulse (8.106) is given by $1/\sigma$.

Evaluate and sketch the squared magnitude $|P(j\omega)|^2$ of the Fourier transform of the pulse $p(t)$ given by (8.106). Find the ratio of its bandwidth to its center frequency $\omega_c = 4\pi/T$. Is it an ultrawideband pulse?

With M arbitrary, verify that

$$|\Phi_Q(\omega)|^2 = \frac{\sin^2(\omega M\Delta/2)}{M^2\sin^2(\omega\Delta/2)}$$

$$= \frac{\sin^2(\omega T/2)}{M^2\sin^2(\omega T/(2M))} \,. \tag{8.108}$$

Now, set $M = 4$. For $M = 4$ and the pulse $p(t)$ of (8.106), sketch the discrete and continuous spectral components $S_d(\omega)$ and $S_c(\omega)$ of the power spectral density $S_Z(\omega)$ of the signal $Z(t)$.

The objective of UWB wireless communications is to communicate at very high rate with low power at short ranges, say within one or several offices on the same floor of a building. Since the bandwidth employed is very large, the power is spread over a large frequency band and is well below the noise level of established users within the range of frequencies employed by this modulation scheme. Unfortunately, in order to ensure that UWB devices will not interfere with GPS systems, the Federal Communications Commission (FCC) has imposed a spectral mask requirement which imposes that the PSD $S_Z(\omega)$ of UWB signals should be very small at GPS frequencies. This in turn imposes challenging design constraints for the pulse $p(t)$, so that the modulated Gaussian pulse employed in this problem needs to be replaced by a specially designed pulse $p(t)$.

8.13 Consider the Miller code Markov chain modulated signal of Problems 5.3 and 5.12. It was shown that the steady-state probability distribution of the one-step transition probability matrix

$$\mathbf{P} = \begin{bmatrix} 0 & 1/2 & 0 & 1/2 \\ 0 & 0 & 1/2 & 1/2 \\ 1/2 & 1/2 & 0 & 0 \\ 1/2 & 0 & 1/2 & 0 \end{bmatrix}$$

is

$$\pi = \frac{1}{4}\begin{bmatrix} 1 & 1 & 1 & 1 \end{bmatrix},$$

and that

$$\mathbf{P} = \mathbf{V}\mathbf{D}\mathbf{V}^{-1},$$

where

$$\mathbf{V} = \begin{bmatrix} 1 & 1 & 1 & 0 \\ 1 & -1 & 0 & 1 \\ 1 & -1 & 0 & -1 \\ 1 & 1 & -1 & 0 \end{bmatrix}$$

and

$$\mathbf{D} = \mathrm{diag}\,\{1, 0, \mathbf{D}_c\} \quad \text{with} \quad \mathbf{D}_c = \frac{1}{2}\begin{bmatrix} -1 & 1 \\ -1 & -1 \end{bmatrix}.$$

Note that since the columns of $\mathbf{V}$ are orthogonal

$$\mathbf{V}^{-1} = \mathrm{diag}\,\{1/4, 1/4, 1/2, 1/2\}\mathbf{V}^T .$$

a) Use the expressions above to show that

$$\mathbf{P}_c = \mathbf{V}\begin{bmatrix} \mathbf{0} & \mathbf{0} \\ \mathbf{0} & \mathbf{D}_c \end{bmatrix}\mathbf{V}^{-1}$$

and if $z = e^{j\omega T}$

$$(\mathbf{I}_4 - \mathbf{P}_c z^{-1})^{-1} = \mathbf{V}\begin{bmatrix} \mathbf{I}_2 & \mathbf{0} \\ \mathbf{0} & (\mathbf{I}_2 - \mathbf{D}_c z^{-1})^{-1} \end{bmatrix}\mathbf{V}^{-1} .$$

Evaluate $(\mathbf{I}_2 - \mathbf{D}_c z^{-1})^{-1}$.

b) Consider now the four waveforms

$$s_1(t) = -s_4(t) = A \ 0 \le t \le T$$

$$s_2(t) = -s_3(t) = \begin{cases} A & 0 \le t < T/2 \\ -A & T/2 \le t < T . \end{cases}$$

Conclude from the symmetries existing between these four waveforms that

$$\sum_{i=1}^{4} \pi_i S_i(j\omega) = 0 ,$$

so that the discrete component of the spectrum is zero. Observe also that

$$\mathbf{V}^{-1}\mathbf{S}(j\omega) = \begin{bmatrix} 0 \\ 0 \\ S_1(j\omega) \\ S_2(j\omega) \end{bmatrix} ,$$

so that if

$$\mathbf{S}_c(j\omega) \triangleq \begin{bmatrix} S_1(j\omega) \\ S_2(j\omega) \end{bmatrix} ,$$

show that the PSD of $Z(t)$ can be expressed as

$$S_Z(\omega) = \frac{1}{T}\Re\{\mathbf{S}_c^H(j\omega)(\mathbf{I}_2 - \mathbf{D}_c e^{-j\omega T})^{-1}\mathbf{S}_c(j\omega)\} - \frac{1}{2T}[|S_1(j\omega)|^2 + |S_2(j\omega)|^2] .$$

c) By observing that

$$s_1(t) = p(t) + p(t - T/2)$$
$$s_2(t) = p(t) - p(t - T/2),$$

where

$$p(t) = \begin{cases} A & 0 \le t \le T/2 \\ 0 & \text{otherwise} , \end{cases}$$

denotes the half-period pulse, show that the PSD of $Z(t)$ can be expressed as [21, p.44]

$$S_Z(\omega) = \frac{2|P(j\omega)|^2}{T}\frac{3 + \cos(\phi) + 2\cos(2\phi) - \cos(3\phi)}{9 + 12\cos(2\phi) + 4\cos(4\phi)} \tag{8.109}$$

where $\phi = \omega T/2$, and

$$P(j\omega) = A\frac{T}{2}\text{sinc}(\omega T/4)\exp(-j\omega T/4)$$

denotes the Fourier transform of $p(t)$. For $A = 1$, plot $S_Z(\omega)$ as a function of ω/ω_b where $\omega_b = 2\pi/T$.

8.14 In mixed signal circuits, the switching of digital gates constitutes a significant source of signal interference for analog components of the circuit, such as analog-to-digital converters (ADCs), phase-locked loops (PLLs), or frequency synthesizers. The interference propagates primarily through the common circuit substrate and is therefore often referred to as substrate noise or switching noise. In [22] a Markov chain modulation model was proposed to describe substrate noise. Suppose most of the noise at a circuit location originates from the switching activity of a single large buffer, which switches from low to high with probability α and from high to low with probability β. If $X(k)$ denotes the state of the buffer in the k-th clock cycle, and if the states of the buffer are labeled as $\{L, H\}$, the one-step transition probability matrix describing the buffer state is therefore

$$\mathbf{P} = \begin{bmatrix} 1 - \alpha & \alpha \\ \beta & 1 - \beta \end{bmatrix} .$$

a) Verify that the steady-state probability distribution of the buffer state is

$$\boldsymbol{\pi} = \begin{bmatrix} \pi_L & \pi_H \end{bmatrix} = \begin{bmatrix} \dfrac{\beta}{\alpha + \beta} & \dfrac{\alpha}{\alpha + \beta} \end{bmatrix} .$$

Unfortunately, the signal generated by the buffer as it switches depends not only on its current state $X(k)$ but also on its new state $X(k + 1)$, in the sense that if $S(k) = (X(k), X(k + 1))$, the substrate noise generated by the switching buffer is

$$Z(t) = \sum_{k=-\infty}^{\infty} s_{S(k)}(t - kT - D),$$

where T denotes the clock period and D denotes a synchronization delay uniformly distributed over $[0, T]$. Four switching waveforms $s_{LL}(t)$, $s_{LH}(t)$, $s_{HL}(t)$, and $s_{HH}(t)$ appear in the model, but typically, since no switching signal is generated if the state remains the same

$$s_{LL}(t) = s_{HH}(t) = 0 ,$$

so only the low-to-high and high-to-low switches generate substrate noise.

b) To recast the switching noise in the Markov chain modulation format (8.58), we can use an expanded Markov chain describing the switching of the pairs $S(k) = (X(k), X(k + 1))$ of two consecutive states. If the expanded states are labeled as $\{LL, LH, HL, HH\}$, where the first label denotes the state $X(k)$ and the second label the state $X(k + 1)$, verify that the one-step transition matrix of the expanded Markov chain can be written as

$$\mathbf{P}_e = \begin{bmatrix} 1 - \alpha & \alpha & 0 & 0 \\ 0 & 0 & \beta & 1 - \beta \\ 1 - \alpha & \alpha & 0 & 0 \\ 0 & 0 & \beta & 1 - \beta \end{bmatrix} .$$

Compute the steady-state probability distribution

$$\boldsymbol{\pi}_e = \begin{bmatrix} \pi_{LL} & \pi_{LH} & \pi_{HL} & \pi_{HH} \end{bmatrix}$$

of the expanded Markov chain. *Hint:* If the two state Markov chain is in steady state

$$P(X(k) = L, X(k + 1) = L) = P(X(k) = L)P(X(k + 1) = L | X(k) = L)$$

$$= \pi_L p_{LL} ,$$

where p_{LL} denotes the $(1, 1)$ element of $\mathbf{P}$.

c) Verify that

$$P_e V = V \Lambda \ , \tag{8.110}$$

where the columns of

$$V = \begin{bmatrix} 1 & \alpha & \alpha & 0 \\ 1 & -\beta & -(1-\alpha) & 0 \\ 1 & \alpha & 0 & 1-\beta \\ 1 & -\beta & 0 & -\beta \end{bmatrix}$$

are right eigenvectors of P_e and

$$\Lambda = \text{diag}\,\{1, 1 - (\alpha + \beta), 0, 0\}$$

is the corresponding diagonal eigenvalue matrix. In the following it is assumed that $0 < |1 - (\alpha + \beta)| < 1$. Let

$$D = \text{diag}\,\{\frac{1}{\alpha + \beta}, \frac{1}{(\alpha + \beta)(1 - (\alpha + \beta))}, \frac{1}{1 - (\alpha + \beta)}, \frac{1}{1 - (\alpha + \beta)}\} \ .$$

By direct multiplication, check that

$$V^{-1} = D \begin{bmatrix} \beta(1-\alpha) & \alpha\beta & \alpha\beta & \alpha(1-\beta) \\ 1-\alpha & \alpha & -\beta & -(1-\beta) \\ -\beta & -(1-\beta) & \beta & 1-\beta \\ -(1-\alpha) & -\alpha & 1-\alpha & \alpha \end{bmatrix} \ .$$

d) If

$$P_c = P_e - v\pi_e,$$

where v is the vector of dimension 4 with all unit entries, use the eigenvalue decomposition (8.109) to show that if $z = e^{j\omega T}$

$$(I_4 - P_c z^{-1})^{-1} = V \text{diag}\,\{1, \frac{1}{1 - (1 - \alpha + \beta)z^{-1}}, 1, 1\} V^{-1} \ .$$

Then use expressions (8.71) and (8.72) to compute the continuous and discrete components of the power spectral density $S_Z(\omega)$ of $Z(t)$. Express your answers in terms of the Fourier transforms $S_{LH}(j\omega)$ and $S_{HL}(j\omega)$ of the switching waveforms $s_{LH}(t)$ and $s_{HL}(t)$.

8.15 $1/f$ noise in metal oxide field effect transistors (MOSFETs), which used to be of concern primarily in low-frequency devices, has become noticeable at higher frequencies due to the progressive shrinking of CMOS circuits. It is now generally thought that $1/f$ noise in MOSFETS is due to traps in the gate oxide. These traps successively capture and then release electrons in the channel, thus creating some current fluctuations. If we consider a single trap, the presence or absence of an electron in the trap can be modeled [23, 24] by a random telegraph wave process $X(t)$, where the customary

$\{-1, 1\}$ states are replaced by $\{0, 1\}$, where $X(t) = 0$ when no electron is trapped, and $X(t) = 1$ when one electron is trapped, so that

$$X(t) = \frac{1}{2}(1 + (-1)^{N(t)}),$$

where $N(t)$ denotes a Poisson process of rate λ. The parameter λ models here the capture rate and emission/release rate of the trap, which are approximately equal when the trap energy level is close to the Fermi level of the bulk. Recall that as noted in Example 4.4, in steady state

$$P[X(t) = 1] = P[X(t) = 0] = \frac{1}{2}$$

and

$$P[X(t + \tau) = X(t)|X(t)] = \frac{1}{2}[1 + \exp(-2\lambda\tau)]$$

$$P[X(t + \tau) = -X(t)|X(t)] = \frac{1}{2}[1 - \exp(-2\lambda\tau)].$$

Since we are interested in fluctuations in the number of trapped electrons, we consider the centered process

$$\tilde{X}(t) = X(t) - \frac{1}{2} = \frac{1}{2}(-1)^{N(t)}.$$

a) Verify that in steady state, the autocorrelation and PSD of the centered single trap process $\tilde{X}(t)$ are

$$R_{\tilde{X}}(\tau|\lambda) = \frac{1}{4}\exp(-2\lambda|\tau|)$$

and

$$S_{\tilde{X}}(\omega|\lambda) = \frac{1}{8}\frac{\lambda}{\lambda^2 + (\omega/2)^2}.$$

In the above analysis, we considered a single trap with capture and emission rate λ, but suppose there are K independent traps where the the capture/emission rate Λ of each trap is random and distributed according the log-uniform PDF

$$f_\Lambda(\lambda) = \begin{cases} \frac{1}{\lambda \ln(\lambda_H/\lambda_L)} & \lambda_L \le \lambda \le \lambda_H \\ 0 & \text{otherwise}. \end{cases}$$

Then let $X_T(t)$ denote the total number of trapped electrons, and

$$\tilde{X}_T(t) = X_T(t) - \frac{K}{2}$$

the corresponding centered process.

b) Verify that the autocorrelation of $\tilde{X}_T(t)$ is

$$R_{\tilde{X}_T}(\tau) = K \int_0^\infty R_{\tilde{X}}(\tau|\lambda) f_\Lambda(\lambda) d\lambda \, .$$

c) Compute the power spectral density of $\tilde{X}_T(t)$ and verify that

$$S_{\tilde{X}_T}(\omega) \approx \begin{cases} A & \omega \ll \lambda_L \\ \dfrac{B}{\omega} & \lambda_L \ll \omega \ll \lambda_H \\ \dfrac{C}{\omega^2} & \lambda_H \ll \omega \, , \end{cases}$$

where A, B, and C are positive constants. This shows that if the range $[\lambda_L, \lambda_H]$ of the capture/emission parameter distribution is sufficiently large, the PSD of the centered trapped electron process has a $1/\omega$ decay rate over a wide frequency range. *Hint:* To prove this result, recall that

$$\int_0^x \frac{1}{u^2 + a^2} du = \frac{1}{a} \arctan(x/a) \, .$$

References

1. H. Nyquist, "Thermal agitation of electric charge in conductors," *Physical Review*, vol. 32, pp. 110–113, July 1928.
2. S. M. Kay, *Modern Spectral Estimation: Theory and Application*. Englewood Cliffs, NJ: Prentice-Hall, 1988.
3. P. Stoica and R. Moses, *Spectral Analysis of Signals*. Upper Saddle River, NJ: Prentice-Hall, 2005.
4. C. W. Helstrom, *Probability and Stochastic Processes for Engineers, 2nd edition*. Englewood Cliffs, NJ: Prentice-Hall, 1991.
5. J. W. B. Davenport and W. L. Root, *An Introduction to the Theory of Random Signals and Noise*. New York, NY: McGraw-Hill, 1958. Reprinted by IEEE Press, New York, NY, in 1987.
6. J. G. Kemeny and J. L. Snell, *Finite Markov Chains, 3rd edition*. New York, NY: Springer Verlag, 1983.
7. C. E. Shannon, "Communication in the presence of noise," *Proc. of the IRE*, vol. 37, pp. 10–21, Jan. 1949.
8. A. V. Oppenheim and R. W. Schafer, *Discrete-Time Signal Processing, 3rd edition*. Upper Saddle River, NJ: Prentice-Hall, 2010.
9. A. Papoulis and S. U. Pillai, *Probability, Random Variables and Stochastic Processes, 4th edition*. New York, NY: McGraw-Hill, 2001.
10. W. A. Gardner, *Introduction to Random Processes with Applications to Signals and Systems, 2nd edition*. New York, NY: McGraw-Hill, 1990.
11. J. J. Shynk, *Probability, Random Variables, and Random processes: Theory and Signal Processing Applications*. Hoboken, NJ: Wiley-Interscience, 2012.
12. A. V. Balakrishnan, "A note on the sampling principle for continuous signals," *IRE Trans. on Information Theory*, vol. 3, pp. 143–146, 1957.
13. E. Parzen, "A simple proof and some extensions of the sampling theorem," Tech. Rep. 7, Department of Statistics, Stanford University, Stanford, CA, Dec. 1956.
14. F. Beutler, "Sampling theorems and bases in a Hilbert space," *Informat. Control*, vol. 4, pp. 97–117, 1961.
15. R. C. Titsworth and L. R. Welch, "Power spectra of signals modulated by random and pseudorandom sequences," Tech. Rep. 32–140, Jet Propulsion Laboratory, Pasadena, CA, Oct. 1961.
16. G. Bilardi, R. Padovani, and G. L. Pierobon, "Spectral analysis of functions of Markov chains with applications," *IEEE Trans. Commun.*, vol. 31, pp. 853–861, July !983.
17. S. O. Rice, "Mathematical analysis of random noise, parts I and II," *Bell System Tech J.*, vol. 23, pp. 282–332, July 1944.
18. S. O. Rice, "Mathematical analysis of random noise, parts III and IV," *Bell System Tech J.*, vol. 24, pp. 46–156, Jan. 1945.

19. S. O. Rice, "Statistical properties of a sine wave plus random noise," *Bell System Tech J.*, vol. 27, pp. 109–157, Jan. 1948.
20. S. O. Rice, "Noise in FM receivers," in *Time Series Analysis* (M. Rosenblatt, ed.), pp. 395–422, New York: John Wiley, 1963.
21. S. Benedetto, E. Biglieri, and V. Castellani, *Digital Transmission Theory*. Englewood Cliffs, NJ: Prentice-Hall, 1987.
22. A. Demir and P. Feldman, "Modeling and simulation of the interference due to digital switching in mixed-signal ICs," in *Proc. of the 1999 IEEE/ACM Internat. Conf. on Computer-aided Design (ICCAD '99)*, pp. 70–75, Nov. 1999.
23. H. Tian and A. El Gamal, "Analysis of $1/f$ noise in switched MOSFET circuits," *IEEE Trans. Circuits Syst. II*, vol. 48, pp. 151–157, Feb. 2001.
24. A. Van Der Ziel, "Unified presentation of $1/f$ noise in electronic devices: fundamental $1/f$ noise sources," *Proceedings of the IEEE*, vol. 76, pp. 233–258, Mar. 1988.

Part III
Advanced Topics

Ergodicity

9

9.1 Introduction

The concept of ergodicity finds its origin in the statistical mechanics work of Boltzmann who proposed the ergodic hypothesis, whereby ensemble averages could be replaced by time averages. A further observation was provided by Gibbs who observed that for Hamiltonian systems, Liouville's theorem ensures that for a system in equilibrium with a large number of particles, the probability density of particles in phase space is constant, thus ensuring that the probability distribution is SSS of order 1. Accordingly, ergodicity is analyzed in the context of either SSS or WSS processes. Unfortunately, ergodicity is not always guaranteed for SSS processes. Indeed, whereas for an iid process $X(t)$, the SLLN ensures that the time average

$$\frac{1}{t} \sum_{s=0}^{t-1} X(s) \overset{a.s.}{\to} m_X = E[X(t)] \,,$$

if we consider a process $X(t) = X$, where X is a fixed random variable, $X(t)$ is SSS, but the average converges to X, which is a random variable, but not to its mean m_X. It was shown by Birkhoff [1] and Von Neumann [2] that time averages of SSS processes converge almost surely and in the mean square, respectively, to a random variable which in some circumstances equals m_X, the mean of $X(t)$. In this chapter, we present several criteria under which time averages are equal asymptotically to ensemble averages.

In Sect. 9.2, the ergodicity of finite-state Markov chains is examined, and it is shown that when the MC is irreducible and aperiodic, time-averages of the form

$$\frac{1}{t} \sum_{s=0}^{t-1} g(X(s), X(s+1), \ldots, X(s+k))$$

converge almost surely to $E[g(X(t), X(t+1), \ldots, X(t+k))]$. This is why irreducible aperiodic finite-state MCs are referred to as ergodic MCs. In Sect. 9.3, the mean-square ergodicity of WSS processes is analyzed and a necessary and sufficient criterion is provided which ensures that

© Springer Nature Switzerland AG 2020
B. C. Levy, *Random Processes with Applications to Circuits and Communications*,
https://doi.org/10.1007/978-3-030-22297-0_9

$$\frac{1}{t} \sum_{s=0}^{t-1} X(s) \overset{m.s.}{\rightarrow} m_X \, .$$

It is also explained how this criterion can be used to analyze not only the mean-square ergodicity of the mean, but of other statistical quantities, like the sampled autocorrelation or sampled CDF.

9.2   Finite Markov Chains

Consider a homogeneous irreducible aperiodic Markov chain with N states and one-step transition matrix $\mathbf{P}$. It was shown in Chap. 5 that under these assumptions, the eigenvalue $\lambda = 1$ of $\mathbf{P}$ has multiplicity one, and the entries of the corresponding left eigenvector $\boldsymbol{\pi}$ of $\mathbf{P}$ are all positive. When this eigenvector is normalized so that its row sum equals one, it represents the steady-state probability distribution of the Markov chain, since starting from an arbitrary initial distribution $\boldsymbol{\pi}(0)$, the state probability distribution $\boldsymbol{\pi}(t)$ converges to $\boldsymbol{\pi}$ as t tends to infinity. To keep the analysis simple, we assume that the initial state $\boldsymbol{\pi}(0) = \boldsymbol{\pi}$, which ensures that the Markov chain is strict-sense stationary, so that its joint PMFs of arbitrary order are invariant under time shifts. If $g(i)$ denotes an arbitrary function of state $i \in \{1, 2, \ldots, N\}$, the ensemble average of $g(X(t))$ can be expressed as

$$E[g(X(t))] = \sum_{i=1}^{N} g(i)\pi_i \, , \tag{9.1}$$

and the ergodic theorem of Markov chains requires proving that

$$\lim_{t \to \infty} \frac{1}{t} \sum_{s=0}^{t-1} g(X(s)) = E[g(X(t))] \tag{9.2}$$

almost surely. In other words time and ensemble averages need to coincide. In this respect, by observing that $g(X(t))$ can be expressed as

$$g(X(t)) = \sum_{i=1}^{N} g(i)1_i(X(t)), \tag{9.3}$$

where $1_i(x)$ denotes the indicator function of state i, after exchanging summations, we find that the time average

$$\frac{1}{t} \sum_{s=0}^{t-1} g(X(s)) = \sum_{i=1}^{N} g(i)q_i(t) \, , \tag{9.4}$$

where

$$q_i(t) = \frac{1}{t} \sum_{s=0}^{t-1} 1_i(X(s)) = \frac{N_i(t)}{t} \tag{9.5}$$

represents the empirical probability of state i based on the first t samples $X(s)$, $0 \le s \le t - 1$. Specifically, if $N_i(t)$ denotes the number of visits to state i until time $t - 1$, $q_i(t)$ represents the fraction of time that $X(s)$ is in state i among the first t samples. The row vector

$$\mathbf{q}(t) = \left[q_1(t) \ \ldots \ q_i(t) \ \ldots \ q_N(t) \right] \tag{9.6}$$

satisfies

$$\sum_{i=1}^{N} q_i(t) = 1,$$

and $\mathbf{q}(t)$ denotes the empirical PMF of the Markov chain based on its first t states. Since the function g in the ergodic theorem (9.2) is arbitrary, we deduce that the theorem holds if and only if the empirical PMF $\mathbf{q}(t)$ converges almost surely to probability distribution π as t tends to infinity.

9.2.1 Mean Duration of State Excursions

To prove the convergence of the empirical PMF, we need first to evaluate the mean duration of excursions from an arbitrary state, say state i. Suppose that $X(0) = i$ and that

$$T = \min\{t \geq 1 : X(t) = i\} \tag{9.7}$$

represents the first time that $X(t)$ returns to state i. To evaluate $d_i = E[T]$, we can leverage the analysis presented in Sect. 5.5 to evaluate the average number of steps u_j needed to reach state i, starting from an arbitrary state j. Clearly $u_i = 0$, and if $\mathbf{u}$ denotes the column vector with entries u_j, $1 \leq j \leq N$, it was shown that $\mathbf{u}$ satisfies equation

$$\mathbf{u} = \tilde{\mathbf{P}}\mathbf{u} + \mathbf{r}, \tag{9.8}$$

where $\tilde{\mathbf{P}}$ is the matrix obtained by replacing the i-th row of $\mathbf{P}$ by δ_{ij} with $1 \leq j \leq N$, and where reward vector $\mathbf{r}$ has all its entries equal to 1, except its i-th entry $r_i = 0$. The modification of the i-th row in $\tilde{\mathbf{P}}$ has for effect to make state i an absorbing state. By performing a first-step analysis where the first state is $X(1) = j$, we find

$$d_i = \sum_{j=1}^{N} E[T, X(1) = j | X(0) = i]$$

$$= \sum_{j=1}^{N} p_{ij} E[T | X(1) = j], \tag{9.9}$$

where by taking into account the fact that one transition takes place between times 0 and 1 we have with

$$E[T | X(1) = j] = 1 + u_j . \tag{9.10}$$

Substituting (9.10) inside (9.9) and using the unit row sum property of transition probability matrices, we find

$$d_i = \sum_{j=1}^{N} p_{ij} u_j + 1 . \tag{9.11}$$

Then by replacing the trivial equation $0 = 0$ corresponding to the ith row of (9.8) by (9.11), we obtain equation

$$\tilde{\mathbf{u}} = \mathbf{Pu} + \mathbf{v}, \tag{9.12}$$

where $\tilde{\mathbf{u}}$ is the column vector obtained by replacing $u_i = 0$ by d_i in $\mathbf{u}$, and $\mathbf{v}$ is the column vector with entries all equal to one. Multiplying (9.12) on the left by $\boldsymbol{\pi}$, we obtain

$$\pi_i d_i + \sum_{j \neq i} \pi_j u_j = \sum_{j \neq i} \pi_j u_j + 1, \tag{9.13}$$

so that

$$d_i = 1/\pi_i \, . \tag{9.14}$$

Thus the mean duration of excursions from state i is the inverse of the probability of state i.

9.2.2 Convergence of Empirical Distributions

We are now in a position to establish the convergence of empirical distributions. Consider state i, and let $N_i(t) \leq t$ denote the number of visits to state i prior to time t. If F denotes the time of the first visit, and T_k denotes the duration of the excursion between the k-th and $k+1$-th visit to state i, we have

$$F + \sum_{k=1}^{N_i(t)-1} T_k \leq t \leq F + \sum_{k=1}^{N_i(t)} T_k \, . \tag{9.15}$$

Note that since the process is Markov, all excursions T_k are independent of one another, given the knowledge of state i at the beginning and end of the excursion. Since all states are recurrent, the number of visits $N_i(t)$ to state i tends to infinity almost surely at t tends to infinity. By the strong law of large numbers, we have

$$\lim_{t \to \infty} \frac{1}{N_i(t)} \sum_{k=1}^{N_i(t)} T_k = E[T] = \frac{1}{\pi_i} \tag{9.16}$$

almost surely. Then by dividing (9.15) by $N_i(t)$ and noting that $t/N_i(t)$ is bracketed between two sequences which converge almost surely to $1/\pi_i$ as t tends to infinity, we deduce that

$$\frac{t}{N_i(t)} \xrightarrow{a.s.} \frac{1}{\pi_i} \, ,$$

or equivalently

$$q_i(t) = \frac{N_i(t)}{t} \xrightarrow{a.s.} \pi_i \tag{9.17}$$

which establishes the almost sure convergence of the empirical distribution to the state probability distribution.

9.2.3 Convergence of Joint Empirical Distributions

For some applications, it is of interest to apply the ergodic theorem to averages of functions depending on several successive values of state $X(t)$. For example, suppose we seek to prove

$$\frac{1}{t} \sum_{s=0}^{t-1} g(X(s), X(s+1)) \overset{a.s.}{\to} E[g(X(t), X(t+1))]. \tag{9.18}$$

We have

$$E[g(X(t), X(t+1))] = \sum_{i=1}^{N} \sum_{j=1}^{N} g(i, j) \pi_i p_{ij}$$

and

$$\frac{1}{t} \sum_{s=0}^{t-1} g(X(s), X(s+1)) = \sum_{i=1}^{N} \sum_{j=1}^{N} g(i, j) q_{ij}(t)$$

where, if $N_{ij}(t)$ denotes the number of transitions from state i to state j observed among the first t transitions,

$$q_{ij}(t) = \frac{1}{t} \sum_{s=0}^{t-1} 1_i(X(t)) 1_j(X(t+1)) = \frac{N_{ij}(t)}{t} \tag{9.19}$$

denotes the empirical joint distribution of $X(t)$ and $X(t+1)$. From these expressions, we deduce that (9.18) will hold for g arbitrary if and only if

$$q_{ij}(t) \overset{a.s}{\to} \pi_i p_{ij} \tag{9.20}$$

for all $1 \leq i\ j \leq N$. In other words, the joint empirical PMF $q_{ij}(t)$ of two successive states needs to converge to the statistical PMF $\pi_i p_{ij}$. To prove this result, note that $q_{ij}(t)$ can be factored as

$$q_{ij}(t) = q_{j|i}(t) q_i(t), \tag{9.21}$$

where

$$q_{j|i}(t) = \frac{q_{ij}(t)}{q_i(t)} = \frac{N_{ij}(t)}{N_i(t)} \tag{9.22}$$

can be viewed as the empirical transition probability from state i to state j. Since we have already proved that $q_i(t)$ converges almost surely to π_i as t tends to infinity, all that needs to be proved is that $q_{j|i}(t)$ converges almost surely to p_{ij} as t tends to infinity. In expression (9.22), we have already observed that since state i is recurrent, the number $N_i(t)$ of visits to i tends to infinity as t tends to infinity. Then, given that we start in state i, all $N_i(t)$ transitions out of state i are independent since $X(t)$ is a Markov process. Thus, by the strong law of large numbers

$$q_{j|i}(t) \overset{a.s.}{\to} E[1_j(X(s+1))|X(s) = i] = p_{ij}.$$

This completes the proof of the version (9.18) of the ergodic theorem for irreducible aperiodic finite Markov chains. Obviously, the above argument could be extended to the computation of averages depending on more than two successive Markov chain states.

Example 9.1 (Entropy rate of a Markov chain) The identity (9.18) can be used to evaluate the entropy rate of a Markov chain. Following [3], the entropy rate of a random source taking values in $\{1, \ldots, i, \ldots, N\}$ is defined as the limit as t tends to infinity of

$$-\frac{1}{t} \log_2 p(X(0), X(1), \ldots, X(t)),$$

where $p(X(0), X(1), \ldots, X(t))$ denotes the joint PMF of $t + 1$ successive source symbols. For a stationary Markov chain

$$-\log_2 p(X(0), X(1), \ldots, X(t)) = -\log_2 \pi_{X(0)} - \sum_{s=0}^{t-1} \log_2 p_{X(s)X(s+1)} \cdot$$

As t tends to infinity, the term $-\log_2 \pi_{X(0)}/t$ will tend to zero, and by using the ergodic theorem of Markov chains, we obtain

$$-\frac{1}{t} \log_2 p(X(0)X(1), \ldots, X(t)) \stackrel{a.s.}{\to} E[-\log_2 p_{X(s)X(s+1)}]$$

$$= \sum_{i=1}^{N} \pi_i \sum_{j=1}^{N} p_{ij} \log_2 p_{ij} \stackrel{\triangle}{=} H(\mathbf{P}). \tag{9.23}$$

The intuition behind this result is that as t becomes large, most observed sequences have a probability close to $2^{-H(\mathbf{P})/t}$. Such sequences are called "typical sequences" in information theory [4], and the asymptotic convergence result (9.23) is usually referred to as the Asymptotic Equipartition Property.

9.3   Mean-Square Ergodicity

In situations where only the first and second moments of a random process are known, and the process is WSS, it is more convenient to study ergodicity in the mean-square sense, instead of the almost-sure approach used for Markov chains. A DT WSS process $X(t)$ with mean m_X and autocovariance $K_X(\ell)$ is said to be mean-square ergodic in the mean if the sample mean

$$M_X(t) = \frac{1}{t} \sum_{s=0}^{t-1} X(s) \tag{9.24}$$

converges to m_X in the mean-square sense as t tends to infinity, i.e., if

$$\lim_{t \to \infty} E[(M_X(t) - m_X)^2] = 0. \tag{9.25}$$

This property extends to dependent random variables the weak law of large numbers of iid sequences.

9.3.1   Mean-Square Ergodicity Criterion

It turns out that $X(t)$ is mean-square ergodic in the mean if and only if

$$\lim_{t \to \infty} \bar{K}_X(t) = 0, \tag{9.26}$$

where

$$\bar{K}_X(t) \triangleq \frac{1}{t} \sum_{s=0}^{t-1} K_X(s) . \tag{9.27}$$

Before proving this result, it is worth pointing out that while the summability of the autocovariance function $K_X(s)$, i.e.

$$\sum_{s=0}^{\infty} |K_X(s)| < \infty , \tag{9.28}$$

ensures that criterion (9.26) is satisfied, the converse is not true. Specifically (9.26) may hold even in situations where the process $X(t)$ has a long memory in the sense that

$$\sum_{s=0}^{t-1} |K(s)|$$

diverges, as will be shown in Example 9.2 below. For zero mean processes, (so that the autocorrelation and autocovariance coincide), the sufficiency of the summability criterion (9.28) indicates that if $X(t)$ has a power spectral density, it is necessarily mean-square ergodic.

Proof Our derivation of criterion of (9.26) follows [5]. To prove sufficiency, consider the random variables

$$Z_1 = X(0) - m_X .$$

and

$$Z_2 = M_X(t) - m_X = \frac{1}{t} \sum_{s=0}^{t-1} (X(s) - m_X) .$$

We have

$$E[Z_1 Z_2] = \bar{K}(t) \quad \text{and} \quad E[Z_1^2] = K_X(0) ,$$

so by applying the Cauchy–Schwartz inequality

$$(E[Z_1 Z_2])^2 \leq E[Z_1^2] E[Z_2^2] = K_X(0) E[(M_X(t) - m_X)^2] ,$$

we deduce that the mean-square convergence property (9.25) implies that criterion (9.26) holds. To prove necessity, note that

$$E[(M_X(t) - m_X)^2] = \frac{1}{t^2} \sum_{r=0}^{t-1} \sum_{s=0}^{t-1} E[(X(r) - m_x)(X(s) - m_X)]$$

$$= \frac{1}{t^2} \sum_{r=0}^{t-1} \sum_{s=0}^{t-1} K_X(r - s)$$

$$= \frac{2}{t^2} \sum_{k=1}^{t} k \bar{K}_X(k) - \frac{K_X(0)}{t} . \tag{9.29}$$

In this expression the second term tends to zero as t tends to infinity. Also since $\bar{K}(k)$ tends to zero, for every $\epsilon > 0$ there exists k_0 such that $|\bar{K}(k)| < \epsilon$ for all $k \geq k_0$. The magnitude of the first term in (9.29) can be upper bounded by

$$\frac{2}{t^2} \sum_{k=0}^{k_0} k\bar{K}(k) + \epsilon,$$

where the first term tends to zero as t tends to infinity, and ϵ is arbitrarily small, which proves (9.25). $\square$

The above analysis can be extended in a straightforward manner to the continuous-time case. In this case, the sample mean is

$$M_X(t) = \frac{1}{t} \int_0^t X(s)ds , \tag{9.30}$$

and the mean-square ergodicity criterion (9.26) holds with

$$\bar{K}_X(t) = \frac{1}{t} \int_0^t K_X(s)ds . \tag{9.31}$$

Example 9.2 Consider a CT random phase process

$$X(t) = A\cos(\omega_c t + \Theta), \tag{9.32}$$

where the phase Θ is uniformly distributed over $[0, 2\pi]$ and the amplitude A and frequency ω_c are known. It was shown in Example 4.1 that $X(t)$ is WSS with zero mean and autocovariance

$$K_X(\tau) = \frac{A^2}{2} \cos(\omega_c \tau) .$$

Clearly the autocovariance $K_X(\tau)$ is not summable. Yet, we have

$$\bar{K}_X(t) = \frac{1}{t} \int_0^t K_X(s)ds = \frac{A^2}{2\omega_c t} \sin(\omega_c t) ,$$

which tends to zero as t tends to infinity. Thus, rather surprisingly, $X(t)$ is mean-square ergodic in the mean.

Example 9.3 Let $X(t)$ be a CT Ornstein–Uhlenbeck process, so it is Gaussian with zero mean and autocovariance

$$K_X(\tau) = P_X \exp(-a|\tau|)$$

with $a > 0$. As was noted earlier, this process has the feature that whenever $X(t)$ and $X(s)$ are separated by $\tau = t - s \gg 1/a$, the random variables $X(t)$ and $X(s)$ become approximately uncorrelated and thus independent. Thus we would expect $X(t)$ to be ergodic in the mean. This is confirmed by observing that

$$\bar{K}_X(t) = \frac{1}{t} \int_0^t K_X(s)ds = \frac{P_X}{at}[1 - \exp(-at)]$$

converges to zero as t tends to infinity.

9.3.2   Mean-Square Ergodicity of the Autocorrelation

The previous analysis concerns only mean-square ergodicity in the mean. To determine whether other averages are mean-square ergodic, one needs only to identify the mean of the averaged quantity. For example, for a DT WSS process $X(t)$ with mean m_X and autocorrelation $R_X(k) = E[X(t)X(t+k)]$, suppose we seek to estimate $R_X(k)$ by performing the average

$$M_Y(t) = \frac{1}{t} \sum_{s=0}^{t-1} X(s)X(s+k) . \tag{9.33}$$

The previous result can be applied as long as we recognize that the process of interest is now

$$Y(t) = X(t)X(t+k)$$

with k fixed, since $m_Y = E[Y(t)] = R_X(k)$. In order to be able to apply the previous result, $Y(t)$ needs to be WSS, so that the fourth-order moment

$$E[Y(t)Y(t+\ell)] = E[X(t)X(t+k)X(t+\ell)X(t+k+\ell)]$$

needs to be independent of t. If we denote this moment as $R_Y(\ell, k)$, the covariance function

$$K_Y(\ell, k) = R_Y(\ell, k) - R_X^2(k) \tag{9.34}$$

must then satisfy the mean-square ergodicity criterion

$$\lim_{t \to \infty} \frac{1}{t} \sum_{\ell=0}^{t-1} K_Y(\ell, k) = 0 . \tag{9.35}$$

Example 9.2 (Continued) For the CT random phase signal (9.32), we have

$$X(t)X(t+\tau)X(t+\sigma)X(t+\tau+\sigma) = \frac{A^4}{4}[\cos(\omega_c(2t+\tau) + 2\Theta) + \cos(\omega_c\tau)]$$

$$\times [\cos(\omega_c(2(t+\sigma) + \tau) + 2\Theta) + \cos(\omega_c\tau)] ,$$

and by taking the expectation of this expression with respect to Θ and using

$$\cos(\omega_c(2t+\tau) + 2\Theta)\cos(\omega_c(2(t+\sigma) + \tau) + 2\Theta) = \frac{1}{2}[\cos(2\omega_c\sigma)$$

$$+ \cos(\omega_c(4t + 2(\sigma + \tau)) + 4\Theta) ,$$

we find that $Y(t) = X(t)X(t + \tau)$ is WSS with

$$R_Y(\sigma, \tau) = E[Y(t)Y(t + \sigma)] = \frac{A^4}{8}\cos(2\omega_c\sigma) + R_X^2(\tau) \,.$$

Then

$$K_Y(\sigma, \tau) = \frac{A^2}{8}\cos(2\omega_c\sigma)$$

is such that

$$\bar{K}_Y(t) = \frac{1}{t}\int_0^t K_Y(\sigma, \tau)d\sigma = \frac{A^2}{16\omega_c t}\sin(2\omega_c t)$$

converges to zero as t tends to infinity, so that $Y(t)$ is mean-square ergodic in the mean, in which case $X(t)$ is said to be mean-square ergodic of the autocorrelation.

Gaussian Processes The computation of the fourth order moment $E[X(t)X(t + k)X(t + \ell)X(t + k + \ell)]$ can be simplified significantly if $X(t)$ is Gaussian. In this case $X(t)$ is SSS, so we already know that $Y(t) = X(t)X(t + k)$ will be WSS, and if $X(t)$ has zero mean, by using Isserlis's formula (see Problem 4.21) we find

$$\begin{aligned} R_Y(\ell, k) &= E[X(t)X(t + k)X(t + \ell)X(t + k + \ell)] \\ &= (R_X(k))^2 + (R_X(\ell))^2 + R_X(\ell + k)R_X(\ell - k) \,, \end{aligned}$$

so the autocovariance of $Y(t)$ is

$$\begin{aligned} K_Y(\ell, k) &= R_Y(\ell, k) - (R_X(k))^2 \\ &= (R_X(\ell))^2 + R_X(\ell + k)R_X(\ell - k) \,. \end{aligned} \tag{9.36}$$

Accordingly, the ergodicity criterion (9.35) becomes

$$\lim_{t\to\infty} \frac{1}{t}\sum_{\ell=0}^{t-1}(R_X(\ell))^2 + R_X(\ell + k)R_X(\ell - k) = 0 \,. \tag{9.37}$$

But since

$$|R_X(\ell + k)R_X(\ell - k)| \le \frac{1}{2}[(R_X(\ell + k))^2 + (R_X(\ell - k))^2]$$

and

$$\lim_{t\to\infty} \frac{1}{t}\sum_{\ell=0}^{t-1}(R_X(\ell \pm k))^2 = 0$$

whenever

$$\lim_{t \to \infty} \frac{1}{t} \sum_{\ell=0}^{t-1} (R_X(\ell))^2 = 0, \qquad (9.38)$$

we deduce that in the Gaussian case $X(t)$ is mean-square ergodic of the autocorrelation if and only if condition (9.38) is satisfied.

Example 9.3 (Continued) The Ornstein–Uhlenbeck process is a zero-mean CT WSS Gaussian process with

$$R_X^2(\tau) = P_X^2 \exp(-2a\tau)$$

and

$$\frac{1}{t} \int_0^t R_X^2(\tau) d\tau = \frac{P_X^2}{2at} (1 - \exp(-2at)),$$

which tends to zero as t tends to infinity, so it is mean-square ergodic of the autocorrelation.

9.3.3 Mean-Square Ergodicity of the CDF

Consider a DT process $X(t)$ which is SSS of order 2. If we observe t samples $\{X(s), 0 \le s \le t - 1\}$ of this process, the CDF

$$F_X(x) = P(X(t) \le x) = E[u(x - X(t)]$$

for a fixed x can be estimated by using the empirical CDF

$$\hat{F}_X(t, x) = \frac{1}{t} \sum_{s=0}^{t-1} u(x - X(s)) = \frac{N(t, x)}{t}, \qquad (9.39)$$

where $N(t, x)$ denotes the number of samples $\{X(s), 0 \le s \le t - 1\}$ which are less than or equal to x. If $\hat{F}_X(t, x)$ converges in the mean-square sense to $F_X(x)$ for all x as t tends to infinity, the process $X(t)$ is said to be mean-square ergodic of the autocorrelation.

Since $F_X(x)$ is the mean of $Z(t) = u(x - X(t))$, we can apply the mean-square ergodicity criterion derived earlier to $Z(t)$. This requires that $Z(t)$ should be WSS, but since $X(t)$ has been assumed SSS of order 2, if $F_{X(t)X(s)}(x_t, x_s, t - s)$ denotes the joint CDF of $X(t)$ and $X(s)$, we have

$$E[Z(t)Z(s)] = E[u(x - X(t))u(x - X(s))] = P(X(t) \le x, X(s) \le x)$$

$$= F_{X(t)X(s)}(x, x, t - s), \qquad (9.40)$$

which depends only on $\ell = t - s$. The autocovariance of $Z(t)$ is then given by

$$K_Z(\ell, x) = F_{X(t)X(s)}(x, x, \ell) - (F_X(x))^2, \qquad (9.41)$$

and $X(t)$ will be mean-square ergodic of the autocorrelation if and only if

$$\lim_{t \to \infty} \frac{1}{t} \sum_{\ell=0}^{t-1} K_Z(\ell, x) = 0 \,. \tag{9.42}$$

Example 9.4 Consider the random telegraph wave process $X(t)$, and assume that it is initialized with the distribution

$$P[X(0) = 1] = P[X(0) = -1] = \frac{1}{2} \,,$$

so that the initial CDF coincides with the stationary CDF

$$F_X(x) = \frac{1}{2}(u(x - 1) + u(x + 1))$$

which is sketched in Fig. 9.1. Accordingly, $X(t)$ is SSS and its CDF $F_{X(t)}(x) = F_X(x)$ for all t. For $t \geq s$, the joint CDF of $X(t)$ and $X(s)$ is given by

$$F_{X(t),X(s)}(x_t, x_s, \tau) = \frac{1}{4}(1 + \exp(-2a\tau))u(x_t - 1)u(x_s - 1)$$

$$+\frac{1}{4}(1 + \exp(-2a\tau))u(x_t + 1)u(x_s + 1)$$

$$+\frac{1}{4}(1 - \exp(-2a\tau))u(x_t - 1)u(x_s + 1)$$

$$+\frac{1}{4}(1 - \exp(-2a\tau))u(x_t + 1)u(x_s - 1)$$

with $\tau = t - s$, where $a > 0$ denotes the switching rate between the two states. It can be rewritten as

$$F_{X(t)X(s)}(x_t, x_s, \tau) = F_X(x_t)F_X(x_s)$$

$$+\frac{\exp(-2a\tau)}{4}\big[u(x_t - 1)u(x_s - 1) + u(x_t + 1)u(x_s + 1)$$

$$-u(x_t - 1)u(x_s + 1) - u(x_t + 1)u(x_s - 1)\big] \,,$$

Fig. 9.1 CDF of the random telegraph wave process

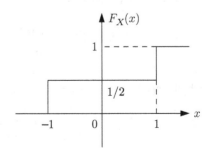

so that the autocovariance of $Z(t) = u(x - X(t))$ is

$$K_Z(\tau, x) = F_{X(t, X(s))}(x, x, \tau) - (F_X(x))^2$$
$$= \frac{\exp(-2a\tau)}{4}[u(x+1) - u(x-1)]^2 .$$

We find that

$$\frac{1}{t}\int_0^t K_Z(\tau, x)d\tau = \frac{1}{8at}(1 - \exp(-2at))[u(x+1) - u(x-1)]^2$$

converges to zero as t tends to infinity, so $X(t)$ is mean-square ergodic of the CDF.

9.4 Bibliographical Notes

Although Birkhoff's ergodic theorem establishing the almost sure convergence of the sample mean of a SSS process is of general interest, it was not included since its proof relies on measure-theoretic concepts beyond the scope of this book. Interested readers are referred to [6, Chap. 8]. Likewise the mean-square convergence of the sample mean of WSS processes is omitted, but a proof can be found in [5, Thm.5.3]. The ergodicity of Markov chains is analyzed in detail in [7, Sec. 3.4] and [8]. The analysis of the mean-square ergodicity of WSS processes appearing in Sect. 9.3 is based in part on the presentation given in [5].

9.5 Problems

9.1 Consider a stationary irreducible aperiodic Markov chain with one step transition matrix $\mathbf{P}$. When computing the entropy rate of a Markov chain, the averaging operation (9.23) used the correct Markov chain statistics. Assume that the true one-step transition matrix $\mathbf{P}$ is unknown, and that the averaging is performed by using transition matrix $\mathbf{R}$.

a) If $r(X(0), X(1), \ldots, X(t))$ denotes the probability of sequence $\{X(s), 0 \le s \le t\}$ based on one-step transition matrix $\mathbf{R}$ and initial distribution $r(X(0))$, verify that

$$\frac{1}{t} - \log_2 r(X(0), X(1), \ldots, X(t)) = -\log_2(r(X(0)))/t$$
$$-\sum_{i=1}^N \sum_{j=1}^N q_{ij}(t)\log_2 r_{ij},$$

where $q_{ij}(t) = N_{ij}(t)/t$ denotes the empirical joint probability distribution of states $X(s) = i$ and $X(s+1) = j$. Here $N_{ij}(t)$ denotes again the number of observed transitions from state i to state j among the first t transitions.

b) Use the convergence of joint empirical distributions to the true statistical distribution to show that

$$\frac{1}{t} - \log_2 r(X(0)X(1), \ldots, X(t)) \overset{a.s.}{\to} H(\mathbf{P}) + D(\mathbf{P}|\mathbf{R}),$$

where if π denotes the true steady-state Markov chain distribution,

$$D(\mathbf{P}|\mathbf{R}) = \sum_{i=1}^{N} \pi_i \sum_{j=1}^{N} p_{ij} \log_2(p_{ij}/r_{ij})$$

is the Kullback Leibler divergence of Markov chain transition matrices $\mathbf{P}$ and $\mathbf{R}$. It can be interpreted as a weighted sum (with weights π_i, $1 \leq i \leq N$) of the Kullback–Leibler divergences of the one-step transition probability distributions $p_{i\cdot}$ and $r_{i\cdot}$ corresponding to the rows of $\mathbf{P}$ and $\mathbf{R}$. Since all entries π_i of the steady-state MC distribution are positive, the divergence $D(\mathbf{P}|\mathbf{R}) \geq 0$ with equality if and only if $\mathbf{P} = \mathbf{R}$.

c) The convergence result of part b) provides the motivation for the maximum likelihood (ML) [9] estimation technique for evaluating the transition matrix of a Markov chain. Since the limit of the negative log-likelihood function is minimized by $\mathbf{R} = \mathbf{P}$, minimizing the finite sample negative log-likelihood function

$$-\ln r(X(0), X(1), \ldots, X(t)) = \sum_{i=1}^{N} \sum_{j=1}^{N} N_{ij}(t) \ln(r_{ij}) + C(t) \tag{9.43}$$

over r_{ij} with $1 \leq i, j \leq N$ should yield a minimum which is close to the correct matrix $\mathbf{P}$ when t is sufficiently large. In (9.43), $C(t)$ denotes a constant independent of $\mathbf{R}$. Also, since logarithms in different bases differ only by a constant scaling factor, the minimization of the negative log-likelihood function is performed in base e, instead of base 2. To do so, note that the entries of $\mathbf{R}$ need to satisfy the row-sum constraints

$$\sum_{j=1}^{N} r_{ij} = 1 \tag{9.44}$$

for $1 \leq i \leq N$. By using the method of Lagrange multipliers to minimize the negative log-likelihood function (9.43) under constraints (9.44), show that the ML estimates of the $\mathbf{R}$ entries are given by

$$\hat{r}_{ij} = \frac{N_{ij}(t)}{N_i(t)} = q_{j|i}(t),$$

i.e., the ML estimates are the conditional empirical transition probabilities (9.22).

9.2 Let $X_0(t)$ and $X_1(t)$ be two independent Markov chains with states $\{0, 1\}$ and one-step transition matrices

$$\mathbf{P}_0 = \begin{bmatrix} 1/2 & 1/2 \\ 1/2 & 1/2 \end{bmatrix} \quad \text{and} \quad \mathbf{P}_1 = \begin{bmatrix} 1/2 & 1/2 \\ 1/4 & 3/4 \end{bmatrix}.$$

a) Find the steady-state probability distributions π_0 and π_1 of these MCs, and assume that the two chains are initialized with these distributions, so that they are SSS.

$X_0(t)$ and $X_1(t)$ can be viewed as the bit sequences produced by two random number generators, where the first number generator producing $X_0(t)$ is working properly, but the second one producing bits stream $X_1(t)$ is defective since bits 0 and 1 do not have probability $1/2$, and successive bits are not independent.

Let B be a binary random variable independent of both $X_0(t)$ and $X_1(t)$ with

$$P(B = 0) = P(B = 1) = \frac{1}{2}$$

and consider the random process

$$Y(t) = \begin{cases} X_0(t) \text{ if } B = 0 \\ X_1(t) \text{ if } B = 1 \end{cases}$$
$$= X_1(t)B + X_0(t)(1 - B).$$

In other words when $B = 0$, we select bit stream $X_0(t)$ and when $B = 1$, we select $X_1(t)$.

b) Let

$$M_Y(t) = \frac{1}{t} \sum_{s=0}^{t-1} Y(s) \tag{9.45}$$

denote the sample mean of $Y(t)$. Show that $M_Y(t)$ converges almost surely to a random variable Y as t tends to infinity. Can you use the limit Y to determine if B was zero or one?

Instead of using a single random variable B to select one of the two bit streams once and for all, we select between the two streams at each time t by using a bit sequence $B(t)$ independent of $X_0(t)$ and $X_1(t)$, resulting in

$$Y(t) = X_1(t)B(t) + X_0(t)(1 - B(t)).$$

The bit sequence $B(t)$ takes values $\{0, 1\}$ and is generated by a MC with one-step transition probability matrix $\mathbf{Q} = \mathbf{P}_0$.

c) Show that in this case, the sample mean $M_Y(t)$ given by (9.45) converges almost surely to the mean m_Y of $Y(t)$. To prove this result, you will need to use the fact that if $X(t)$ and $Z(t)$ are two independent MCs with one-step transition probability matrices $\mathbf{P}$ and $\mathbf{Q}$, then the tensor product of the two chains defined by $(X(t), Z(t))$ has for one-step transition probability matrix $\mathbf{P} \otimes \mathbf{Q}$, where

$$\mathbf{P} \otimes \mathbf{Q} = \begin{bmatrix} p_{11}\mathbf{Q} & \cdots & p_{1j}\mathbf{Q} & \cdots & p_{1m}\mathbf{Q} \\ \vdots & \ddots & & & \vdots \\ p_{i1}\mathbf{Q} & & p_{ij}\mathbf{Q} & & p_{im}\mathbf{Q} \\ \vdots & & & \ddots & \vdots \\ p_{m1}\mathbf{Q} & \cdots & p_{mj}\mathbf{Q} & \cdots & p_{mm}\mathbf{Q} \end{bmatrix}$$

denotes the Kronecker product of matrices $\mathbf{P} \in \mathbb{R}^{m \times m}$ and $\mathbf{Q} \in \mathbb{R}^{n \times n}$. Furthermore, if π and μ are the steady-state probability distributions of $\mathbf{P}$ and $\mathbf{Q}$, the steady-state distribution of $\mathbf{P} \otimes \mathbf{Q}$ is $\pi \otimes \mu$.

9.3 Consider the random urn selection scheme of Problems 5.14 and 5.15, which is a variant of a model used by Paul and Tatyana Ehrenfest to illustrate the convergence of a system to thermodynamic equilibrium. One objection to this model was that while convergence to thermodynamic equilibrium is considered as irreversible in thermodynamics, in an irreducible Markov chain, all states are visited infinitely often, thus suggesting that return to an initial state far from equilibrium is always possible. The purpose of this problem is to demonstrate that while return to an initial state far from equilibrium always occurs, the time needed is so large that convergence to equilibrium can be viewed as irreversible on a reasonable time scale.

a) Prove that the average duration of excursions away from state i is $d_i = 1/\pi_i$, where

$$\pi_i = \binom{N}{i}(1/2)^N$$

is the invariant distribution of the Markov chain. Note that because the MC has period 2, the result (9.14) is not directly applicable. Instead, observe that the Markov chains modeling 2-step transitions from even indexed states $i = 2k$ to even states, and the 2-step transitions of odd states $2k + 1$ to odd states are irreducible and aperiodic. Use the result (9.14) for these chains, and remember that the duration of excursions for these chains needs to be multiplied by 2, since each transition of the modified chains is the aggregation of two transitions of the original chain.

b) For $N = 100$, plot $\log_{10} d_i$ as a function of state i and verify that for states i beyond a few standard deviations from the mean $N/2 = 50$, d_i is extremely large, even though the number $N = 100$ of balls is quite small from a statistical physics viewpoint. Recall that it was shown in Problem 5.14 that the variance of the number of white balls in the urn is $N/4$.

9.4 Consider the process

$$X(t) = A_c \cos(\omega_c t) - A_s \sin(\omega_c t),$$

where A_c and A_s are two independent zero mean Gaussian random variables with variance σ^2, and the frequency ω_c is known. Note that $X(t)$ can also be be written as

$$X(t) = A \cos(\omega_c t + \Theta),$$

where A is Rayleigh distributed and independent of Θ which is uniformly distributed over $[0, 2\pi)$. Determine whether this process has mean-square ergodicity of the mean and of the autocorrelation. Compare your results with those obtained in Example 9.2 for the case where A is known and Θ is uniformly distributed.

9.5 Let $X(t)$ be a WSS random process whose mean is *nonzero* and such that $X(t)$ has mean-square ergodicity of both the mean and the autocorrelation. Let

$$Y(t) = AX(t),$$

where the amplitude A is random, independent of $X(t)$, with probability density $f_A(a)$. Determine whether $Y(t)$ has mean-square ergodicity of the mean and of the autocorrelation.

9.6 Consider the moving average process

$$X(t) = \sum_{s=t-(L-1)}^{t} V(s),$$

where $V(t)$ is a zero-mean white Gaussian noise process with variance $P_V = E[V(t)^2]$.

a) Evaluate the mean and autocorrelation of $X(t)$.
b) Determine whether $X(t)$ has the mean-square ergodicity property of the mean.
c) Find whether $X(t)$ has the mean-square ergodicity property of the autocorrelation.

9.7 Consider the random signal

$$X(t) = A\cos(\Omega t + \Theta),$$

where the amplitude A is known, the frequency Ω is a Gaussian random variable with mean ω_0 and standard deviation Δ, i.e., its probability density is given by

$$f_\Omega(\omega) = \frac{1}{(2\pi\Delta^2)^{1/2}} \exp(-\frac{(\omega - \omega_0)^2}{2\Delta^2}),$$

and the phase Θ is independent of Ω and uniformly distributed over $[0, 2\pi)$.

a) Evaluate the mean and autocorrelation of $X(t)$. To do so, you may want to condition first with respect to Ω and take repeated expectations. To evaluate expressions

$$m_X = E[E[X(t)|\Omega]]$$
$$R_X(\tau) = E[E[X(t + \tau)X(t)|\Omega]],$$

note that it was shown in class that when A and Ω are known, the process $X(t)$ is zero-mean WSS with autocorrelation

$$\frac{A^2}{2}\cos(\Omega\tau).$$

After conditioning with respect to Ω, to evaluate the autocorrelation $R_X(\tau)$, you may want to use the fact that the characteristic function of the Gaussian distributed frequency Ω is

$$\Phi_\Omega(u) = E[e^{iu\Omega}] = \exp(i\omega_0 u - \frac{(\Delta u)^2}{2}).$$

b) Determine whether $X(t)$ has the mean-square ergodicity property of the mean.
c) Determine whether $X(t)$ has the mean-square ergodicity property of the autocorrelation.

To solve this problem, you may need the trigonometric identities

$$\cos(x)\cos(y) = \frac{1}{2}[\cos(x - y) + \cos(x + y)]$$

$$\cos^2(x) = \frac{1}{2}[1 + \cos(2x)].$$

9.8 It was shown in Problems 3.17 and 3.18 that given a sequence of independent binary random variables $B_k, k \geq 1$ such that

$$P[B_k = 0] = P[B_k = 1] = \frac{1}{2},$$

the random variable

$$X = \sum_{k=1}^{\infty} 2^{-k} B_k$$

is uniformly distributed over $[0, 1]$. In a sequential quantizer, the bits B_k appearing in the binary expansion of X can be computed by performing the recursion

$$X_{t+1} = 2X_t - B_t$$

with

$$B_t = \lfloor 2X_t \rfloor = \begin{cases} 0 \text{ if } X_t < 1/2 \\ 1 \text{ if } X_t \geq 1/2 \end{cases},$$

starting from $X_1 = X$.

a) Verify that

$$X_t = \sum_{k=1}^{\infty} 2^{-k} B_{t+k-1}$$

for $t \geq 1$.

b) Evaluate the mean and variance of B_k and then the mean and autocovariance of X_t. Show that X_t is WSS.

c) Is X_t mean-square ergodic in the mean?

References

1. G. D. Birkhoff, "Proof of the ergodic theorem," *Proc. Nat. Acad. Sciences USA*, vol. 17, pp. 656–550, 1931.
2. J. V. Neumann, "Proof of the quasi-ergodic hypothesis," *Proc. Nat. Acad. Sciences USA*, vol. 18, pp. 70–82, 1932.
3. C. E. Shannon, "Communication in the presence of noise," *Proc. of the IRE*, vol. 37, pp. 10–21, Jan. 1949.
4. T. M. Cover and J. A. Thomas, *Elements of Information Theory, second edition*. Hoboken, NJ: Wiley-Interscience, 2006.
5. S. Karlin and H. M. Taylor, *A First Course in Stochastic Processes, 2nd edition*. New York, NY: Academic Press, 1975.

6. H. McKean, *Probability: The Classical Limit Theorems*. Cambridge, United Kingdom: Cambridge University Press, 2014.

7. P. Brémaud, *Markov Chains: Gibbs Fields, Monte Carlo Simulations, and Queues*. New York, NY: Springer Verlag, 1999.

8. J. R. Norris, *Markov Chains*. Cambridge, United Kingdom: Cambridge University Press, 1997.

9. B. C. Levy, *Principles of Signal Detection and Parameter Estimation*. New York, NY: Springer Verlag, 2008.

Scalar Markov Diffusions and Ito Calculus

10

10.1 Introduction

In this chapter, we examine scalar Markov diffusions. These processes have "Brownian-motion-like" properties in the sense that their sample paths are continuous almost surely, but their dynamics are richer in the sense that unlike the Wiener process, which is homogeneous in both time and space, the process dynamics depend on both the current time t and position $X(t) = x$ of the process. Thus diffusion processes can be used to describe a great variety of physical phenomena, such as oscillators in noise, the dynamics of airplanes subjected to turbulence, or the evolution of the prices of financial securities, like stock options.

A Markov diffusion $X(t)$ can be viewed in some way as a continuous variable and continuous time version of a Markov chain, where the one-step transition matrix $\mathbf{P}$ is replaced by a second-order differential operator parametrized by two parameters: the drift and diffusion coefficients $a(x, t)$ and $d(x, t)$, and where the state probability vector $\pi(t)$ at time t is replaced by the PDF $f(x, t)$ of state $X(t)$. In Sect. 10.2, Markov diffusions are introduced by placing two requirements on the transition density of a Markov process $X(t)$: a continuity requirement, as well as a condition on first and second conditional moments of infinitesimal increments $X(t + h) - X(t)$ of the process given $X(t) = x$. These moment conditions define the drift and diffusion coefficients of the process. Given this parametrization, it is then shown that the PDF $f(x, t)$ can be propagated forward in time by a parabolic space–time PDE called the Fokker–Planck equation (or forward Kolmogorov equation). This equation is the analog for Markov diffusions of the right multiplication by $\mathbf{P}$ used to advance the state probability distribution $\pi(t)$ of a Markov chain by one time step. It is also shown that the conditional expectation of a function $g(X(t))$ conditioned on a previous value $X(s) = x_s$ can be computed backward in time (in s and x_s) by propagating the backward Kolmogorov equation. This allows in particular the solution of exit time problems, like finding the distribution of the time when $X(t)$ will exit a certain set, or the probability that the exit will take place on one side or the other of the set. These problems are the analog for diffusions of exit times and probabilities for the class of transient states of a Markov diffusion.

An important feature of Markov diffusions is that when the drift and square-root of the diffusion coefficient obey a Lipschitz continuity condition and a slow growth condition, they can be viewed as the solution of a stochastic differential equation (SDE). The construction of a solution to this equation makes use of the Ito stochastic integral. This integral, which is defined in Sect. 10.3, can be viewed as a generalization of the Wiener integral defined in Chap. 6. Specifically, while the integrand in the Wiener

© Springer Nature Switzerland AG 2020
B. C. Levy, *Random Processes with Applications to Circuits and Communications*,
https://doi.org/10.1007/978-3-030-22297-0_10

integral was deterministic, it is random in Ito integration theory. This creates obviously an important difficulty: if the value $X(t)$ of the integral at time t is allowed to depend on future values of X, the self-referencing structure of the integral makes it impossible to define $X(t)$ properly. This problem was solved by Ito by requiring that the integral should be *non-anticipative*. Then the construction technique employed by Wiener can be extended, and once again gives rise to an isometry, called the Ito isometry, between integral values, and the expectation of the inner product of the corresponding random integrands. An important feature of processes constructed in this manner is that functions of these processes satisfy a differentiation rule which differs from the standard chain rule of deterministic calculus. An extra term, due to the finiteness of the quadratic variation of the Wiener process, is introduced, which gives rise to the Ito calculus. This means that the use of integrating factors, which is common for solving deterministic differential equations, needs to be modified appropriately for stochastic differential equations. This chapter concludes in Sect. 10.4 with an analysis of the Picard iteration method used to construct a unique solution to SDEs. Under appropriate conditions for the coefficients $a(x, t)$ and $b(x, t)$ of the SDE, its solution is a Markov diffusion, so that under appropriate conditions for the drift and diffusion coefficients of a Markov diffusion, the diffusion can be viewed as solving a SDE driven by a Wiener process. This indicates that SDEs can be viewed as a mechanism for constructing a wide class of processes with properties similar to the Wiener process.

Our discussion focuses on scalar diffusions, but the extension of the results presented to the vector case is straightforward. Although the material presented in this chapter is rather challenging, it will be needed for the analysis of phase noise in oscillators presented in Chap. 13.

10.2   Diffusion Processes

10.2.1 Diffusion Parametrization

A Markov diffusion $X(t)$ with $0 \leq t \leq T$ is a Markov process with continuous sample paths. To rule out the presence of jumps, the transition density $q(x_t, t; x_s, s)$ is required to satisfy the following two conditions for all $x_t \in \mathbb{R}$.

i) Continuity condition: for all $\epsilon > 0$ and $h > 0$ small

$$P(|X(t + h) - X(t)| > \epsilon | X(t) = x_t)$$

$$= \int_{|x_{t+h} - x_t| > \epsilon} q(x_{t+h}, t + h; x_t, t) dx_{t+h} = o(h). \tag{10.1}$$

ii) Drift and diffusion coefficients definition: for small $h > 0$

$$E[X(t + h) - X(t) | X(t) = x_t]$$

$$= \int_{-\infty}^{\infty} (x_{t+h} - x_t) q(x_{t+h}, t + h; x_t, t) dx_{t+h} = a(x_t, t)h + o(h) \tag{10.2}$$

and

$$E[(X(t + h) - X(t))^2 | X(t) = x_t]$$

$$= \int_{-\infty}^{\infty} (x_{t+h} - x_t)^2 q(x_{t+h}, t + h; x_t, t) dx_{t+h} = d(x_t, t)h + o(h). \tag{10.3}$$

The coefficients $a(x, t)$ and $d(x, t)$ are called the drift and diffusion coefficients of $X(t)$, respectively. To illustrate these conditions, we consider several examples.

Example 10.1 (Wiener Process) For the Wiener process, the transition density

$$q(x_{t+h}, t + h; x_t, t) = \frac{1}{(2\pi h)^{1/2}} \exp(-(x_{t+h} - x_t)^2/2).$$

In this case

$$P(|X(t + h) - X(t)| > \epsilon|X(t) = x_t) = 2Q(\epsilon/h^{1/2})$$

$$= \left(\frac{2h}{\pi}\right)^{1/2} \frac{1}{\epsilon} \exp\left(-\frac{\epsilon^2}{2h}\right) = o(h),$$

so that the continuity condition (10.1) holds. The moments of the conditional density are given by

$$E[(X(t + h) - X(t))^n|X(t) = x_t] = 0$$

for n odd, and

$$E[(X(t + h) - X(t))^n|X(t) = x_t] = h^{n/2}(n - 1)!$$

for n even. In particular, for $n = 2$

$$E[(X(t + h) - X(t))^2|X_t = x_t] = h.$$

Thus $a(x, t) = 0$ and $d(x, t) = 1$.

Example 10.2 (Ornstein–Uhlenbeck Process) The Ornstein–Uhlenbeck (OU) process is a zero-mean WSS Gaussian process with autocorrelation $K_X(\tau) = \exp(-a\tau)P_X$ for $\tau > 0$, with $a > 0$ and where $P_X = E[X^2(t)]$ denotes the power of $X(t)$. The conditional distribution of $X(t + h)$ given $X(t)$ is therefore Gaussian, and by using expressions (2.105) and (2.106) we find that the conditional mean and variance are

$$E[X(t + h)|X(t) = x_t] = K_X(h)K_X^{-1}(0)x_t = \exp(-ah)x_t \tag{10.4}$$

and

$$E[(X(t + h) - E[X(t + h)|X(t) = x_t])^2|X(t) = x_t]$$

$$= K_X(0) - K_X(h)K_X^{-1}(0)K_X(h) = (1 - \exp(-2ah))P_X. \tag{10.5}$$

Accordingly, the transition density of the OU process is

$$q(x_{t+h}, t + h; x_t, t) = \frac{1}{(2\pi(1 - \exp(-2ah)P_X)^{1/2}} \exp\left(-\frac{(x_{t+h} - \exp(-ah)x_t)^2}{2(1 - \exp(-2ah))P_X}\right).$$

From (10.4) and (10.5) we find

$$E[X(t + h) - X(t)|X(t) = x_t] = (\exp(-ah) - 1)x_t = -ax_t h + O(h^2)$$

and

$$E[(X(t+h) - X(t))^2 | X(t) = x_t]$$
$$= E[(X(t+h) - E[X(t+h) | X(t) = x_t])^2 | X(t) = x_t]$$
$$+ (E[X(t+h) | X(t) = x_t] - x_t)^2 = 2a P_X h + O(h^2),$$

so that the drift coefficient $a(x, t) = -ax$ depends linearly on x and the diffusion coefficient $d(x, t) = 2a P_X$ is constant.

Example 10.3 (Geometric Brownian Motion) It was shown in Problem 6.4 that the transition density of the geometric Brownian motion $X(t) = \exp(\mu t + \sigma W(t))$, where $W(t)$ is a standard Wiener process, is the log-normal density

$$q(x_t, t; x_s, s) = \frac{1}{(2\pi(t-s)\sigma^2)^{1/2} x_t} \exp\left(-\frac{1}{2\sigma^2(t-s)} (\ln(x_t/x_s) - \mu(t-s))^2\right) \qquad (10.6)$$

for $x_t > 0$ and

$$q(x_t, t; x_s, s) = 0$$

for $x_t \leq 0$. In addition, it was shown that the conditional mean and variance of the geometric Brownian motion satisfy

$$E[X(t) | X(s)] = \exp(\mu(t-s) + \sigma^2(t-s)/2) X(s),$$
$$E[(X(t) - E[X(t) | X(s)])^2 | X(s)] = \exp(\sigma^2(t-s)) X^2(s),$$

so that

$$E[X(t+h) - X(t) | X(t) = x_t] = (\mu + \sigma^2/2) x_t h + o(h)$$
$$E[(X(t+h) - X(t))^2 | X(t) = x_t] = (\sigma x_t)^2 h + o(h),$$

for small h. Accordingly, $X(t)$ is a diffusion process with linear drift coefficient $a(x, t) = (\mu + \sigma^2/2)x$ and diffusion coefficient $d(x, t) = (\sigma x)^2$.

When the drift and diffusion coefficients of a diffusion equation do not depend on time t, the diffusion is called homogeneous.

10.2.2 Fokker–Planck Equation

The drift and diffusion coefficients $a(x, t)$ and $d(x, t)$ have the feature that they are sufficient to compute the transition density $q(x_t, t; x_s, s)$ of the corresponding Markov diffusion by solving a second-order partial differential equation called the *Fokker–Planck or forward Kolmogorov equation*. This equation is derived only in a weak sense by considering an arbitrary infinitely continuously differentiable test function $g(x)$ with compact support. Then for $s < t < t + h$ with h small, by iterated expectations and using the Markov property of $X(t)$, we find

$$E[g(X(t+h))|X(s) = x_s] = \int_{-\infty}^{\infty} g(x)q(x, t+h; x_s, s)dx$$

$$= E[E[g(X(t+h)|X(t)]|X(s) = x_s], \tag{10.7}$$

where

$$E[g(X(t+h))|X(t) = x] = \int_{-\infty}^{\infty} g(x_{t+h})q(x_{t+h}, t+h; x, t)dx . \tag{10.8}$$

Since the function g is smooth, it admits a Taylor series expansion of the form

$$g(x_{t+h}) = g(x) + (x_{t+h} - x)\frac{dg}{dx}(x) + \frac{(x_{t+h} - x)^2}{2}\frac{d^2g}{dx^2}(x) + O((x_{t+h} - x)^3)$$

in the vicinity of x, which after substitution in (10.8) gives

$$E[g(X(t+h))|X(t) = x] = g(x) + [a(x, t)\frac{dg}{dx}(x) + \frac{d(x, t)}{2}\frac{d^2g}{dx^2}(x)]h + o(h). \tag{10.9}$$

Using this expression in (10.7) and rearranging terms, we obtain

$$\int_{\infty}^{\infty} g(x)[q(x, t+h; x_s, s) - q(x, t; x_s, s)]dx$$

$$= \int_{-\infty} [a(x, t)\frac{dg}{dx} + \frac{d(x, t)}{2}\frac{d^2g}{dx^2}]q(x, t; x_s, t)dzh + o(h). \tag{10.10}$$

Dividing both sides of (10.10) by h and letting h tend to zero yield

$$\int_{-\infty}^{\infty} g(x)\frac{\partial}{\partial t}q(x, t; x_s, t)]dz = \int_{-\infty}^{\infty} (L_t g(x))q(x, t; x_s, s)dx, \tag{10.11}$$

where if $u(x)$ is an arbitrary function twice differentiable in x, L_t denotes the second-order differential operator

$$L_t u(x) \stackrel{\triangle}{=} a(x, t)\frac{\partial u}{\partial x} + \frac{d(x, t)}{2}\frac{\partial^2 u}{\partial x^2}. \tag{10.12}$$

Then by integration by parts, we find

$$\int_{-\infty}^{\infty} g(x)[\frac{\partial q}{\partial t} - (L_t^* q)(x, t; x_s, s)]dx = 0, \tag{10.13}$$

where if we assume that the drift and diffusion coefficients admit one and two derivatives in x, respectively, the dual operator L_t^* of L_t is given by

$$L_t^* u(x) = -\frac{\partial}{\partial x}(a(x, t)u(x)) + \frac{1}{2}\frac{\partial^2}{\partial x^2}(d(x, t)u(x)). \tag{10.14}$$

Since the test function $g(x)$ in (10.13) is arbitrary, $q(x, t; x_s, s)$ must satisfy the differential equation

$$\frac{\partial q}{\partial t} = L_t^* q(x, t; x_s, s) \tag{10.15}$$

in a weak sense, i.e., in the sense of distributions, for $t > s$ with initial condition

$$\lim_{t \to s} q(x, t; x_s, s) = \delta(x - x_s). \tag{10.16}$$

Equation (10.15) with initial condition (10.16) is the Fokker–Planck (FP) equation. This shows that the transition density $q(x, t; x_s, s)$ can be interpreted as the Green's function of parabolic operator

$$\Lambda_t = -\frac{\partial}{\partial t} + L_t^*,$$

so that the pair $(a(x, t), d(x, t))$ formed by the drift and diffusion coefficients, together with boundary conditions discussed below, parametrize entirely the transition density of the diffusion process $X(t)$.

In this context, it is worth observing that if the diffusion is homogeneous, i.e., $a(x), d(x)$, and the boundary conditions are time invariant, then the transition density $q(x, t; x_s, s)$ depends only on $t - s$. This is in particular the case for the Wiener, OU, and geometric Brownian motion processes examined in Examples 10.1–10.3, but as shown in Problem 10.1, the Brownian bridge is not homogeneous.

Probability Density Propagation   Since the PDF $f_{X(t)}(z, t)$ of the diffusion at time t can be expressed in terms of the density $f_{X(s)}(x_s, s)$ at time $s < t$ as

$$f_{X(t)}(x, t) = \int_{-\infty}^{\infty} q(x, t; x_s, s) f_{X(s)}(x_s, s) dx_s, \tag{10.17}$$

by applying parabolic operator Λ_t to both sides of (10.17) we deduce that the PDF of $X(t)$ itself obeys the FP equation

$$\frac{\partial f}{\partial t} + \frac{\partial}{\partial x}(a(x, t) f(x, t)) - \frac{1}{2} \frac{\partial^2}{\partial x^2}(d(x, t) f(x, t)) = 0, \tag{10.18}$$

where for convenience we have suppressed the subscript $X(t)$ of the PDF $f_{X(t)}(x)$.

The FP equation (10.18) admits a simple interpretation as a conservation law. Let

$$j(x, t) \overset{\triangle}{=} a(x, t) f(x, t) - \frac{1}{2} \frac{\partial}{\partial x}(d(x, t) f(x, t)) \tag{10.19}$$

denote the probability flux at point x and time t. Then (10.18) can be rewritten as

$$\frac{\partial f}{\partial t} + \frac{\partial j}{\partial x} = 0. \tag{10.20}$$

To interpret this equation, consider an arbitrary interval $[a, b]$ of $\mathbb{R}$. The probability that $X(t)$ belongs to this interval is

$$P(t) = P(a \leq X(t) \leq b) = \int_a^b f(x, t) dx.$$

The rate of change of this probability is

$$\frac{dP}{dt} = \int_a^b \frac{\partial f(x, t)}{\partial t} dx$$
$$= -\int_a^b \frac{\partial j(x, t)}{\partial x} dx = j(a, t) - j(b, t), \tag{10.21}$$

which corresponds to the probability flux entering interval $[a, b]$ at a minus the flux leaving at b. Thus the FP equation is just an expression of the conservation of probability mass.

Boundary Conditions Up to this point, we have ignored the need to specify boundary conditions for the FP equation. When the support of the PDF $f_{X(t)}(x, t)$ or transition density $q(x, t; x_s, s)$ is the real line, or the half line $[0, \infty)$, the requirement that the total probability mass should equal one, so that the density should be summable, implies that the PDF or transition density should satisfy a decay condition of the form

$$f_{X(t)}(x, t) = o(1/|x|) \tag{10.22}$$

as $|x|$ tends to infinity. Conditions of this type are called a *natural boundary conditions* since they are satisfied automatically if the initial density has compact support.

However, over finite domains or for the half line, other types of boundary conditions often arise. An *absorbing boundary condition* at $x = a$ takes the form

$$f_{X(t)}(a, t) = 0 \tag{10.23}$$

for all $t \geq 0$. In this case the total probability mass is not conserved, since the probability flux $-j(a, t)$ across the boundary $x = a$ is removed and never returns. Absorbing boundary conditions play therefore the same role for Markov diffusions as absorbing states for Markov chains. In particular, they are very useful to study exit problems for an interval $[a, b]$ of $\mathbb{R}$, such as the mean time to exit, or the probability that starting from a point x between a and b, the first exit takes place at a

For a *reflecting boundary condition* at $x = a$, an obstacle exists at $x = a$ which ensures that the probability flux

$$j(a, t) = 0 \tag{10.24}$$

for all $t \geq 0$.

Example 10.4 (Reflected Brownian Motion) Consider the reflected Brownian motion process $X(t) = |W(t)|$. Since $W(t)$ is Markov and symmetric about the origin (i.e., the processes $W(t)$ and $-W(t)$ are identical), $X(t)$ is Markov. To evaluate the conditional CDF $F_R(x_t, t|x_s, s)$, the symmetry of $W(t)$ with respect to the origin implies that we can assume $X(s) = W(s) = x_s$. Then

$$F_R(x_t, t|x_s, s) = P(-x_t \leq W(t) \leq x_t | W(s) = x_s)$$

$$= \frac{1}{(2\pi(t-s))^{1/2}} \int_{-x_t}^{x_t} \exp\left(-\frac{(u - x_s)^2}{2(t-s)}\right) du.$$

Differentiating with respect to x_t, we find that the transition density of $X(t)$ is given by

$$q_R(x_t, t; x_s, s) = \frac{1}{(2\pi(t-s))^{1/2}} \left[\exp\left(-\frac{(x_t - x_s)^2}{2(t-s)}\right) \right.$$

$$\left. + \exp\left(\left(-\frac{(x_t + x_s)^2}{2(t-s)}\right)\right) \right]. \tag{10.25}$$

Interestingly, we see that $q_R(x_t, t; x_s, s)$ is the superposition of the Wiener process transition densities $q(x_t, t; x_s, s)$ and $q(x_t, t; -x_s, s)$ corresponding to initial positions at $\pm x_s$. Since each of these transition densities obeys the continuity condition (10.1) it is also satisfied by q_R. Furthermore, it is easy to verify that $a_R(x, t) = 0$ and $d_R(x, t) = 1$, so that the drift and diffusion coefficients are the same as for the Wiener process. On the other hand, since all sample paths of $W(t)$ are reflected at $t = 0$, the probability flux must be zero at $x = 0$, so that

$$j(0, t) \equiv 0.$$

Since the drift is zero and the diffusion coefficient is constant, the expression (10.19) for the flux implies

$$\frac{\partial q_R}{\partial x}(0, t; x_s, s) = 0$$

for all $t \geq s$. From this perspective, the form (10.25) of the transition density can be interpreted as obtained by using the method of images to cancel the probability flux across the $x = 0$ boundary. Specifically, since $x = 0$ is located in the middle of sources placed at x_s and $-x_s$, the probability fluxes due to each component Green's functions cancel each other across the boundary.

For some problems, periodic boundary conditions arise. For example if we consider the phase process

$$X(t) = X(0) + d^{1/2}W(t) \quad \text{mod } (2\pi)$$

examined in Problem 6.7, because $X(t)$ can be viewed as obtained by wrapping the scaled Wiener process $d^{1/2}W(t)$ around a cylinder of unit radius starting from a random initial position $X(0)$, the periodic conditions

$$f(-\pi, t) = f(\pi, t) \quad \text{and} \quad j(-\pi, t) = j(\pi, t) \tag{10.26}$$

are satisfied. Note indeed that the probability flux exiting interval $[-\pi, \pi]$ at π must be recycled at $-\pi$. Otherwise, the drift and diffusion coefficients of $X(t)$ are those of the Wiener process.

Stationary Solutions and Convergence When the drift and diffusion coefficients, as well as the boundary conditions satisfied by the FP equation do not depend on t, the FP equation admits a stationary solution $f(x)$ independent of t. Specifically, when the initial PDF of the Markov diffusion is $f_{X(0)}(x) = f(x)$, the PDF of $X(t)$ remains $f(x)$, so that the diffusion is SSS. By setting the derivative $\frac{\partial f}{\partial t}$ equal to zero in (10.20), we find that a stationary distribution needs to satisfy

$$\frac{dj}{dx}(x) = 0,$$

so that the probability flux $j(x) = C$ is constant. But when the x domain is either $\mathbb{R}$ or the half-line, if we assume that the drift coefficient grows slower than the rate of decay of $f(x)$ at infinity, then for large x, the condition $j(x) = C$ becomes

$$\frac{d}{dx}(d(x)f(x)) \approx -2C$$

so that

$$f(x) \approx \frac{-2Cx + B}{d(x)},$$

where B denotes an integration constant. Then if $d(x)$ is bounded, the boundary condition (10.22) implies that we must have $B = C = 0$. Thus under the assumption of modest growth for $a(x)$ and boundedness of $d(x)$, the conservation of probability mass implies that the probability flux is zero for all x. In this case the flux equation

$$j(x) = a(x)f(x) - \frac{1}{2}\frac{d}{dx}(d(x)f(x)) = 0$$

can be rewritten as

$$\frac{d}{dx}\ln(d(x)f(x)) = \frac{2a(x)}{d(x)} = -\frac{dV}{dx}(x), \qquad (10.27)$$

where the potential

$$V(x) \stackrel{\triangle}{=} -\int^x \frac{2a(z)}{d(z)}dz. \qquad (10.28)$$

Integrating (10.27) gives the stationary solution

$$f(x) = \frac{1}{Zd(x)}\exp(-V(x)), \qquad (10.29)$$

where the integration constant

$$Z = \int_{-\infty}^{\infty}\frac{1}{d(z)}\exp(-V(z))dz$$

ensures that the probability mass of stationary density $f(x)$ is normalized.

In this context, it is of particular interest to consider the case of *Langevin diffusions*, where if $U(x)$ denotes a potential function, the drift

$$a(x) = -\frac{dU}{dx}$$

and $d(x) = d$ is a constant. In this case if we let $\beta \stackrel{\triangle}{=} 2/d$, the stationary solution

$$f(x) = \frac{1}{Z(\beta)}\exp(-\beta U(x)) \qquad (10.30)$$

takes the form of a Gibbs distribution which we denote in the following as $f_G(x)$. Furthermore, the convergence of the solution $f(x, t)$ of the FP equation towards the Gibbs distribution $f_G(x)$ can be analyzed by considering the relative entropy (or Kullback–Leibler divergence)

$$D(f(\cdot, t)|f_G) = \int_{-\infty}^{\infty}\ln\left(\frac{f(x, t)}{f_G(x)}\right)f(x, t)dx \qquad (10.31)$$

of $f(\cdot, t)$ with respect to the Gibbs PDF f_G. The relative entropy is not a true distance since it does not satisfy the triangle inequality, but it has the property that $D(f_1|f_2) \geq 0$ with equality if and only if $f_1 = f_2$[1]. Assume that the initial PDF $f(x, 0)$ of $X(0)$ is such that $D(f(\cdot, 0), f_G)$ is finite. Then, following [2], we have

$$\frac{d}{dt}D(f(\cdot, t)|f_G) = -\frac{1}{\beta}I(f(\cdot, t)|f_G), \qquad (10.32)$$

where

$$I(f(\cdot, t)|f_G) \triangleq \int_{-\infty}^{\infty} |\frac{\partial}{\partial x} \ln\left(\frac{f(x, t)}{f_G(x)}\right)|^2 f(x, t)dx \qquad (10.33)$$

denotes the relative Fisher information of $f(\cdot, t)$ and f_G.

To prove (10.32), note that by differentiating (10.31) and using the conservation law form (10.20) of the FP equation, we obtain

$$\frac{d}{dt} D(f(\cdot, t)|f_G) = \int_{-\infty}^{\infty} \frac{\partial f}{\partial t}\left[\ln\left(\frac{f}{f_G}\right) + 1\right]dx = -\int_{-\infty}^{\infty} \frac{\partial j}{\partial x}\left[\ln\left(\frac{f}{f_g}\right) + 1\right]dx$$

$$= \int_{-\infty}^{\infty} j(x, t)\frac{\partial}{\partial x} \ln\left(\frac{f}{f_G}\right)dx, \qquad (10.34)$$

where we have used integration by parts to go from the first to the second line, as well as the fact that as long as $a(x)$ does not grow too fast, the probability flux $j(x, t)$ tends to zero as $|x|$ tends to infinity. In this expression

$$j(x, t) = [-\frac{dU}{dx} + \frac{1}{\beta}\frac{\partial}{\partial x} \ln(f(x, t)]f(x, t)$$

$$= \frac{1}{\beta}[\frac{\partial}{\partial x}(\ln(f_G(x)) - \ln(f(x, t)))]f(x, t),$$

which proves identity (10.32). This identity is quite interesting, since it shows that the relative entropy of $f(\cdot, t)$ and the Gibbs distribution f_G are strictly monotone decreasing along trajectories of the FP equation, so that the solution $f(x, t)$ becomes progressively closer to $f_G(x)$. Convergence occurs when the relative Fisher information $I(f(\cdot, t)|f_G) = 0$, which implies

$$\frac{\partial}{\partial x} \ln\left(\frac{f(x, t)}{f_G(x)}\right) = 0$$

for all x or equivalently

$$\ln\left(\frac{f(x, t)}{f_G(x)}\right) = C, \qquad (10.35)$$

where the constant $C = 0$ since both PDFs $f(\cdot, t)$ and f_G are normalized. Thus at convergence, we must have $f(x, t) = f_G(x)$ for all x, which in turn ensures that the relative entropy is zero.

Example 10.2 (Continued) The Ornstein–Uhlenbeck process $X(t)$ is probably the simplest Langevin diffusion since in this case the potential $U(x) = ax^2/2$ is quadratic with $a > 0$, so that

$$a(x) = -\frac{dU}{dx} = -ax$$

and $d = 2P_X a$. Then $\beta = 1/(aP_X)$, and the Gibbs distribution (10.29) takes the form of the Gaussian distribution

$$f(x) = \frac{1}{Z(\beta)} \exp\left(-\frac{x^2}{2P_X}\right),$$

where $Z(\beta) = (2\pi P_X)^{1/2}$.

Example 10.5 Consider the phase process

$$X(t) = X(0) + d^{1/2}W(t) \quad \text{mod } (2\pi),$$

where the density $f(x, 0)$ of the initial phase $X(0)$ is nonzero over $[-\pi, \pi]$ and such that the periodic boundary conditions

$$f(-\pi, 0) = f(\pi, 0) \quad \text{and} \quad \frac{\partial}{\partial x}f(-\pi, 0) = \frac{\partial}{\partial x}f(\pi, 0).$$

In this case the potential $U(x)$ and drift coefficient $a(x)$ are both zero, and the diffusion coefficient d is constant. The Gibbs distribution $f_G(x)$ reduces therefore to a uniform distribution over $[-\pi, \pi]$, i.e.,

$$f_G(x) = \begin{cases} \frac{1}{2\pi} & -\pi \le x \le \pi \\ 0 & \text{otherwise .} \end{cases}$$

In the definition (10.31) of the relative entropy of $f(\cdot, t)$ and f_G, the domain of integration becomes $[-\pi, \pi]$. By observing that the boundary conditions

$$f(-\pi, t) = f(\pi, t) \quad \text{and} \quad \frac{\partial}{\partial x}f(-\pi, t) = \frac{\partial}{\partial x}f(\pi, t)$$

are satisfied, we find that the relative entropy identity (10.32) still holds, since in the integration by parts step needed to go from the first to the second line of (10.34) we have

$$-j(\pi, t)f(\pi, t) + j(-\pi, t)f(-\pi, t) = 0.$$

Hence we conclude that, starting from an arbitrary phase PDF which is nonzero over $[-\pi, \pi]$ so that $D(f(\cdot, 0), f_G)$ is finite, the relative entropy of $f(\cdot, t)$ and f_G is monotone decreasing and the solution of the FP equation converges to the uniform density $f_G(x)$.

10.2.3 Backward Kolmogorov Equation

The transition density $q(x_t, t; x_s, s)$ of a Markov diffusion can also be propagated backwards in time. Let $g(x)$ be a twice continuously differentiable bounded function. Then conditioned on $X(s) = x_s$, the expectation of $g(X(t))$ can be expressed as

$$u(x_s, s) \overset{\triangle}{=} E[g(X(t))|X(s) = x_s] = \int_{-\infty}^{\infty} g(x)q(x, t; x_s, s)dx. \tag{10.36}$$

In addition, by using iterated expectation and the Markov property of $X(t)$, we find that for $r < s < t$

$$u(x_r, r) = E[E[g(X(t))|X(s)]|X(r) = x_r] = E[u(X(s), s)|X(r) = x_r]$$

$$= \int_{-\infty}^{\infty} u(z, s)q(z, s; x_r, r)dz.$$

Setting $s \rightarrow s + h$ and $r \rightarrow s$, this gives

$$u(x_s, s) = \int_{-\infty}^{\infty} u(z, s + h) q(z, s + h; x_s, s) dz. \tag{10.37}$$

By performing a Taylor series expansion of $u(z, s + h)$ in the vicinity of (x_s, s), we have

$$u(x_s, s) = u(x_s, s) + h \frac{\partial u}{\partial s}(x_s, s) + (z - x_s) \frac{\partial u}{\partial x_s}(x_s, s)$$

$$+ \frac{(z - x_s)^2}{2} \frac{\partial^2 u}{\partial x_s^2}(x_s, s) + O((z - x_s)^3) + O((z - x_s)h) + O(h^2). \tag{10.38}$$

Substituting this expression inside (10.37) and taking into account the definition (10.2) and (10.3) of the drift and diffusion coefficients yield

$$0 = h[(\frac{\partial u}{\partial s} + L_s u(x_s, s)] + o(h),$$

where L_s denotes the second-order differential operator (10.12), so that by dividing by h and letting h tend to zero, we find that $u(x_s, s)$ obeys the *backward Kolmogorov equation*

$$\frac{\partial u}{\partial s} + L_s(u_x, s) = 0 \tag{10.39}$$

for $s < t$, with initial value $u(x, t) = g(x)$. In light of the definition (10.36) of $u(x_s, s)$ this implies that $q(x_t, t; x_s, s)$ viewed as a function of x_s and s is the the Green's function of operator

$$\frac{\partial}{\partial s} + L_s,$$

i.e., it satisfies the backward partial differential equation

$$(\frac{\partial}{\partial s} + L_s) q(x_t, t; x_s, s) = 0 \tag{10.40}$$

for $s \leq t$, where the operator L_s operates on the x_s variable, with terminal condition

$$q(x_t, t; x_s, s) = \delta(x_s - x_t). \tag{10.41}$$

Remark Unlike the Fokker–Planck equation which is defined only in the sense of distributions, the backward Kolmogorov equation for $u(x_s, s)$ is an ordinary PDE. It is also worth pointing out that the conditions needed for its existence are weaker, since whereas the FP equation required the differentiability of $a(x, t)$ and $d(x, t)$, these conditions are not required for the backward Kolmogorov equation. Attentive readers will also notice that the argument employed to derive this equation, which to evaluate a conditional mean based on $X(s) = x_s$ introduces an intermediate conditioning with respect to $X(t + h)$, is similar to the first-step analysis used to analyze exit problems for the transient class of a Markov chain. It is therefore not surprising that exit problems for Markov diffusions can be analyzed in the same manner.

Distribution of the Exit Time from a Finite Interval Consider a homogeneous Markov diffusion with drift and diffusion parameters $a(x)$ and $d(x)$. Then suppose that the diffusion is initially at a point $X(0) = x_0$ such that $a \le x_0 \le b$, where $[a, b]$ is a finite interval of $\mathbb{R}$. Let

$$T = \inf\{t : X(t) \notin (a, b)\}$$

denote the exit time of the Markov diffusion from interval $[a, b]$, i.e., the first time that $X(t)$ leaves this interval. T is a random variable, and we assume in the following that T is finite with probability one. We are interested in finding its CDF

$$F_T(t, x_0) = P(T \le t | X(0) = x_0).$$

It turns out that it is more convenient to evaluate its complement

$$G(t, x_0) = 1 - F_T(t, x_0) = P(T > t | X(0) = x_0). \tag{10.42}$$

Since the exit problem terminates as soon as $X(t)$ hits either $x = a$ or $x = b$, we consider absorbing boundary conditions

$$q(a, t; x_0, 0) = q(b, t : x_0, 0)$$

at $x = a$ and $x = b$. Then the probability that $T > t$ can be identified with the remaining probability mass contained in interval $[a, b]$ at time t, so that

$$G(t, x_0) = \int_a^b q(x, t; x_0, 0)dx. \tag{10.43}$$

But since the diffusion is homogeneous, we have $q(x, t; x_0, 0) = q(x_0, 0; x_0, -t)$ where $q(x, 0; x_0, -t)$ satisfies the backward Kolmogorov equation. Thus $G(t, x_0)$ satisfies the PDE

$$\frac{\partial G}{\partial t} = a(x_0)\frac{\partial G}{\partial x}(x_0, t) + \frac{d(x_0)}{2}\frac{\partial^2 G}{\partial x_0^2}(x_0, t) \tag{10.44}$$

with boundary conditions

$$G(t, a) = G(t, b) = 0 \tag{10.45}$$

and initial condition

$$G(0, x_0) = 1 \tag{10.46}$$

for $a < x_0 < b$.

Let

$$u(x_0) = E[T | X(0) = x_0]$$

denote the mean exit time. We have

$$u(x_0) = \int_0^\infty s f_T(s, x_0)ds = \int_0^\infty f_T(s, x_0))[\int_0^s dt]ds$$

$$= \int_0^\infty [\int_t^\infty f_T(s, x_0)ds]dt = \int_0^\infty G(t, x_0)dt, \tag{10.47}$$

where $f_T(t, x0)$ denotes the PDf of T given $X(0) = x_0$, and where to go from the first to the second line, we have exchanged the order of integration. Then by integrating the PDE (10.44) with respect to t, we find that the mean exit time $u(x)$ obeys the second-order differential equation

$$\int_0^\infty \frac{\partial G}{\partial t}(t, x)dt = -1 = a(x)\frac{du}{dx} + \frac{d(x)}{2}\frac{d^2u}{dx^2} \qquad (10.48)$$

with boundary conditions

$$u(a) = u(b) = 0. \qquad (10.49)$$

This differential equation is the counterpart for Markov diffusions of matrix equation (5.43) for the average exit time of a Markov chain from its transient class.

Exit Probabilities Suppose that we consider again a diffusion starting at point x_0 in the interior of a finite interval $[a, b]$. We are interested in finding the probability that $X(t)$ will hit $x = a$ before $x = b$. Specifically, let

$$T(a) = \inf\{t : X(t) = a\} \quad \text{and} \quad T(b) = \inf\{t : X(t) = b\}.$$

We are interested in evaluating

$$v(x_0) = P(T(a) < T(b)|X(0) = x_0).$$

We employ again a first-step analysis argument. By conditioning with respect to $X(h) = x_h$ and using the Markov property of $X(t)$, we find by the principle of total probability that

$$v(x_0) = \int_a^b P(T(a) < T(b)|X(h) = x_h)q(x_h, h; x_0, 0)dx_h, \qquad (10.50)$$

where the homogeneity of the diffusion ensures

$$P(T(a) < T(b)|X(h) = x_h) = v(x_h).$$

Then if we assume that $v(x)$ is twice continuously differentiable, it admits a Taylor series expansion

$$v(x_h) = v(x_0) + (x_h - x_0)\frac{dv}{dx}(x_0) + \frac{(x_h - x_0)^2}{2}\frac{d^2v}{dx^2}(x_0) + O((x_h - x_0)^3).$$

By substituting this expression inside (10.50) and using the definition (10.2) and (10.3) of the drift and diffusion coefficients, we obtain

$$v(x_0) = v(x_0) + \left[a(x_0)\frac{dv}{dx}(x_0) + \frac{d(x_0)}{2}\frac{d^2v}{dx^2}(x_0)\right]h + o(h).$$

Dividing by h and letting h tend to zero, we find that $v(x)$ satisfies the second-order differential equation

$$0 = a(x)\frac{dv}{dx} + \frac{d(x)}{2}\frac{d^2v}{dx^2}. \qquad (10.51)$$

Also, since the probability that $T(a) < T(b)$ is one if the initial diffusion position is $X(0) = a$ and zero if it is $X(0) = b$, the boundary conditions

$$v(a) = 1 \quad \text{and} \quad v(b) = 0 \tag{10.52}$$

are satisfied.

Example 10.1 (Continued) For the Wiener process $a(x) = 0$ and $d(x) = 0$, so the differential equation (10.51) reduces to

$$\frac{d^2x}{dx^2} = 0.$$

After applying the boundary conditions (10.52) to the solution $v(x) = Ax + B$, we find

$$v(x) = \frac{b - x}{b - a},$$

so that the probability of exit from $x = a$ depends linearly on the position x of $X(0)$ relative to a in the interval $[a, b]$. This result is consistent with the linear expression derived in Example 5.5 for the exit probability of a random walk on a linear graph, as expected since the Wiener process is just a scaled random walk.

10.3 Ito Calculus

A Markov diffusion can also be interpreted as the solution of a stochastic differential equation (SDE) of the form

$$dX(t) = a(X(t), t)dt + b(X(t), t)dW(t), \tag{10.53}$$

where $b^2(x, t) = d(x, t)$, which can be rewritten in integral form as

$$X(t) - X(s) = \int_s^t a(X(u), u)du + \int_s^t b(X(u), u)dW(u). \tag{10.54}$$

However, even before being able to discuss the existence of solutions to nonlinear equations of the form (10.54), the problem of defining the second integral appearing in this equation arises. Specifically, like the Wiener integral examined in Chap. 6, it has the feature that the Wiener process does not have a finite variation, so it cannot be defined as a Stieltjes integral. In addition, while the integrand $b(u)$ appearing in the Wiener integral is deterministic, we need now to contend with the fact that $b(X(u), u)$ is random. It turns out that this problem was solved by Ito, who started just where Wiener had stopped.

10.3.1 Ito Integral

Consider the problem of defining an integral of the form

$$I(h)(\omega) = \int_0^t h(s, \omega)dW(s), \tag{10.55}$$

where $h(t, \omega)$ is a random function defined over interval $0 \leq t \leq T$. Three conditions will be imposed on h: nonanticipativeness, almost sure square integrability over $[0, T]$, and integrability of its second moment. To define nonanticipativeness, we introduce first the sets of events (usually called sigma fields) $\mathcal{W}[0, t]$ and $\mathcal{W}_t^+$ spanned by intersection, union and complementation of events of the form $W(s)^{-1}(B)$ and $(W(s) - W(t))^{-1}(B)$ for $s \leq t$ and $s > t$, respectively, where B denotes an arbitrary set of the Borel field. Intuitively, $\mathcal{W}[0, t]$ and $\mathcal{W}_t^+$ are the sets of events depending on the past of $W(t)$ and its future, respectively. Sometimes it is necessary to augment $\mathcal{W}[0, t]$ by considering additional events independent of $\mathcal{W}_t^+$ pertaining for example to the initial condition $X(0)$ of differential equation (10.53). Thus we consider a sequence $\{\mathcal{F}_t, 0 \leq t \leq T\}$ of sigma fields satisfying the following three conditions:

  i) The sequence is increasing: $\mathcal{F}_s \subset \mathcal{F}_t$ for $s \leq t$;
 ii) Events in the past of $W(t)$ are included in $\mathcal{F}_t$: $\mathcal{W}[0, t] \subset \mathcal{F}_t$;
iii) $\mathcal{F}_t$ is independent of $\mathcal{W}_t^+$, i.e., arbitrary events A and B of these two sets are independent.

Then the function $h(t, \cdot)$ is *nonanticipative* if for all t in $[0, T]$, $h(t, \cdot)$ is $\mathcal{F}_t$ measurable, i.e.,

$$h(t)^{-1}(B) \in \mathcal{F}_t$$

for all sets B of the Borel field.

The second condition placed on $h(t, \cdot)$ is that it should be square integrable over $[0, T]$ with probability 1. In other words, if

$$A = \{\omega : \int_0^T h^2(t, \omega)dt < \infty\},$$

then $P(A) = 1$. Finally, we also assume that

$$\int_0^T E[h^2(t)]dt$$

is finite.

To define the Ito integral, we consider first the case where $h(t, \cdot)$ is a right continuous piecewise constant function, so that there exists a partition $t_0 = 0 < t_1 < \cdots < t_N = T$ of $[0, T]$ such that $h(t, \omega) = h(t_k, \omega)$ for $t_k \leq t < t_{k+1}$. Then the Ito integral is defined as

$$I(h) \stackrel{\triangle}{=} \int_0^T h(t)dW(t) = \sum_{k=0}^{N-1} h(t_k)(W(t_{k+1}) - W(t_k)). \qquad (10.56)$$

Properties The Ito integral has the following properties:

a) Linearity: for arbitrary real numbers a and b and arbitrary right continuous piecewise continuous functions h_1 and h_2

$$I(ah_1 + bh_2) = aI(h_1) + bI(h_2); \qquad (10.57)$$

b) Zero mean:

$$E[I(h)] = 0; \qquad (10.58)$$

c) Variance:

$$E[I(h)^2] = \int_0^T E[h^2(t)]dt. \tag{10.59}$$

Proof

a) is proved by noting that the linear combination $ah_1 + bh_2$ is piecewise right continuous but with respect to a partition of $[0, T]$ formed by the union of the breakpoints of each partition. Then the linearity is due to the linearity of expression (10.56) with respect to h. The property b) is due to the fact that since $h(t_k)$ and $W(t_{k+1}) - W(t_k)$ are independent, then

$$E[I(h)] = \sum_{k=0}^{N-1} E[h(t_k)]E[W(t_k) - W(t_{k-1})] = 0,$$

where the last equality uses the fact that Wiener process increments have zero mean. Finally,

$$E[I(h)^2] = \sum_{k=0}^{N-1}\sum_{\ell=0}^{N-1} E[h(t_k)h(t_\ell)(W(t_{k+1}) - W(t_k))(W(t_{\ell+1}) - W(t_\ell))]$$

$$= \sum_{k=0}^{N-1} E[h^2(t_k)(W(t_{k+1}) - W(t_k))^2]$$

$$+2\sum_{k<\ell} E[h(t_k)h(t_\ell)(W(t_{k+1}) - W(t_k))(W(t_{\ell+1}) - W(t_\ell))], \tag{10.60}$$

where by observing that $W(t_{\ell+1}) - W(t_\ell)$ is independent of the other terms and has zero mean, the term on the third line is zero, and by using the independence of $h^2(t_k)$ and $(W(t_{k+1}) - W(t_k))^2$, we find

$$E[I(h)^2] = \sum_{k=0}^{N-1} E[h^2(t_k)]E[(W(t_k) - W(t_{k-1})^2]$$

$$= \sum_{k+1}^{N-1} E[h^2(t_k)](t_{k+1} - t_k) = \int_0^T E[h^2(t)]dt. \tag{10.61}$$

To define the Ito integral for the general case, we use the same approach as for the Wiener integral and observe that if $h(t)$ is square integrable, it can be approximated by a sequence of piecewise constant function $h_n(t)$ such that

$$\lim_{n\to\infty} \int_0^T |h(t) - h_n(t)|^2 dt = 0.$$

Then Ito integral $I(h)$ is defined as the mean-square limit of integrals $I(h_n)$ as n tends to infinity, i.e.

$$I(h_n) \overset{m.s.}{\to} I(h) = \int_0^T h(t)dW(t). \tag{10.62}$$

Then it is not difficult to prove that the limit $I(h)$ inherits the properties (10.57)–(10.59) satisfied by Ito integrals of piecewise constant functions.

In addition, if h_1 and h_2 are two nonanticipating functions, we have

$$E[I(h_1)I(h_2)] = \frac{1}{2}[E[I(h_1 + h_2)^2] - E[I(h_1)^2] - E[I(h_2)^2]]$$

$$= \int_0^T E[h_1(t)h_2(t)]dt, \tag{10.63}$$

which defines the Ito isometry between Ito integrals and the inner product of random nonanticipating square-integrable functions. This isometry is remarkably similar to the Wiener isometry introduced in Chap. 6 between Wiener integrals and the inner product of square integrable deterministic functions over $[0, T]$. From this perspective, we see that the Ito integral can be viewed as a generalization of the Wiener integral where the key idea is the nonanticipativeness of integrand $h(t)$. Yet, this definition yields astonishing results as can be seen in the following example.

Example 10.6 Consider the Ito integral

$$\int_0^T W(t)dW(t),$$

where $W(t)$ is continuous and thus square-integrable over $[0, T]$. Consistently with the nonanticipativeness assumption, we partition the interval $[0, T]$ into N subintervals with spacing $h = T/N$, so that if $t_k = kh$, and approximate the integral by the partial sum

$$\sum_{k=0}^{N-1} W(t_k)(W(t_{k+1}) - W(t_k)),$$

where the increment $W(t_{k+1}) - W(t_k)$ is independent of $W(t_k)$. This sum can be rewritten as

$$\frac{1}{2}\sum_{k=0}^{N-1}[W^2(t_{k+1}) - W^2(t_k) - (W(t_{k+1}) - W(t_k))^2],$$

where

$$\frac{1}{2}\sum_{k=0}^{N-1}(W^2(t_{k+1}) - W^2(t_k)) = \frac{W^2(T)}{2},$$

and by using the fact that the quadratic variation of the Wiener process over $[0, T]$ is T, we find

$$\lim_{N\to\infty}\sum_{k=0}^{N-1}(W(t_{k+1}) - W(t_k))^2 = T,$$

so that the Ito integral yields

$$\int_0^T W(t)dW(t) = \frac{1}{2}(W^2(T) - T). \tag{10.64}$$

In contrast, for an ordinary differentiable function $w(t)$, the Riemann integral would yield

$$\int_0^T w(t)\frac{dw}{dt}dt = \frac{w^2(T)}{2}. \tag{10.65}$$

The difference between (10.64) and (10.65) is due to two causes: first, unlike ordinary functions, the Wiener process has a finite quadratic variation, and second, the nonanticipatory definition of the Ito integral. If we had selected the average

$$\frac{1}{2}(W(t_{k+1}) + W(t_k))$$

of the values of W at the two end point of interval $[t_k, t_{k+1})$ to approximate $W(t)$, the result would have been identical to the conventional integral. This choice corresponds to the Stratonovich integral [3, 4]. However, it has the serious disadvantage that with this convention, the two moment properties (10.58) and (10.59) of the Ito integral no longer hold. So in practice, the Ito integral is preferred, even though it leads to unconventional results.

Ito Integral Process By considering the integral

$$X(t) = \int_0^t h(s)dW(s) = \int_0^T h(s)1_{[0,t]}(s)dW(s), \tag{10.66}$$

where

$$1_{[0,t]}(s) = \begin{cases} 1 & 0 \le s \le t \\ 0 & \text{otherwise}, \end{cases}$$

denotes the indicator function of set $[0, t]$, it is possible to generate a random process by Ito integration. This process has several important properties:

i) $X(t)$ is $\mathcal{F}_t$ measurable, so it is nonanticipating;
ii) $E[X(t)] = 0$ and

$$E[X(t)X(s)] = \int_0^{\min(t,s)} E[h^2(u)]du. \tag{10.67}$$

iii) $X(t)$ has continuous sample paths almost surely.

The reader is referred to [3, Sec. 5.1] for a proof. Note that (10.67) is a consequence of the Ito isometry. A consequence of this property is that the process $X(t)$ has orthogonal increments. Specifically if $0 \le s_1 < t_1 \le s_2 < t_2 \le T$, a direct substitution of (10.66) implies

$$E[(X(t_2) - X(s_2))(X(t_1) - X(s_1))] = 0.$$

However, the fact that the increments are orthogonal does not mean they are independent.

10.3.2 The Ito Formula

Consider a random process $X(t)$ satisfying

$$X(t) = X(0) + \int_0^t g(s, \omega)ds + \int_0^t h(s, \omega)dW(s) \tag{10.68}$$

for $0 \leq t \leq T$, where the second integral is defined as an Ito integral and the first integral is a Riemann integral. The function $g(t, \cdot)$ is nonanticipative with respect to the sequence $\{\mathcal{F}_t, 0 \leq t \leq T\}$ used to define the Ito integral, i.e., $g(t, \cdot)$ is $\mathcal{F}_t$ measurable for all t, and with probability one

$$\int_0^T |g(t)|dt < \infty$$

Finally, it is also assumed that $X(0)$ is $\mathcal{F}_0$ measurable. In this case, we say that $X(t)$ has the stochastic differential

$$dX(t) = g(t)dt + h(t)dW(t). \tag{10.69}$$

The *Ito formula* is the analog for random processes of the form (10.68) of the chain rule of ordinary calculus. Specifically, let $\phi(x, t)$ be a continuous function of $\mathbb{R} \times [0, T]$ such that

$$\frac{\partial \phi}{\partial t} \;,\quad \frac{\partial \phi}{\partial x} \quad \text{and} \quad \frac{\partial^2 \phi}{\partial x^2}$$

exist and are continuous. Then the random process $Y(t) = \phi(X(t), t)$ has the stochastic differential

$$dY(t) = \frac{\partial \phi}{\partial t}dt + \frac{\partial \phi}{\partial x}dX(t) + \frac{1}{2}\frac{\partial^2 \phi}{\partial x^2}dt$$

$$= [\frac{\partial \phi}{\partial t} + \frac{\partial \phi}{\partial x}g + \frac{1}{2}\frac{\partial^2 \phi}{\partial x^2}]dt + \frac{\partial \phi}{\partial x}dW(t). \tag{10.70}$$

A very rough way of interpreting Ito's formula is to view it as a Taylor series expansion where the $(dW(t))^2$ term is replaced by dt, so that the corresponding second-order term which would normally be neglected in ordinary calculus becomes a first-order term in Ito calculus. Like all nonlinear transformations which are often used to solve nonlinear differential equation, the Ito formula can be employed to solve stochastic differential equations.

Example 10.7 Let $X(t)$ be a process with differential

$$dX(t) = aXdt + \sigma XdW \tag{10.71}$$

and initial condition $X(0) = 1$. Consider the transformation $Y(t) = \ln(X(t))$, so that $\phi(x) = \ln(x)$. Since

$$\frac{d\phi}{dx} = \frac{1}{x} \quad \text{and} \quad \frac{d^2\phi}{dx^2} = -\frac{1}{x^2},$$

by applying the formula (10.70) we find that $Y(t)$ has for differential

$$dY(t) = \mu dt + \sigma dW(t)$$

with $\mu = a - \sigma^2/2$, where the right-hand side is independent of $Y(t)$ and can be easily integrated. Taking into account the initial condition $Y(0) = 0$, this gives

$$Y(t) = \mu t + \sigma W(t)$$

for $0 \leq t \leq T$, which shows that the solution of the stochastic differential equation (10.71) is the geometric Brownian motion

$$X(t) = \exp(Y(t)) = \exp(\mu t + \sigma W(t)).$$

Ito's Product Rule Suppose now that $X_1(t)$ and $X_2(t)$ are two stochastic processes with differentials

$$dX_i(t) = g_i(t)dt + h_i(t)dW(t)$$

for $i = 1, 2$, where as before we assume that g_i and h_i are $\mathcal{F}_t$ measurable, with g_i integrable and h_i square-integrable over $[0, T]$ almost surely for $i = 1, 2$. Then the product $X_1 X_2$ has the differential

$$d(X_1 X_2) = X_2 dX_1 + X_1 dX_2 + h_1 h_2 dt, \tag{10.72}$$

which yields the stochastic integration by parts formula

$$\int_0^t X_2 dX_1 = X_1(t)X_2(t) - X_1(0)X_2(0) - \int_0^t X_1 dX_2 - \int_0^t h_1 h_2 dt, \tag{10.73}$$

which differs from the usual formula by the appearance of the last term. Like Ito's formula, the product rule (10.72) can be interpreted as obtained by replacing $(dW(t))^2$ by dt in ordinary differential calculus.

The product rule (10.72) can often be used to solve stochastic differential equations by eliminating the variable of interest.

Example 10.8 Suppose we want to solve the linear stochastic differential equation

$$dX(t) = (\alpha dt + \beta dW(t))X(t) + (\gamma dt + \delta dW(t)) \tag{10.74}$$

with initial condition $X(0)$, where α, β, γ, and δ are real constants. The case where $\gamma = \delta = 0$ corresponds to the geometric Brownian motion process examined in Example 10.7. This suggests introducing an integrating factor of the form

$$Y(t) = \exp(-\mu t - \beta W(t)), \tag{10.75}$$

where μ remains to be determined. It has the stochastic differential

$$dY(t) = [(-\mu + \beta^2/2)dt - \beta dW(t)]Y(t).$$

Then if $Z(t) = Y(t)X(t)$, by using the product rule (10.72), we obtain

$$dZ = [\alpha + \beta dW(t) - \mu + \beta^2/2 - \beta dW(t) - \beta^2]X(t)Y(t)$$
$$+ [(\gamma - \beta\delta)dt + \delta dW(t)]Y(t). \tag{10.76}$$

By selecting

$$\mu = \alpha - \beta^2/2$$

the XY term of (10.76) is zero, so that we obtain

$$dZ = (\gamma - \beta\delta)Y(t)dt + \delta Y(t)dW(t)$$

or equivalently

$$Z(t) = Z(0) + (\gamma - \beta\delta)\int_0^t Y(s)ds + \delta\int_0^t Y(s)dW(s). \tag{10.77}$$

Since $Y(0) = 1$, we have $Z(0) = X(0)$, so that the solution of the stochastic differential equation (10.74) is given by

$$X(t) = Y^{-1}(t)[X(0) + (\gamma - \beta\delta)\int_0^t Y(s)ds + \delta\int_0^t Y(s)dW(s)], \tag{10.78}$$

where $Y(t)$ is given by (10.75) with $\mu = \alpha - \beta^2/2$.

10.4   Stochastic Differential Equations

When it is not possible to solve stochastic differential equations by nonlinear transformation or the use of integrating factors, a Picard iteration technique similar to the one used to solve ordinary differential equations [5, Chap. 1] needs to be employed. Specifically, consider the SDE

$$dX(t) = a(X(t), t)dt + b(X(t), t)dW(t) \tag{10.79}$$

with initial condition $X(0)$, or equivalently, in integral form

$$X(t) = X(0) + \int_0^t a(X(s), s)ds + \int_0^t b(X(s), s)dW(s). \tag{10.80}$$

To ensure the existence and uniqueness of a solution to the SDE (10.80) over interval $[0, T]$, two conditions need to be placed on coefficients $a(x, t)$ and $b(x, t)$.

i) Lipschitz continuity: for all $t \in [0.T]$ and $x, y \in \mathbb{R}$,

$$|a(x, t) - a(y, t)| + |b(x, t) - b(y, t)| \le C|x - y| \tag{10.81}$$

for some constant C.

ii) Slow growth condition: for all $t \in [0, T]$ and $x \in \mathbb{R}$, the coefficients $a(x, t)$ and $b(x, t)$ satisfy

$$|a(x, t)| + |b(x, t)| \le D(1 + |x|^2)^{1/2} \tag{10.82}$$

for some constant D.

In addition, it is assumed that the random initial condition $X(0)$ has finite second moment, i.e., $E[X(0)^2]$ is finite, and is independent of the Wiener process $W(t)$. Furthermore, the increasing sequence $\{\mathcal{F}_t, t \geq 0\}$ of sigma fields used to define the second integral of (10.80) in the Ito sense is such that $X(0)$ is measurable with respect to $\mathcal{F}_0$. To establish the uniqueness of a solution to the SDE (10.80), we will use the following inequality, which is proved in [5, Chap. 1].

Grönwall's Inequality Let $h(t)$ be a nonnegative continuous function over $[0, T]$ such that

$$h(t) \leq K + L \int_0^t h(s)ds \tag{10.83}$$

for $0 \leq t \leq T$, where K and L denote nonnegative constants. Then

$$h(t) \leq K \exp(Lt) \tag{10.84}$$

for $0 \leq t \leq T$.

Uniqueness Assume that $X(t)$ and $\check{X}(t)$ are two solutions of (10.80) with initial condition $X(0)$. Then

$$X(t) - \check{X}(t) = \int_0^t [a(X(s), s) - a(\check{X}(s), s)]ds + \int_0^t [b(X(s), s) - b(\check{X}(s), s)]dW(s)$$

for $0 \leq t \leq T$, and by observing that $(\alpha + \beta)^2 \leq 2(\alpha^2 + \beta^2)$, we deduce that

$$E[|X(t) - \check{X}(t)|^2] \leq 2E\left[\left|\int_0^t [a(X(s), s) - a(\check{X}(s), s)]ds\right|^2\right]$$

$$+ 2E\left[\left|\int_0^t [b(X(s), s) - b(\check{X}(s), s)]dW(s)\right|^2\right]. \tag{10.85}$$

By using the Cauchy–Schwartz inequality

$$\left|\int_0^t v(s)ds\right|^2 \leq t \int_0^t v^2(s)ds$$

and the Lipschitz continuity condition (10.81), we find

$$E\left[\left|\int_0^t [a(X(s), s) - a(\check{X}(s), s)]ds\right|^2\right] \leq TE\left[\int_0^t \left|a(X(s), s) - a(\check{X}(s), s)\right|^2 ds\right]$$

$$\leq C^2 T \int_0^t E[|X(s) - \check{X}(s)|^2]ds. \tag{10.86}$$

Likewise, by using the Ito isometry and condition (10.81), we have

$$E\left[\left|\int_0^t [b(X(s), s) - b(\check{X}(s), s)]dW(s)\right|^2\right] = E\left[\int_0^t \left|b(X(s), s) - b(\check{X}(s), s)\right|^2 ds\right]$$

$$\leq C^2 \int_0^t E[|X(s) - \check{X}(s)|^2]ds. \tag{10.87}$$

Combining (10.85)–(10.87) gives

$$E[|X(t) - \check{X}(t)|^2] \le 2C^2(T + 1) \int_0^t E[|X(s) - \check{X}(s)|^2]ds \tag{10.88}$$

and by applying Grönwall's inequality with $K = 0$ and $L = 2C^2(T + 1)$, we deduce

$$E[|X(t) - \check{X}(t)|^2] = 0$$

so $X(t) = \check{X}(t)$ almost surely for $0 \le t \le T$.

Existence The existence of a solution to the SDE (10.80) is established by proving the convergence of the Picard iteration

$$X^{k+1}(t) = X(0) + \int_0^t a(X^k(s), s)ds + \int_0^t b(X^k(s), s)dW(s) \tag{10.89}$$

for $0 \le t \le T$ with initial iterate $X^0(t) = X(0)$ for $0 \le t \le T$.

By using the same bounding approach as for uniqueness, we find

$$E[|X^{k+1}(t) - X^k(t)|^2] \le 2C^2(T + 1) \int_0^t E[|X^k(s) - X^{k-1}(s)|^2]ds \tag{10.90}$$

for $k \ge 1$. For $k = 0$ we have

$$E[|X^1(t) - X^0(t)|^2] \le 2E\left[\left|\int_0^t a(X(0), s)ds\right|^2\right]$$

$$+2E\left[\left|\int_0^t b(X(0), s)dW(s)\right|^2\right], \tag{10.91}$$

where

$$E\left[\left|\int_0^t a(X(0), s)ds\right|^2\right] \le TE\left[\int_0^t \left|a(X(0), s)\right|^2 ds\right]$$

$$\le D^2 Tt E[1 + |X(0)|^2], \tag{10.92}$$

and

$$E\left[\left|\int_0^t b(X(0), s)dW(s)\right|^2\right] = E\left[\int_0^t \left|b(X(s), s)\right|^2 ds\right]$$

$$\le D^2 t E[1 + |X(0)^2]. \tag{10.93}$$

Combining (10.91)–(10.93) gives the bound

$$E[|X^1(t) - X^0(t)|^2] \le 2D^2(T + 1)E[1 + |X(0)|^2]t \tag{10.94}$$

Then from (10.94) and (10.90) we find by induction that

$$E[|X^{k+1}(t) - X^k(t)|^2] \le \frac{M^{k+1}t^{k+1}}{(k+1)!} \tag{10.95}$$

with

$$M = \max(2C^2(T+1), 2D^2(T+1)E[1 + |X(0)|^2]).$$

Accordingly for $k > \ell$ we deduce that

$$E[(X^k(t) - X^\ell(t))^2] = E[(\sum_{m=\ell}^{k-1} (X^{m+1}(t) - X^m(t)))^2]$$

$$\le 2^{-\ell} \sum_{m=\ell}^{k-1} 2^m E[(X^{m+1}(t) - X^m(t))^2]$$

$$\le 2^{-\ell} \sum_{m=\ell}^{k-1} \frac{(2Mt)^m}{m!} \le 2^{-\ell} \exp(2Mt), \tag{10.96}$$

where $2^{-\ell} \exp(2Mt)$ tends to zero for $0 \le t \le T$ as ℓ tends to infinity. This shows that $X^k(t)$ is a Cauchy sequence in the space of square integrable random variables, so it admits a limit $X(t)$ in the mean-square sense. When k tends to infinity in (10.96) we also obtain the inequality

$$E[(X(t) - X^\ell(t))^2] \le 2^{-\ell} \exp(2Mt) \tag{10.97}$$

for $0 \le t \le T$, which will be used below.

It remains to show that the limit $X(t)$ obeys the SDE (10.80) in the mean-square sense. This requires showing

$$\int_0^t a(X^k(s), s)ds \overset{m.s.}{\to} \int_0^t a(X(s), s)ds \tag{10.98}$$

$$\int_0^t b(X^k(s), s)dW(s) \overset{m.s.}{\to} \int_0^t b(X(s), s)dW(s) \tag{10.99}$$

uniformly in t. By using the Cauchy–Schwartz inequality, the Lipschitz condition (10.81), and (10.97) we find

$$E\left[\left|\int_0^t [a(X(s), s) - a(X^k(s), s)]ds\right|^2\right]$$

$$\le T \int_0^T E[[a(X(s), s) - a(X^k(s), s)]^2]ds$$

$$\le C^2 T \int_0^T E[(X(s) - X^k(s))^2]ds \le C^2 T^2 2^{-k} \exp(2MT), \tag{10.100}$$

where the upper bound on the right-hand side is independent of t and tends to zero as k tends to infinity. This proves (10.98). Likewise, by using Ito's isometry and the Lipschitz condition (10.81) as well as inequality (10.97), we find

$$E\left[\left|\int_0^t [b(X(s), s) - b(X^k(s), s)]dW(s)\right|^2\right]$$

$$= \int_0^t E[[b(X(s), s) - b(X^k(s), s)]^2]ds$$

$$\leq C^2 \int_0^t E[(X(s) - X^k(s))^2]ds \leq C^2 T 2^{-k} \exp(2MT), \qquad (10.101)$$

where the upper bound on the right-hand side is independent of t and tends to zero as k tends to infinity, which proves (10.99).

This proves the existence of a solution to the stochastic differential equation (10.80) in the mean-square sense. It turns out [3, p. 118] that this solution is mean-square continuous. In addition, the solution can also be constructed in the almost sure sense and has continuous sample paths almost surely [3, 6].

Markov Property When the conditions (10.80)–(10.82) (as well as the assumptions that $E[X(0)^2]$ is finite and $X(0)$ is $\mathcal{F}_0$ measurable) ensuring the existence and uniqueness of a solution to the SDE (10.80 are satisfied, its solution $X(t)$ is a Markov process. Furthermore, if $a(x, t)$ and $b(x, t)$ are continuous functions of t over $[0, T]$, then $X(t)$ is a Markov diffusion with parameters $a(x, t)$ and $d(x, t) = b^2(x, t)$.

These results are proved in Sects. 9.2 and 9.3 of [3]. They establish that under modest smoothness and slow growth conditions for $a(x, t)$ and $b(x, t) = d^{1/2}(x, t)$, diffusion processes can be viewed as the solution of a stochastic differential equation, whose properties such as almost sure continuity of the sample paths are similar to those of the Wiener process. In this sense, through the mechanism of Ito integration, the Wiener process can be viewed as the key building block used to construct an entire family of random processes, the Markov diffusion processes, with "Brownian motion-like" properties.

10.5   Bibliographical Notes

Due to the important role played by stochastic calculus in modern mathematical finance, the last 20 years have seen a proliferation of books focused on stochastic differential equations. Yet, Arnold's 1974 book [3] probably remains the easiest introduction to this general topic, and our presentation of Ito calculus is largely based on it. Jazwinski's book [7] on nonlinear filtering also contains an accessible presentation of stochastic differential equations targeted at engineers. More recent books which require only a moderate level of mathematical training include those by Kloeden and Platen [4], Oksendal [6], and Evans [8]. The books by Gardiner [9], Risken and Frank [10], and Pavliotis [11] contain detailed discussions of the Fokker–Planck and Langevin equations. Exit problems of Markov diffusions are discussed in detail in [9], as well as in the second volume of Karlin and Taylor's stochastic processes treatise [12].

10.6   Problems

10.1 The Brownian bridge process $X(t)$ is a Wiener process $W(t)$ conditioned on $W(T) = 0$. It can be represented as

$$X(t) = W(t) - \frac{t}{T}W(T),$$

and it was shown in Problem 6.2 that this process is Gaussian with zero mean and covariance function

$$K_X(t, s) = \min(s, t)(1 - \frac{\max(s, t)}{T}).$$

a) Evaluate the transition density $q(x_t, t; x_s, s)$ of the Brownian bridge.
b) Specify the drift and diffusion coefficients $a(x, t)$ and $d(x, t)$ of $X(t)$.

10.2 In the discussion of reflected Brownian motion in Example 10.4, we saw that the method of images can be used to find the transition density of a Wiener process satisfying a reflecting boundary condition. The same approach can also be used to obtain the transition density of a Wiener process satisfying absorbing boundary conditions. Specifically, consider a diffusion $X(t)$ with $a(x) = 0$ and $d(x) = 1$ defined over the half space $(-\infty, a]$ with $a > 0$ satisfying the absorbing condition

$$f(a, t) = 0 \tag{10.102}$$

for all $t \geq 0$.

a) For $x_s < a$, show that the transition density of $X(t)$ is given by

$$q_A(x_t, t; x_s, s) = \frac{1}{(2\pi(t - s))^{1/2}}\left[\exp\left(-\frac{(x_t - x_s)^2}{2(t - s)}\right)\right.$$
$$\left. - \exp\left(-\frac{(x_t - (2a - x_s))^2}{2(t - s)}\right)\right]. \tag{10.103}$$

This density can be interpreted as obtained by pairing the source at x_s with a source of opposite polarity at the mirror image location $2a - x_s$ of x_s with respect to a. On the line $x_t = a$, these two opposite sources located at the same distance of a cancel each other, so that the boundary condition

$$q(a, t; x_s, s) = 0$$

is satisfied.
b) The transition density (10.103) can be used to derive the distribution of the first passage time

$$T = \inf\{t : W(t) = a\}$$

of the Wiener process through level $a > 0$. Since the problem terminates as soon as level a is crossed, we apply the boundary condition (10.102) to the Wiener diffusion. Then by observing that

$$P(T > t) = \int_{-\infty}^{a} q_A(x, t; 0, 0)dt,$$

show that

$$P(T > t) = 2F_{W(t)}(a) - 1, \tag{10.104}$$

where $F_{W(t)}(w)$ denotes the CDF of Wiener process $W(t)$.

c) Obtain the PDF $f_T(t)$ of T. This distribution is called the Lévy distribution. Together with the Gaussian and Cauchy distributions, it is one of three stable distributions which admit a closed form.

10.3 The method of images can be extended to more complicated boundary conditions. For example consider a Brownian motion process $X(t)$ over interval $[a, b]$ with absorbing boundary conditions

$$f(a, t) = f(b, t) = 0 \tag{10.105}$$

at both $x = a$ and $x = b$. In this case, given an initial value $X(s) = x_s$, the transition density $q_I(x_t, t; x_s, s)$ can be obtained by the method of images. This is done by mirroring a source placed at x_s by sources of opposite polarities located at the images $2b - x_s$ and $2a - x_s$ of source x_s with respect to boundaries a and b. However, after this mirroring, it becomes necessary to mirror, again with opposite polarity, the new sources with respect to the boundaries with respect to which they do not yet have a matching image. For example, the image of the source at $2a - x_s$ with respect to the boundary at b is located at $2(b - a) + x_s$. This process needs to be repeated until complete mirroring is achieved by an infinite sequence of positive sources located at $x_s + 2n(b - a)$ with $n \in \mathbb{Z}$ and an infinite sequence of negative sources located at $2a - x_s + 2m(b - a)$.

a) Use the above mirroring technique to evaluate the transition density
   $q(x_t, t; x_s, s)$ of the Brownian motion process $X(t)$ defined over interval $[a, b]$ with boundary conditions (10.105).
b) Use the transition density of part a) to find an expression for the exit time probability

$$G(x_0, t) = P(T > t | X(0) = x_0)$$

defined in (10.42).

10.4 Because it evaluates the expectation of a function $g(\cdot)$ of a future diffusion value $X(T)$ given its current value $X(t)$, the backward Kolmogorov equation can be used to analyze mathematical finance problems such as computing the value of stock options. Specifically, consider a European stock option which, unlike American options, can only be exercised on the closing date T of option. For a call option with strike price S, if $X(t)$ denotes the price of the corresponding stock at time t, the option holder has the right to purchase the stock at price S at time T, so that the value of the option at time T is

$$g(X(T)) = \max(X(T) - S, 0).$$

In other words, the profit on the option is $X(T) - S$ if the stock price exceeds S at time T, and zero if it is less than S, since in this case the option holder does not exercise the option. We assume that the stock price $X(t)$ is modeled by a geometric Brownian motion with drift parameter rx and diffusion parameter $(\sigma x)^2$. To exclude arbitrage opportunities, i.e., the opportunity to achieve profits based on riskless strategies, the coefficient r is equal to the rate of return on riskless investment instruments, such as Treasury Bills. Then conditioned on the value $X(t) = x$ of the stock at time t, the value of the option at time t is

$$V(x, t) = \exp(-r(T - t))E[g(X(T))|X(t) = x].$$

The discount rate $\exp(-r(T-t))$ is introduced here to evaluate the option appreciation with respect to a benchmark represented by the rate of return of a riskless investment. Use the backward Kolmogorov equation to show that the option value $V(x, t)$ obeys the partial differential equation

$$\frac{\partial V}{\partial t} + rx\frac{\partial V}{\partial t} + \frac{1}{2}(\sigma x)^2\frac{\partial^2 V}{\partial x^2} - rV = 0.$$

This equation is known as the Black–Scholes formula [13,14]. Although it has a number of flaws, such as the assumption that the volatility σ remains constant and that stock price fluctuations are Gaussian, it is a remarkable analytical tool, since it allows the systematic computation of the option value for all possible current stock prices $X(t) = x$. In particular, it shows that even if the current stock price x is below the strike price S, an option is not worthless.

10.5 Consider a shifted Wiener process $X(t) = x + W(t)$ whose initial position $X(0) = x$ is located in the interior of interval $[a, b]$. Evaluate the mean exit time $u(x)$ of $X(t)$ from interval (a, b). For what value of x is it maximized?

10.6 For a diffusion with parameters $a(x)$ and $d(x)$, the exit probability $v(x)$ satisfying (10.51) with boundary conditions (10.52) can be evaluated in closed form. Let

$$V(x) = -\int^x \frac{2a(z)}{d(z)}dz$$

be the potential introduced in (10.27). If

$$S(x) = \int^x \exp(V(u))du,$$

verify that

$$v(x) = \frac{S(b) - S(x)}{S(b) - S(a)}$$

satisfies (10.51) and (10.52).

10.7 Consider a first-order analog phase-locked loop (PLL) such that the quiescent frequency of the voltage controlled oscillator is already tuned to the frequency of the received signal. Then the phase error can be described by the first order stochastic differential equation [15,16]

$$\frac{d\Phi}{dt} = -a\sin(\Phi(t)) + bdW(t), \tag{10.106}$$

with $a > 0$. In this equation, the phase error is unwrapped, so that it takes values in $\mathbb{R}$, but suppose we consider the wrapped phase $X(t) = \Phi(t) \mod 2\pi$, which takes values over in $[-\pi, \pi)$.

a) Find the stationary PDF $f_G(x)$ of $X(t)$.
b) Show that if the initial PDF $f_{X(0)}(x)$ is nonzero for all $x \in [-\pi, \pi)$, the PDF $f_{X(t)}(x)$ converges to $f_G(x)$ as t tends to infinity.

10.8 Consider the unwrapped first-order PLL phase satisfying (10.106). The stable equilibria of the PLL occur at $n2\pi$ with n integer, since in the absence of noise, if the phase $\Phi(t)$ is slightly perturbed away from $n2\pi$, it will return to $n2\pi$ due to the effect of the nonlinear feedback term $-a\sin(\phi)$. If the PLL is initially in the zero error position ($n = 0$), a cycle slip occurs when due to the presence of the driving noise, $\Phi(t)$ hits either 2π or -2π since the PLL is again in stable equilibrium in these two positions, but the phase error is now $\pm 2\pi$. The problem of finding the mean time to loss of lock or frequency of cycle slips can be formulated as an exit problem for the unwrapped phase equation (10.106). Since an exit occurs whenever $\Phi(t)$ hits $\pm 2\pi$, we can apply the boundary conditions

$$f(\pm 2\pi, t) = 0.$$

Specify the differential equation satisfied by the mean exit time $u(\phi)$ and obtain a closed-form solution. To do so, you may want to use the change of variable

$$s(\phi) = \exp(\beta \cos(\phi)) \frac{du}{d\phi}$$

in the differential equation satisfied by $u(\phi)$, where $\beta = 2a/b^2$.

10.9 Since the Markov property is independent of the time direction, it is reasonable to expect that if $X(t)$ is a Markov diffusion in forward time, it is also be a diffusion in reverse time. Let $X(t)$ be a Markov diffusion with drift $a(x, t)$ and diffusion coefficient $d(x, t)$ defined over $[0, T]$ and let $f(x, t)$ denote its PDF at time t, so that $f(x, t)$ obeys the Fokker–Planck equation (10.18). Then consider the time-reversed process $\check{X}(t) = X(T - t)$.

a) Show that the transition density of $\check{X}(t)$ can be expressed as

$$\check{q}(x_t, t; x_s, s) = q(x_s, T - s; x_t, T - t) \frac{f(x_t, T - t)}{f(x_s, T - s)}. \tag{10.107}$$

b) Then consider the second-order differential operator

$$\check{L}_t u(x) = \check{a}(x, t) \frac{du}{dx} + \frac{\check{d}(x, t)}{2} \frac{d^2 u}{dx^2},$$

where the reverse-time drift and diffusion coefficients are given by

$$\check{a}(x, t) = -a(x, T - t) + \frac{1}{f(x, T - t)} \frac{\partial}{\partial x}(d(x, T - t)f(x, T - t))$$

$$\check{d}(x, t) = d(x, T - t). \tag{10.108}$$

By using the backward Kolmogorov equation satisfied by $q(x_s, T - s; x_t, T - t)$ and the Fokker–Planck equation satisfied by $f(x_t, T - t)$, verify that the reverse-time transition density $\check{q}(x_t, t; x_s, s)$ obeys the Fokker–Planck equation

$$\frac{\partial \check{q}}{\partial t} - \check{L}_t^* \check{q}(x, t; x_s, s) = 0$$

with initial condition

$$\check{q}(x_t, s; x_s, s) = \delta(x_t - x_s),$$

which indicates that the time-reversed process $\check{X}(t)$ is a diffusion with parameters $\check{a}(x, t)$ and $\check{d}(x, t)$.

c) Consider now the case where the diffusion $X(t)$ is homogeneous, so that $a(x)$ and $d(x)$ do not depend on time, and the PDF of $X(t)$ is stationary ($f(x)$ does not depend on t). Prove that $\check{L} = L$, so that in this case the Markov diffusion is reversible.

A complete analysis of the time reversal of diffusions is given in [17–20].

10.10 Hermite polynomials are defined by

$$H_n(x, t) = \frac{(-t)^n}{n!} \exp\left(\frac{x^2}{2t}\right) \frac{d^n}{dx^n} \left(\exp\left(-\frac{x^2}{2t}\right)\right).$$

They satisfy the three terms recursion

$$H_{n+1} = \frac{x}{n+1} H_n - \frac{t}{n+1} H_{n-1}$$

with initial conditions $H_0 = 1$ and $H_1 = x$, so that

$$H_2(x, t) = \frac{x^2}{2} - \frac{t}{2} \quad, \quad H_3(x, t) = \frac{x^3}{6} - \frac{tx}{2}.$$

They have also the properties

$$\frac{\partial H_n}{\partial x} = H_{n-1}(x, t) \quad \text{and} \quad \frac{\partial H_n}{\partial t} = -\frac{1}{2} H_{n-2}(x, t).$$

Prove that $H_n(W(t), t)$ satisfies

$$H_{n+1}(W(t), t) = \int_0^t H_n(W(t), t) dW(t),$$

so that Hermite polynomials play for the successive integration of the Wiener process the same role as $t^n/n!$ for ordinary integration.

10.11 In 1838, the Belgian mathematician Pierre Verhulst proposed the differential equation

$$\frac{dX}{dt} = \lambda X(t)(1 - X(t)),$$

which is now known as the logistic equation, to describe population dynamics. This equation has the feature that for small $X(t)$, the population grows exponentially with rate λ, but the rate decreases progressively as $X(t)$ approaches 1 (the normalized carrying capacity of the environment) and is negative if $X(t)$ exceeds 1. Suppose that parameter λ is actually random and that the equation is actually

$$dX = \lambda X(t)(1 - X(t)) + \sigma X(t) dW(t).$$

Solve this equation with initial condition $X(0)$ by performing the transformation $Y(t) = 1/X(t)$. To solve the resulting stochastic differential equation for $Y(t)$, you may want to use the approach of Example 10.8.

References

1. S. Kullback, *Information Theory and Statistics*. New York: J. Wiley & Sons, 1959. Reprinted by Dover Publ., Mineola, NY, 1968.
2. P. A. Markowich and C. Villani, "On the trend to equilibrium of the Fokker-Planck equation: an interplay between physics and functional analysis," *Mathematica Contoporeana*, vol. 19, pp. 1–29, 2000.
3. L. Arnold, *Stochastic Differential Equations*. New York, NY: J. Wiley, !974.
4. P. E. Kloeden and E. Platen, *Numerical Solution of Stochastic Differential Equations, 2nd edition*. Berlin, Germany: Springer Verlag, 1992.
5. E. A. Coddington and N. Levinson, *Theory of Ordinary Differential Equations*. New York: McGraw-Hill, 1955.
6. B. Oksendal, *Stochastic Differential Equations: An Introduction with Applications, 6th edition*. Berlin, Germany: Springer, 2003.
7. A. H. Jazwinski, *Stochastic Processes and Filtering Theory*. New York, NY: Academic Press, 1970. Reprinted in 2007 by Dover Publications, Mineola NY.
8. L. C. Evans, *An Introduction to Stochastic Differential Equations*. Providence, RI: American Math. Society, 2013.
9. C. W. Gardiner, *Handbook of Stochastic Methods, 3rd edition*. Berlin, Germany: Springer, 2004.
10. H. Risken and T. Frank, *The Fokker-Planck Equation: Method of Solution and Applications, 2nd edition*. Berlin, Germany: Springer, 1996.
11. G. A. Pavliotis, *Stochastic Processes and Applications: Diffusion Processes, the Fokker-Planck and Langevin Equations*. New York: Springer Verlag, 2014.
12. S. Karlin and H. M. Taylor, *A Second Course in Stochastic Processes*. New York, NY: Academic Press, 1981.
13. F. Black and M. Scholes, "The pricing of options and corporate liabilities," *The Journal of Political Economy*, vol. 81, pp. 637–654, 1973.
14. S. Shreve, *Stochastic Calculus for Finance II: Continuous-Time Models*. New York: Springer Verlag, 2010.
15. A. J. Viterbi, "Phase-locked loop dynamics in the presence of noise by Fokker-Planck techniques," *Proceedings of the IEEE*, vol. 51, pp. 1737–1753, Dec. 1963.
16. W. C. Lindsey, *Synchronization Systems in Communication and Control*. Englewood-Cliffs, NJ: Prentice-Hall, 1972.
17. B. D. O. Anderson, "Reverse-time diffusion equation models," *Stochastic Processes and their Applications*, vol. 12, pp. 313–326, 1982.
18. D. A. Castanon, "Reverse-time diffusion processes," *IEEE Trans. Informat. Theory*, vol. 28, pp. 1953–1956, Nov. 1982.
19. R. J. Elliott and B. D. O. Anderson, "Reverse time diffusions," *Stochastic Processes and Applications*, vol. 19, pp. 327–339, 1985.
20. U. G. Haussmann and E. Pardoux, "Time reversal of diffusions," *Annals of Probability*, vol. 14, pp. 1186–1205, 1986.

Part IV
Applications

Wiener Filtering

11

11.1 Introduction

An important application of linear time-invariant filtering of WSS random processes is for the design of optimum estimation filters. In this problem, given some observations $Y(t)$ related to a process $X(t)$ of interest, one seeks to design a filter which is optimal in the sense that if $Y(t)$ is the filter input, the estimate $\hat{X}(t)$ produced at the filter output minimizes the mean-square estimation error with the desired process $X(t)$. The level of difficulty of this problem depends on whether the estimation filter is required to be causal or not. The first complete solution of this problem for both the noncausal and causal cases was proposed by Wiener [1], and optimal estimation filters are therefore often referred to as Wiener filters.

This chapter starts in Sect. 11.2 with a discussion of the linear least-squares estimation problem for a single random variable based on a set, possibly infinite, of observations. It is assumed that both the variable to be estimated and observations have finite second-order moments. Random variables with finite second-order moments can be viewed as elements of a Hilbert space, and in this context, it is shown that the linear least-squares estimation problem can be formulated geometrically as a projection problem onto the linear subspace spanned by the available observations. It is also shown that the optimum linear estimate is characterized uniquely by an orthogonality property between the estimation error and the observations. In spite of its innocuous appearance, this orthogonality property plays a key role in the derivation of all results of this chapter.

While comprehensive studies of optimal filtering [2, 3] start typically with the derivation of finite impulse response (FIR) Wiener filters, which are useful in situations where process statistics are not known exactly, so that optimum filters need to be designed adaptively, since the purpose of this chapter is illustrative rather than comprehensive, we focus exclusively on the derivation of infinite impulse response (IIR) Wiener filters. The noncausal Wiener filter and its associated mean-square error (MSE) are derived in Sect. 11.3. Unlike the noncausal case, the causal Wiener filter involves the solution of a difficult class of equations named after Wiener and Hopf, who were the first to propose a solution technique. The approach we employ uses the spectral factorization of the power spectrum of the observations process (when this spectrum is rational and free of zeros on the unit circle) into minimum phase and maximum phase spectral factors. This spectral factorization is described in Sect. 11.4 and is used to construct a white random process, called the innovations process, which can be constructed causally from the observations, and in return can be used to recover the observations causally. The innovation process [4] represents the new information contained in each successive observation. It

B. C. Levy, *Random Processes with Applications to Circuits and Communications*,
https://doi.org/10.1007/978-3-030-22297-0_11

is used in Sect. 11.5 to obtain a simple solution to the Wiener–Hopf equation. Then in Sect. 11.6 several special cases are considered: the case when the signal $X(t)$ is observed directly in white noise, the k-step ahead prediction problem, and the fixed-lag smoothing problem. This last problem corresponds to situations where a fixed lag k is introduced between the latest observation $Y(t)$ and the variable $X(t - k)$ to be estimated. This problem combines the advantages of causal filtering, where estimates are computed online, as observations are received, and smoothing, i.e., noncausal filtering, which allows looking ahead. In particular, when the lag k exceeds the correlation length between the observations process $Y(t)$ and process $X(t)$, the fixed lag smoother achieves the same performance as the noncausal filter. Since modern applications of Wiener filters are typically implemented in discrete-time, our discussion in this chapter centers on the DT case, but for completeness, CT expressions are presented briefly in Sect. 11.7.

11.2   Linear Least-Squares Estimation

The set H of random variables X defined on probability space $(\Omega, \mathcal{A}, P)$ with finite second moment $E[X^2]$ can be viewed as an infinite dimensional vector space with respect to the scalar multiplication cX with $c \in \mathbb{R}$ and random variable addition $X + Y$. Note indeed that $E[(cX)^2] = c^2 E[X^2]$, so cX has a finite second moment, and since

$$(x + y)^2 \leq 2(x^2 + y^2)$$

for any two real numbers x and y, we have

$$E[(X + Y)^2] \leq 2(E[X^2] + E[Y^2]) < \infty$$

if X and Y have finite second moments. Observe also that if $X \in H$, it has necessarily a finite mean, since by applying the Cauchy–Schwartz inequality for X and $Y = 1$, we have

$$|E[X]| = |E[X.1]| \leq (E[X^2]E[1^2])^{1/2} = (E[X^2])^{1/2}.$$

In the space H, two random variables X_1 and X_2 are considered equal if they are identical except on a set of measure zero, i.e., $P[X_1 = X_2] = 1$. The space H can be endowed with an inner product defined by

$$< X, Y >= E[XY]. \tag{11.1}$$

The inner product is clearly symmetric, i.e.

$$< X, Y >=< Y, X >$$

and for an arbitrary linear combination $aX_1 + bX_2$ of random variables X_1 and X_2 of H with $a, b \in \mathbb{R}$, we have

$$< (aX_1 + bX_2), Y >= a < X_1, Y > +b < X_2, Y > .$$

The corresponding vector norm is then

$$||X||^2 =< X, X >= E[X^2] \geq 0,$$

and since $X = 0$ almost surely when $E[X^2] = 0$, we conclude that $||X||^2 = 0$ implies $X = 0$. Thus H is an inner product space.

If X_n is a sequence of random variables of H and $X \in H$, the convergence of X_n to X in the mean-square can be expressed in terms of the H norm as

$$\lim_{n \to \infty} ||X_n - X|| = 0.$$

Finally, if X_n is a Cauchy sequence of H, i.e.,

$$\lim_{n, m \to \infty} ||X_n - X_m|| = 0,$$

its limit is in H [5, Sec. 6.10], so the inner product space H is complete, which makes it a Hilbert space [6].

Consider a set of random variables $\{Y_j, \ j \in \mathcal{J}\}$ where $\mathcal{J}$ is not required to be finite. The linear space L spanned by these random variables contains all the random variables which can be written as

$$Z = \sum_{j \in \mathcal{J}} a_j Y_j. \tag{11.2}$$

When the set $\mathcal{J}$ is infinite, the infinite sum on the right-hand side of (11.2) is defined as the limit as n tends to infinity of the partial sum formed by the first n terms of (11.2). Thus the linear space L is itself closed. Note also that by selecting $Y_0 = 1$, the random variable Z in (11.2) takes the form of an affine linear combination

$$Z = a_0 + \sum_{j \in \mathcal{J} - \{0\}} a_j Y_j, \tag{11.3}$$

so that affine linear spaces are contained in the family of subspaces L.

11.2.1 Orthogonality Property of Linear Least-Squares Estimates

Given a random variable X in H and a linear subspace L spanned by observations $\{Y_j, \ j \in \mathcal{J}\}$, the linear least-squares estimate (LLSE) $\hat{X}$ of X is defined as the solution of the minimization problem

$$\hat{X} \overset{\triangle}{=} \arg \min_{\check{X} \in L} ||X - \check{X}||^2$$

$$= \arg \min_{\check{X} \in L} E[(X - \check{X})^2]. \tag{11.4}$$

The estimate $\hat{X}$ is *unique* and can be characterized by the orthogonality property

$$E[(X - \hat{X})Z] = < X - \hat{X}, Z >= 0 \tag{11.5}$$

for all random variables $Z \in L$. In other words, $\hat{X}$ solves the minimization problem (11.4) if and only if it satisfies (11.5).

Proof To prove that (11.4) implies (11.5), let $\hat{X}$ denote a solution of the minimization problem (11.4), and let Z be an arbitrary vector of L. Then, consider the function

$$J(s) \stackrel{\triangle}{=} ||X - \hat{X} - sZ||^2$$

of $s \in \mathbb{R}$. We have

$$J(s) = J(0) - 2s < X - \hat{X}, Z > + s^2 ||Z||^2, \qquad (11.6)$$

and since $\hat{X} + sZ$ belongs to L, $J(0) = ||X - \hat{X}||^2$ is the minimum of $J(s)$. Since $J(s)$ is minimized at $s = 0$, its derivative

$$\frac{dJ}{ds}(0) = -2 < X - \hat{X}, Z > \qquad (11.7)$$

must be zero, so that (11.5) is satisfied. Conversely, assume that $\hat{X}$ is a random variable of L satisfying the orthogonality property (11.5). Then, if S is an arbitrary random variable of L, we have

$$||X - S||^2 = ||X - \hat{X} + \hat{X} - S||^2$$
$$= ||X - \hat{X}||^2 + ||\hat{X} - S||^2 \geq ||X - \hat{X}||^2, \qquad (11.8)$$

where to go from the first to the second line, we have used the fact that

$$< X - \hat{X}, \hat{X} - S > = 0,$$

since $\hat{X} - S$ belongs to L. The inequality (11.8) shows that the orthogonality property (11.5) implies the minimization property (11.4).

Finally, suppose there exists two solutions $\hat{X}_1$, $\hat{X}_2 \in L$ of the linear least-squares minimization problem (11.4). Since $\hat{X}_2 - \hat{X}_1$ belongs to L, we have

$$||X - \hat{X}_2||^2 = ||X - \hat{X}_1||^2 + ||\hat{X}_1 - \hat{X}_2||^2. \qquad (11.9)$$

But since $||X - \hat{X}_2||^2 = ||X - \hat{X}_1||^2$, this implies

$$||\hat{X}_2 - \hat{X}_1||^2 = 0,$$

and thus $\hat{X}_2 = \hat{X}_1$, so the LLSE is unique.   □

In conclusion, we have established that the LLSE $\hat{X}$ of X given a set of observations $\{Y_j, , \in \mathcal{J}\}$ coincides with the *orthogonal projection* $P_L X$ of X onto the linear space L spanned by the given observations, as depicted in Fig. 11.1. Furthermore, for any $Z \in L$, Pythagoras's theorem

$$||X - Z||^2 = ||X - \hat{X}||^2 + ||\hat{X} - Z||^2 \qquad (11.10)$$

holds.

Fig. 11.1 Interpretation of the LLS estimate $\hat{X}$ as the orthogonal projection of X on the subspace L spanned by the observations

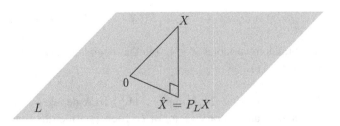

11.2.2 Finite Dimensional Case

When the number n of observations $\{Y_j, 1 \leq j \leq n\}$ is finite, it is possible to express the LLSE estimate explicitly in terms of the given observations and the first- and second-order moments of X and the Y_js. Let

$$\mathbf{Y} = \begin{bmatrix} Y_1 \ldots Y_j \ldots Y_n \end{bmatrix}^T$$

denote the n-dimensional vector of observations, to which we add the deterministic observation $Y_0 = 1$ to keep track of the first moments of X and $\mathbf{Y}$, thus yielding the expanded observation vector

$$\mathbf{Y}_e = \begin{bmatrix} 1 \\ \mathbf{Y} \end{bmatrix}.$$

Then if L is the linear space spanned by the entries of $\mathbf{Y}_e$, the LLSE $\hat{X}$ of random variable X given the observations $\{Y_j, 0 \leq j \leq n\}$ can be expressed as

$$\hat{X} = \mathbf{a}_e^T \mathbf{Y}_e = a_0 + \mathbf{a}^T \mathbf{Y},$$

where

$$\mathbf{a}_e = \begin{bmatrix} a_0 \, a_1 \ldots a_j \ldots a_n \end{bmatrix}^T = \begin{bmatrix} a_0 \, \mathbf{a} \end{bmatrix}^T$$

is the vector of observation weights used to form the LLSE. If $E = X - \hat{X}$ is the estimation error, the orthogonality property $< E, Y_j >= 0$ for $0 \leq j \leq n$ can be expressed in vector form as

$$E[(X - \hat{X})\mathbf{Y}_e^T] = \mathbf{0}_{n+1}^T, \tag{11.11}$$

where $\mathbf{0}_{n+1}$ denotes the $n + 1$-dimensional zero vector, which yields

$$\mathbf{R}_{XY_e} = \mathbf{a}_e^T \mathbf{R}_{Y_e}, \tag{11.12}$$

where

$$\mathbf{R}_{XY_e} = E[X\mathbf{Y}_e^T] = \begin{bmatrix} m_X \, \mathbf{R}_{XY} \end{bmatrix} \tag{11.13}$$

with $\mathbf{R}_{XY} = E[X\mathbf{Y}^T]$, and

$$\mathbf{R}_{Y_e} = E[\mathbf{Y}_e \mathbf{Y}_e^T] = \begin{bmatrix} 1 & \mathbf{m}_Y^T \\ \mathbf{m}_Y & \mathbf{R}_Y \end{bmatrix} \tag{11.14}$$

with $\mathbf{R}_Y = E[\mathbf{Y}\mathbf{Y}^T]$. If the matrix $\mathbf{R}_{Y_e}$ is invertible, the weight vector satisfying equation (11.12) is given by

$$\mathbf{a}_e^T = \mathbf{R}_{XY_e} \mathbf{R}_{Y_e}^{-1}, \tag{11.15}$$

so the LLSE can be expressed as

$$\hat{X} = \mathbf{R}_{XY_e} \mathbf{R}_{Y_e}^{-1} \mathbf{Y}_e, \tag{11.16}$$

which can be viewed as a parametric form of the projection operator P_L. Let $\mathbf{K}_Y = \mathbf{R}_Y - \mathbf{m}_Y \mathbf{m}_Y^T$ denote the covariance matrix of observation vector $\mathbf{Y}$. The expanded correlation matrix $\mathbf{R}_{Y_e}$ admits the block lower-diagonal-upper (LDU) factorization

$$\mathbf{R}_{Y_e} = \begin{bmatrix} 1 & 0 \\ \mathbf{m}_Y & \mathbf{I}_n \end{bmatrix} \begin{bmatrix} 1 & 0 \\ 0 & \mathbf{K}_Y \end{bmatrix} \begin{bmatrix} 1 & \mathbf{m}_Y^T \\ 0 & \mathbf{I}_n \end{bmatrix}, \tag{11.17}$$

where $\mathbf{I}_n$ denotes the identity matrix of size N. The factorization (11.17) indicates that $\mathbf{R}_{Y_e}$ is invertible if and only if the covariance matrix $\mathbf{K}_Y$ of observations vector $\mathbf{Y}$ is invertible, i.e., if and only if the centered observations $Y_j - m_{Y_j}$ with $1 \leq j \leq n$ are linearly independent. Then by substituting the LDU factorization (11.17) inside expression (11.16) for the LLSE and performing simple algebraic manipulations, we obtain the familiar expression

$$\hat{X} = m_X + \mathbf{K}_{XY}\mathbf{K}_Y^{-1}(\mathbf{Y} - \mathbf{m}_Y) \tag{11.18}$$

for the LLSE, where

$$\mathbf{K}_{XY} = \mathbf{R}_{XY} - m_X \mathbf{m}_Y.$$

To compute the mean-square error (MSE) of the linear-least squares estimate $\hat{X}$, we apply Pythagoras's identity (11.10) with $Z = 0$. This gives

$$E[X^2] = E[(X - \hat{X})^2] + E[\hat{X}^2],$$

so that

$$MSE \stackrel{\triangle}{=} E[(X - \hat{X})^2] = R_X - E[\hat{X}^2], \tag{11.19}$$

where $R_X = E[X^2]$ and

$$E[\hat{X}^2] = \mathbf{R}_{XY_e}\mathbf{R}_{Y_e}^{-1} E[\mathbf{Y}_e \mathbf{Y}_e^T] \mathbf{R}_{Y_e}^{-1} \mathbf{R}_{Y_e X}$$

$$= \mathbf{R}_{XY_e}\mathbf{R}_{Y_e}^{-1} \mathbf{R}_{Y_e X}. \tag{11.20}$$

This yields

$$MSE = R_X - \mathbf{R}_{XY_e}\mathbf{R}_{Y_e}^{-1}\mathbf{R}_{Y_e X}. \tag{11.21}$$

By observing that

$$\mathbf{R}_{Y_e X}^T = \mathbf{R}_{XY_e} = \begin{bmatrix} m_X & \mathbf{R}_{XY} \end{bmatrix},$$

and using the LDU factorization (11.17) to compute the inverse of $\mathbf{R}_Y$, we find

$$\mathbf{R}_{XY_e}\mathbf{R}_Y^{-1}\mathbf{R}_{Y_e X} = m_X^2 + \mathbf{K}_{XY}\mathbf{K}_Y^{-1}\mathbf{K}_{YX},$$

which after substitution in (11.21) yields the usual expression

$$MSE = K_X - \mathbf{K}_{XY}\mathbf{K}_Y^{-1}\mathbf{K}_{YX}, \tag{11.22}$$

where $K_X = R_X - m_X^2$. The expressions (11.18) and (11.22) for the LLSE and its MSE have the feature that they depend exclusively on the first- and second-order statistics of X and $\mathbf{Y}$. Since higher-order statistics are often difficult to estimate, this feature is one of the main advantages of linear estimation.

At this point, it is worth observing that the geometric interpretation of linear least-squares estimation as an orthonormal projection on the space spanned by the given observations applies only to scalar random variables with finite second moments. The LLSE of a random vector

$$\mathbf{X} = \begin{bmatrix} X_1 \ldots X_j \ldots X_m \end{bmatrix}^T$$

with mean

$$\mathbf{m}_X = \begin{bmatrix} m_{X_1} \ldots m_{X_j} \ldots m_{X_m} \end{bmatrix}^T$$

and finite autocorrelation matrix $\mathbf{R}_X = E[\mathbf{XX}^T]$ is obtained by evaluating separately the LLSE $\hat{X}_j$ of each entry of $\mathbf{X}$ and then forming the vector

$$\hat{\mathbf{X}} = \begin{bmatrix} \hat{X}_1 \\ \vdots \\ \hat{X}_j \\ \vdots \\ \hat{X}_m \end{bmatrix} = \mathbf{m}_X + \mathbf{K}_{XY}\mathbf{K}_Y^{-1}(\mathbf{Y} - \mathbf{m}_Y), \tag{11.23}$$

which is the LLSE of $\mathbf{X}$. For this vector if $\mathbf{E} = \mathbf{X} - \hat{\mathbf{X}}$ denotes the error vector, the correlation matrix

$$\mathbf{R}_E = E[\mathbf{EE}^T]$$

can be obtained by noting that the orthogonality property of each component $E_j = X_j - \hat{X}_j$ of $\mathbf{E}$ implies

$$\begin{aligned} \mathbf{R}_E &= E[\mathbf{E}(\mathbf{X} - \hat{\mathbf{X}})^T] = E[\mathbf{EX}^T] \\ &= E[\mathbf{XX}^T] - E[\hat{\mathbf{X}}\mathbf{X}^T] \\ &= \mathbf{K}_X - \mathbf{K}_{XY}\mathbf{K}_Y^{-1}\mathbf{K}_{YX}. \end{aligned} \tag{11.24}$$

Note that since

$$\mathbf{m}_E = E[\mathbf{E}] = E[\mathbf{X}] - E[\hat{\mathbf{X}}] = \mathbf{0},$$

the correlation matrix $\mathbf{R}_E$ of the error vector is also its covariance matrix $\mathbf{K}_E$.

Remarks

1. The LLSE $\hat{X}$ of a random variable X minimizes the mean-square error $E[(X - \hat{X})^2]$ only within the class of linear estimators. If the estimator is not required to be linear, the estimator minimizing the mean-square error is the conditional mean

$$\hat{X}_{MSE}(\mathbf{Y}) = E[X|\mathbf{Y}],$$

which is characterized by the much stronger orthogonality property

$$E[(X - \hat{X}_{MSE})g(\mathbf{Y})] = 0$$

for all nonlinear observation functions $g(\mathbf{Y})$ [7, Sec 4.2]. In this respect, it is interesting to note that when $\mathbf{X}$ and $\mathbf{Y}$ are jointly Gaussian, the LLSE $\hat{\mathbf{X}}$ given by (11.23) coincides with the conditional mean $E[\mathbf{X}|\mathbf{Y}]$ evaluated in (2.106), so the LLSE and MSE estimates coincide in the Gaussian case. However, this is not the case in general, since the conditional mean $E[\mathbf{X}|\mathbf{Y}]$ is usually not a linear function of $\mathbf{Y}$.

2. In the geometric formulation of linear least-squares estimation presented above, we added the extra deterministic observation $Y_0 = 1$ as a device to keep track of the means of X and observation vector $\mathbf{Y}$. This scheme allowed us to convert affine spaces of observations into linear spaces. This approach becomes unnecessary if all random variables X and Y_j with $1 \leq j \leq n$ have zero mean, since in this case

$$< Y_0, X >=< Y_0, Y_j >= 0$$

for $1 \leq j \leq n$. This means that all random variables of interest are located in the subspace H_c of H formed by centered (zero-mean) random variables, which is the orthogonal complement of the one-dimensional subspace spanned by Y_0 in H. In this case, all orthogonal projections can be performed inside H_c, and the orthogonality of the LLSE estimates reduces to

$$< X - \hat{X}, Y_j >= 0$$

for $1 \leq j \leq n$. Since it is easy to center random variables by subtracting their means, it will be assumed that all random processes of interest have zero mean in the Wiener filtering analysis of the following sections.

11.2.3 Linear Observation Model

In the previous subsection, the exact relationship between random vector $\mathbf{X}$ and observations vector $\mathbf{Y}$ was not described. Assume that $\mathbf{Y}$ is related to $\mathbf{X}$ through a linear model of the form

$$\mathbf{Y} = \mathbf{HX} + \mathbf{V}, \tag{11.25}$$

where $\mathbf{H}$ is a known $n \times m$ matrix and the observation noise vector $\mathbf{V} \in \mathbb{R}^n$ has zero mean and autocorrelation/autocovariance matrix $\mathbf{R}_V$, and is uncorrelated with $\mathbf{X}$.

In this case the mean of $\mathbf{Y}$ satisfies

$$\mathbf{m}_Y = \mathbf{Hm}_X$$

and after subtracting it from (11.25), we find

$$\mathbf{Y} - \mathbf{m}_Y = \mathbf{H}(\mathbf{X} - \mathbf{m}_X) + \mathbf{V}.$$

Since $\mathbf{V}$ and $\mathbf{X}$ are uncorrelated, this implies

$$\mathbf{K}_{YX} = \mathbf{H}\mathbf{K}_X$$
$$\mathbf{K}_Y = \mathbf{H}\mathbf{K}_X\mathbf{H}^T + \mathbf{R}_V.$$

Substituting these expressions inside (11.23) and (11.24) yields

$$\hat{\mathbf{X}} = \mathbf{m}_X + \mathbf{K}_X\mathbf{H}^T(\mathbf{H}\mathbf{K}_X\mathbf{H}^T + \mathbf{R}_V)^{-1}(\mathbf{Y} - \mathbf{H}\mathbf{m}_X) \tag{11.26}$$

$$\mathbf{K}_E = \mathbf{K}_X - \mathbf{K}_X\mathbf{H}^T(\mathbf{H}\mathbf{K}_X\mathbf{H}^T + \mathbf{R}_V)^{-1}\mathbf{H}\mathbf{K}_X. \tag{11.27}$$

In the case when the covariance $\mathbf{K}_X$ of random vector $\mathbf{X}$ and noise covariance matrix $\mathbf{R}_V$ are invertible, these expressions can be written more compactly by using the ABCD matrix inversion lemma (also known as the Sherman–Morrison–Woodbury matrix inversion identity [8, p. 48])

$$(\mathbf{A} + \mathbf{B}\mathbf{C}\mathbf{D})^{-1} = \mathbf{A}^{-1} - \mathbf{A}^{-1}\mathbf{B}(\mathbf{C}^{-1} + \mathbf{D}\mathbf{A}^{-1}\mathbf{B})^{-1}\mathbf{D}\mathbf{A}^{-1} \tag{11.28}$$

with $\mathbf{A} = \mathbf{K}_X^{-1}$, $\mathbf{B} = \mathbf{H}^T$, $\mathbf{C} = \mathbf{R}_V^{-1}$ and $\mathbf{D} = \mathbf{H}$. Then the error covariance expression (11.27) can be rewritten as

$$\mathbf{K}_E = (\mathbf{K}_X^{-1} + \mathbf{H}^T\mathbf{R}_V^{-1}\mathbf{H})^{-1} \tag{11.29}$$

and after simple manipulations [7, pp. 130–131], the LLS estimate can be rewritten as

$$\hat{\mathbf{X}} = \mathbf{K}_E\left(\mathbf{K}_X^{-1}\mathbf{m}_X + \mathbf{H}^T\mathbf{R}_V^{-1}\mathbf{Y}\right) \tag{11.30}$$

Together expressions (11.29) and (11.30) are sometimes referred to as the information form of the linear estimator.

Example 11.1 Consider the DT observations

$$Y(t) = X_c\cos(\omega_c t) + X_s\sin(\omega_c t) + V(t)$$

for $1 \le t \le T$ where

$$\mathbf{X} = \begin{bmatrix} X_c \\ X_s \end{bmatrix}$$

is a zero-mean random vector with covariance matrix $\mathbf{K}_X$ and $V(t)$ is a white noise uncorrelated with $\mathbf{X}$ with variance σ_V^2, so that $\mathbf{R}_V = \sigma_V^2\mathbf{I}_T$ where $\mathbf{I}_T$ denotes the identity matrix of dimension T. We seek to compute the LLSE $\hat{\mathbf{X}}$ of vector $\mathbf{X}$ and its error covariance. This can be accomplished by writing the observations in the vector form (11.25) with

$$\mathbf{Y}^T = \begin{bmatrix} Y(1) \ldots Y(t) \ldots Y(T) \end{bmatrix}$$
$$\mathbf{V}^T = \begin{bmatrix} V(1) \ldots V(t) \ldots V(T) \end{bmatrix}$$

and

$$\mathbf{H}^T = \begin{bmatrix} \cos(\omega_c) \ldots \cos(\omega_c t) \ldots \cos(\omega_c T) \\ \sin(\omega_c) \ldots \sin(\omega_c t) \ldots \sin(\omega_c T) \end{bmatrix}.$$

The two columns of the matrix $\mathbf{H}$ are approximately orthogonal to each other in general, and are exactly orthogonal if $\omega_c = k\pi/T$ for some integer k. So to simplify the analysis we assume that ω_c satisfies this constraint. In this case

$$\mathbf{H}^T \mathbf{R}_V^{-1} \mathbf{H} = \frac{T}{2\sigma_V^2} \mathbf{I}_2,$$

so the error covariance matrix is given by

$$\mathbf{K}_E = \left(\mathbf{K}_X^{-1} + \frac{T}{2\sigma_V^2} \mathbf{I}_2\right)^{-1}.$$

In this expression the term $\mathbf{K}_X^{-1}$ becomes negligible when T becomes large. Furthermore if the random amplitudes X_c and X_s are uncorrelated with the same power P_X, so that

$$\mathbf{K}_X = P_X \mathbf{I}_2,$$

then the error covariance is also diagonal, i.e.

$$\mathbf{K}_E = P_E \mathbf{I}_2,$$

with

$$P_E = \frac{\sigma_V^2}{T/2 + SNR^{-1}},$$

where $SNR = P_X/\sigma_V^2$ denotes the signal to noise ratio. In this case, the LLS estimates of X_c and X_s can be expressed as

$$\hat{X}_c = \frac{1}{T/2 + SNR^{-1}} \sum_{t=1}^{T} Y(t) \cos(\omega_c t)$$

$$\hat{X}_s = \frac{1}{T/2 + SNR^{-1}} \sum_{t=1}^{T} Y(t) \sin(\omega_c t).$$

Intuitively, we see that the amplitude estimates $\hat{X}_c$ and $\hat{X}_s$ are obtained by correlating the observation with the in-phase and quadrature components of the modulated signal and scaling with a weight depending on the length T of the observation record and SNR.

11.3 Noncausal Wiener Filter

Consider two jointly WSS zero-mean DT random processes $X(t)$ and $Y(t)$, where $Y(t)$ is the observed process and $X(t)$ is the process we seek to estimate. The matrix autocorrelation function of the vector process

$$\mathbf{Z}(t) = \begin{bmatrix} X(t) \\ Y(t) \end{bmatrix} \tag{11.31}$$

is given by

$$\mathbf{R}_Z(m) = E[\mathbf{Z}(t+m)\mathbf{Z}^T(t)]$$

$$= \begin{bmatrix} E[X(t+m)X(t)] & E[X(t+m)Y(t)] \\ E[Y(t+m)X(t)] & E[Y(t+m)Y(t)] \end{bmatrix}$$

$$= \begin{bmatrix} R_X(m) & R_{XY}(m) \\ R_{YX}(m) & R_Y(m) \end{bmatrix},$$ (11.32)

where the definition of $\mathbf{R}_Z(m)$ implies

$$\mathbf{R}_Z^T(-m) = E[\mathbf{Z}(t)\mathbf{Z}^T(t-m)] = \mathbf{R}_Z(m),$$ (11.33)

which is just a compact way of writing the even property

$$R_X(-m) = R_X(m) \ , \ R_Y(-m) = R_Y(m)$$

of autocorrelation functions $R_X(m)$ and $R_Y(m)$, together with the symmetry property

$$R_{YX}(m) = R_{XY}(-m)$$

of the cross correlation function of $Y(t)$ and $X(t)$. We assume that the matrix autocorrelation function $\mathbf{R}_Z(m)$ is summable, so it admits the 2×2 matrix PSD

$$\mathbf{S}_Z(e^{j\omega}) = \sum_{m=-\infty}^{\infty} \mathbf{R}_Z(m)e^{-jm\omega} = \begin{bmatrix} S_X(e^{j\omega}) & S_{XY}(e^{j\omega}) \\ S_{YX}(e^{j\omega}) & S_Y(e^{j\omega}) \end{bmatrix}.$$ (11.34)

Since the matrix autocorrelation function $\mathbf{R}_Z(m)$ is real, the symmetry condition (11.33) implies

$$\mathbf{S}_Z^H(e^{j\omega}) = \mathbf{S}_Z(e^{j\omega}),$$ (11.35)

where H denotes the Hermitian transpose (the complex conjugate transpose) of a complex matrix.

Since $\mathbf{R}_Z(m)$ is real, the Hermitian symmetry of the PSD just expresses the fact that the scalar PSDs $S_X(e^{j\omega})$ and $S_Y(e^{j\omega})$ are real and even, and the off-diagonal cross-spectral densities satisfy

$$S_{YX}(e^{j\omega}) = S_{XY}^*(e^{j\omega}) = S_{XY}(e^{-j\omega}).$$

In the following, it will also be useful to consider the z-transform

$$\mathbf{S}_Z(z) = \sum_{m=-\infty}^{\infty} \mathbf{R}_Z(m)z^{-m} = \begin{bmatrix} S_X(z) & S_{XY}(z) \\ S_{YX}(z) & S_Y(z) \end{bmatrix}$$ (11.36)

of the 2×2 matrix autocorrelation function $\mathbf{R}_Z(m)$. This transform will be referred to as the power spectrum of $\mathbf{Z}(t)$. Because $\mathbf{R}_Z(m)$ is summable, the region of convergence of $\mathbf{S}_Z(z)$ includes the unit circle, and by setting $z = e^{j\omega}$, the power spectrum reduces on the unit circle to the power spectral density matrix $\mathbf{S}_Z(e^{j\omega})$. Furthermore, the symmetry (11.33) implies

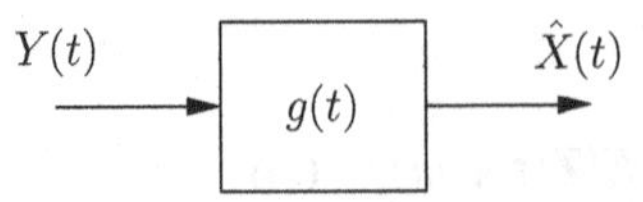

Fig. 11.2 Estimate $\hat{X}(t)$ obtained by passing observations $Y(t)$ through noncausal filter $g(t)$

$$\mathbf{S}_Z(z) = \mathbf{S}_Z^T(z^{-1}), \tag{11.37}$$

so that the components of $\mathbf{S}_Z(z)$ satisfy

$$S_X(z) = S_X(z^{-1}) \quad , \quad S_Y(z) = S_Y(z^{-1})$$

and

$$S_{YX}(z) = S_{XY}(z^{-1}).$$

The property (11.37) implies also that the region of convergence of $\mathbf{S}_Z(z)$ is of the form $c < |z| < c^{-1}$ with $c < 1$.

Noncausal Filter We seek to find the linear least-squares estimate $\hat{X}(t)$ of $X(t)$ given the entire observations record $\{Y(t - m), -\infty < m < \infty\}$. Since the estimate $\hat{X}(t)$ is allowed to depend on future observations, this estimate is noncausal and is called a *smoothed estimate*. Because the processes $X(t)$ and $Y(t)$ are jointly WSS, the estimation problem is invariant under a time shift, so the estimate $\hat{X}(t)$ can be obtained by passing the observations process $Y(t)$ through a noncausal LTI filter with impulse response $g(t)$ as shown in Fig. 11.2.

The resulting estimate $\hat{X}(t)$ can therefore be expressed as

$$\hat{X}(t) = g(t) * Y(t) = \sum_{m=-\infty}^{\infty} g(m)Y(t - m), \tag{11.38}$$

and the error $E(t)$ is given by

$$E(t) = X(t) - \hat{X}(t) = X(t) - g(t) * Y(t). \tag{11.39}$$

The filter $g(t)$ is selected to minimize the mean-square error

$$MSE = E[E^2(t)] = R_E(0), \tag{11.40}$$

where

$$R_E(n) = E[E(t + n)E(t)]$$

denotes the autocorrelation of the error process. To find the optimum estimation filter, we rely on the orthogonality property

$$E[E(t)Y(t - n)] = 0 \tag{11.41}$$

of linear least-squares estimates, where $n \in \mathbb{Z}$ since it is assumed that the entire observations record is available.

By substituting (11.39) inside (11.41), we obtain

$$R_{XY}(n) = E[\hat{X}(t)Y(t-n)] = \sum_{m=-\infty}^{\infty} g(m)E[Y(t-m)Y(t-n)]$$

$$= \sum_{m=-\infty}^{\infty} g(m)R_Y(n-m) = g(n) * R_Y(n). \tag{11.42}$$

By z-transforming (11.42), we find

$$S_{XY}(z) = G(z)S_Y(z),$$

so that the optimum noncausal filter is given by

$$G(z) = \frac{S_{XY}(z)}{S_Y(z)}. \tag{11.43}$$

By observing that the error process $E(t)$ can be expressed in terms of vector process $\mathbf{Z}(t)$ defined in (11.31) as

$$E(t) = \big[\delta(t) - g(t) \big] * \mathbf{Z}(t),$$

where $*$ denotes the convolution operation, we deduce that its power spectrum takes the form

$$S_E(z) = \big[1 - G(z) \big] \mathbf{S}_Z(z) \begin{bmatrix} 1 \\ -G(z^{-1}) \end{bmatrix}, \tag{11.44}$$

which after substituting (11.36) and taking into account the form (11.43) of the filter $G(z)$ gives

$$S_E(z) = S_X(z) - G(z)S_{YX}(z) = S_X(z) - \frac{S_{XY}(z)S_{XY}(z^{-1})}{S_Y(z)}. \tag{11.45}$$

The MSE $R_E(0)$ can then be evaluated by computing the inverse z-transform $R_E(n)$ of $S_E(z)$ and setting $n = 0$. This gives

$$MSE = \frac{1}{2\pi} \int_{-\pi}^{\pi} S_E(e^{j\omega})d\omega$$

$$= R_X(0) - (g * R_{YX})(0). \tag{11.46}$$

Noisy Deconvolution In the previous analysis, the relationship existing between WSS processes $X(t)$ and $Y(t)$ was not specified, but it was assumed that the auto- and cross-correlations $R_X(m)$, $R_Y(m)$, and $R_{YX}(m)$ were available. In practical situations, the signals $X(t)$ and $Y(t)$ are related through an observation model such as the one depicted in Fig. 11.3, where the observation $Y(t)$ can be interpreted as obtained by passing the signal of interest $X(t)$ through a convolution filter $h(t)$ (corresponding to the channel impulse response for a communications channel) and observing the resulting signal $S(t)$ in the presence of an additive observation noise $V(t)$. Given the observations $Y(t-m)$, $m \in \mathbb{Z}\}$, the problem of estimating $X(t)$ can be viewed as a deconvolution problem (undoing the filtering effects of $h(t)$) in the presence of noise $V(t)$.

Fig. 11.3 Deconvolution
of a LTI system with
additive noise

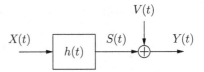

It is assumed that the signal $X(t)$ and noise $V(t)$ are zero-mean, WSS, and uncorrelated. Furthermore the autocorrelations

$$R_X(m) = E[X(t+m)X(t)] \quad \text{and} \quad R_V(m) = E[V(t+m)V(t)]$$

of $X(t)$ and $V(t)$ are known. In this case, to compute the optimum filter $G(z)$, we need to evaluate the cross-spectrum $S_{YX}(z)$ and spectrum $S_Y(z)$. We note first that the autocorrelation of $S(t) = h(t)*X(t)$ is

$$R_S(m) = h(m) * R_X(m) * h(-m)$$

and its cross-correlation with $X(t)$ is

$$R_{SX}(m) = h(m) * R_X(m).$$

Then since $Y(t) = S(t) + V(t)$ where $V(t)$ and $X(t)$ are uncorrelated, the cross-correlation of $Y(t)$ and $X(t)$ is

$$R_{YX}(m) = R_{SX}(m) = h(m) * R_X(m), \tag{11.47}$$

and the autocorrelation of $Y(t)$ is

$$R_Y(m) = R_S(m) + R_V(m) = h(m) * R_X(m) * h(-m) + R_V(m). \tag{11.48}$$

If $H(z)$ denotes the z-transform of filter $h(t)$ and $S_X(z)$ and $S_V(z)$ are the power spectra of signal $X(t)$ and noise $V(t)$, the z-transform of expressions (11.47) and (11.48) takes the form

$$S_{YX}(z) = H(z)S_X(z)$$
$$S_Y(z) = H(z)S_X(z)H(z^{-1}) + S_V(z), \tag{11.49}$$

and by substituting (11.49) inside expression (11.43) for the optimum filter, we find that it can be expressed as

$$G(z) = \frac{S_X(z)H(z^{-1})}{H(z)S_X(z)H(z^{-1}) + S_V(z)}$$
$$= \frac{1}{H(z) + \dfrac{S_V(z)}{S_X(z)H(z^{-1})}}. \tag{11.50}$$

The second line of this expression shows that in the absence of noise, i.e., for $S_V(z) = 0$, the filter $G(z) = H^{-1}(z)$ reduces to a straightforward deconvolution operation, but when noise is present, the second term in the denominator of $G(z)$ has the effect of de-emphasizing the filter $G(z)$, i.e., reducing its magnitude, whenever the noise power spectrum $S_V(z)$ is large.

By substituting (11.50) inside (11.45) and performing simplifications, we also find that the error power spectrum is given by

$$S_E(z) = \frac{S_X(z)S_V(z)}{S_Y(z)} \tag{11.51}$$

$$= (S_X^{-1}(z) + H(z)S_V^{-1}(z)H(z^{-1}))^{-1}, \tag{11.52}$$

where the second expressions are usually called the "information form" of the error power spectrum.

Example 11.2 Consider the model of Fig. 11.3, where $X(t)$ and $V(t)$ are zero-mean white noises with power spectra

$$S_X(z) = 0.72 \quad \text{and} \quad S_V(z) = 2$$

and $h(t) = (0.8)^t u(t)$, or equivalently

$$H(z) = \frac{1}{1 - 0.8z^{-1}}.$$

By substituting these expressions inside (11.48), we find

$$S_{XY}(z) = S_X(z)H(z^{-1}) = \frac{0.72}{1 - 0.8z}$$

$$S_Y(z) = \frac{0.72}{(1 - 0.8z^{-1})(1 - 0.8z)} + 2$$

$$= \frac{3.2(1 - z^{-1}/2)(1 - z/2)}{(1 - 0.8z^{-1})(1 - 0.8z)},$$

so that the optimum smoothing filter

$$G(z) = \frac{S_{XY}(z)}{S_Y(z)} = \frac{0.72(1 - 0.8z^{-1})}{3.2(1 - z^{-1}/2)(1 - z/2)}$$

$$= \frac{-0.45(1 - 0.8z^{-1})z^{-1}}{(1 - z^{-1}/2)(1 - 2z^{-1})},$$

where to go from the first to the second line, the numerator and denominator were multiplied by $-2z^{-1}$ in order to obtain a rational function of z^{-1}. Since numerator and denominator polynomials of $G(z)$ have both degree 2, $G(z)$ has for partial fraction expansion

$$G(z) = A + \frac{B}{1 - z^{-1}/2} + \frac{C}{1 - 2z^{-1}},$$

where

$$A = \lim_{z^{-1} \to \infty} G(z) = 0.45 \times 0.8 = 0.36$$

$$B = (1 - z^{-1}/2)G(z)\,|_{z^{-1}=2} = \frac{-0.45(1 - 0.8z^{-1})z^{-1}}{1 - 2z^{-1}}\,\bigg|_{z^{-1}=2}$$

$$= \frac{-0.45(1 - 1.6)2}{1 - 4} = -0.18$$

$$C = (1 - 2z^{-1})G(z)\,|_{z^{-1}=1/2} = \frac{-0.45(1 - 0.8z^{-1})z^{-1}}{1 - z^{-1}/2}\,|_{z^{-1}=1/2}$$

$$= \frac{-0.45(1 - 0.4)/2}{1 - 1/4} = -0.18.$$

Inverse z-transforming, we find

$$g(t) = 0.18(2)^t u(-t) - 0.18(1/2)^t u(t - 1),$$

which is plotted in Fig. 11.4.

To compute the MSE of the noncausal estimator, we use expression (11.51) to evaluate the error spectrum. This gives

$$S_E(z) = \frac{0.72(1 - 0.8z^{-1})(1 - 0.8z)}{1.6(1 - z^{-1}/2)(1 - z/2)}$$

$$= \frac{0.72(1 - 4/5z^{-1})(1 - 5/4z^{-1})}{(1 - z^{-1}/2)(1 - 2z^{-1})}$$

$$= \alpha + \frac{\beta}{1 - z^{-1}/2} + \frac{\gamma}{1 - 2z^{-1}}.$$

In this expression, since $1/(1 - 2z^{-1})$ is the z transform of a strictly anticausal sequence, and since we are only interested in $R_E(0)$, we only need to evaluate α and β. We have

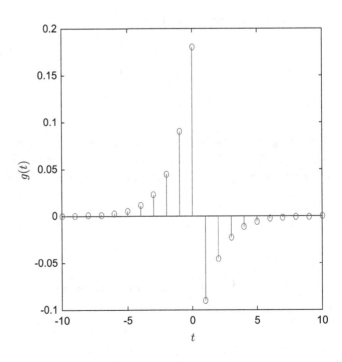

Fig. 11.4 Impulse response $g(t)$ of the optimal noncausal filter

$$\alpha = \lim_{z^{-1} \to \infty} S_E(z) = 0.72$$

$$\beta = (1 - z^{-1}/2) S_E(z) \mid_{z^{-1}=2} = \frac{0.72(1 - 4/5z^{-1})(1 - 5/4z^{-1})}{1 - 2z^{-1}} \mid_{z^{-1}=2}$$

$$= \frac{0.72 \times -3/5 \times -3/2}{-3} = -0.3 \times 0.72,$$

so that

$$MSE = \alpha + \beta = 0.7 \times 0.72.$$

The relatively small 30% reduction in the MSE (compared to $R_X(0) = 0.72$) obtained by noncausal estimation is due to the fact that the variance $E[V^2(t)] = 2$ of noise $V(t)$ is large, which reduces the benefits of estimation.

11.4 Spectral Factorization

Up to this point it has been assumed that the autocorrelation matrix $\mathbf{R}_Z(m)$ of the zero-mean DT WSS vector process

$$\mathbf{Z}(t) = \begin{bmatrix} X(t) \\ Y(t) \end{bmatrix}$$

is summable, which ensures that the region of convergence of the matrix power spectrum $\mathbf{S}_Z(z)$ is of the form $c < |z| < c^{-1}$ with $c < 1$, so that it includes the unit circle. This implies that $S_X(z)$ or $S_Y(z)$ have no pole on the unit circle. To discuss the causal filtering problem consisting of estimating $X(t)$ from the current and previous observations $\{Y(t - m), \ m \geq 0\}$, we need to make two additional assumptions:

  i) $\mathbf{S}_Z(z)$ is a rational matrix function of z, i.e., all its entries are ratios of polynomials in z. Note that rationality in z implies rationality in in z^{-1}, and vice versa.
  ii) $S_Y(z)$ has no zero on the unit circle, or equivalently the PSD $S_Y(e^{j\omega}) > 0$ for all ω.

The property $S_Y(z) = S_Y(z^{-1})$ implies that each real pole or zero z_0 of $S_Y(z)$ inside the unit circle is matched by a corresponding real pole of zero z_0^{-1} outside the unit circle. Likewise every pair (z_0, z_0^*) of complex conjugate poles or zeros of $S_Y(z)$ inside the unit circle is matched by a corresponding pair $((z_0^*)^{-1}, z_0^{-1})$ of poles and zeros outside the unit circle. This ensures [9, Sec. 5.4] that $S_Y(z)$ admits a factorization of the form

$$S_Y(z) = L(z) P_v L(z^{-1}), \tag{11.53}$$

where $L(z)$ is real and has all its poles and zeros inside the unit circle, and is thus a minimum phase digital filter. The conjugate filter $L(z^{-1})$ includes all the poles and zeros of $S_Y(z)$ outside the unit circle, and P_v is a positive constant. Since $Y(z)$ has no pole or zero on the unit circle and $L(z)$ is is required to be real, this factorization is unique, except for a multiplication by a real constant, which is fixed by requiring $L(\infty) = 1$.

Because all the poles of $L(z)$ are inside the unit circle, the region of convergence of $L(z)$ is of the form $|z| > c$ with $c < 1$, so it includes the unit circle. Likewise, since all the zeros of $L(z)$ are inside the unit circle, the region of convergence of $M(z) = L^{-1}(z)$ has also the form $|z| > c'$ with $c' < 1$. This implies that the impulse response $\ell(t)$ of filter $L(z)$ is causal and stable, i.e., $\ell(t) = 0$ for $t < 0$ and

$$\sum_{t=0}^{\infty} |\ell(t)| < \infty,$$

and similarly, the impulse response of the inverse filter $L^{-1}(z)$ is causal and stable. Note also that the condition $L(\infty) = 1$ is equivalent to assuming that the first term of the impulse response satisfies $\ell(0) = 1$. Since $L(\infty) = 1$ implies $M(\infty) = 1$, if $m(t)$ with $t \geq 0$ denotes the impulse response of $M(z)$, its first term admits also the normalization $m(0) = 1$.

Innovations Process The properties of $L(z)$ and $L^{-1}(z)$ indicate that the filter $L(z)$ is causal and causally invertible. Consider the process $v(t)$ obtained by passing the observations $Y(t)$ through filter $M(z) = L^{-1}(z)$, as shown in Fig. 11.5. Since $Y(t)$ is zero-mean WSS, $v(t)$ is zero mean WSS, and its power spectrum is given by

$$S_v(z) = L^{-1}(z)S_Y(z)L^{-1}(z^{-1}) = P_v, \tag{11.54}$$

so it is a white noise process with variance $P_v > 0$, i.e.,

$$R_v(m) = E[v(t+m)v(t)] = P_v\delta(m).$$

Since $M(z)$ is causal, and its impulse response $m(t)$ satisfies $m(0) = 1$, $v(t)$ can be expressed in terms of the previous observations as

$$v(t) = Y(t) + \sum_{k=1}^{\infty} m(k)Y(t-k). \tag{11.55}$$

Conversely $Y(t)$ can be recovered by passing $v(t)$ through the causal and stable filter $L(z)$, so

$$Y(t) = v(t) + \sum_{k=1}^{\infty} \ell(k)v(t-k).$$

This shows that the information contained in observations $\{Y(t-k),\ k \geq 0\}$ and innovations $\{v(t-k),\ k \geq 0\}$ up to time t is exactly the same, since these two sets of observations can be constructed causally from each other. Furthermore, the whiteness of process $v(t)$ and the equivalence of $\{v(t-k),\ k \geq 1\}$ and $\{Y(t-k),\ k \geq 1\}$ imply that $v(t)$ has the orthogonality property

$$E[v(t)Y(t-k)] = 0 \tag{11.56}$$

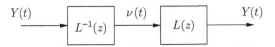

Fig. 11.5 Generation of innovations process $v(t)$ by passing observations $Y(t)$ through whitening filter $L^{-1}(z)$

for $k \geq 1$. This implies that $v(t)$ can be interpreted as the estimation error for estimating $Y(t)$ based on prior observations $\{Y(t-k),\ k \geq 1\}$, which in turn indicates that

$$\hat{Y}_p(t) = -\sum_{k=1}^{\infty} m(k)Y(t-k) \tag{11.57}$$

is the one step-ahead prediction estimate of $Y(t)$ given $\{Y(t-k),\ k \geq 1\}$. From this perspective, the innovations process can be interpreted as the new information contained in $Y(t)$ not already included in prior observations $\{Y(t-k),\ k \geq 1\}$. Also, the computation of $v(t)$ can be viewed as performing a Gram–Schmidt orthogonalization of observations $Y(t)$ for successive values of t. In this context, since

$$v(t) = Y(t) - \hat{Y}_p(t).$$

can be interpreted as the one-step ahead causal prediction error, its variance $P_v = MSE_p$ represents the MSE of the one-step ahead prediction problem.

MSE of One-Step Ahead Prediction Under the assumption that $S_Y(z)$ is rational and has no pole and zero on the unit circle, $S_Y(z)$ is free of poles and zeros over an annulus $\gamma < |z| < \gamma^{-1}$ such that $\gamma < 1$, so that $C(z) = \ln S_Y(z)$ is analytic over this region. The function $C(z)$ is called the cepstrum of $Y(t)$. The analyticity property implies that $C(z)$ admits a convergent Laurent series expansion

$$C(z) = \sum_{k=-\infty}^{\infty} c_k z^{-k} \tag{11.58}$$

over $\gamma < |z| < \gamma^{-1}$, which reduces to the Fourier series

$$C(e^{j\omega}) = \sum_{k=-\infty}^{\infty} c_k e^{-j\omega k}$$

on the unit circle, where

$$c_k = \frac{1}{2\pi} \int_{-\pi}^{\pi} \ln(S_Y(e^{j\omega})) e^{j\omega k} d\omega \tag{11.59}$$

denotes the k-th cepstral coefficient. Note that the property $C(z) = C(z^{-1})$ of the cepstrum implies $c_k = c_{-k}$. The cepstrum can be decomposed additively as

$$C(z) = Q(z) + c_0 + Q(z^{-1}), \tag{11.60}$$

where the region of convergence of the z transform

$$Q(z) \overset{\Delta}{=} \sum_{k=1}^{\infty} c_k z^{-k}$$

includes $\gamma < |z|$ since the sequence $\{c_k,\ k \geq 1\}$ is summable and causal. By exponentiating (11.60) we obtain

$$S_Y(z) = \exp(Q(z)) \exp(c_0) \exp(Q(z^{-1})) \tag{11.61}$$

where the function $\exp(Q(z))$ and its inverse are analytic over $\gamma < |z|$. Since $Q(\infty) = 0$, this implies that $L(z) = \exp(Q(z))$, so by comparing (11.61) with the spectral factorization (11.53), we find

$$P_v = \exp(c_0) = \exp\left(\frac{1}{2\pi} \int_{-\pi}^{\pi} \ln S_Y(e^{j\omega}) d\omega\right), \tag{11.62}$$

which provides a closed-form expression for MSE_p in terms of $S_Y(e^{j\omega})$. This formula is known as the Kolmogorov–Szegö identity.

11.5   Causal Wiener Filter

Consider now the problem of computing the linear least squares estimate

$$\hat{X}_f(t) = \sum_{m=0}^{\infty} k(m) Y(t - m) \tag{11.63}$$

of $X(t)$ given the current and prior observations $\{Y(t - m), m \geq 0\}$. The corresponding error is denoted by

$$E_f(t) = X(t) - \hat{X}_f(t)$$

and by using the orthogonality property

$$E[E_f(t) Y(t - n)] = 0 \tag{11.64}$$

of the error for $n \geq 0$, we obtain the Wiener–Hopf equation

$$R_{XY}(n) = k(n) * R_Y(n) \tag{11.65}$$

for $n \geq 0$. Even though this equation has a form similar to the noncausal estimation equation (11.42), the constraint $n \geq 0$ makes it much harder to solve, except if the available observations process happens to be white. Specifically, since the knowledge of current and past observations $\{Y(t-m), m \geq 0\}$ is strictly equivalent to the knowledge of current and past innovations $\{v(t-m), m \geq 0\}$, the filtered estimate $\hat{X}_f(t)$ can also be written in terms of the innovations as

$$\hat{X}_f(t) = \sum_{m=0}^{\infty} \bar{k}(m) v(t - m), \tag{11.66}$$

where because $v(t)$ is white, Eq. (11.65) for filter $\bar{k}(n)$ can be expressed as

$$R_{Xv}(n) = P_v \bar{k}(n) \tag{11.67}$$

for $n \geq 0$.

To solve this equation, we use the fact that a noncausal function can be decomposed additively into causal and anticausal parts. Specifically, let $f(t)$ be a summable noncausal function, so that its z transform

$$F(z) = \sum_{t=-\infty}^{\infty} f(t)z^{-t}$$

has for region of convergence $a < |z| < b$ where $a < 1 < b$, since $F(z)$ is defined on the unit circle because $f(t)$ is summable. Then the causal and strictly anticausal parts of $f(t)$ have for z transforms

$$\{F(z)\}_+ = \sum_{t=0}^{\infty} f(t)z^{-t}$$

$$\{F(z)\}_- = \sum_{t=-\infty}^{0} f(t)z^{-t},$$

and have respectively for regions of convergence $a < |z|$ and $|z| < b$, which both include the unit circle. Note that the operations consisting of extracting the causal and strictly anticausal parts of $f(t)$ are not symmetric in time since $f(0)$ is allocated to $\{F(z)\}_+$, whereas $\{F(z)\}_-$ includes only the response for $t \leq -1$. Then the solution of (11.67) can be expressed in the z-domain as

$$\bar{K}(z) = \frac{1}{P_\nu}\{S_{X\nu}(z)\}_+, \tag{11.68}$$

where since the innovations process $\nu(t)$ is obtained by passing $Y(t)$ through filter $L^{-1}(z)$, we have

$$S_{X\nu}(z) = \frac{S_{XY}(z)}{L(z^{-1})}.$$

In addition, by noting that the optimum filter $k(m)$ satisfying (11.65) can be implemented by first whitening $Y(t)$ by passing it through $L^{-1}(z)$ and then applying filter $\bar{k}(m)$ as shown in Fig. 11.6, we deduce that the optimum causal filter $K(z)$ is given by

$$K(z) = \frac{1}{L(z)P_\nu}\left\{\frac{S_{XY}(z)}{L(z^{-1})}\right\}_+. \tag{11.69}$$

By using the orthogonality property (11.64) of the filtering error $E_f(t)$, we find that the MSE can be expressed as

$$MSE_f = E[E_f^2(t)] = E[E_f(t)X(t)]$$
$$= R_{E_fX}(0) = R_X(0) - k * R_{YX}(0). \tag{11.70}$$

Fig. 11.6 Implementation of optimum filter $K(z)$ by whitening $Y(t)$ and applying filter $\bar{K}(z)$

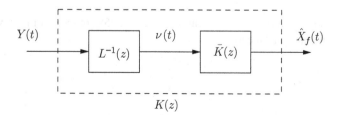

Example 11.2 (Continued) For this case, the power spectrum $S_Y(z)$ admits the factorization (11.53) with

$$L(z) = \frac{1 - z^{-1}/2}{1 - 0.8z^{-1}}$$

and $P_v = 3.2$. In this case

$$\frac{S_{XY}(z)}{L(z^{-1})} = \frac{0.72}{1 - z/2}$$

is anticausal and its causal part reduces to

$$\left\{ \frac{S_{XY}(z)}{L(z^{-1})} \right\}_+ = 0.72,$$

so that the optimum causal filter is

$$K(z) = \frac{0.72}{3.2L(z)} = \frac{0.45(1 - 0.8z^{-1})}{2(1 - z^{-1}/2)}$$

$$= a + \frac{b}{1 - z^{-1}/2},$$

with

$$a = \lim_{z^{-1} \to \infty} K(z) = 0.35 \times 0.8 = 0.36$$

$$b = (1 - z^{-1/2})K(z)\ |_{z^{-1}=2} = \frac{0.45(1 - 0.8z^{-1})}{2}\ |_{z^{-1}=2}$$

$$= \frac{0.45 \times -0.6}{2} = -0.135.$$

The impulse response of $K(z)$ is therefore

$$k(t) = 0.36 - 0.135(1/2)^t u(t).$$

By setting $t = 0$ in the inverse z-transform of

$$K(z)S_{YX}(z) = \frac{0.45 \times 0.72}{2(1 - z^{-1}/2)}$$

and using (11.70), we find that the MSE is given by

$$MSE_f = 0.72(1 - 0.225),$$

so that causal filtering achieves only a 22.5% reduction in the variance of $X(t)$, compared to 30% in the noncausal case.

Relation Between Noncausal and Causal MSEs We already saw that the noncausal MSE is $R_E(0)$ where $S_E(z)$ is given (11.45). But the causal MSE is $R_{E_f X}(0)$ where

$$S_{E_f X}(z) = S_X(z) - K(z) S_{YX}(z)$$

$$= S_X(z) - \frac{1}{L(z) P_v} \left\{ \frac{S_{XY}(z)}{L(z^{-1})} \right\}_+ S_{YX}(z). \tag{11.71}$$

Comparing $S_E(z)$ and $S_{E_f X}(z)$, we find

$$S_{E_f X}(z) = S_E(z) + \frac{1}{P_v} \left\{ \frac{S_{XY}(z)}{L(z^{-1})} \right\}_- \frac{S_{YX}(z)}{L(z)}. \tag{11.72}$$

But

$$\frac{S_{YX}(z)}{L(z)} = \left\{ \frac{S_{YX}(z)}{L(z)} \right\}_+ + \left\{ \frac{S_{YX}(z)}{L(z)} \right\}_-,$$

and by observing that

$$R_{E_f X}(0) = \frac{1}{2\pi j} \int_{|z|=1} S_{E_f X}(z) dz,$$

where the unit circle integration is performed in the counter-clockwise direction, by Cauchy's theorem we deduce that the contribution of

$$\left\{ \frac{S_{XY}(z)}{L(z^{-1})} \right\}_- \times \left\{ \frac{S_{YX}(z)}{L(z)} \right\}_-$$

will be zero since this function is analytic inside the unit circle. This implies

$$MSE_f = MSE + \frac{1}{2\pi P_v} \int_{-\pi}^{\pi} \left| \left\{ \frac{S_{XY}(e^{j\omega})}{L(e^{-j\omega})} \right\}_- \right|^2 d\omega. \tag{11.73}$$

Example 11.2 (Continued) In this case

$$J(z) = \left\{ \frac{S_{XY}(z)}{L(z^{-1})} \right\}_- = \frac{0.72}{1 - z/2} - 0.72 = \frac{0.36z}{1 - z/2}$$

and

$$W(z) = J(z) J(z^{-1}) = \frac{(0.36)^2}{(1 - z/2)(1 - z^{-1}/2)}$$

$$= \frac{-(0.72)^2 z^{-1}/2}{(1 - z^{-1}/2)(1 - 2z^{-1})} = \frac{c}{1 - z^{-1}/2} + \frac{d}{1 - 2z^{-1}}$$

with

$$c = (1 - z^{-1}/2) W(z) \mid_{z^{-1}=2} = \frac{-(0.72)^2}{-3} = 0.72 \times 0.24.$$

Thus if $w(t)$ denotes the inverse z-transform of $W(z)$, $w(0) = c$ and the formula (11.72) yields

$$MSE_f - MSE = \frac{c}{P_v} = \frac{0.24}{3.2} \times 0.72 = 0.075 \times 0.72,$$

which coincides with the 7.5% performance difference already observed between the causal and noncausal Wiener filters.

11.6 Special Cases

In this section we consider several special cases of interest.

11.6.1 Signal in White Noise

Suppose that

$$Y(t) = X(t) + V(t), \tag{11.74}$$

where $V(t)$ is a zero-mean white noise process with variance P_V which is uncorrelated with $X(t)$. In this case $R_{XY}(m) = R_X(m)$ and $S_{XY}(z) = S_X(z)$. Furthermore

$$S_Y(z) = S_X(z) + P_V,$$

since $X(t)$ and $V(t)$ are uncorrelated. If we consider the spectral factorization (11.53), we have

$$\frac{S_{XY}(z)}{L(z^{-1})} = \frac{S_X(z)}{L(z^{-1})} = \frac{S_Y(z) - P_V}{L(z^{-1})}$$

$$= L(z)P_v - \frac{P_V}{L(z^{-1})}. \tag{11.75}$$

In this expression $L(z)P_v$ is the z-transform of a causal sequence, so

$$\{L(z)P_V\}_+ = L(z)P_v$$

and

$$\frac{P_V}{L(z^{-1})} = P_V M(z^{-1}) = P_V\left(1 + \sum_{t=1}^{\infty} m(t)z^t\right)$$

is anticausal, so that its causal part reduces to

$$\left\{\frac{P_V}{L(z^{-1})}\right\}_+ = P_V.$$

Accordingly, the optimal causal filter given by (11.69) can be expressed as

$$K(z) = \frac{1}{L(z)P_v}(L(z)P_v - P_V) = 1 - \frac{P_V}{L(z)P_v}. \tag{11.76}$$

This result admits a simple geometric interpretation. According to (11.74)

$$X(t) = Y(t) - V(t), \tag{11.77}$$

where $Y(t)$ is observed, so

$$\hat{X}_f(t) = Y(t) - \hat{V}_f(t), \tag{11.78}$$

where $\hat{V}_f(t)$ denotes the linear least-squares estimate of $V(t)$ given the current and previous observations, i.e.,

$$\hat{V}_f(t) = P_{L_t} V(t), \tag{11.79}$$

where L_t is the linear space of random variables spanned by $\{Y(t), \ m \geq 0\}$. But L_t is also spanned by innovations $\{v(t - m), \ m \geq 0\}$ and it admits the orthogonal decomposition

$$L_t = \{v(t)\} \oplus L_{t-1}.$$

But since $V(t)$ is a white noise process, it is uncorrelated with all prior observations, so it is orthogonal to L_{t-1}. Accordingly, as shown in Fig. 11.7, the orthogonal projection (11.79) reduces to the projection

$$\hat{V}_f(t) = P_{v(t)} V(t)$$

onto the single random variable $v(t)$, which can be expressed as

$$P_{v(t)} V(t) = E[V(t)v(t)](E[v^2(t)])^{-1} v(t),$$

where $E[v^2(t)] = P_v$ and

$$E[V(t)v(t)] = E[V(t)(Y(t) - \hat{Y}_p(t))] = E[V(t)Y(t)] = P_V$$

because $V(t)$ is uncorrelated with the one-step ahead predicted estimate $\hat{Y}_p(t)$ since it belongs to L_{t-1}. Thus

$$\hat{V}_f(t) = \frac{P_V}{P_v} v(t), \tag{11.80}$$

and by combining (11.78) and (11.80) we obtain

$$\hat{X}_f(t) = Y(t) - \frac{P_V}{P_v} v(t), \tag{11.81}$$

which is the time domain equivalent of (11.76).

By subtracting (11.78) from (11.77), we find also that the filtering error can be expressed as

$$E_f(t) = X(t) - \hat{X}_f(t) = -(V(t) - \hat{V}_f(t)). \tag{11.82}$$

But the MSE associated with estimating noise $V(t)$ from innovation value $v(t)$ is given by

$$MSE_f = P_V - \frac{P_V^2}{P_v}, \tag{11.83}$$

which can be viewed as obtained by applying Pythagoras's theorem to the orthogonal triangle formed by 0, $\hat{V}_f(t)$, and $V(t)$ in Fig. 11.7.

Fig. 11.7 Projection of $V(t)$ onto L_t by orthogonal decomposition of subspace L_t into $\{v(t)\}$ and L_{t-1}

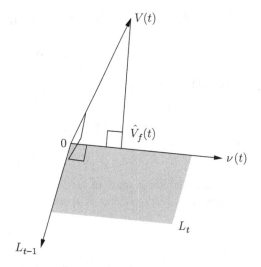

Example 11.3 In the additive observations model (11.74), $X(t)$ has for power spectrum

$$S_X(z) = \frac{1}{(1 - z^{-1}/2)(1 - z/2)}$$

and white noise $V(t)$ has variance $P_V = 5$ and is uncorrelated with $X(t)$. The power spectrum of $Y(t)$ is therefore

$$S_Y(z) = S_X(z) + P_V = \frac{1}{(1 - z^{-1}/2)(1 - z/2)} + 5$$

$$= \frac{25(1 - 2/5z^{-1})(1 - 2/5z)}{4(1 - z^{-1}/2)(1 - z/2)},$$

which can be factored in the form (11.53) where

$$L(z) = \frac{1 - 2/5z^{-1}}{1 - z^{-1}/2}$$

and $P_v = 25/4$. In this case, the optimal causal filter for estimating $X(t)$ from the current and prior observations is

$$K(z) = 1 - \frac{5}{25/4L(z)} = 1 - \frac{4(1 - z^{-1}/2)}{5(1 - 2/5z^{-1})}$$

$$= \frac{1}{5(1 - 2/5z^{-1})}.$$

Its impulse response is

$$k(t) = \frac{1}{5}\left(\frac{2}{5}\right)^t u(t),$$

and the resulting MSE is

$$MSE_f = 5 - \frac{25}{25/4} = 1.$$

Since the autocorrelation corresponding to $S_X(z)$ is

$$R_X(m) = \frac{4}{3}(1/2)^{|m|},$$

we find therefore that causal filtering achieves a 25% reduction in the variance $P_X = R_X(0) = 4/3$ of $X(t)$.

11.6.2 k-Steps Ahead Prediction

Consider a zero-mean WSS random process $Y(t)$ with autocorrelation $R_Y(m) = E[Y(t+m)Y(t)]$ and power spectrum $S_Y(z)$ satisfying the spectral factorization conditions of Sect. 11.4, i.e., $R_Y(m)$ is summable, with $S_Y(z)$ rational and free of zeros on the unit circle. In this case if $L(z)$ denotes the minimum phase spectral factor of $S_Y(z)$ and $M(z) = L^{-1}(z)$ denotes its inverse, it was found in (11.57) that the optimal one step ahead prediction filter for estimating $Y(t)$ given $\{Y(t-m), m \geq 1\}$ is

$$K_p(z) = -(M(z) - 1) = 1 - \frac{1}{L(z)},$$

and since the corresponding error is $v(t)$, the associated MSE is the innovations variance P_v. If we shift the problem forward by one time unit, the optimum filter for estimating $Y(t + 1)$ based on the current and past observations $\{Y(t - m), m \geq 0\}$ is

$$K_{p,1}(z) = zK_p(z) = z(1 - M(z)). \tag{11.84}$$

Sometimes, it is necessary to predict further in time, so we consider the k-step ahead prediction problem with $k \geq 1$, in which case the process to be estimated is $X(t) = Y(t + k)$. Since $X(t)$ can be viewed as obtained by passing $Y(t)$ through advance filter z^k, the cross-spectrum of $X(t)$ and $Y(t)$ is therefore

$$S_{XY}(z) = z^k S_Y(z),$$

and the optimum k-step ahead prediction filter

$$K_{p,k}(z) = \frac{1}{L(z)P_v}\left\{\frac{z^k S_Y(z)}{L(z^{-1})}\right\}_+$$

$$= \frac{1}{L(z)}\left\{z^k L(z)\right\}_+. \tag{11.85}$$

By observing that

$$z^k L(z) = \sum_{t=0}^{k-1} \ell(t)z^{k-t} + \sum_{t=k}^{\infty} \ell(t)z^{k-t},$$

where the terms appearing in the first sum on the right-hand side contain positive powers of z, we deduce that

$$\left\{z^k L(z)\right\}_+ = \sum_{t=k}^{\infty} \ell(t) z^{k-t}$$

$$= z^k (L(z) - L_k(z)) \tag{11.86}$$

with

$$L_k(z) \triangleq \sum_{t=0}^{k-1} \ell(t) z^{-t}.$$

Thus the optimum k-step ahead prediction filter is given by

$$K_{p,k}(z) = z^k \left(1 - \frac{L_k(z)}{L(z)}\right). \tag{11.87}$$

By noting that $\ell(0) = 1$, this filter reduces to (11.84) when $k = 1$.

To evaluate the mean-square error, observe that when $X(t) = Y(t + k)$, $S_X(z) = S_Y(z)$, and $S_{YX}(z) = z^{-k} S_Y(z)$, so the cross-spectrum of the k-step ahead prediction error $E_{p,k}(t)$ and $Y(t)$ is given by

$$S_{E_{p,k}Y}(z) = S_Y(z) - K_{p,k}(z) S_{XY}(z) = (1 - (1 - \frac{L_k(z)}{L(z)})) S_Y(z)$$

$$= L_k(z) P_v L(z^{-1}), \tag{11.88}$$

which yields

$$MSE_{p,k} = R_{E_{p,k}X}(0) = \frac{P_v}{2\pi} \int_{-\pi}^{\pi} |L_k(e^{j\omega})|^2 d\omega. \tag{11.89}$$

As expected, since $L_k(z)$ converges pointwise to $L(z)$ as the estimation gap k becomes infinite,

$$\lim_{k \to \infty} MSE_{p,k} \to P_Y = E[Y^2(t)],$$

which indicates that the benefits of prediction vanish as k becomes large.

Example 11.4 Consider a WSS random process with power spectrum

$$S_Y(z) = \frac{(1 - z^{-1}/4)(1 - z/4)}{(1 - z^{-1}/2)(1 - z/2)},$$

so that its minimum phase spectral factor

$$L(z) = \frac{1 - z^{-1}/4}{1 - z^{-1}/2}$$

and the innovations process has variance $P_v = 1$. The impulse response of $L(z)$ is

$$\ell(t) = \delta(t) + \frac{1}{2}(1/2)^t u(t-1).$$

Suppose we seek to compute the two-step ahead predicted estimate of $Y(t+2)$ given all observations up to time t. By retaining the first two terms of impulse response $\ell(t)$, we find

$$L_2(z) = 1 + z^{-1}/4,$$

so the optimum prediction filter is given by

$$K_{p,2}(z) = z^2(1 - \frac{L_2(z)}{L(z)})$$

$$= z^2 \frac{(1 - z^{-1}/4) - (1 + z^{-1}/4)(1 - z^{-1}/2)}{1 - z^{-1}/4} = \frac{1/8}{1 - z^{-1}/4}.$$

Since

$$L_2(z)L_2(z^{-1}) = (1 + z^{-1}/4)(1 + z/4) = \frac{17}{16} + (z + z^{-1})/4,$$

we deduce that

$$MSE_{p,2} = 17/16,$$

so that predicting the observations two steps ahead increases the MSE by $1/16$ compared to one step ahead.

11.6.3 Fixed-Lag Smoothing

Consider again the signal in white noise model (11.74). In some communications applications, instead of attempting to estimate the current signal $X(t)$ based on observations $\{Y(t-m), m \geq 0\}$, we seek to estimate a lagged value $X(t-k)$ of the signal with $k \geq 0$ based on all observations up to time t. This problem is similar to the smoothing problem of Sect. 11.3, except that instead of having access to the entire observations record in the future of the variable $X(t-k)$ of interest, we have access only to k future values $Y(s)$, with $t \geq s \geq t-k$, as well as to all current and past values. This problem is called a fixed-lag smoothing problem. It has the advantage of allowing the online computation of estimates of $X(t-k)$ without having to wait until all observations have been recorded. It is worth noting that the cross-correlation $R_{XY}(m)$ between $X(t+m)$ and $Y(t)$ has usually a finite length ℓ in the sense that

$$R_{XY}(m) \approx 0,$$

for $|m| \geq \ell$. In such cases, a fixed lag smoother will have the same performance as a smoother using the entire observations record as long as the lag k exceeds the cross-correlation length ℓ.

If the variable to be estimated is $S(t) = X(t-k)$, the cross-spectral density of $S(t)$ and $Y(t)$ is

$$S_{SY}(z) = z^{-k}S_X(z) = z^{-k}(S_Y(z) - P_V),$$

so the optimum k-th lag smoothing filter is given by

$$
\begin{aligned}
K_{s,k}(z) &= \frac{1}{L(z)P_v}\left\{\frac{z^{-k}S_X(z)}{L(z^{-1})}\right\}_+ \\
&= \frac{1}{L(z)P_v}\left\{z^{-k}(L(z)P_v - \frac{P_V}{L(z^{-1})})\right\}_+ \\
&= z^{-k} - \frac{P_V}{L(z)P_v}\left\{z^{-k}M(z^{-1})\right\}_+.
\end{aligned}
\tag{11.90}
$$

The filter $z^{-k}M(z^{-1}))$ can be decomposed into causal and anticausal parts as

$$
z^{-k}M(z^{-1}) = z^{-k}M_k(z^{-1}) + \sum_{t=k+1}^{\infty} m(t)z^{t-k},
$$

where

$$
M_k(z) \triangleq \sum_{t=0}^{k} m(t)z^{-t}
$$

retains only the first $k+1$ terms of $M(z)$. Accordingly, the optimum k-th lag smoothing filter can be expressed as

$$
K_{s,k}(z) = z^{-k}\left(1 - \frac{M_k(z^{-1})}{L(z)P_v}\right).
\tag{11.91}
$$

In this case the cross-spectrum of the k-th lag smoothing error $E_{s,k}(t)$ and $S(t) = X(t-k)$ is given by

$$
\begin{aligned}
S_{E_{s,k}S}(z) &= S_X(z) - K_{s,k}(z)S_{YZ}(z) \\
&= S_X(z) - K_{s,k}(z)z^k S_X(z) \\
&= S_X(z)P_V\frac{M_k(z^{-1})}{L(z)P_v}.
\end{aligned}
\tag{11.92}
$$

The MSE of the k-th lag smoother can therefore be expressed as

$$
MSE_{s,k} = R_{E_{s,k}S}(0) = \frac{1}{2\pi}\int_{-\pi}^{\pi} S_X(e^{j\omega})P_V\frac{M_k^*(e^{j\omega})}{L(e^{j\omega})P_v}d\omega.
\tag{11.93}
$$

In this expression, by observing that $M_k(z)$ converges pointwise to $M(z) = L^{-1}(z)$ in the domain of convergence of $S_Y(z)$ as k tends to infinity, we deduce that

$$
\frac{M_k(z^{-1})}{L(z)P_v} \to \frac{1}{S_Y(z)}
$$

pointwise as k tends to infinity, so $MSE_{s,k}$ converges asymptotically to the noncausal Wiener filter MSE as k tends to infinity.

Example 11.3 (Continued) Suppose we seek to compute the lag-2 estimate of $X(t-2)$ based on all observations up to time t. We have

$$M(z) = L^{-1}(z) = \frac{1 - z^{-1}/2}{1 - 2/5z^{-1}} = A + \frac{B}{1 - 2/5z^{-1}}$$

with

$$A = \lim_{z^{-1} \to \infty} M(z) = \frac{5}{4}$$

$$B = (1 - 2/5z^{-1})M(z) \mid_{z^{-1}=5/2} = (1 - 5/4) = -1/4,$$

so the impulse response of $M(z)$ is

$$m(t) = \frac{5}{4}\delta(t) - \frac{1}{4}(2/5)^t u(t).$$

Truncating this response to its first three terms gives

$$M_2(z) = 1 - z^{-1}/10 - z^{-2}/25.$$

Since $P_V = 5$ and $P_v = 25/4$, the optimum lag-2 smoothing filter can be expressed as

$$Ks, 2(z) = z^{-2}\left(1 - \frac{4(1 - z/10 - z^2/25)(1 - z^{-1}/2)}{5(1 - 2/5z^{-1})}\right).$$

11.7 CT Wiener Filtering

If $X(t)$ and $Y(t)$ are two jointly WSS zero-mean CT random processes, the 2×2 autocorrelation matrix

$$\mathbf{R}_Z(\tau) = E[\mathbf{Z}(t + \tau)\mathbf{Z}^T(t)] = \begin{bmatrix} R_X(\tau) & R_{XY}(\tau) \\ R_{YX}(\tau) & R_Y(\tau) \end{bmatrix}$$

of the vector process $\mathbf{Z}(t)$ given by (11.31) has the symmetry property

$$\mathbf{R}_Z^T(-\tau) = \mathbf{R}_Z(\tau). \tag{11.94}$$

If $\mathbf{R}_Z(\tau)$ is summable, the region of convergence (ROC) of the Laplace transform

$$\mathbf{S}_Z(s) = \int_{-\infty}^{\infty} \mathbf{R}_Z(\tau) \exp(-s\tau)d\tau = \begin{bmatrix} S_X(s) & S_{XY}(s) \\ S_{YX}(s) & S_Y(s) \end{bmatrix}$$

includes the $j\omega$ axis. $\mathbf{S}_Z(s)$ is the power spectrum of $\mathbf{Z}(t)$, and for $s = j\omega$ it coincides with its power spectral density. The symmetry condition (11.94) implies

$$\mathbf{S}_Z^T(-s) = \mathbf{S}_Z(s), \tag{11.95}$$

so that the region of convergence of the spectrum is of the form $-c < \Re s < c$ with $c > 0$, which is symmetric about the $j\omega$ axis. In the following, it will be assumed that $\mathbf{S}_Z(s)$ is a rational function of s and that $S_Y(s)$ has no finite or infinite zero on the $j\omega$ axis (since the ROC of $\mathbf{S}_Z(s)$ includes the $j\omega$ axis, we already know that $S_Y(s)$ has no pole on this axis). The assumption that $S_Y(s)$ has no infinite zero is equivalent to requiring that that $S_Y(\infty) > 0$. Indeed, if we consider the rational expression

$$S_Y(s) = \frac{N(s)}{D(s)},$$

where $N(s)$ $D(s)$ are polynomials of s, $S_Y(s)$ is free of zeros at infinity if and only if $N(s)$ has the same degree as $D(s)$, so that $S_Y(\infty) > 0$.

Then consider the CT estimate $\hat{X}(t)$ of $X(t)$ obtained by passing observation process $Y(t)$ through a CT filter $g(t)$ as shown in Fig. 11.2. When the filter $g(t)$ is noncausal, if $E(t) = X(t) - \hat{X}(t)$ denotes the estimation error, the orthogonality condition

$$E[E(t)Y(t-\tau)] = 0 \qquad (11.96)$$

with τ arbitrary yields equation

$$R_{XY}(\tau) = g(\tau) * R_Y(\tau) \qquad (11.97)$$

for the optimum filter, whereas when the filter $g(\tau)$ is required to be causal, the orthogonality condition (11.96) holds only for $\tau \geq 0$, and the filter equation (11.97) holds only for $\tau \geq 0$. In the noncausal case, Laplace transforming (11.97) gives

$$G(s) = \frac{S_{XY}(s)}{S_Y(s)}, \qquad (11.98)$$

and following the same steps as in the DT case, we find that the error spectrum

$$S_E(s) = S_X(s) - G(s)S_{YX}(s)$$

and the MSE is given by

$$MSE = E[E^2(t)] = R_X(0) - (g * R_{YX})(0). \qquad (11.99)$$

To solve the Wiener–Hopf equation (11.97) where $\tau \geq 0$, we perform a spectral factorization

$$S_Y(s) = L(s)P_v L(-s) \qquad (11.100)$$

of $S_Y(s)$ where $L(s)$ has all its poles and zeros in domain $\Re s \leq -\gamma$ with $\gamma > 0$ and $L(\infty) = 1$. This means that $L(s)$ and $M(s) = L^{-1}(s)$ are both causal and stable. The assumption $S_Y(\infty) > 0$ ensures that in the rational representation

$$L(s) = \frac{B(s)}{A(s)} \quad M(s) = \frac{A(s)}{B(s)}$$

of $L(s)$ and its inverse $M(s)$, the numerator and denominator polynomials $B(s)$ and $A(s)$ have the same degree, so that the condition $L(\infty) = M(\infty) = 1$ can be satisfied. This condition also ensures that $P_v > 0$. Then, as in the DT case, the innovations process $v(t)$ is obtained by passing the observations $Y(t)$ through the LTI whitening filter $M(s)$. Its power spectrum

$$S_\nu(s) = M(s)S_Y(s)M(-s) = P_\nu$$

is constant, so $\nu(t)$ is a white noise with intensity P_ν. The innovations and observations processes $\nu(t)$ and $Y(t)$ can be obtained causally from one another, so the information contained in observations $\{Y(t - \tau), \tau \geq 0\}$ and innovations $\{\nu(t - \tau), \tau \geq 0\}$ is the same. Following the same steps as in the DT case, we then find that the optimal causal estimation filter is given by

$$K(s) = \frac{1}{L(s)P_\nu} \left\{ \frac{S_{XY}(s)}{L(-s)} \right\}_+, \qquad (11.101)$$

where if

$$F(s) = \int_{-\infty}^{\infty} f(t) \exp(-st) dt$$

denotes the two-sided Laplace transform of $f(t)$, then $\{F(s)\}_+$ is the Laplace transform of $f(t)u(t)$, i.e.

$$\{F(s)\}_+ = \int_{0_-}^{\infty} f(t) \exp(-st) dt.$$

By using the orthogonality property of linear least-squares estimates, the causal MSE can be expressed as

$$MSE_f = E[(X(t) - \hat{X}_f(t))X(t)] = R_X(0) - (k * R_{YX})(0). \qquad (11.102)$$

Example 11.4 Consider the signal in white noise model (11.74), where $V(t)$ is a white noise with unit intensity, so that

$$E[V(t)V(s)] = \delta(t - s)$$

and $S_V(s) = 1$. The signal $X(t)$ is zero mean WSS with autocorrelation

$$R_X(\tau) = \frac{5}{4} \exp(-2|\tau|)$$

and is uncorrelated with $V(t)$. Its power spectrum is

$$S_X(s) = \frac{5}{-s^2 + 4}.$$

The power spectrum of $Y(t)$ is

$$S_Y(s) = S_X(s) + S_V(s) = \frac{-s^2 + 9}{-s^2 + 4},$$

which is rational and such that $S_Y(\infty) = 1$. It admits therefore a spectral factorization of the form (11.100) with

$$L(s) = \frac{s + 3}{s + 2}$$

and $P_\nu = 1$. Then by observing that $S_{XY}(s) = S_X(s)$ and following the steps used to derive expression (11.76) in the DT case for estimating causally a signal in white noise, we find

$$K(s) = 1 - \frac{1}{L(s)} = \frac{1}{s+3}.$$

We then have

$$K(s)S_{YX}(s) = K(s)S_X(s) = \frac{5}{(s+3)(s+2)(-s+2)}$$

$$= \frac{A}{s+3} + \frac{B}{s+2} + \frac{C}{-s+2}$$

with

$$A = (s+3)K(s)|_{s=-3} = -1$$

$$B = (s+2)K(s)|_{s=-2} = \frac{5}{4}$$

$$C = (-s+2)K(s)|_{s=2} = \frac{1}{4},$$

so

$$k * R_{YX}(0) = A + B + C = \frac{1}{2}$$

and

$$MSE_f = R_X(0) - k * R_{YX}(0) = \frac{5}{4} - \frac{1}{2} = \frac{3}{4}.$$

11.8 Bibliographical Notes

Although Wiener's original derivation [1] of optimum estimation filters is of historical interest, it probably does not represent the easiest introduction to the topic of optimal filtering, since it is expressed in terms of time averages instead of the ensemble average approach adopted by most modern probability textbooks. A concise and insightful presentation of Wiener filtering is given in [10]. Wiener filtering has been superseded by Kalman filtering [11], which is more flexible since it can handle transients and time varying processes. This approach to optimal filtering relies on employing a linear state-space representation of the process to be estimated and observations. In this context the computation of optimal filtering and smoothing estimates can be mechanized easily by employing simple state-space recursions. Comprehensive presentations of Kalman and Wiener filtering are given in [3, 12]. One limitation of both Wiener and Kalman filtering is that they assume that the parameters and statistics of the process to be estimated and observations are known exactly. But in practice, some parameters are often unknown. Also, due to the presence of unmodeled dynamics and perturbations, models are only approximations of reality. Adaptive filtering [13] applies optimal filtering techniques to situations where some parameters are unknown. This requires the availability of training data to allow adaptive algorithms to learn the values of the unknown system parameters. Robust filtering [14–

16] deals with estimation problems where unstructured modeling errors are present. In this context, a common but conservative approach consists of solving a minimax problem, where the goal is to find the best filter for the worst model in the class of allowable models. Usually, the presentation of Kalman filtering, adaptive filtering, and possibly nonlinear and robust filtering requires a full semester course.

11.9 Problems

11.1 In fitting a line through experimental data, we assume

$$Y(t) = A + Bt + V(t) \ 0 \le t \le T,$$

where $V(t)$ is a discrete-time zero-mean white noise with variance $\sigma_V^2 = 1$. The intercept A and slope B are uncorrelated, have means m_A and m_B, and variances K_A and K_B, respectively. Find the the linear least-squares estimates of A and B based on all observations $Y(t), 0 \le t \le T$. Assume that A and B are uncorrelated with $V(t)$. Which of the two parameters benefit the most from the prior knowledge? *Hint:* You may want to compute the mean-square errors for the parameters A and B and compare them to the a priori variances K_A and K_B.

11.2 Consider the discrete-time observations

$$Y(t) = h(t) * X(t) + V(t)$$

with $-\infty < t < \infty$, where $X(t)$ and $V(t)$ are two independent zero-mean white noise processes with variances $\sigma_X^2 = 0.75$ and $\sigma_V^2 = 1$, i.e.,

$$E[X(t)X(s)] = 0.75\delta(t-s)$$
$$E[V(t)V(s)] = \delta(t-s).$$

The LTI filter

$$h(t) = (0.5)^t u(t),$$

where $u(t)$ denotes the discrete-time unit-step function and $*$ denotes the convolution operation. We seek to find the linear least-squares estimate (LLSE) $\hat{X}(t)$ of $X(t)$ by passing the observations $Y(t)$ through a linear time-invariant filter with impulse response $g(t)$ and z-transform $G(z)$.

a) Find the optimum noncausal Wiener filter $G(z)$.
b) By performing a partial fraction expansion of $G(z)$, find the impulse response $g(t)$ and verify that it is noncausal.

To solve this problem, you may need the following Z-transform pairs

$$a^t u(t) \longleftrightarrow \frac{1}{1 - az^{-1}} \quad |a| < |z|$$

$$-b^t u(-t-1) \longleftrightarrow \frac{1}{1 - bz^{-1}} \quad |z| < |b|$$

$$\delta(t) \longleftrightarrow 1.$$

11.3 Consider a zero mean WSS random process $Y(t)$ with autocorrelation $R_Y(m) = E[Y(t + m)Y(t)]$ and power spectrum $S_Y(z)$. We consider a noncausal interpolation problem consisting of finding the linear least squares estimate of $Y(t)$ given the observations $Y(t - m)$ for all $m \in \mathbb{Z}$ with $m \neq 0$. This problem arises for example when one observation is missing from a data record and we seek to recover the missing observation from the surrounding observations. If $g(t)$ denotes the optimum interpolation filter, the interpolated estimate can be expressed as

$$\hat{Y}(t) = g(t) * Y(t) = \sum_{m \in \mathbb{Z} - \{0\}} g(m)Y(t - m),$$

and the corresponding error takes the form

$$E(t) = Y(t) - \hat{Y}(t) = f(t) * Y(t),$$

where if $\delta(t)$ denotes the DT unit impulse function, the filter

$$f(t) = \delta(t) - g(t).$$

Let $MSE_I = E[E^2(t)]$ denote the mean-square noncausal interpolation error.

a) Use the orthogonality property $E[E(t)Y(t - n)] = 0$ for $n \neq 0$ to show that the error filter $f(t)$ satisfies the identity
$$f(n) * R_Y(n) = MSE_I \delta(n). \tag{11.103}$$

b) By z-transforming (11.103), verify that

$$F(z) = \frac{MSE_I}{S_Y(z)},$$

and use the fact that $f(0) = 1$ to show that

$$MSE_I = \left[\frac{1}{2\pi} \int_{-\pi}^{\pi} S_Y^{-1}(e^{j\omega}) d\omega \right]^{-1}. \tag{11.104}$$

c) Find the power spectrum $S_E(z)$ of the error process $E(t)$.
d) Since noncausal interpolation uses two-sided data, it is expected that MSE_I should be less than the one-step ahead mean-square prediction error MSE_p for estimating $Y(t)$ based on the prior values $\{Y(t - m), m \geq 1\}$, since the prediction problem uses only one-sided data. Use the Kolmogorov–Szegö formula (11.61) and Jensen's inequality to show that $MSE_I \leq MSE_p$.
e) Evaluate $g(t)$ and MSE_I when $R_Y(m) = (1/2)^{|m|}$, or equivalently

$$S_Y(z) = \frac{3/4}{(1 - z^{-1}/2)(1 - z/2)}.$$

11.4 Consider the noncausal Wiener filtering problem for a signal $X(t)$ in white noise $V(t)$, so that the observations can be expressed as

$$Y(t) = X(t) + V(t),$$

where $V(t)$ has zero-mean, variance P_V, and is uncorrelated with $X(t)$. This corresponds to the case where $H(z) = 1$ and $S_V(z) = P_V$ in the noisy deconvolution problem considered in Sect. 11.3.

a) By observing that

$$X(t) = Y(t) - V(t),$$

where $Y(t)$ is given, note that the optimal estimate $\hat{X}(t)$ of $X(t)$ can be written as

$$\hat{X}(t) = Y(t) - \hat{V}(t), \tag{11.105}$$

where $\hat{V}(t)$ denotes the linear least-squares estimate of $V(t)$ based on observations $\{Y(t-m),\ m \in \mathbb{Z}\}$. To evaluate $\hat{V}(t)$, note that the Hilbert space L of random variables spanned by observations $\{Y(t-m),\ m \in \mathbb{Z}\}$ admits the orthogonal decomposition

$$L = \{E_I(t)\} \oplus L_t^m,$$

where L_t^m denotes the Hilbert space spanned by removing $Y(t)$ from L, and

$$E_I(t) = Y(t) - \hat{Y}_I(t)$$

is the interpolation error for the optimal interpolation problem considered in Problem 11.3. $\hat{Y}_I(t)$ denotes here the optimal interpolation estimate of $Y(t)$ given observations $Y(t-m)$, $m \in \mathbb{Z} - \{0\}$. But since $V(t)$ is white and uncorrelated with signal $X(t)$, it is orthogonal to L_t^m, so

$$\hat{V}(t) = P_{E_I(t)} V(t) = K E_I(t), \tag{11.106}$$

where $P_{E_I(t)}$ denotes the projection on the scalar random variable $E_I(t)$. Evaluate the scalar gain K, and by noting that

$$E_I(t) = f(t) * Y(t),$$

where it was found in Problem 11.3 that the optimum interpolation filter $F(z)$ is given by

$$F(z) = \frac{MSE_I}{S_Y(z)},$$

combine identities (11.105) and (11.106) to obtain the optimum noncausal estimation filter $G(z)$.

b) Verify that the optimum filter of part a) coincides with expression

$$G(z) = \frac{S_X(z)}{S_Y(z)}$$

derived in Sect. 11.3.

c) Use the geometric interpretation of the noncausal Wiener filtering problem derived in part a) to show that the noncausal MSE can be expressed as

$$MSE = P_V - \frac{P_V^2}{MSE_I}, \tag{11.107}$$

where

$$MSE_I = \left[\frac{1}{2\pi} \int_{-\pi}^{\pi} S_Y^{-1}(e^{j\omega}) d\omega\right]^{-1}$$

is the mean-square interpolation error for estimating $Y(t)$ given Y_t^m which was obtained in Problem 11.3.

11.5 Consider the observations

$$Y(t) = X(t) + V(t)$$

of a zero-mean DT WSS random signal $X(t)$ with autocorrelation

$$R_X(m) = 2(0.8)^{|m|},$$

where $V(t)$ is a white noise with variance $\sigma_V^2 = 2$, which is uncorrelated with $X(t)$.

a) Find the optimal *noncausal* Wiener filter for estimating $X(t)$ based on all observations $\{Y(s)\ ;\ -\infty < s < \infty\}$. Evaluate the corresponding MSE.
b) Find the optimal *causal* Wiener filter for estimating $X(t)$ based on $\{Y(s)\ ;\ -\infty < s \le t\}$. Evaluate the MSE.

11.6 A zero-mean WSS process $Y(t)$ admits the autoregressive moving average model

$$Y(t) - 0.8Y(t-1) = V(t) - 0.6V(t-1) + 0.36V(t-2),$$

where $V(t)$ is a zero-mean white noise with unit variance. Find the optimal one-step ahead prediction filter for estimating $Y(t)$ based on the observations $\{Y(t-m),\ m \ge 1\}$, and evaluate the corresponding MSE.

11.7 We seek to use a Wiener filter to solve the *noise cancellation problem* depicted in Fig. 11.8. The signal received by a microphone is given by

$$Y_1(t) = X(t) + N_1(t),$$

where $X(t)$ is the desired signal, and $N_1(t)$ is a random noise that we seek to remove. The procedure that we employ consists in locating a second microphone close to the noise source, which yields the measurement

$$Y_2(t) = N_2(t),$$

where the noise $N_2(t)$ is of course not identical to the noise $N_1(t)$ at the first microphone, but is correlated with it. Then, given the measurements $\{Y_2(s)\ ;\ -\infty < s \le t\}$, we design a *causal* Wiener filter $G(z)$ which estimates the noise N_1 at the first microphone. The estimate $\hat{N}_1(t)$ is then subtracted from the signal $Y_1(t)$, resulting in the improved signal

$$Z(t) = Y_1(t) - \hat{N}_1(t),$$

Fig. 11.8 Noise cancellation system

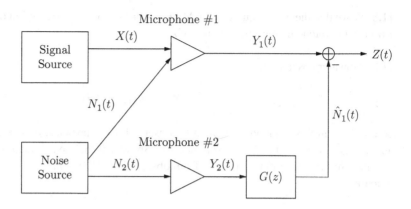

which can be used for amplification or broadcasting purposes.

For the case we consider here, we assume that the noises $N_1(t)$ and $N_2(t)$ admit first-order autoregressive models

$$N_1(t) = a_1 N_1(t-1) + V(t)$$
$$N_2(t) = a_2 N_2(t-1) + V(t),$$

which are driven by the *same* zero-mean white noise process $V(t)$ with intensity σ_V^2, so that N_1 and N_2 are correlated. The coefficients a_1 and a_2 have a magnitude less than 1, so that the above models are stable, and are assumed to be in steady state. We also assume that the signal of interest $X(t)$ is a sinusoidal waveform

$$X(t) = A\cos(\omega_0 t + \Theta),$$

where Θ is uncorrelated with $V(t)$ and uniformly distributed over the interval $[0, 2\pi)$. The amplitude A and frequency ω_0 are known.

a) Show that the autocorrelations and cross-correlation of the noises $N_1(t)$ and $N_2(t)$ are given by

$$R_{N_i}(k) = a_i^{|k|}\frac{\sigma_V^2}{1-a_i^2} \quad \text{for } i = 1, 2$$

$$R_{N_1 N_2}(k) = \begin{cases} a_1^k \dfrac{\sigma_V^2}{1-a_1 a_2} & \text{for } k \geq 0 \\[2mm] a_2^{-k} \dfrac{\sigma_V^2}{1-a_1 a_2} & \text{for } k \leq 0. \end{cases}$$

b) Show that the impulse response $k(t)$ of the IIR causal Wiener filter for estimating $N_1(t)$ based on the values of $\{Y_2(t-m),\ m \geq 0\}$ is given by

$$k(t) = \begin{cases} 1 & \text{for } t = 0 \\ (a_1 - a_2)\,a_1^{t-1} & \text{for } t \geq 1. \end{cases}$$

11.8 Prove that the mean-square error $MSE_{p,k}$ for k-step ahead prediction of a WSS random process $Y(t)$ is a monotone increasing function of k.

11.9 Consider observations

$$Y(t) = X(t) + V(t)$$

of a zero-mean WSS random process $X(t)$ observed in the presence of a zero-mean white noise $V(t)$ uncorrelated with $X(t)$ and with variance P_V. Show that the mean-square-error $MSE_{s,k}$ of the k-th lag smoother for estimating $X(t - k)$ from observations $\{Y(t - m),\ m \geq 0\}$ is monotone decreasing function of k.

11.10 A zero mean WSS process $X(t)$ is observed in the presence of additive zero-mean white noise $V(t)$ with variance $P_V = E[V^2(t)]$, so that the resulting observations take the form

$$Y(t) = X(t) + V(t).$$

We assume that the signal power $P_X = E[X^2(t)] > 0$ and that $X(t)$ and $V(t)$ are uncorrelated.

a) Consider first the filtering problem where $X(t)$ is estimated based on observations $\{Y(t-m),\ m \geq 0\}$. It was shown that MSE_f satisfies (11.93). Use the Kolmogorov–Szegö formula to show that the innovations variance P_ν appearing in this expression satisfies $P_\nu \geq P_V$, which ensures $MSE_f \geq 0$.
b) For the noncausal estimation problem where $X(t)$ is estimated based on all observations $\{Y(t - m),\ m \in \mathbb{Z}\}$, it was shown in Problem 11.4 that the noncausal estimation error MSE satisfies (11.107). Prove that $MSE_I \geq P_V$, which ensures $MSE \geq 0$.

11.11 Let $Z(t)$ be a CT zero-mean WSS random process with PSD $S_Z(\omega)$. Although $Z(t)$ is not necessarily bandlimited, it is sampled with sampling period T, and the samples $Z(kT)$ are used to form the pulse amplitude modulated process

$$Y(t) = \sum_{n=-\infty}^{\infty} Z(nT)p(t - nT - D),$$

where $p(t)$ is an arbitrary signaling pulse and D is a synchronization delay uniformly distributed over $[-T/2, T/2]$. We seek to reconstruct

$$X(t) = Z(t - D)$$

from the observations $\{Y(t - \tau), \tau \in \mathbb{R}\}$ by using a noncausal Wiener filter $g(t)$. The estimate produced by the filter is

$$\hat{X}(t) = g(t) * Y(t),$$

and the corresponding MSE is $E[(X(t) - \hat{X}(t))^2]$.

a) Verify that $X(t)$ and $Y(t)$ are jointly WSS, and evaluate the cross-spectral density $S_{XY}(\omega)$ and PSD $S_Y(\omega)$ in terms of the PSD $S_Z(\omega)$ of $Z(t)$ and the Fourier transform $P(j\omega)$ of $p(t)$. To do so, you are reminded that the CT Fourier transform of the sampled autocorrelation

$$R_S(\tau) = \sum_{m=-\infty}^{\infty} R_Z(mT)\delta(\tau - mT)$$

is [9, Sec. 4.2]

$$S_S(\omega) = \frac{1}{T} \sum_{k=-\infty}^{\infty} S_Z(\omega - k\omega_s),$$

where $\omega_s = 2\pi/T$ denotes the sampling frequency.

b) Compute the optimum noncausal CT Wiener filter

$$G(j\omega) = \frac{S_{XY}(\omega)}{S_Y(\omega)}$$

and its associated MSE.

c) If the filter $P(j\omega)$ is nonzero whenever $S_X(\omega)$ is nonzero, verify that the MSE is independent of the filter $P(j\omega)$, and verify that the MSE is zero if $Z(t)$ is bandlimited with bandwidth less than $\omega_s/2 = \pi/T$.

References

1. N. Wiener, *Extrapolation, Interpolation, and Smoothing of Stationary Time Series with Engineering Applications.* Cambridge, MA: MIT Press, !964 (Paperback Edition).
2. M. H. Hayes, *Statistical Digital Signal Processing and Modeling.* New York, NY: John Wiley, 1996.
3. T. Kailath, A. H. Sayed, and B. Hassibi, *Linear Estimation.* Upper Saddle River, NJ: Prentice-Hall, 2000.
4. T. Kailath, "A view of three decades of linear filtering theory," *IEEE Trans. Informat. Theory*, vol. 20, pp. 146–181, Mar. 1974.
5. D. Williams, *Probability with Martingales.* Cambridge, United Kingdom: Cambridge University Press, 1991.
6. N. Young, *An Introduction to Hilbert Space.* Cambridge, United Kingdom: Cambridge University Press, 1988.
7. B. C. Levy, *Principles of Signal Detection and Parameter Estimation.* New York, NY: Springer Verlag, 2008.
8. A. J. Laub, *Matrix Analysis for Scientists & Engineers.* Philadelphia, PA: Soc. Industrial and Applied Math., 2005.
9. A. V. Oppenheim and R. W. Schafer, *Discrete-Time Signal Processing, 3rd edition.* Upper Saddle River, NJ: Prentice-Hall, 2010.
10. P. Whittle, *Prediction and Regulation by Linear Least-Squares Methods.* Minneapolis, MN: Univ. Minnesota Press, 1983.
11. R. E. Kalman, "A new approach to linear filtering and prediction problems," *Trans. ASME– J. of Basic Engineering*, vol. 82, pp. 35–45, 1960.
12. B. D. O. Anderson and J. B. Moore, *Optimal Filtering.* Englewood Cliffs, NJ: Prentice-Hall, 1979. Republished in 2005 by Dover Publications, Mineola NY.
13. A. H. Sayed, *Adaptive Filters.* Hoboken, NJ: J. Wiley– IEEE Press, 2008.
14. P. Whittle, *Risk-sensitive Optimal Control.* Chichester, UK: J. Wiley, 1990.
15. B. Hassibi, A. H. Sayed, and T. Kailath, *Indefinite-Quadratic Estimation and Control– A Unified approach to H2 and H-infinity Theories*, vol. 16 of *Studies in Applied and Numerical Math.* Philadelphia, PA: Soc. Indust. and Applied Math., 1999.
16. L. P. Hansen and T. J. Sargent, *Robustness.* Princeton, NJ: Princeton Univ. Press, 2008.

Quantization Noise and Dithering 12

12.1 Introduction

While various circuit noise sources, such as thermal noise, shot noise, or $1/f$ noise, have a physical origin, other types of noise, like quantization or roundoff noise, occur as a consequence of computational operations performed by circuits. Specifically, the digitization of bandlimited analog signals involves two operations which are typically combined in analog-to-digital converters (ADCs). The first is signal sampling in time, and the other is the quantization of the amplitude of each signal sample. Since the quantization operation requires approximating an analog amplitude by a digital representation, it necessarily involves errors. The error arising in this approximation is called the quantization noise, and the characterization of its properties plays an important role in analyzing DSP systems performance and in ADC design. Bennett [1] observed in 1948 that for a sufficiently small discretization step, quantization noise can be viewed approximately as uniformly distributed and white. In his exhaustive study of quantization noise, Widrow [2, 3] was able to obtain precise conditions for these properties to hold, and necessary and sufficient conditions were later presented by Sripad and Snyder [4].

Since the topic of signal quantization represents a vast area of investigation (see [5] for a book-length coverage), we focus in this chapter on quantization noise, thereby ignoring the analysis of quantizer outputs. The presentation employed here is a variant of classical studies of quantization, which typically consider infinite quantizers and rely on the characteristic function of the input signal distribution to analyze quantization noise properties. Instead, following [6], we assume that amplitude of the quantizer input is confined to a fixed dynamic range, normalized here to $[-1, 1]$. This amounts to assuming that the quantizer never overloads, an assumption which is also made by other quantization studies. This allows the representation of the input PDF as a Fourier series defined over $[-1, 1]$, where the Fourier coefficients can in fact be interpreted as samples of the characteristic function (with a scaling constant). The advantage of this approach is that when a signal is passed through an M-level quantizer, the quantization operation can be interpreted in the Fourier domain as a downsampling operation of its Fourier coefficient sequence by a factor of M (combined with a polar modulation operation). This property can be used to derive easily Sripad and Snyder's conditions for quantization noise uniformity and whiteness. It also allows the interpretation of quantization as a contraction operation on the PDF of the input signal. The uniform distribution is the unique fixed point of this contraction. Since many quantizers (like pipelined ADCs) are implemented in stages, one can think of each stage as performing a flattening operation on the input PDF, until the quantization

B. C. Levy, *Random Processes with Applications to Circuits and Communications*,
https://doi.org/10.1007/978-3-030-22297-0_12

residual is uniformly distributed at the end of the quantization chain. This insight is in fact used by a class of random number generators [7] which operate by feeding back the output of a quantizer to its input, and by using each successive quantization decision to generate the random number sequence.

In some signal processing systems, such as audio or imaging systems, one unsatisfactory aspect of signal quantization systems is that they produce an error which is not independent of the input signal, since it is in fact a deterministic function of this signal. This lack of independence is often detected by the human auditory or vision system, which is an undesirable aspect of quantization [8]. To alleviate this limitation, Roberts [9] and later Schuchman [10] proposed to add a small dither signal to the quantizer input. For subtractive dither systems, this dither is removed from the quantizer output, whereas for nonsubtractive dither [11, 12], it is not removed. For simplicity, we focus here on subtractive dither. The Fourier series analysis of PDFs with finite support is extended to the case where an additive analog dither is added to the quantizer, and dither signals which ensure the independence of the quantization error from the input signal are characterized. It turns out that in this class, the dithers with the smallest amplitude are uniformly distributed with half of a least-significant bit (LSB) amplitude.

This chapter is organized as follows. Section 12.2 describes the operation of single and multistage quantizers. Section 12.3 characterizes the quantization error statistics and analyzes the properties of the Frobenius–Perron operator which converts the quantizer input PDF into its output PDF. The effect of quantization on the joint statistics of the quantization error is studied in Sect. 12.4, where conditions for whiteness are presented. Finally, the effect of subtractive dither signals on the quantization error is analyzed in Sect. 12.5.

12.2   Quantization

Consider the quantization of a random variable X with CDF $F_X(x)$ and PDF $f_X(x)$. Unlike the ideal quantizer considered in Problem 2.23 which had an infinite number of steps, in this section we consider quantizers with a finite number of discretization steps. We consider two architectures: single-stage quantizers, which discretize X in one step, or multistage quantizers which perform the discretization gradually. These two types of quantizers can be implemented by flash and pipelined analog-to-digital converters (ADCs), respectively [13].

12.2.1 Single-Stage Quantizer

The quantizer we consider has M steps evenly distributed over a $[-1, 1]$ normalized dynamic range, so that the step size is $\Delta = 2/M$. To simplify the analysis, we assume that the quantizer does not overload, so that the input random variable X takes values in the range $[-1, 1]$. This is typically ensured by scaling the input signal to ensure that its amplitude rarely exceeds 1. We consider a uniform roundoff quantizer $q(\cdot)$ which, given input X, produces the quantized value

$$q(X) = \left(\lfloor \frac{(X+1)}{\Delta} \rfloor + \frac{1}{2} \right) \Delta - 1$$

$$= \left(\lfloor \frac{(X+1)}{\Delta} \rfloor - \frac{(M-1)}{2} \right) \Delta, \tag{12.1}$$

where the floor function $\lfloor z \rfloor$ denotes the largest integer smaller than or equal to z. Thus if

$$m = \lfloor \frac{(X+1)}{\Delta} \rfloor, \tag{12.2}$$

we have $0 \leq m \leq M - 1$ and the corresponding quantized value of X is

$$Q(X) = x_m = \left(m - \frac{(M-1)}{2}\right)\Delta. \tag{12.3}$$

The quantizer input–output characteristic is plotted in Fig. 12.1 for $M = 4$.

The number M of quantization steps is often a power of 2, and if $M = 2^L$, the quantizer is called an L-bits quantizer. Note that M varies exponentially with L, and since the quantizer implementation requires $M - 1$ comparators, its complexity grows very quickly with L.

The corresponding quantization error

$$E = X - q(X) \tag{12.4}$$

is such that $-\Delta/2 \leq X \leq \Delta/2$. It is a piecewise linear function of X, which is plotted in Fig. 12.2 for $M = 4$. The comparator values used in the discretization of X correspond to the vertical lines in this plot.

To analyze the quantizer, we consider the joint PMF/CDF of the quantized value Q and error E. We have

Fig. 12.1 Input–output characteristic of a 4 level roundoff quantizer

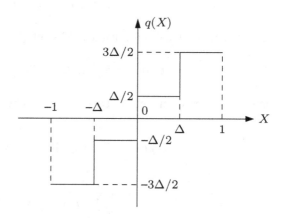

Fig. 12.2 Error function for a $M = 4$ level quantizer

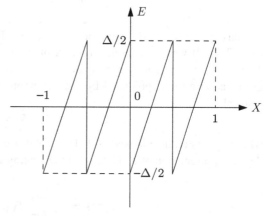

$$F_{QE}(m, e) = P(Q = x_m, E \le e)$$

$$= \begin{cases} 0 & e \le -\Delta/2 \\ F_X(x_m + e) - F_X(x_m - \Delta/2) & -\Delta/2 < e \le \Delta/2 \\ F_X(x_m + \Delta/2) - F_X(x_m - \Delta/2) & e > \Delta/2 \end{cases}$$

$$= [F_X(x_m + e) - F_X(x_m - \Delta/2)]u(e + \Delta/2)$$

$$+ [F_X(x_m + \Delta/2) - F_X(x_m + e)]u(e - \Delta/2), \tag{12.5}$$

where

$$u(z) = \begin{cases} 1 & z \ge 0 \\ 0 & z < 0 \end{cases}$$

denotes the unit step function. Differentiating with respect to e, we find that the joint PMF/PDF of Q and E is specified by

$$f_{QE}(m, e) = \frac{d}{de} F_{QE}(m, e)$$

$$= f_X(x_m + e)[u(e + \Delta/2) - u(e - \Delta/2)]. \tag{12.6}$$

From (12.5), we find that the marginal PMF of Q is

$$p_Q(m) = F_Q(m, \infty) = F_X(x_m + \Delta/2) - F_X(x_m - \Delta/2), \tag{12.7}$$

and from (12.6), the marginal PDF of E is given by

$$f_E(e) = \sum_{m=0}^{M-1} f_{QE}(m, e)$$

$$= \left[\sum_{m=0}^{M-1} f_X(x_m + e) \right] (u(e + \Delta/2) - u(e - \Delta/2)). \tag{12.8}$$

Thus, given an input distributed over interval $[-1, 1]$, the quantizer error is distributed over $[-\Delta/2, \Delta/2]$ which represents $1/M$-th of the dynamic range of X. This suggests that to digitize X, instead of proceeding in a single step, we can actually use a multistage quantizer, where the quantization error E produced by each component quantizer is first amplified by M, producing a residual

$$R = ME \tag{12.9}$$

whose dynamic range is now $[-1, 1]$, so that it can be passed to another quantizer for refining the digital representation of X. Note that after scaling E by M, the PDF of R becomes

$$f_R(r) = \frac{1}{M} \sum_{m=0}^{M-1} f_X\left(x_m + \frac{r}{M}\right)$$

$$= \frac{1}{M} \sum_{m=0}^{M-1} f_X\left(\frac{2m - (M - 1) + r}{M}\right). \tag{12.10}$$

for $-1 \le r \le 1$ and $f_R(r) = 0$ otherwise.

Fig. 12.3 Multistage quantizer

Fig. 12.4 Stage of a
multistage quantizer

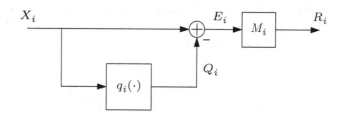

12.2.2 Multistage Quantizer

A multistage quantizer is obtained by applying a chain of I uniform quantizers to the input X as shown in Fig. 12.3. The i-th quantizer, with $1 \leq i \leq I$, discretizes the residual R_{i-1} produced by the previous quantizer with M_i discretization steps, and its quantization error E_i is scaled by M_i to produce the residual R_i which will be passed to the next quantizer as shown in Fig. 12.4.

The input of the first stage of the multistage quantization chain is $X_1 = X$. By eliminating E_1 from the quantization identities

$$X = Q_1 + E_1$$

$$M_1 E_1 = R_2 = Q_2 + E_2$$

of the first two stages, we obtain

$$X = Q_1 + \frac{1}{M_1}(Q_2 + E_2), \tag{12.11}$$

and proceeding by induction, we find

$$X = Q + E, \tag{12.12}$$

where

$$Q = Q_1 + \sum_{i=2}^{I} \frac{1}{\prod_{j=1}^{i-1} M_j} Q_j \tag{12.13}$$

$$E = \frac{1}{\prod_{i=1}^{I-1} M_i} E_I \tag{12.14}$$

denote respectively the quantized approximation and error of the multistage quantizer. Since E_I is distributed over interval $[-\Delta_I/2, \Delta_I/2]$ with $\Delta_I = 2/M_I$, the expression (12.14) indicates that the error E is distributed over $[-\Delta/2, \Delta/2]$ where

$$\Delta = \frac{2}{M} \quad \text{with} \quad M = \prod_{i=1}^{I} M_i, \tag{12.15}$$

so that a multistage quantizer with I stages, each with M_i quantization levels, $1 \leq i \leq I$, is equivalent to a single stage quantizer with $M = \prod_{i=1}^{I} M_i$ levels. Suppose that the numbers $M_i = 2^{L_i}$ are powers of two. Then the representation (12.13) indicates that the first L_1 significant bits of the binary representation of Q are computed by the first stage, the next L_2 bits are computed by the second stage, and finally the L_I least significant bits are computed by the last stage of the quantizer chain.

What are the advantages and disadvantages of using a multistage quantizer? One advantage is that since each stage requires only $M_i - 1$ comparators, the total number of comparators needed by a multistage implementation is

$$\sum_{i=1}^{I}(M_i - 1)$$

instead of $M - 1$. This difference is substantial if one-bit quantization stage is employed. Indeed, suppose that $L_i = \log_2 M_i = 1$ for all i. Then only one comparator per stage is needed, so that $I = \log_2 M$ comparators are needed for a multistage quantizer with I stages, instead of $M - 1 = 2^I - 1$ for a single stage quantizer. The downside of this clear advantage for a multistage quantizer implementation is that an amplifier with gain M_i is required to generate the residual R_i at each stage, which increases the power consumption of the quantizer, and an increase in latency since I quantization operations are needed to produce the quantized value Q of X in (2.13). While an increase in latency is unavoidable, a multistage quantizer does not necessarily have a lower throughput than a single-stage quantizer if pipelining is employed, i.e., while Stage 2 of the quantizer computes quantization component Q_2 of X, Stage 1 can already be computing quantization component Q_1 of the next analog sample that needs to be digitized.

12.3   Quantization Noise Statistics

Fourier Series Model   To analyze the effect of uniform quantization on a random input X, we follow the approach presented in [6]. Since the PDF $f_X(x)$ has for support $[-1, 1]$, it admits a Fourier series expansion of the form

$$f_X(x) = \sum_{k=-\infty}^{\infty} F_k^X \exp(j\pi k x) \tag{12.16}$$

for $-1 \leq x \leq 1$, where the Fourier coefficients F_k^X are given by

$$F_k^X = \frac{1}{2} \int_{-1}^{1} f_X(x) \exp(-j\pi k x) dx. \tag{12.17}$$

Note that since the total probability mass of X is one, the zero-th order Fourier coefficient

$$F_0^X = \frac{1}{2} \int_{-1}^{1} f_X(x) dx = \frac{1}{2}.$$

The Fourier coefficients F_k^X can be interpreted as the scaled sampled values $F_k^X = \Phi_X(-k\pi)/2$ of the characteristic function

$$\Phi_X(u) = \int_{-1}^{1} f_X(x) \exp(jux) dx \tag{12.18}$$

of X. Note that since $f_X(x)$ is space-limited to interval $[-1, 1]$, by applying a space-frequency dual of Nyquist's sampling theorem, it is not surprising that $\Phi_X(u)$ can be reconstructed exactly from its Nyquist rate samples $\Phi_X(-k\pi)$. To evaluate the probability distribution of the quantization error E, we examine instead the distribution of the residual $R = ME$ which has the advantage of having exactly the same support $[-1, 1]$ as X. We found in (12.10) that the density $f_R(x)$ can be expressed as

$$f_R(x) = \mathbf{P} f_X(x), \tag{12.19}$$

where the Frobenius–Perron operator $\mathbf{P}$ is defined by the right-hand side of (12.10). Substituting the Fourier series expansion (12.16) inside (12.10) gives

$$f_R(x) = \frac{1}{M} \sum_{m=0}^{M-1} \sum_{k=-\infty}^{\infty} F_k^X \exp(j\pi kx/M) \exp(j\pi k(2m - (M-1))/M)$$

$$= \sum_{k=-\infty}^{\infty} S(k) \exp(-j\pi k(M-1)/M) F_k^X \exp(j\pi kx/M), \tag{12.20}$$

where the order of summation is exchanged to go from the first to the second line, and

$$S(k) \triangleq \frac{1}{M} \sum_{m=0}^{M-1} \exp(j2\pi km/M)$$

$$= \begin{cases} 1 \text{ for } k \mod M = 0 \\ 0 \text{ for } k \mod M \neq 0. \end{cases} \tag{12.21}$$

Then if

$$\epsilon \triangleq M - 1 \mod 2 = \begin{cases} 1 & M \text{ even} \\ 0 & M \text{ odd}, \end{cases} \tag{12.22}$$

substituting (12.21) inside (12.20) gives

$$f_R(x) = \sum_{\ell=-\infty}^{\infty} (-1)^{\epsilon\ell} F_{M\ell}^X \exp(j\pi\ell x), \tag{12.23}$$

which is the Fourier series representation of the residual PDF $f_R(x)$. Thus, the Fourier coefficients F_k^R of the PDF of R satisfy

$$F_k^R = (-1)^{\epsilon k} F_{Mk}^X \tag{12.24}$$

so they can be viewed as obtained by downsampling (see [14, Sec. 4.6] for a discussion of the downsampling operation for discrete-time signals) the input Fourier coefficient sequence F_k^X by a factor M, followed when M is even by a polar modulation operation consisting of flipping the sign of odd elements of the resulting downsampled sequence. The equivalence between the Frobenius–Perron operator $\mathbf{P}$ applied on the input PDF and a combined downsampling and modulation operation in the Fourier coefficient domain is illustrated in Fig. 12.5.

If we consider a multistage quantizer with I stages, the PDF of the residual R_i at the output of the i-th stage is related to the PDF of the input residual R_{i-1} through

$$f_{R_i}(x) = \mathbf{P}_i f_{R_{i-1}}(x), \tag{12.25}$$

Fig. 12.5 Equivalence between **P** and decimation followed by sign modulation in the Fourier coefficient domain

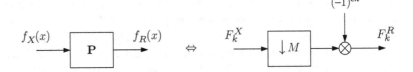

where

$$\mathbf{P}_i f(x) = \frac{1}{M_i} \sum_{m=0}^{M_i-1} f\left(x_m^i + \frac{x}{M_i}\right)$$

denotes the Frobenius–Perron operator of the i-th quantization stage, with

$$x_m^i = \left(m - \frac{(M_i - 1)}{2}\right)\Delta_i \quad \text{and} \quad \Delta_i = \frac{2}{M_i}.$$

Accordingly, the Frobenius–Perron operator of the entire multistage quantizer is

$$\mathbf{P} = \mathbf{P}_I \dots \mathbf{P}_i \dots \mathbf{P}_1 \tag{12.26}$$

and can be expressed in the form (12.10) with $M = \prod_{i=1}^{I} M_i$. The Fourier coefficient representation of the i-th stage Frobenius–Perron operator (12.25) is

$$F_k^{R_i} = (-1)^{\epsilon_i k} F_{M_i k}^{R_{i-1}}, \tag{12.27}$$

so if

$$N_i = \prod_{\ell=1}^{i} M_\ell$$

$$\eta_i = \sum_{j=1}^{i} \left(\prod_{\ell=j+1}^{i} M_j\right)\epsilon_j \quad \text{mod } 2,$$

the cumulative effect of the first i quantization stages can be represented in the Fourier domain as

$$F_k^{R_i} = (-1)^{\eta_i k} F_{N_i k}^{X}, \tag{12.28}$$

where F_k^X denotes the kth Fourier coefficient of the input PDF $f_X(x)$. It can be proved by induction that η_i satisfies

$$\eta_i = N_i - 1 \quad \text{mod } 2. \tag{12.29}$$

At the output of stage 1, we have trivially $N_1 = M_1$ and

$$\eta_1 = \epsilon_1 = M_1 - 1 \quad \text{mod } 2.$$

Then, suppose that (12.29) holds for index i. This implies

$$\eta_{i+1} = \epsilon_{i+1} + M_{i+1}\eta_i \quad \text{mod } 2$$

$$= M_{i+1} - 1 + M_{i+1}(N_i - 1) \quad \text{mod } 2$$

$$= M_{i+1}N_i - 1 \quad \text{mod } 2 = N_{i+1} - 1 \quad \text{mod } 2,$$

so identity (12.29) holds for $i + 1$. This verifies that the Fourier coefficient representations of the overall multistage quantizer $\mathbf{P}$ and of its stages $\mathbf{P}_i$ are consistent. Thus whereas a single-stage quantizer with $M = \prod_{i=1}^{I} M_i$ steps can be represented in the Fourier coefficient domain by a downsampling operation by M combined with a polar modulation whose index ϵ is given by (12.22), when this quantizer is implemented by elementary quantization stages in series, the downsampling and modulation operations are implemented incrementally, so that the i-th stage $1 \leq i \leq I$ applies downsampling factor M_i and modulation index ϵ_i. From an analytical point of view, a single stage uniform quantizer or a multistage one with uniform stages are entirely equivalent as long as they have the same number M of quantization steps and the same dynamic range $[-1, 1]$.

Let

$$v(x) = \begin{cases} 1/2 & -1 \leq x \leq 1 \\ 0 & \text{otherwise} \end{cases} \tag{12.30}$$

denote the uniform distribution over interval $[-1, 1]$ whose Fourier coefficients are

$$F_k^v = \begin{cases} 1/2 & k = 0 \\ 0 & k \neq 0. \end{cases}$$

By observing that for any M, the decimation and modulation operations of Fig. 12.5 do not alter F_k^v, we deduce that the uniform density $v(x)$ is invariant under $\mathbf{P}$, i.e.,

$$\mathbf{P}v(x) = v(x) \tag{12.31}$$

so that quantizing a uniform density produces a uniform residual. Even when $f_X(x)$ is not uniform, the quantization residual R (or equivalently the quantization error E) may be uniformly distributed.

Theorem 1 *If X is quantized with an M-level quantizer, the residual R is uniformly distributed over $[-1, 1]$ (or equivalently E is uniformly distributed over $[-\Delta/2, \Delta/2]$) if and only if the Fourier coefficients F_k^X of $f_X(x)$ satisfy*

$$F_{Mk}^X = 0 \tag{12.32}$$

for $k \neq 0$.

This result relies on the representation (12.24) of the residual Fourier coefficients and the observation that in order for R to be uniformly distributed, its Fourier coefficients must satisfy

$$F_k^R = \begin{cases} 1/2 & k = 0 \\ 0 & k \neq 0. \end{cases}$$

It turns out that several classes of PDFs satisfy condition (12.32).

Example 12.1 The most important class, which was first identified by [2] and further studied in [3], consists of input PDFs $f_X(x)$ which are *spatially bandlimited*, i.e., such that

$$F_k^X = 0 \quad \text{for} \quad |k| \geq k_0 \tag{12.33}$$

Then as long as the number M of quantization steps exceeds k_0, the condition (12.32) holds.

However, as noted in [4], the class of spatially bandlimited PDFs is not the only one for which R is uniformly distributed.

Example 12.2 Consider the triangular PDF

$$f_T(x) = \begin{cases} 1 - |x| & |x| \leq 1 \\ 0 & \text{otherwise} \end{cases}$$

plotted in Fig. 12.6, or the raised cosine PDF

$$f_{RC}(x) = \begin{cases} \frac{1}{2}[1 + \cos(\pi x)] & |x| \leq 1 \\ 0 & \text{otherwise} \end{cases}$$

plotted in Fig. 12.7. Both of these densities have the feature that they can be expressed as

$$f(x) = v(x) + d(x),$$

Fig. 12.6 Triangular input probability density

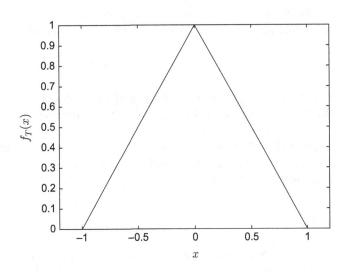

Fig. 12.7 Raised cosine probability density

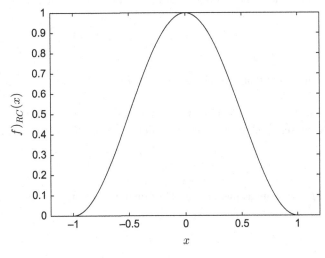

where the function $d(x)$ is such that it has the half-wave symmetry when it is extended periodically outside interval $[-1, 1]$. Periodic functions with this property have the feature that all their even-indexed Fourier coefficients are zero [15, pp. 766–767]. Accordingly, the residual R will be uniformly distributed if the number M of quantization steps is even.

For M-level quantizers, it is also shown in Problem 12.2 that PDFs which are piecewise constant over quantization intervals $[x_m - \Delta/2, x_m + \Delta/2]$ with x_m given by (12.3) have a uniformly distributed residual.

Contraction Property of P To explain why the sequential application of quantization operators $\mathbf{P}_i$, possibly with different numbers M_i of quantization levels, typically produces a uniform density, it is useful to observe that $\mathbf{P}$ is *contractive* for both the $L^1[-1, 1]$ and $L^2[-1, 1]$ norms. Let f and g be two PDFs with support $[-1, 1]$. Then

$$
\begin{aligned}
\|\mathbf{P}(f - g)\|_1 &= \int_{-1}^{1} |\mathbf{P}(f(x) - g(x))| dx \\
&= \frac{1}{M} \int_{-1}^{1} |\sum_{m=0}^{M-1} f\left(x_m + \frac{x}{M}\right) - g\left(x_m + \frac{x}{M}\right)| dx \\
&\leq \frac{1}{M} \sum_{m=0}^{M-1} \int_{-1}^{1} |f\left(x_m + \frac{x}{M}\right) - g\left(x - m + \frac{x}{M}\right)| dx \\
&= \sum_{m=0}^{M-1} \int_{x_m - \Delta/2}^{x_m + \Delta/2} |f(z) - g(z)| dz \\
&= \int_{-1}^{1} |f(z) - g(z)| dz = \|f - g\|_1,
\end{aligned}
\tag{12.34}
$$

where the discretization levels x_m, $0 \leq m \leq M - 1$ are given by (12.3), $\Delta = 2/M$ and to go from the third to fourth line we have performed the change of variable $z = x_m + x/M$.

Likewise, if we consider the L^2 norm, by using Parseval's identity and the downsampling interpretation of $\mathbf{P}$ in the Fourier coefficient domain, we obtain

$$
\begin{aligned}
\|\mathbf{P}(f - g)\|_2^2 &= \int_{-1}^{1} |\mathbf{P}(f(x) - g(x))|^2 dx = \frac{1}{2} \sum_{k=-\infty}^{\infty} |F_{Mk} - G_{Mk}|^2 \\
&\leq \frac{1}{2} \sum_{k=-\infty}^{\infty} |F_k - G_k|^2 \\
&= \int_{-1}^{1} |f(x) - g(x)|^2 dx = \|f - g\|_2^2.
\end{aligned}
\tag{12.35}
$$

Since the uniform density $v(x)$ is an invariant density of $\mathbf{P}$, the contraction identities (12.34) and (12.35) imply

$$
\|Pf - v\| \leq \|f - v\|
\tag{12.36}
$$

for both the L^1 and L^2 norms, so typically, the quantization operator $\mathbf{P}$ brings the PDF $f(x)$ closer to uniform density $v(x)$. Unfortunately, the inequality (12.34) is not strict, as can be seen from the following example.

Example 12.3 Consider a quantizer with $M \geq 2$ steps. Its input X has for PDF

$$f_X(x) = \frac{1}{2}(1 + \cos(\pi M x))$$

for $0 \leq |x| \leq 1$ and $f_X(x) = 0$ otherwise. By using the downsampling model of Fig. 12.5, we find

$$\mathbf{P}f_X(x) = \frac{1}{2}(1 + (-1)^\epsilon \cos(\pi x)).$$

But since

$$f_X(x) - v(x) = \frac{1}{2} \cos(\pi M x)$$

$$\mathbf{P}f_X(x) - v(x) = \frac{(-1)^\epsilon}{2} \cos(\pi x)$$

have the same $L^2[-1, 1]$ norm, we deduce that for this example

$$\|\mathbf{P}f_X - v\|_2 = \|f_X - v\|_2,$$

so that the distance from v is not affected by the quantization operator $\mathbf{P}$. However, this example is rather artificial, since when a quantizer with $M - 1$ quantization levels or a number of quantization levels greater than M is applied to $f_X(x)$, the residual R is uniformly distributed. So this example just involves a poor match between the quantizer and its input density.

Since we have seen earlier that the Fourier coefficients F_k^X of PDF $f_X(x)$ are obtained by sampling the characteristic function $\Phi_X(u)$ with sampling period π, the sampling model of Fig. 12.5 suggests that L^2 distance between $\mathbf{P}f_X$ and v will decrease rapidly as long as $\Phi_X(u)$ decays sufficiently rapidly as u becomes large.

Example 12.4 To illustrate the last comment, consider a multistage quantizer with $I = 3$ stages, where each stage has $M_i = 2$ quantization levels, resulting in a three-bits quantizer. The quantizer input is an iid zero-mean Gaussian sequence with standard deviation $\sigma = 1/3$ truncated to interval $[-1, 1]$. The input PDF is therefore

$$f_X(x) = \frac{1}{C\sigma(2\pi)^{1/2}} \exp\left(-\frac{x^2}{2\sigma^2}\right)$$

for $0 \leq |x| \leq 1$ and $f_X(x) = 0$ otherwise, where the normalization constant

$$C = \frac{1}{\sigma(2\pi)^{1/2}} \int_{-1}^{1} \exp\left(-\frac{x^2}{2\sigma^2}\right) dx$$

$$= \text{erf}\left(\frac{1}{\sigma 2^{1/2}}\right)$$

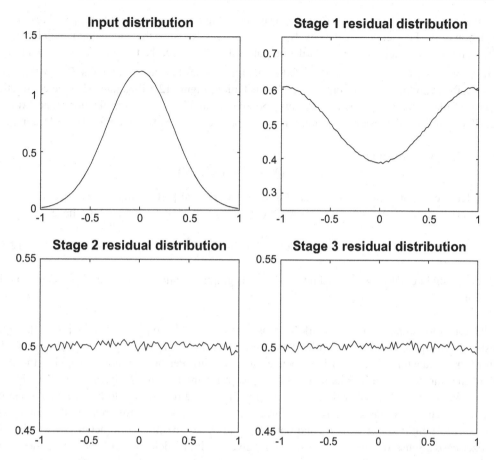

Fig. 12.8 Empirical input and residual distributions at stages 1–3 for an iid truncated Gaussian input sequence of length $K = 10^7$ with $\sigma = 1/3$

ensures that the total probability mass is unity. With $\sigma = 1/3$, the maximum allowed amplitude of input X is 3 times the standard deviation, so that the amount of truncation is negligible ($C = 0.9973$). A simulation was performed by passing a sequence of length $K = 10^7$ through the quantizer. The empirical distributions of the residuals at the outputs of each quantization stage are shown in Fig. 12.8. While the quantization residual at the output of the first stage is not uniform, the residuals at the outputs of the second and third stage are approximately uniformly distributed.

To interpret this experimental result, note that the characteristic function of an untruncated zero-mean Gaussian variable with standard deviation σ is

$$\Phi_X(u) = \exp(-\sigma^2 u^2/2).$$

Neglecting truncation effects, the Fourier coefficients of the PDF of X are therefore approximately

$$F_k^X = \frac{1}{2}\Phi(-k\pi) = \frac{1}{2}\exp(-(k\sigma\pi)^2/2).$$

Since passing X and then residuals R_i through successive quantization stages with $M_i = 2$ results in a decimation of the Fourier coefficients by M_i, at each successive stage the rate of exponential decay of $|F_k^{R_i}|$ increases by a factor $M_i^2 = 4$. At the output of the first stage, the first Fourier coefficients $F_{\pm 1}^{R_1}$ are still non-negligible, but after the second stage, all residual Fourier coefficients $F_k^{R_2}$ with $k \neq 0$ are negligible, resulting in a uniform residual distribution. Figure 12.8 illustrates also the contraction property of the quantization operator $\mathbf{P}_i$ corresponding to an elementary quantization stage. At each stage, $\mathbf{P}_i$ compresses the residual distribution $f_{R_i}(x)$ progressively closer to uniform distribution $\nu(x)$.

Remark The transformation

$$R = M(X - q(X)) = \tau(X) \tag{12.37}$$

has the feature that it is piecewise linear and maps the interval $[-1, 1]$ into itself. Specifically over each subinterval $I_m = [x_m - \Delta/2, x_m + \Delta/2]$ with $0 \leq m \leq M - 1$ of $[-1, 1]$, we have

$$\tau(x) = M(x - x_m), \tag{12.38}$$

so that each subinterval I_m is mapped into $[-1, 1]$. Its graph is plotted in Fig. 12.9 for a quantizer with $M = 5$ levels.

The transformation $\tau(x)$ is a Markov transformation of the type studied in [16, 17]. These transformations play an important role in the design of chaos and random number generators. To each Markov transformation $\tau(x)$ is associated a Frobenius–Perron operator $\mathbf{P}$ mapping the input PDF of the transformation into its output PDF. Except for the fact that $\mathbf{P}$ operates on PDFs instead of finite dimensional probability distribution vectors, it plays a role similar to the one-step transition probability matrix of a Markov chain. The convergence of iterates $\mathbf{P}^m f_X$ towards an invariant density as m tends to infinity under certain conditions on τ has been investigated in detail [16, 17]. For the quantizer case discussed here, we have seen that, given an initial density $f_X(x)$ whose characteristic function $\Phi_X(u)$ decays sufficiently rapidly with u, $\mathbf{P}^m f_X(x)$ approaches a uniform density $\nu(x)$ rapidly. If we assume that $\mathbf{P}$ is a one-bit ($M = 2$) quantizer, once convergence has been achieved, the successive quantization bits $B_i = Q_i + 1/2$ generated by the quantizer will be iid with binary distribution

Fig. 12.9 Transformation $\tau(x)$ for $M = 5$

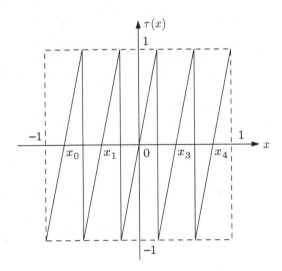

Fig. 12.10 Random
number generator
implementation by a
one-bit quantizer in
feedback configuration

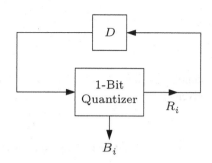

$$P[B_i = 0] = P[B_i = 1] = 1/2.$$

Thus consider the system shown in Fig. 12.10 obtained by feeding back the output residual of a binary quantizer to its input after a unit delay D. After a short transient period, an ideal binary random sequence is generated [7]. Of course, this scheme assumes that the quantizer is ideal, which is difficult to ensure in practice.

12.4 Random Process Quantization

Up to this point, we have considered the quantization of a single random variable X. In practice, analog-to-digital converters (ADCs) quantize random processes. Specifically, given a CT random process $X_c(t)$, an ADC will generate quantized values $Q(X(n))$ of the DT sampled process $X(n) = X_c(nT)$ obtained by sampling $X_c(t)$ with period T. In other words ADCs combine time sampling and quantization. Consider N samples $X(n_1), X(n_2), \ldots, X(n_N)$ with $n_1 < n_2 < \cdots < n_N$ and assume that they admit a joint PDF $f_{\mathbf{X}}(\mathbf{x})$, where

$$\mathbf{X} = \left[X(n_1) \, X(n_2) \, \ldots \, X(n_N) \right]^T$$
$$\mathbf{x} = \left[x_1 \, x_2 \, \ldots \, x_N \right]^T \in \mathbb{R}^N.$$

This assumption rules out situations where the input signal is not sufficiently rich. For example, a constant signal $X(n) = A$ with A random is completely known from one time sample, so it does not admit joint PDFs at two or more times. Likewise a sinusoidal signal $X(n) = A \cos(\omega_0 n + \Theta)$, with random amplitude and phase A and Θ, is perfectly known up to a phase symmetry from two different time samples, so it does not admit joint densities at three different times.

We assume again that $X(n)$ takes values in $I = [-1, 1]$ for all n, so that the quantizer does not overload. Accordingly the support of $f_{\mathbf{X}}(\mathbf{x})$ is contained in I^N, and $f_{\mathbf{X}}(\mathbf{x})$ admits an N-dimensional Fourier series representation

$$f_{\mathbf{X}}(\mathbf{x}) = \sum_{\mathbf{k} \in \mathbb{Z}^N} F_{\mathbf{k}}^{\mathbf{X}} \exp(j\pi \mathbf{k}.\mathbf{x}), \tag{12.39}$$

where

$$\mathbf{k} = \left[k_1 \, k_2 \, \ldots \, k_N \right]^T$$

denotes the Fourier coefficient vector index and

$$\mathbf{k}.\mathbf{x} = \sum_{i=1}^{N} k_i x_i$$

denotes the inner product of vectors $\mathbf{k}$ and $\mathbf{x}$. In (12.39), the N-dimensional Fourier coefficients are given by

$$F_{\mathbf{k}}^{\mathbf{X}} = \frac{1}{2^N} \int_{I^N} f_{\mathbf{X}}(\mathbf{x}) \exp(-j\pi \mathbf{k}.\mathbf{x}) d\mathbf{x}. \tag{12.40}$$

To each sample $X(n)$ corresponds a quantization error $E(n) = X(n) - Q(X(n))$ and residual

$$R(n) = ME(n) = \tau(X(n)),$$

where the transformation $\tau(\cdot)$ is given by (12.37). Since $R = \tau(X)$ is M-to-1 and piecewise linear with inverse

$$X = x_m + \frac{R}{M}$$

in the m-th linearity region $I_m = [x_m - \Delta/2, x_m + \Delta/2]$, the joint PDF $f_{\mathbf{R}}(\mathbf{r})$ of the residual vector

$$\mathbf{R} = \begin{bmatrix} R(n_1) \\ R(n_2) \\ \vdots \\ R(n_N) \end{bmatrix} = \begin{bmatrix} \tau(X(n_1)) \\ \tau(X(n_2)) \\ \vdots \\ \tau(X(n_N)) \end{bmatrix}$$

can be expressed in terms of the PDF $f_{\mathbf{X}}(\mathbf{x})$ as

$$f_{\mathbf{R}}(\mathbf{x}) = (\mathbf{P} \otimes \mathbf{P} \ldots \otimes \mathbf{P}) f_{\mathbf{X}}(\mathbf{x})$$
$$= \frac{1}{M^N} \sum_{\mathbf{m} \in [0, M-1]^N} f_{\mathbf{X}}\left(\mathbf{x_m} + \frac{\mathbf{x}}{M}\right). \tag{12.41}$$

In this first line of this expression, $\mathbf{P} \otimes \mathbf{P} \ldots \otimes \mathbf{P}$ denotes the N-fold tensor product of the one-dimensional Frobenius–Perron operator $\mathbf{P}$. In the second line

$$[0, M - 1] = \{0, 1, \ldots, M - 1\}$$

denotes the set of integers modulo M,

$$\mathbf{m} = \begin{bmatrix} m_1 & m_2 & \ldots m_N \end{bmatrix}^T$$

is a vector of $[0, M - 1]^N$, and

$$\mathbf{x_m} = \begin{bmatrix} x_{m_1} & x_{m_2} & \ldots & x_{m_N} \end{bmatrix}^T,$$

where each x_{m_i} has the form (12.3).

Then by following steps similar to those used in the previous section to derive a Fourier coefficient model of the one-dimensional Frobenius–Perron operator $\mathbf{P}$, it is not difficult to verify that for joint

PDFs of order N, the N-fold tensor product $\mathbf{P} \otimes \mathbf{P} \ldots \otimes \mathbf{P}$ admits the equivalent Fourier coefficient representation

$$F_{\mathbf{k}}^{\mathbf{R}} = (-1)^{\epsilon \sum_{i=1}^{N} k_i} F_{M\mathbf{k}}^{\mathbf{X}}, \qquad (12.42)$$

so that the Fourier coefficients of the joint PDF of order N of the residuals $R(n_i)$, $1 \leq i \leq N$ are obtained by first downsampling the Fourier coefficients $F_{\mathbf{k}}^{\mathbf{X}}$ of the joint PDF of the $X(n_i)$s with downsampling matrix $M\mathbf{I}_N$ (see [18, Chap. 12] for an introduction to multirate DSP in several dimensions) and then applying a polar modulation operation to the downsampled sequence.

Theorem 2 *The residuals $R(n_i)$, $1 \leq i \leq N$ are independent and uniformly distributed over $[-1, 1]$ if and only if*

$$F_{M\mathbf{k}}^{\mathbf{X}} = 0 \quad for \quad \mathbf{k} = \begin{bmatrix} k_1 \\ k_2 \\ \vdots \\ k_N \end{bmatrix} \neq \mathbf{0} = \begin{bmatrix} 0 \\ 0 \\ \vdots \\ 0 \end{bmatrix}. \qquad (12.43)$$

Proof The condition (12.43) implies that the Fourier coefficients of the joint PDF of order N of residuals $R(n_i)$, $1 \leq i \leq N$ are all zero except

$$F_0^{\mathbf{R}} = F_0^{\mathbf{X}}$$
$$= \frac{1}{2^N} \int_{I^N} f_{\mathbf{X}}(\mathbf{x}) d\mathbf{x} = \frac{1}{2^N}, \qquad (12.44)$$

so that the joint PDF of the residuals $R(n_i)$, $1 \leq i \leq N$ is given by

$$f_{\mathbf{R}}(\mathbf{x}) = F_0^{\mathbf{R}} 1_{I^N}(\mathbf{x}) = \frac{1}{2^N} 1_{I^N}(\mathbf{x})$$
$$= \prod_{i=1}^{N} \nu(x_i), \qquad (12.45)$$

where

$$1_{I^N}(\mathbf{x}) = \begin{cases} 1 & \mathbf{x} \in I^N \\ 0 & \text{otherwise} \end{cases}$$

denotes the characteristic function of the subset I^N of $\mathbb{R}^N$ and $\nu(x)$ denotes the uniform density over $I = [-1, 1[$ defined in (12.30).

Remark The condition (12.43) is quite restrictive, since in order to ensure that residuals $R(n_i)$, $1 \leq i \leq N$ are independent and uniformly distributed for all choices of times n_i, $1 \leq i \leq N$, all the Fourier coefficients of the joint PDFs at the corresponding times must obey (12.43). These conditions can be reduced if we assume that the input process $X(n)$ is SSS of an appropriate order. For example, if $X(n)$ is SSS of order 2, with

$$\mathbf{X} = \begin{bmatrix} X(n) \\ X(n+m) \end{bmatrix},$$

the joint PDF $f_{\mathbf{X}}(\mathbf{x})$ depends on m only, so that the condition (12.43) needs to be satisfied for each $m \neq 0$ to ensure that residuals $R(n)$ and $R(n + m)$ are independent and uniformly distributed over $[-1, 1]$ for each n and m. In this case, $R(n)$ and $R(n+m)$ are uncorrelated, so that the residual process $R(n)$ (and thus the quantization error $E(n) = R(n)/M$) is a white uniformly distributed process.

Example 12.5 Consider a discrete-time Ornstein–Uhlenbeck (OU) process $X(n)$ with autocorrelation function

$$R_X(m) = E[X(n + m)X(n)] = P_X a^m$$

with $P_X = (0.3)^2$ (so that $\sigma_X = 0.3$, as in Example 12.4) and $a = 0.8$. This process can be generated by using the recursion

$$X(n + 1) = aX(n) + V(n),$$

where $V(n)$ is a WGN with intensity $q = P_X/(1 - a^2)$. Since $X(n)$ is Gaussian and SSS, the joint characteristic function of $X(n)$ and $X(n + m)$ can be expressed as

$$\Phi_{\mathbf{X}}(\mathbf{u}) = \exp\left(- P_X \mathbf{u}^T \begin{bmatrix} 1 & a^m \\ a^m & 1 \end{bmatrix} \mathbf{u}/2\right) \tag{12.46}$$

for $m \neq 0$, where

$$\mathbf{X} = \begin{bmatrix} X(n) \\ X(n + m) \end{bmatrix} \quad \text{and} \quad \mathbf{u} = \begin{bmatrix} u_1 \\ u_2 \end{bmatrix}.$$

Then by comparing the definition (12.40) of the Fourier coefficients $F_{\mathbf{k}}^X$ of the joint PDF $f_{\mathbf{X}}(\mathbf{x})$ of $X(n + m)$ and $X(n)$ with the definition of the joint characteristic function $\Phi_{\mathbf{X}}(\mathbf{u})$, we deduce that

$$F_{\mathbf{k}}^X = \frac{1}{4}\Phi_{\mathbf{X}}(-\mathbf{k}\pi) \quad \text{with} \quad \mathbf{k} = \begin{bmatrix} k_1 \\ k_2 \end{bmatrix}. \tag{12.47}$$

Substituting (12.46) inside (12.47), we find that the Fourier coefficients of the joint density $f_{\mathbf{X}}(\mathbf{x})$ of the OU process are given by

$$F_{\mathbf{k}}^X = \frac{1}{4} \exp\left(- P_X \pi^2 \mathbf{k}^T \begin{bmatrix} 1 & a^m \\ a^m & 1 \end{bmatrix} \mathbf{k}/2\right) \quad \text{for} \quad m \neq 0,$$

and it is then easy to check that the condition (12.43) is approximately satisfied for $M = 8$. Thus the quantization residual $R(n)$ obtained by passing the OU process $X(n)$ through an $M = 8$-levels quantizer should be approximately white. To verify this result, the periodogram of the residuals $R(n)$ obtained by passing an OU process segment of length $N = 10^6$ through an 8-steps quantizer is shown in Fig. 12.11. As expected, the PSD is approximately flat, so that the quantization noise can be modeled as white.

•

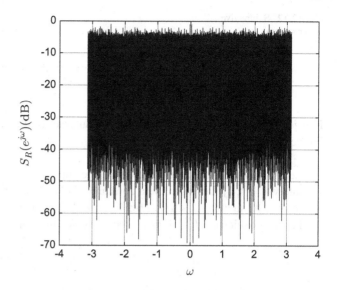

Fig. 12.11 Power spectral density of the quantization residuals obtained by passing a block of length $N = 10^6$ of an OU process with $P_X = (0.3)^2$ and $a = 0.8$ through an 8-levels quantizer

12.5 Dithering

One limitation of the quantization model

$$Q(X) = X - E \tag{12.48}$$

is that although we can ensure that the error E is uniformly distributed by selecting a small enough step size Δ, E is not independent of X, since it is in fact a deterministic function of X. Thus (12.48) cannot be treated as a conventional signal plus independent noise communications model. To overcome this defect, Roberts [9] proposed the addition of a random dither signal D to X, so that after quantization, the new error can be expressed as

$$\check{E} = X + D - Q(X + D). \tag{12.49}$$

Then if the known dither signal is subtracted from the quantized signal $Q(X + D)$, we obtain the observation model

$$\hat{X} = Q(X + D) - D = X - \check{E} \tag{12.50}$$

of quantization. The complete subtractive dither quantization model is shown in Fig. 12.12. Since $\check{E}$ is the quantization error corresponding to input $X + D$, it has exactly the same range of values $[-\Delta/2, \Delta/2]$ as the error E of an undithered quantizer. On the other hand, since the maximum amplitude of the overall input $Z = X + D$ cannot exceed 1, the introduction of dither D necessarily restricts the range of allowable amplitudes of X. Thus dithering has usually the effect of decreasing slightly the quantizer resolution. For the dithering system of Fig. 12.12, Schuchman [10] derived a condition on the PDF $f_D(d)$ of dither signal D which ensures that the error $\check{E}$ is independent of input X, so that observation (12.50) can be viewed as a conventional signal plus independent noise model.

Fig. 12.12 Subtractive
dither system

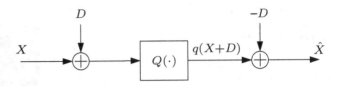

Fourier Series Properties Our analysis of subtractive dither relies on several useful properties of Fourier series.

i) Let X be a random variable such that the support (a, b) of its PDF $f_X(x)$ is contained inside interval $[-1, 1]$. Let also c be a constant such that the random variable $Z = X + c$ is contained in $[-1, 1]$, so that $-1 < a + c < b + c < 1$. Then the PDF of Z is $f_Z(z) = f_X(z - c)$, and the Fourier coefficients F_k^Z of f_Z can be expressed in terms of those of f_X as

$$F_k^Z = F_k^X \exp(-j\pi kc). \tag{12.51}$$

ii) Let X and Y be two independent random variables such that the supports (a, b) and (c, d) of their PDFs f_X and f_Y are contained inside $[-1, 1]$. Then the PDF $f_Z(z)$ of the sum $Z = X + Y$ is the convolution $f_Z(z) = (f_X * f_Y)(z)$ of f_X and f_Y and its support is $(a + c, b + d)$. If this support is contained inside $[-1, 1]$, it is shown in Prob. 12.4 that the Fourier coefficients F_k^Z of f_Z can be expressed in terms of those of f_X and f_Y as

$$F_k^Z = 2F_k^X F_k^Y. \tag{12.52}$$

This is just the Fourier series version of the classical result whereby a convolution in the space domain becomes a product in the Fourier domain.

iii) Let $h(x)$ be a real summable function defined over $[-1, 1]$, so that it admits the Fourier series expansion

$$h(x) = \sum_{k=-\infty}^{\infty} H_k \exp(j\pi kx).$$

Then, by Parseval's identity, if X is random variable contained in $[-1, 1]$ with PDF $f_X(x)$, the expectation of $h(X)$ can be expressed as

$$E[h(X)] = \int_{-1}^{1} h(x) f_X(x) = 2 \sum_{k=-\infty}^{\infty} H_k^* F_k^X, \tag{12.53}$$

where H_k^* denotes the complex conjugate of H_k.

Dithering Analysis Consider a dithered quantizer input $Z = X + D$ where input X and dither D are independent and such that X, D, and Z are all contained in interval $[-1, 1]$. In this case, the PDF $f_Z(z)$ of Z is the convolution of the PDFs f_X and f_D of X and D, respectively, and according to property ii) of Fourier series described above, the Fourier coefficients F_k^Z of $f_Z(z)$ are given by

$$F_k^Z = 2F_k^X F_k^D, \tag{12.54}$$

where F_k^X and F_k^D denote the Fourier coefficients of f_X and f_D, respectively. Likewise, given $X = x$, by property i) of Fourier series, the conditional density $f_{Z|X}(z|x)$ of $Z = D + x$ has for Fourier coefficients

$$F_k^{Z|X}(x) = F_k^D \exp(-j\pi kx). \tag{12.55}$$

Because as noted earlier, the error $\check{E}$ given by (12.49) belongs to interval $[-\Delta/2, \Delta/2]$, the residual

$$\check{R} = M\check{E} = \tau(X + D)$$

is contained in interval $[-1, 1]$. Thus its PDF $f_{\check{R}}(\check{r})$ admits a Fourier series with coefficients $F_k^{\check{D}}$, and the downsampling model (12.24) of quantization implies that if $Q(\cdot)$ is an M-steps quantizer

$$F_k^{\check{R}} = (-1)^{\epsilon k} F_{Mk}^Z = (-1)^{\epsilon k} 2 F_{Mk}^D F_{Mk}^X, \tag{12.56}$$

where ϵ is given by (12.22). Likewise, the conditional density of $\check{R}$ given $X = x$ has for Fourier coefficients

$$F_k^{\check{R}|X}(x) = (-1)^{\epsilon k} F_{Mk}^{Z|X}(x) = (-1)^{\epsilon k} F_{Mk}^D \exp(-j\pi Mkx). \tag{12.57}$$

To ensure that the residual $\check{R}$ (or equivalently, the error $\check{E} = \check{R}/M$) is independent of X, the conditional density $f_{\check{R}|X}(\check{r}|x)$ must be independent of x. Equivalently, the Fourier coefficients $F_k^{\check{R}|X}(x)$ must not depend on x, and expression (12.57) implies therefore

$$F_{Mk}^D = 0 \tag{12.58}$$

for $k \neq 0$.

Theorem 3 *The residual $\check{R}$, or equivalently error $\check{E}$, of the subtractive dither system of Fig. 12.12 is independent of X if and only if the Fourier coefficients F_k^D of the random dither PDF satisfy (12.58). In this case, the residual $\check{R}$ is also uniformly distributed over $[-1, 1]$.*

Since the Fourier coefficients F_k^D of the dither PDF can be expressed in terms of the dither characteristic function $\Phi_D(u)$ as $F_k^D = \Phi_D(-k\pi)/2$, the conditions (12.58) are equivalent to those obtained by Schuchman [10]. By substituting (12.58) inside (12.56), we find

$$F_k^{\check{R}} = \begin{cases} 1/2 & k = 0 \\ 0 & k \neq 0, \end{cases}$$

so that

$$f_{\check{R}}(\check{r}) = \frac{1}{2}$$

for all $-1 \leq \check{r} \leq 1$. Thus, independently of the distribution of X, as long as X, D, and $X + D$ are confined to interval $[-1, 1]$, the dither condition (12.58) ensures uniformity of error $\check{E}$.

It is also worth noting that the statistical independence of $\check{R}$ from X is equivalent to requiring

$$E[h(\check{R})|X] = E[h(\check{R})] \tag{12.59}$$

for all integrable functions h over $I = [-1, 1]$. Indeed, condition (12.59) is implied by

$$f_{\check{R}|X}(\check{r}|x) = f_{\check{R}}(\check{r}). \tag{12.60}$$

Conversely, if (12.59) holds, by selecting $h(\check{r}) = \exp(-j\pi k\check{r})/2$ in (12.59), we obtain $F_k^{\check{R}|X}(x) = F_k^{\check{R}}$ for all k, which in turn implies (12.60).

The condition (12.58) can be recognized as an M-th band filter condition [18, p. 152], and PDFs which satisfy this condition can be characterized as follows. Note first that if

$$f_L(z) = \begin{cases} \frac{1}{\Delta} & -\frac{\Delta}{2} \le z \le \frac{\Delta}{2} \\ 0 & \text{otherwise}, \end{cases} \tag{12.61}$$

is a uniform distribution over an interval $[-\Delta/2, \Delta/2]$ of half a least significant bit (LSB) amplitude, its Fourier series coefficients (see Prob. 12.3) are given by

$$F_k^L = \frac{1}{2}\frac{\sin(k\pi/M)}{k\pi/M}. \tag{12.62}$$

These coefficients satisfy condition (12.58) and have the feature that they are nonzero as long as k is not an integer multiple of M. They represent therefore a least constrained solution of (12.58), in the sense that any other solution may have more zeros. Thus if the dither sequence F_k^D, $k \in \mathbb{Z}$ satisfies (12.58), it can necessarily be expressed as

$$F_k^D = 2F_k^L G_k. \tag{12.63}$$

Assuming that

$$g(z) = \sum_{k=-\infty}^{\infty} G_k \exp(j\pi kz)$$

is nonnegative and has its support in $[-(1 - \Delta/2), (1 - \Delta/2)]$, this implies

$$f_D(z) = f_L(z) * g(z), \tag{12.64}$$

which has a larger support than $f_L(z)$ since when two functions are convolved, the lengths of its supports are added. Thus, among all dithers satisfying condition (12.58), $f_L(z)$ has the smallest support, i.e., the corresponding dither has least amplitude.

Example 12.6 To illustrate the effect of dithering with a half-LSB uniform dither, consider a four-bit quantizer ($M = 16$) with dynamic range $[-1, 1]$, so that $\Delta = 1/8$. For such a quantizer, over interval $I_m = [x_n - \Delta/2, x_m + \Delta/2]$, the squared error $E(x) = (x - x_m)^2$ is quadratic. It equals zero when $X = x_m$ and it reaches its maximum $\Delta^2/4 = 1/256$ at the edges of I_m. The squared error is a periodic piecewise quadratic function as x varies from -1 to 1, as shown in Fig. 12.13.

In contrast, assume that the quantizer is dithered with a half-LSB uniform dither with PDF (12.61). The empirical conditional mean-square-error MSE $< \check{E}^2|x >$ (MSE) obtained by averaging 4×10^6 samples at each discretized value of x in interval $[-(1 - \Delta/2), (1 - \Delta/2)]$ is shown in Fig. 12.14. In

Fig. 12.13 Squared error $E^2(x)$ as a function of x for a four-bit undithered quantizer

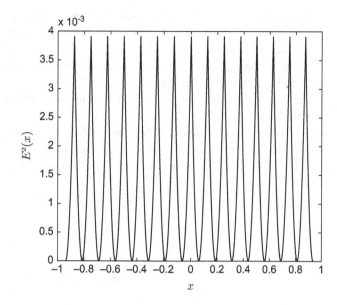

Fig. 12.14 Empirical conditional MSE error $< \check{E}^2 | x >$ of a four-bit dithered quantizer obtained by averaging 4×10^6 samples for each x

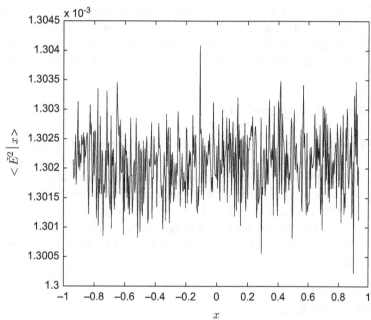

this case, the conditional MSE $< \check{E}^2 | x >$ is almost constant around the variance $\Delta^2/12 = 1.302 \times 10^{-3}$ corresponding to a uniform error distribution over interval $[-\Delta/2, \Delta/2]$, and the remaining random fluctuations are due to the finiteness of the averaging time.

Remarks

a) One drawback of subtractive dither systems is that the dither D must be subtracted from the quantized value $Q(X + D)$. So if the purpose of quantization is signal compression prior to transmission, the analog dither D needs to be approximated and subtracted from the quantized value $Q(X + D)$ prior to transmission. Alternatively, instead of being truly random, D needs to be computed by identical pseudo-random number generators at both the transmitter and receiver. Another approach consists of using a digital dither where D takes uniformly distributed discrete values in interval $[-\Delta/2, \Delta/2]$ (i.e., it takes fraction of a LSB values). For such a dither it is shown in [19] that the conditional distribution of $\check{E}$ given X becomes asymptotically independent of X and converges in distribution to a half-LSB uniform distribution as the number of dithering levels tends to infinity.

b) The dither analysis presented here focused on the independence of the input signal $X(n)$ and quantization error $\check{E}(n)$ at a single time. Sherwood [20] showed that if the Fourier coefficients of the joint PDF of the dither samples $D(n + m)$ and $D(n)$ satisfy the two-dimensional version

$$F^{\mathbf{D}}_{M\mathbf{k}} = 0 \tag{12.65}$$

of condition (12.58), with

$$\mathbf{D} = \begin{bmatrix} D(n) \\ D(n+m) \end{bmatrix} \quad \text{and} \quad \mathbf{k} \in \mathbb{Z}^2,$$

then for $m \neq 0$, $\check{E}(n + m)$ and $\check{E}(n)$ are independent of $X(n + m)$ and $X(n)$, and are independent and uniformly distributed over $[-\Delta/2, \Delta/2]$. Note that condition (12.65) is automatically satisfied if the dither $D(n)$ is an independent sequence of half-a LSB uniformly distributed random variables (or more generally if the $D(n)$s are i.i.d., with a distribution satisfying (12.58)).

12.6   Bibliographical Notes

As mentioned in the introduction, the classical theory of quantization was developed by Widrow [2, 3] and Sripad and Snyder [4], and the effect of subtractive dithering was analyzed by [10, 20]. See also [5] for a book length treatment of quantization and dithering. While conventional analyses of quantization and dithering assume implicitly an infinite quantizer, the presentation given here is based on [6, 19], which applies to a finite quantizer with a fixed dynamic range, which was normalized here to $[-1, 1]$. In contemporary circuit design, quantizer analysis plays an important role in characterizing the behavior of delta–sigma modulator ADCs, and the readers interested by this topic should consult [21–23].

12.7   Problems

12.1 A random variable X has the PDF $f_X(x)$ shown in Fig. 12.15. What is the smallest number M of quantization levels needed to ensure that the residual R (or quantization error E) is uniformly distributed?

12.2 Consider an M-step uniform quantizer and basis functions

$$\phi_m(x) = \begin{cases} 1 & |x - x_m| \leq \Delta/2 \\ 0 & \text{otherwise} \end{cases},$$

Fig. 12.15 PDf $f_X(x)$ of random variable X

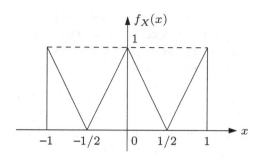

where the center locations x_m with $0 \le m \le M - 1$ are given by (12.3)

a) Consider a piecewise constant PDF of the form

$$g(x) = \sum_{m=0}^{M-1} c_m \phi_m(x) \tag{12.66}$$

with $c_m \ge 0$ for $0 \le m \le M - 1$ and where the constraint

$$\left(\sum_{m=0}^{M-1} c_m \right) \Delta = 1$$

ensures that the total probability mass is one. Prove that

$$\mathbf{P}g(x) = v(x),$$

so that the quantization residual is uniformly distributed.

b) Let $f_X(x)$ be a Lipschitz continuous PDF over $I = [-1, 1]$, so that it satisfies

$$|f_X(x) - f_X(y)| \le C|x - y| \tag{12.67}$$

for $x, y \in I$, where the smallest constant C satisfying this condition is called the Lipschitz modulus of continuity. Functions of this type have the feature that they can be approximated with arbitrary accuracy by piecewise constant functions of the form (12.66). Specifically, consider the piecewise constant approximation

$$f^\Delta(x) = \sum_{m=0}^{M-1} f_X(x_m) \phi_m(x)$$

of $f_X(x)$, where $f_X(x)$ is approximated over each subinterval $I_m = [x_m - \Delta/2, x_m + \Delta/2]$ by its value in the middle of the interval. Show that that for an arbitrary $\epsilon > 0$, the number $M = 2/\Delta$ of approximation subintervals can be selected such that

$$\|f_X - f^\Delta\|_1 \le \epsilon.$$

c) Assume that $f_X(x)$ is Lipschitz continuous and is approximated with accuracy ϵ in the sense of the L_1 norm by piecewise constant function $f^\Delta(x)$ with M approximation subintervals. Use the result of part a) to prove that when $f_X(x)$ is quantized with M levels, if $\mathbf{P}$ denotes the Frobenius–Perron operator of the quantizer

$$\|\mathbf{P}f_X - v\|_1 \leq \epsilon$$

so that the quantization residual is approximately uniform, with the same tolerance ϵ as the approximation of f_X by f^Δ.

12.3 Consider a random dither D with uniform density

$$f_D(x) = \begin{cases} M/2 & |x| \leq 1/M \\ 0 & \text{otherwise} \end{cases}$$

over $[-1/M, 1/M]$.

a) Compute its Fourier coefficients F_k^D and verify that $F_{Mk}^D = 0$ for $k \neq 0$, so that the residual R obtained by passing D through an M-level quantizer is uniformly distributed over $[-1, 1]$.
b) Equivalently, by direct evaluation of $f_R(x) = \mathbf{P}f_D(x)$, where $\mathbf{P}$ is given by (12.10), verify that $f_R(x) = v(x)$ where $v(x)$ is the uniform distribution over $[-1, 1]$ specified by (12.10).

12.4 Let X and Y be two independent random variables with PDFs $f_X(x)$ and $f_Y(y)$ whose supports (a, b) and (c, d) are contained inside interval $I = [-1, 1]$, and let F_k^X and F_k^Y denote the Fourier coefficients of f_X and f_Y, respectively. Then if $Z = X + Y$, the PDF $f_Z(z)$ of Z is the convolution of the densities f_X and f_Y of X and Y and has for support $(a + c, b + d)$. If we assume that this support is again contained in I, so that $-1 < a + c < b + d < 1$, show that the Fourier coefficients F_k^Z of the PDF $f_Z(z)$ satisfy

$$F_k^Z = 2F_k^X F_k^Y,$$

so that as expected, a convolution in the space domain corresponds to a product in the Fourier coefficient domain. *Hint:* Substitute the Fourier series expansion (12.16) of $f_Y(y)$ inside the convolution

$$f_Z(z) = \int_{-1}^{1} f_X(x) f_Y(z - x)dx$$

and then use the definition (12.17) of the Fourier coefficients of $f_X(x)$.

12.5 Consider the average

$$Y = \frac{1}{M} \sum_{i=1}^{M} X_i \tag{12.68}$$

of M independent random variables X_i uniformly distributed over $[-1, 1]$. It can be viewed as the sum of random variables $D_i = X_i/M$, where the D_is are uniformly distributed over $[-1/M, 1/M]$.

a) Use the result of Problem 12.4 to show that the Fourier coefficients of Y are given by

$$F_k^Y = 2^{M-1}(F_k^D)^M,$$

where F_k^D denotes the kth Fourier coefficient of a random variable D uniformly distributed over $[-1/M, 1/M]$.

b) Use the expression for the Fourier coefficients F_k^D obtained in part a) of Problem 12.3 to evaluate the Fourier coefficients F_k^Y of the average Y, and deduce from this expression that the residual R (or equivalently the quantization error E) obtained by passing Y through a M-level quantizer is uniformly distributed over $[-1, 1]$.

c) To illustrate the result of part b), for $M = 4$, generate $K = 10^7$ independent random variables Y_k, $1 \leq k \leq K$ of the form (12.68) and evaluate the corresponding quantization residuals R_k for a 4-level quantizer. Plot the empirical PDFs of the Y_ks and R_ks and verify that the residuals are uniformly distributed over $[-1, 1]$.

12.6 Consider the Perron–Frobenius operator $\mathbf{P}$ of an M-step quantizer. We seek to show that in the class of PDFs $f(x)$ with a summable Fourier coefficient series F_k, i.e., such that

$$0 < f(x) < \sum_{k=-\infty}^{\infty} |F_k| < \infty, \tag{12.69}$$

the uniform PDF $v(x)$ defined by (12.30) is the unique fixed point of $\mathbf{P}$.

a) To explain why the summability condition (12.69) (which guarantees the boundedness of $f(x)$) is required, assume that M is odd and that the quantizer input is $X = 0$. In this case X is deterministic with CDF $F_X(x) = u(x)$ and PDF $f_X(x) = \delta(x)$. Verify that the mapping $\tau(x)$ defined by (12.37) satisfies $\tau(0) = 0$, and conclude that $\delta(x)$ is another fixed point of $\mathbf{P}$.

b) More generally when $X = \xi_0$ is deterministic, verify that $P^n f_X(x) = \delta(x - \xi_n)$ where ξ_n satisfies the recursion $\xi_{n+1} = \tau(\xi_n)$, with initial value ξ_0, so that $P^n f_X$ does not converge to v as n tends to infinity.

c) Under the condition (12.69), prove the pointwise convergence

$$\lim_{n \to \infty} |P^n f(x) - v(x)| = 0$$

for all x in $[-1, 1]$. This proves that v is the unique fixed point of $\mathbf{P}$ in the class of PDFs with a summable Fourier series.

12.7 A misconception appearing often in quantization studies is that if the input distribution f_X satisfies condition (12.32), so that the quantization error E is uniformly distributed over $[-\Delta/2, \Delta/2]$, then the error E and quantizer input X are uncorrelated. Unfortunately, this assertion is incorrect as can be seen from the following counter-example. Consider a one-bit quantizer ($M = 2$) with dynamic range $[-1, 1]$, so that $\Delta = 1$, and the error function

$$E(x) = \begin{cases} x - 1/2 & 0 \leq x \leq 1 \\ x + 1/2 & -1 \leq x < 0 \end{cases}$$

Assume that input X has the uniform PDF

$$f_X(x) = \begin{cases} 1 & 0 \le x \le 1 \\ 0 & \text{otherwise} \end{cases}$$

over range $[0, 1]$.

a) Show that

$$\mathbf{P} f_X(x) = \nu(x),$$

so that the quantization residual R (and thus error E) is uniformly distributed.
b) Evaluate $E[E(X)X]$ and verify it is nonzero.

12.8 Consider a discrete-time sinusoidal signal

$$X(n) = A \cos(\omega_0 n + \Theta),$$

where the amplitude A is Rayleigh distributed with parameter σ_X^2 and independent of the phase Θ, which is uniformly distributed over $[0, 2\pi)$.

a) By observing that $X(n)$ can be rewritten as

$$X(n) = X_c \cos(\omega_0 n) + X_c \sin(\omega_0 n),$$

where X_c and X_s are independent $N(0, \sigma_X^2)$ random variables, verify that

$$\mathbf{X} = \begin{bmatrix} X(n) \\ X(n+m) \end{bmatrix}$$

with $m \ne 0$ is $N(\mathbf{0}, \mathbf{K}_X)$ distributed. Evaluate its covariance matrix $\mathbf{K}_X$ and show it is nonsingular as long as $f_0 = \omega_0/(2\pi)$ is an irrational number, i.e., it cannot be expressed as the ratio of two integers. Observe also that $\mathbf{K}_X$ depends on m only, so that $X(n)$ is SSS of order 2.
b) When f_0 is irrational, use the approach of Example 12.5 to evaluate the 2-D characteristic function $\Phi_{\mathbf{X}}(\mathbf{u})$ of $\mathbf{X}$, and when σ_X is significantly smaller than 1, find the Fourier coefficients $F_{\mathbf{k}}^X$ of the joint PDF $f_{\mathbf{X}}(\mathbf{x})$ of $X(n)$ and $X(n+m)$ over $[-1, 1]^2$.
c) With $f_0 = 1/(2\sqrt{5})$ and $\sigma_X = 0.3$, verify that the condition (12.43) is approximately satisfied for $M = 8$. By passing a block of length $N = 10^6$ of process $X(n)$ through an 8-levels uniform quantizer, compute the periodogram of the residual process $R(n)$. Note that although the periodogram looks approximately flat over $[-\pi, \pi]$, it contains many tones across this band.
d) By observing that $R(n)$ is obtained by passing the sinusoidal signal $X(n)$ through the odd nonlinear function $\tau(x)$, can you interpret the presence of secondary tones in the spectrum of $R(n)$?

12.9

a) Consider an M-level quantizer with step-size $\Delta = 2/M$, and let D be a dither with the triangular PDF

$$f_T(d) = \begin{cases} \frac{1}{\Delta}(1 - |\frac{d}{\Delta}|) & -\Delta \le d \le \Delta \\ 0 & \text{otherwise} . \end{cases}$$

Compute the Fourier coefficients F_k^T of f_T. Do they satisfy condition (12.58)? *Hint:* Note that $f_D(d)$ can be viewed as the convolution of the half-LSB dither PDF $f_L(d)$ with itself.

b) Consider a dither D with the raised cosine PDF

$$f_{RC}(d) = \begin{cases} \frac{1}{2\Delta}(1 + \cos(\pi d/\Delta)) & -\Delta \le d \le \Delta \\ 0 & \text{otherwise} . \end{cases}$$

Compute the Fourier coefficients F_k^{RC} of f_{RC}. Do they satisfy condition (12.58)?

References

1. W. R. Bennett, "Spectra of quantized signals," *Bell System Tech. J.*, vol. 27, pp. 446–472, July 1948.
2. B. Widrow, "A study of rough amplitude quantization by means of Nyquist sampling theory," *IRE Trans. Circuit Theory*, vol. 3, pp. 266–276, Dec. 1956.
3. B. Widrow, "Signal analysis of amplitude-quantized sampled-data systems," *Trans. AIEE, Part II: Applicat. Ind.*, vol. 79, pp. 555–568, Jan. 1961.
4. A. B. Sripad and D. L. Snyder, "A necessary and sufficient condition for quantization errors to be uniform and white," *IEEE Trans. Acoust., Speech, Signal Proc.*, vol. 25, pp. 442–448, Oct. 1977.
5. B. Widrow and I. Kollar, *Quantization Noise– Roundoff Error in Digital Computation, Signal Processing, Control, and Communications*. Cambridge, U. K: Cambridge University Press, 2008.
6. B. C. Levy, "A propagation analysis of residual distributions in pipeline ADCs," *IEEE Trans. Circuits Syst. I*, vol. 58, pp. 2366–2376, Oct. 2011.
7. S. Callegari, R. Rovatti, and G. Setti, "Embeddable ADC-based true random number generator for cryptographic applications exploiting nonlinear signal processing and chaos," *IEEE Trans. Signal Proc.*, vol. 53, pp. 793–805, Feb. 2005.
8. S. P. Lipshitz, R. A. Wannamaker, and I. Vanderkooy, "Quantization and dither: a theoretical survey," *J. Audio Eng. Soc.*, vol. 5, pp. 355–375, May 1992.
9. L. G. Roberts, "Picture coding using pseudo-random noise," *IRE Trans. Information Theory*, vol. 8, pp. 145–154, Feb. 1962.
10. L. Schuchman, "Dither signals and their effect on quantization noise," *IEEE Trans. Commun. Technol.*, vol. 12, pp. 162–165, Dec. 1964.
11. R. M. Gray and T. G. Stockham, Jr., "Dithered quantizers," *IEEE Trans. Informat. Theory*, vol. 39, pp. 805–812, May 1993.
12. R. A. Wannamaker, S. P. Lipshitz, J. Vanderkooy, and J. Nelson Wright, "A theory of nonsubtractive dither," *IEEE Trans. Signal Proc.*, vol. 48, pp. 499–516, Feb. 2000.
13. F. Maloberti, *Data Converters*. New York: Springer Verlag, 2007.
14. A. V. Oppenheim and R. W. Schafer, *Discrete-Time Signal Processing, 3rd edition*. Upper Saddle River, NJ: Prentice-Hall, 2010.
15. J. A. Nilsson and S. A. Riedel, *Electric Circuits, 7th edition*. Upper Saddle River, NJ: Pearson Prentice-Hall, 2005.
16. A. Lasota and M. C. Mackey, *Chaos, Fractals and Noise– Stochastic Aspects of Dynamics, 2nd edition*. New York: Springer Verlag, 1994.
17. A. Boyarsky and P. Gora, *Laws of Chaos– Invariant Measures and Dynamical Systems in One Dimension*. Boston, MA: Birkhauser, 1997.
18. P. P. Vaidyanathan, *Multirate Systems and Filter Banks*. Englewood Cliffs, NJ: Prentice Hall, 1993.

19. B. C. Levy, "A study of subtractive digital dither in single-stage and multi-stage quantizers," *IEEE Trans. Circuits Syst. I*, vol. 60, pp. 2888–2901, Nov. 2013.
20. D. T. Sherwood, "Some theorems on quantization and an example using dither," in *Proc. Conf. Record 19th Asilomar Conf. Circuits, Systems and Computers*, (Pacific Grove, CA), pp. 207–212, Nov. 1986.
21. R. M. Gray, "Quantization noise spectra," *IEEE Trans. Informat. Theory*, vol. 36, pp. 1220–1244, Nov. 1990.
22. I. Galton, "Granular quantization noise in a class of delta-sigma modulators," *IEEE Trans. Informat. Theory*, vol. 40, pp. 848–859, May 1994.
23. S. Pamarti, J. Welz, and I. Galton, "Statistics of the quantization noise in 1-bit dithered single-quantizer digital delta-sigma modulators," *IEEE Trans. Circuits Syst. I*, vol. 54, pp. 492–503, Mar. 2007.

Phase Noise in Autonomous Oscillators

13

13.1 Introduction

Due to its importance in circuit design, the problem of modeling phase noise in autonomous oscillators has attracted considerable attention over the last 50 years. Noteworthy phase noise studies include a spectrum model based on a feedback circuit oscillator parametrization proposed by Leeson [1] and the impulse sensitivity function model proposed by Hajimiri and Lee [2, 3]. Unfortunately, even though these models offer valuable insights for oscillator design, they exhibit significant defects, such as their failure to predict a Lorentzian power spectral density for the oscillator tones affected by noise. The primary reason for this failure is that most early phase noise theories were based on linear perturbations. The first satisfactory nonlinear phase noise theory was proposed by Kaertner [4], with significant later refinements by Demir, Mehrothra, and Roychowdury [5]. The key insight of this approach is that for a stable oscillator, random fluctuations in directions away from the oscillator trajectory remain contained due to the inherent trajectory stability, but fluctuations along the trajectory direction accumulate since they are not subjected to a restoring force. The model constructed by this approach relies on rather simplistic approximations, such as the neglect of amplitude fluctuations about the noiseless oscillator trajectory, but it results in a single scalar diffusion equation for the phase noise. This model can be simplified further by noticing that the phase noise evolves on time scale much slower than the rate of oscillation of the oscillator, so that an averaging method can be employed to describe the slow time-scale evolution of the phase noise. This averaging operation reveals that on a slow time-scale, the phase noise behaves like a scaled Wiener process, where the scaling parameter represents therefore the only and most important parameter of the model. This slow time-scale model can then be used to compute the oscillator autocorrelation and its power spectral density, which shows that the effect of phase noise is to smear the pure impulsive spectral lines of the noiseless oscillator into narrow Lorentzian shaped spectra, which conform with experimental observations.

The analysis presented in this chapter relies on an in-depth understanding of the characteristics of noiseless oscillators. Section 13.2 starts by reviewing conditions which can be used to establish the existence of periodic orbits in nonlinear free-running oscillators. The Floquet theory for analyzing linearized motion in the vicinity of a stable oscillator trajectory is then reviewed. Although perturbations about the periodic trajectory are not periodic, they can be decomposed with respect to a periodic basis, called the Floquet basis, and with respect to this rotating basis, the motion becomes time invariant. Next, we introduce the concept of isochron surface, i.e., the locus of points in the attractor domain of the oscillator with the same asymptotic phase as they reach the oscillator orbit.

© Springer Nature Switzerland AG 2020
B. C. Levy, *Random Processes with Applications to Circuits and Communications*,
https://doi.org/10.1007/978-3-030-22297-0_13

It is shown that the gradient to the isochron surface at a point located on the oscillator orbit is the first vector of the conjugate Floquet basis. With these preliminaries out of the way, the model of the phase noise of noisy oscillators obtained in [4, 5] is derived in Sect. 13.3 under two assumptions. The first assumption is that amplitude fluctuations of the noisy oscillator can be neglected. The second assumption, which seems to have been first noticed in [6], is that the isochron surface is approximately flat in the neighborhood of the current oscillator location. Note that this flatness assumption does not mean that the isochron surface intersects the oscillator orbit perpendicularly. Then by using rules of Ito calculus, we obtain a scalar diffusion equation for the phase noise. This equation is driftless, but the diffusion coefficient is phase noise dependent. Then by observing that for small oscillator noise, the phase noise evolves on a time scale much slower than the oscillator rate of oscillation, it is possible to apply time-averaging methods of the type described in [7, 8] to construct a slow time-scale model of the oscillator phase noise which reduces to a scaled Wiener process. The wrapped phase noise (its evaluation modulo the oscillator period) is therefore uniformly distributed, and the scaled Wiener process characterization of phase noise is used in Sect. 13.4 to evaluate the noisy oscillator autocorrelation and spectrum. It is shown that the oscillator is asymptotically WSS, and its power spectral density is a weighted sum of Lorentzian spectra centered about each of the deterministic oscillator harmonics.

13.2   Oscillator Characteristics

The oscillators we consider in this chapter are described by unforced nonlinear time-invariant differential systems of the form

$$\frac{d\mathbf{x}}{dt} = \mathbf{f}(\mathbf{x}), \tag{13.1}$$

where $\mathbf{x} \in \mathbb{R}^n$ and $\mathbf{f}$ is a vector field mapping $\mathbb{R}^n$ onto itself. The state transition function $\phi(t, \mathbf{x}_0)$ is the solution of (13.1) at time t corresponding to initial state $\mathbf{x}(0) = \mathbf{x}_0$. Because the system is time invariant, the solution at time t corresponding to initial state $\mathbf{x}_s$ at time s is therefore $\phi(t - s, \mathbf{x}_s)$. We say that $\mathbf{x}_0$ is an equilibrium point of system (13.1) if $\mathbf{f}(\mathbf{x}_0) = \mathbf{0}$. Consider the Jacobian

$$\mathbf{J}(\mathbf{x}) = \nabla_{\mathbf{x}}^T \mathbf{f}(\mathbf{x}) = \begin{bmatrix} \dfrac{\partial f_1}{\partial x_1} & \dfrac{\partial f_1}{\partial x_2} & \cdots & \dfrac{\partial f_1}{\partial x_n} \\ \dfrac{\partial f_2}{\partial x_1} & \dfrac{\partial f_2}{\partial x_2} & \cdots & \dfrac{\partial f_2}{\partial x_n} \\ \vdots & & \ddots & \vdots \\ \dfrac{\partial f_n}{\partial x_1} & \dfrac{\partial f_n}{\partial x_2} & \cdots & \dfrac{\partial f_n}{\partial x_n} \end{bmatrix}. \tag{13.2}$$

Then $\mathbf{x}_0$ is a stable equilibrium if $J(\mathbf{x}_0)$ is a stable matrix, i.e., if all its eigenvalues have a negative real part.

13.2.1 Limit Cycles

We are interested in systems of the form (13.1) which have stable periodic trajectories. A trajectory $\mathbf{x}(t)$ is periodic with period T if $\mathbf{x}(t+kT) = \mathbf{x}(t)$ for any $k \in \mathbb{Z}$. It is stable if, after a small perturbation away from the trajectory, it asymptotically reverts to the original trajectory as t tends to ∞. Given a stable periodic trajectory $\mathbf{x}_c(t)$, the curve $C = \{\mathbf{x}_c(t), 0 \le t \le T\}$ is called a *limit cycle* or *periodic orbit* of the system. A system of the form (13.1) may have several limit cycles, but typically, only one is of interest. For planar systems, i.e., for $n = 2$, the following criterion [9, Sec. 3.7], [10, Sec. 3.1] can be employed to establish the existence of a limit cycle.

Poincaré–Bendixson Theorem Consider a domain D enclosed between two simple closed curves C_1 and C_2 such that D does not contain any equilibrium point of (13.1). Recall that a curve is simple if it does not cross itself. Then if along C_1 and C_2, the vector field $f(\mathbf{x})$ points towards the interior of D, the system (13.1) admits a limit cycle C contained in D.

To illustrate this result, we consider two simple examples.

Example 13.1 Consider the dynamical system

$$\frac{dx}{dt} = -y + x(1 - x^2 - y^2)$$
$$\frac{dy}{dt} = x + y(1 - x^2 - y^2). \tag{13.3}$$

The only equilibrium point of this system is the origin. As will be seen below, it is not stable. In the polar coordinates

$$r = (x^2 + y^2)^{1/2} \quad , \quad \theta = \arctan(y/x)$$

the system (13.3) reduces to the two decoupled equations

$$\frac{dr}{dt} = r(1 - r^2)$$
$$\frac{d\theta}{dt} = 1.$$

From these equations we see that $\theta(t) = t + \theta(0)$ and $dr/dt > 0$ for $r < 1$ and $dr/dt < 0$ for $r > 1$. This indicates that for all points (x, y) inside the unit disk, the vector field f is normal to the circle passing through (x, y) and points towards the unit circle $r = 1$. This implies in particular that the equilibrium point at $(0, 0)$ is unstable. Conversely, for points (x, y) outside the unit circle, the vector field is again normal to the circle passing through (x, y) and points towards $r = 1$. This means that the Poincaré–Bendixson theorem is satisfied if we select circles $C_1 = \{(r, \theta) : r = 1/2\}$ and $C_2 = \{(r, \theta) : r = 3/2\}$. For $\theta(0) = 0$, the limit cycle is the periodic trajectory $x_c(t) = \cos(t)$, $y_c(t) = \sin(t)$ which traces the unit circle $r = 1$ and has period $T = 2\pi$. To illustrate this result, the plots of trajectories obtained by selecting initial conditions

$$\begin{bmatrix} x(0) \\ y(0) \end{bmatrix} = \begin{bmatrix} 1/2 \\ 0 \end{bmatrix} \quad \text{and} \quad \begin{bmatrix} 3/2 \\ 0 \end{bmatrix}$$

are shown in Fig. 13.1. The trajectories starting inside and outside the unit circle both converge to the limit cycle tracing the unit circle.

Example 13.2 Consider the Van der Pol oscillator

$$\frac{d^2x}{dt^2} - \mu(1 - x^2)\frac{dx}{dt} + x = 0. \tag{13.4}$$

Fig. 13.1 Trajectories of
oscillator (13.3) with initial
conditions (1/2, 0) and
(3/2, 0)

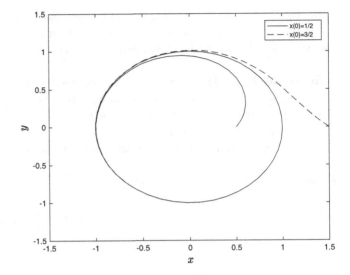

It can be rewritten as the first-order system

$$
\begin{aligned}
\frac{dx}{dt} &= y \\
\frac{dy}{dt} &= -x + \mu(1 - x^2)y.
\end{aligned}
\tag{13.5}
$$

The origin is again the only equilibrium point of this system. To analyze this system, it is again convenient to perform the polar coordinate conversion

$$
x = r \cos(\theta), \; y = r \sin(\theta).
$$

The coordinates are no longer separable since

$$
\frac{dr}{dt} = \mu r (1 - r^2 \cos^2(\theta)) \sin^2(\theta),
$$

but we can deduce that for $r < 1$, $dr/dt > 0$, so that along any circle C_1 of radius less than 1, the vector field points outward. By piecing several curve segments [11], it is also possible to construct a second curve C_2 such that the vector field points towards the interior of the domain D between C_1 and C_2. The equilibrium point at the origin does not belong to this domain and the Poincaré–Bendixson theorem indicates therefore the presence of a limit cycle inside D. To illustrate this limit cycle, let $\mu = 1/4$. Then the trajectories obtained by selecting initial conditions

$$
\begin{bmatrix} x(0) \\ y(0) \end{bmatrix} = \begin{bmatrix} 1 \\ 0 \end{bmatrix} \quad \text{and} \quad \begin{bmatrix} 3 \\ 0 \end{bmatrix}
$$

are plotted in Fig. 13.2. The two trajectories spiral outward and inward, respectively, until they reach the same limit cycle, which unfortunately does not have a closed-form expression.

Fig. 13.2 Trajectories of
the Van der Pol oscillator
with initial conditions
$(1, 0)$ and $(3, 0)$

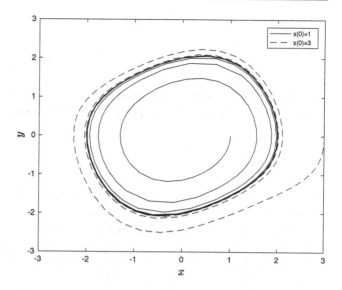

13.2.2 Floquet Theory

When a small perturbation $\delta(t)$ is applied to a periodic orbit $\mathbf{x}_c(t)$ of (13.1) so that the oscillator position is

$$\mathbf{x}(t) = \mathbf{x}_c(t) + \delta(t),$$

by performing a Taylor series expansion of (13.1) in the vicinity of $\mathbf{x}_c(t)$, we obtain

$$\frac{d}{dt}(\mathbf{x}_c(t) + \delta(t)) = \mathbf{f}(\mathbf{x}_c(t)) + \mathbf{A}(t)\delta(t) + O(||\delta(t)||^2), \tag{13.6}$$

where

$$\mathbf{A}(t) = \mathbf{J}(\mathbf{x}_c(t))$$

denotes the vector field Jacobian defined in (13.2) evaluated along the periodic orbit $\mathbf{x}_c(t)$, and where $O(||\delta||^2)$ denotes a term proportional to $||\delta||^2$. This indicates that small trajectory perturbations satisfy the linear differential equation

$$\frac{d\delta}{dt} = \mathbf{A}(t)\delta(t), \tag{13.7}$$

where $\mathbf{A}(t)$ is periodic with period T since $\mathbf{x}_c(t)$ is periodic with the same period.

Linear periodic systems of the form (13.7) were studied in detail by the French mathematician Gaston Floquet in the second half of the 19th century. To analyze system (13.7), we start by noting that since $\mathbf{x}_c(t)$ solves the differential equation (13.1), by differentiating

$$\frac{d\mathbf{x}_c}{dt} = \mathbf{f}(\mathbf{x}_c(t))$$

and denoting $\mathbf{v}(t) = d\mathbf{x}_c/dt$, we obtain

$$\frac{d\mathbf{v}}{dt} = \mathbf{A}(t)\mathbf{v}(t), \tag{13.8}$$

where the velocity vector $\mathbf{v}(t)$ is periodic with period T and is tangent to orbit C at point $\mathbf{x}_c(t)$. This indicates that in (13.7), if the initial perturbation $\boldsymbol{\delta}(0)$ is colinear with vector $\mathbf{v}(0)$, i.e., $\boldsymbol{\delta}(0) = \alpha\mathbf{v}(0)$, then

$$\boldsymbol{\delta}(t) = \alpha\mathbf{v}(t)$$

for all t, so that in this case $\boldsymbol{\delta}(t)$ remains colinear with $\mathbf{v}(t)$ and is periodic with period T. Unfortunately, perturbations $\boldsymbol{\delta}(0)$ in directions other than the tangent to C at $\mathbf{x}_c(0)$ do not give rise to periodic solutions, but the Floquet theory of periodic differential equations shows that they can be characterized in terms of a set periodic vector functions $\{\mathbf{p}_i(t), 1 \leq i \leq n\}$ and characteristic exponents $\{\mu_i, 1 \leq i \leq n\}$. A detailed account of Floquet's results is given in [12, Chap. 4] and [10, Sec. 2.2.], and the main steps are described below.

We say that a $n \times n$ matrix $\mathbf{Y}(t)$ is a fundamental solution of periodic linear system (13.7) if

$$\frac{d\mathbf{Y}}{dt} = \mathbf{A}(t)\mathbf{Y}(t) \tag{13.9}$$

with $\det \mathbf{Y}(0) \neq 0$. In this case the Jacobi–Liouville identity [13, p. 28]

$$\det \mathbf{Y}(t) = \exp\left(\int_0^t \text{tr}(\mathbf{A}(s))ds\right) \det \mathbf{Y}(0)$$

implies that $\mathbf{Y}(t)$ is nonsingular for all t. Also, since $\mathbf{A}(t)$ is periodic with period T, we deduce that $\mathbf{Y}(t+T)$ must be a fundamental solution, and since $\mathbf{Y}(t)$ spans the space of all solutions to differential equation (13.7), there exists an invertible matrix $\mathbf{B}$ such that

$$\mathbf{Y}(t+T) = \mathbf{Y}(t)\mathbf{B}. \tag{13.10}$$

Setting $t = 0$ in (13.10) gives

$$\mathbf{B} = \mathbf{Y}^{-1}(0)\mathbf{Y}(T).$$

In this context, it is important to note that the eigenvalues of $\mathbf{B}$ do not depend on the choice of fundamental matrix $\mathbf{Y}(t)$ of system (13.9). Indeed if $\tilde{\mathbf{Y}}(t)$ is another fundamental matrix, there exists an invertible matrix $\mathbf{S}$ such that

$$\tilde{\mathbf{Y}}(t) = \mathbf{Y}(t)\mathbf{S}. \tag{13.11}$$

Then if we substitute (13.11) in the identity

$$\tilde{\mathbf{Y}}(t+T) = \tilde{\mathbf{Y}}(t)\tilde{B}$$

satisfied by $\tilde{\mathbf{Y}}(t)$ and compare the resulting expression to (13.10), we find

$$\mathbf{B} = \mathbf{S}\tilde{B}\mathbf{S}^{-1}, \tag{13.12}$$

so that matrices $\mathbf{B}$ and $\tilde{\mathbf{B}}$ are related by a similarity transformation, and thus they have the same eigenvalues. This shows that the eigenvalues $\{\rho_i, 1 \le i \le n\}$ of $\mathbf{B}$ characterize the dynamical system under consideration and are independent of the coordinate system we consider. These eigenvalues are the *characteristic multipliers* of system (13.7). If $\mathbf{u}$ is an eigenvector of B corresponding to characteristic multiplier ρ, so that

$$B\mathbf{u} = \rho\mathbf{u}, \tag{13.13}$$

the function

$$\boldsymbol{\phi}(t) = \mathbf{Y}(t)\mathbf{u}$$

is a solution of the differential equation (13.7) such that

$$\boldsymbol{\phi}(t + T) = \rho\boldsymbol{\phi}(t). \tag{13.14}$$

Since we have already observed that the velocity vector $\mathbf{v}(t) = d\mathbf{x}_c/dt$ is periodic and satisfies the differential equation (13.7), we conclude that one of the characteristic multipliers, say ρ_1, must equal one. In addition, since the oscillator orbit we consider is stable, small perturbations in directions other than the tangent to the trajectory must disappear as t becomes large, i.e., if $\boldsymbol{\delta}(0)$ has no component along the $\mathbf{v}(0)$ direction, we must have

$$\lim_{t \to \infty} \boldsymbol{\delta}(t) = 0.$$

Since

$$\boldsymbol{\phi}(mT) = \rho^m\boldsymbol{\phi}(0),$$

this implies that all the characteristic multipliers ρ_i other than ρ_1 must be such that $|\rho_i| < 1$.

Next, we use the fact that since the matrix $\mathbf{B}$ is invertible, there exists a matrix $\mathbf{R}$ such that

$$\mathbf{B} = \exp(\mathbf{R}T).$$

The matrix

$$\mathbf{R} = \frac{1}{T}\log\mathbf{B} \tag{13.15}$$

is just obtained by scaling the logarithm of matrix $\mathbf{B}$. When $\mathbf{B}$ is diagonalizable, i.e.

$$\mathbf{B} = \mathbf{U}\,\mathrm{diag}\{\rho_i, 1 \le i \le n\}\mathbf{U}^{-1},$$

where the i-th column $\mathbf{u}_i$ of $\mathbf{U}$ is an eigenvector corresponding to eigenvalue ρ_i, then

$$\log\mathbf{B} = \mathbf{U}\,\mathrm{diag}\{\log(\rho_i), 1 \le i \le n\}\mathbf{U}^{-1}.$$

In this expression, if $\rho_i = |\rho_i|e^{j\theta_i}$ is the polar coordinate representation of ρ_i, then

$$\log(\rho_i) = \log(|\rho|_i) + j\theta_i.$$

Note that the argument of a complex number is defined modulo 2π, so to ensure that $\log(\rho_i)$ is uniquely defined, we require $-\pi < \theta_i \leq \pi$. When **B** is not diagonalizable, the construction of $\log \mathbf{B}$ is more complicated and relies on the Jordan form of **B** [12].

Floquet's Theorem   If $\mathbf{Y}(t)$ is a fundamental solution of system (13.7), it can be expressed as

$$\mathbf{Y}(t) = \mathbf{P}(t)\exp(\mathbf{R}t), \tag{13.16}$$

where $\mathbf{P}(t)$ is a periodic with period T.

Proof Let $\mathbf{P}(t) \overset{\triangle}{=} \mathbf{Y}(t)\exp(-\mathbf{R}t)$. Then

$$\mathbf{P}(t+T) = \mathbf{Y}(t+T)\exp(-\mathbf{R}(t+T)) = \mathbf{Y}(t)\mathbf{B}\exp(-\mathbf{R}T)\exp(-\mathbf{R}t)$$
$$= \mathbf{Y}(t)\exp(-\mathbf{R}t) = \mathbf{P}(t),$$

so that $\mathbf{P}(t)$ is periodic. $\square$

An important consequence of Floquet's theorem is that with respect to the basis defined by the columns $\mathbf{p}_i(t)$ of $\mathbf{P}(t)$, $1 \leq i \leq n$, which is referred to as the *Floquet frame*, the motion of perturbations $\boldsymbol{\delta}(t)$ is described by a time-invariant linear system. Specifically, if $\boldsymbol{\delta}(t)$ is an arbitrary solution of (13.7), consider the coordinate transformation

$$\boldsymbol{\gamma}(t) = \mathbf{P}^{-1}(t)\boldsymbol{\delta}(t). \tag{13.17}$$

Since $\boldsymbol{\delta}(t)$ can be expressed in terms of the fundamental matrix function $\mathbf{Y}(t)$ as

$$\boldsymbol{\delta}(t) = \mathbf{Y}(t)\mathbf{c}, \tag{13.18}$$

where we recognize that the coefficient vector

$$\mathbf{c} = \mathbf{Y}^{-1}(0)\boldsymbol{\delta}(0) = \mathbf{P}^{-1}(0)\boldsymbol{\delta}(0) = \boldsymbol{\gamma}(0),$$

after substituting (13.18) inside (13.17) and using the Floquet identity (13.16), we obtain

$$\boldsymbol{\gamma}(t) = \exp(\mathbf{R}t)\boldsymbol{\gamma}(0), \tag{13.19}$$

so that $\boldsymbol{\gamma}(t)$ has the linear time-invariant dynamics

$$\frac{d\boldsymbol{\gamma}}{dt} = \mathbf{R}\boldsymbol{\gamma}(t). \tag{13.20}$$

Intuitively, the frame $\mathbf{P}(t)$ follows the periodic motion of the oscillator along its trajectory, and when analyzed with respect to this basis, the motion of oscillator perturbations appears time-invariant. In this respect it is worth noting that both the matrix **B** in (13.10) and the matrix **R** defined by (13.15) depend on the choice of fundamental matrix $\mathbf{Y}(t)$. However, as noted earlier, the eigenvalues ρ_i of **B** and therefore the eigenvalues

$$\mu_i = \frac{1}{T}\log \rho_i,$$

$1 \leq i \leq n$, of R are independent of the choice of $\mathbf{Y}(t)$ and characterize the system dynamics. The eigenvalues μ_i of $\mathbf{R}$ are called the *characteristic exponents* of the oscillator. Since we observed earlier that the exponent ρ_1 corresponding to the velocity function $\mathbf{v}(t)$ equals one, we deduce $\mu_1 = 0$. In addition, since the oscillator orbit was assumed stable, we deduce that $\Re(\mu_i) < 0$ for $2 \leq i \leq n$, which ensures that the time-invariant dynamics (13.20) are stable.

Example 13.1 (Continued) The Jacobian of the oscillator dynamics is given by

$$\mathbf{J}(\mathbf{x}) = \begin{bmatrix} (1 - r^2) - 2x^2 & -1 - 2xy \\ 1 - 2xy & (1 - r^2) - 2y^2 \end{bmatrix},$$

where $r = (x^2 + y^2)^{1/2}$, so that when the Jacobian is evaluated along the limit cycle trajectory

$$\mathbf{x}_c(t) = \begin{bmatrix} \cos(t) & \sin(t) \end{bmatrix}^T$$

we have

$$\mathbf{A}(t) = \begin{bmatrix} 0 & -1 \\ 1 & 0 \end{bmatrix} - 2\mathbf{x}_c(t)\mathbf{x}_c^T(t).$$

Then

$$\mathbf{Y}(t) = \begin{bmatrix} -\sin(t) & \exp(-2t)\cos(t) \\ \cos(t) & \exp(-2t)\sin(t) \end{bmatrix} = \begin{bmatrix} \dot{\mathbf{x}}_c(t) & \exp(-2t)\mathbf{x}_c(t) \end{bmatrix}$$

is a fundamental matrix of the linearized periodic system and

$$\mathbf{Y}(t) = \mathbf{P}(t)\mathrm{diag}\{1, \exp(-2t)\}, \tag{13.21}$$

where

$$\mathbf{P}(t) = \begin{bmatrix} -\sin(t) & \cos(t) \\ \cos(t) & \sin(t) \end{bmatrix} = \begin{bmatrix} \dot{\mathbf{x}}_c(t) & \mathbf{x}_c(t) \end{bmatrix}$$

is periodic with period 2π. The decomposition (13.21) indicates that

$$\mathbf{R} = \mathrm{diag}\{0, -2\}$$

and the characteristic exponents of the linearized system are 0 and -2. In this particular case, the two columns $\mathbf{p}_1(t) = \dot{\mathbf{x}}_c(t)$ and $\mathbf{p}_2(t) = \mathbf{x}_c(t)$ are perpendicular to each other. This is not the case in general.

Example 13.3 Consider the Coram oscillator [14] whose dynamics can be expressed in polar coordinates as

$$\dot{r} = r - r^2$$

$$\dot{\theta} = 1 + r.$$

Its only equilibrium point is $r = 0$, and if $D = \{(r, \theta) : 1/2 \leq r \leq 3/2\}$ is the domain located between circles $C_1 = \{(r, \theta) : r = 1/2\}$ and $C_2 = \{(r, \theta) : r = 3/2\}$, it does not include any equilibrium point and the vector field $\mathbf{f}$ on circles C_1 and C_2 points towards the interior of D. Thus D contains a limit cycle which is given by $r_c(t) = 1$ and $\theta_c(t) = 2t + \theta_c(0) \mod 2\pi$, where the phase $\theta_c(0)$ denotes the initial oscillator phase, assuming that the starting point is on the limit cycle. With

$$\mathbf{z} = \begin{bmatrix} r \\ \theta \end{bmatrix} \quad \text{and} \quad \mathbf{f}(\mathbf{z}) = \begin{bmatrix} r - r^2 \\ 1 + r \end{bmatrix},$$

we have

$$J(\mathbf{z}) = \begin{bmatrix} 1 - 2r & 0 \\ 1 & 0 \end{bmatrix}$$

and

$$\mathbf{A}(t) = \mathbf{J}(\mathbf{z}_c(t)) = \begin{bmatrix} -1 & 0 \\ 1 & 0 \end{bmatrix}.$$

Thus with respect to the rotating coordinate system represented by $(r_c = 1, \theta_c(t))$, the matrix $\mathbf{A}$ is constant, and its eigendecomposition is given by

$$\mathbf{A} = \mathbf{P} \operatorname{diag}\{0, -1\}\mathbf{P}^{-1},$$

where

$$\mathbf{P} = \begin{bmatrix} 0 & 1 \\ 1 & -1 \end{bmatrix} = \begin{bmatrix} \mathbf{p}_1 & \mathbf{p}_2 \end{bmatrix}.$$

Accordingly

$$\mathbf{Y}(t) = \mathbf{P}\operatorname{diag}\{1, \exp(-t)\},$$

so that the characteristic exponents of the oscillator are 0 and -1. Note that in this case the vector $\mathbf{p}_2$ is not orthogonal to the vector $\mathbf{p}_1$ which is tangent to the oscillator's limit cycle, as shown in Fig. 13.3.

Although the vectors $\{\mathbf{p}_j(t), 1 \leq j \leq n\}$ forming the Floquet frame are not necessarily orthogonal, it is easy to construct the orthogonal complement of any one of these vectors. Specifically, the rows $\mathbf{q}_i^T(t)$ of the matrix inverse

$$\mathbf{Q}(t) = \begin{bmatrix} \mathbf{q}_1^T(t) \\ \vdots \\ \mathbf{q}_i^T(t) \\ \vdots \\ \mathbf{q}_n^T(t) \end{bmatrix} = \mathbf{P}^{-1}(t) \tag{13.22}$$

satisfy

$$\mathbf{q}_i^T(t)\mathbf{p}_j(t) = \delta_{ij}, \tag{13.23}$$

Fig. 13.3 Floquet vectors
of the Coram oscillator

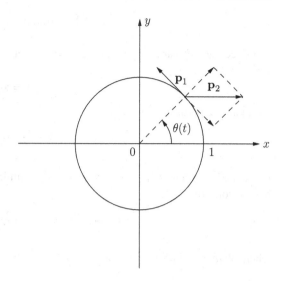

so that $\{\mathbf{q}_i(t), i \neq j\}$ spans the orthogonal complement of $\mathbf{p}_j(t)$. For this reason, the bases $\{\mathbf{p}_i(t), 1 \leq j \leq n\}$ and $\{\mathbf{q}_i(t), 1 \leq i \leq n\}$ are said to be *biorthogonal*.

13.2.3 Isochrons

Consider a stable periodic orbit $\mathbf{x}_c(t)$ with period T. If $\mathbf{x}_0 = \mathbf{x}_c(0)$, then a point $\mathbf{x}_1 = \mathbf{x}_c(\phi)$ also located on the periodic orbit is said to have phase (measured in unit of time) ϕ, where $0 \leq \phi < T$. Unfortunately, when oscillators are subjected to perturbations, the trajectory does not stay on the stable orbit but evolves in its basin of attraction. However, because we know that any trajectory will ultimately converge to the stable orbit, it is possible to assign to each point $\mathbf{x}$ of the attraction basin the phase it will assume asymptotically when the trajectory reaches the limit cycle. Specifically, we say that a point $\mathbf{x}$ has asymptotic phase $\phi(\mathbf{x})$ if, given initial condition $\mathbf{x}(0) = \mathbf{x}$, the oscillator trajectory $\mathbf{x}(t)$ is such that

$$\lim_{t \to \infty} ||\mathbf{x}(t) - \mathbf{x}_c(t + \phi(\mathbf{x}))|| = 0. \tag{13.24}$$

Then for an oscillator evolving in $\mathbb{R}^n$, an isochron [6, 15–17] is an $n - 1$ dimensional surface satisfying

$$\phi(\mathbf{x}) = \phi_0,$$

i.e., it is the surface formed by all points whose asymptotic phase is ϕ_0. In other words, all points on the same isochron can be viewed as asymptotically synchronized.

The isochrons of a dynamical system of the form (13.1) do not usually admit a closed form and are rather difficult to compute. However, isochrons admit several important properties which will be exploited to constructing a model of phase noise. First, given a point $\mathbf{x}_0 = \mathbf{x}_c(0)$ on the oscillator's periodic orbit, the tangent space to the isochron passing through point $\mathbf{x}_0$ is spanned by the vectors $\{p_j(0), 2 \leq j \leq n\}$ obtained by removing the tangent vector $\mathbf{p}_1(0) = \mathbf{v}(0)$ from the Floquet frame. Note that we are assuming here that the coordinate system has been selected so that $\mathbf{R}$ has the diagonal structure

$$\mathbf{R} = \text{diag}\{0, \mu_2, \ldots, \mu_n\},$$

and since the μ_i's can be complex the Floquet vectors $\mathbf{p}_j(t)$ may be complex. Then at time 0, consider a point

$$\mathbf{x}(0) = \mathbf{x}_0 + \boldsymbol{\delta} \tag{13.25}$$

such that the perturbation vector

$$\boldsymbol{\delta} = \sum_{j=2}^{n} \gamma_j \mathbf{p}_j(0) \tag{13.26}$$

is spanned by vectors $\{p_j(0), 2 \le j \le n\}$. Assuming that $\boldsymbol{\delta}$ is small, the linearized equation (13.7) has for solution

$$\mathbf{x}(t) - \mathbf{x}_c(t) = \boldsymbol{\delta}(t) = \sum_{j=2}^{N} \gamma_j \mathbf{p}_j(t) \exp(\mu_j t), \tag{13.27}$$

where $\Re(\mu_j) < 0$ for $2 \le j \le n$, so

$$\lim_{t \to \infty} ||\mathbf{x}(t) - \mathbf{x}_c(t)|| = 0.$$

This shows that when $\boldsymbol{\delta}$ is small, all initial points $\mathbf{x}(0)$ of the form (13.25) and (13.26) belong to the same isochron as $\mathbf{x}_0$, and accordingly the tangent space to the isochron at point $\mathbf{x}_0 = \mathbf{x}_c(0)$ is spanned by vectors $\{\mathbf{p}_j(0), 2 \le j \le n\}$. The biorthogonality property (13.23) of the basis $\{\mathbf{q}_i(t), 1 \le i \le n\}$ implies therefore that the vector $\mathbf{q}_1(0)$ is perpendicular to the isochron tangent space at point $\mathbf{x}_0$, since all perturbations of the form (13.26) satisfy

$$\mathbf{q}_1^T(0)\boldsymbol{\delta} = 0. \tag{13.28}$$

In the above discussion we have considered the isochron surface passing through the orbit location $\mathbf{x}_0 = \mathbf{x}_c(0)$ at time $t = 0$, but the analysis is independent of the time considered, so the tangent space to the isochron surface passing through orbit point $\mathbf{x}_c(t)$ at time t is perpendicular to vector $\mathbf{q}_1(t)$. The local geometry of isochrons in the vicinity of an orbit point $\mathbf{x}_c(t)$ is illustrated in Fig. 13.4 for the case of a planar oscillator.

It is useful to note that the phase gradient $\nabla_\mathbf{x}\phi(\mathbf{x})$ on the orbit $x_c(t)$ is the vector $\mathbf{q}_1(t)$. Since isochrons are level set surfaces for the phase and $\mathbf{q}_1(t)$ is perpendicular to the tangent hyperplane to the isochron at point $\mathbf{x}_c(t)$, we know that

$$\nabla_\mathbf{x}\phi(\mathbf{x})\,|_{\mathbf{x}=\mathbf{x}_c(t)} = a\mathbf{q}_1(t), \tag{13.29}$$

ı.e., $\nabla_\mathbf{x}\phi(\mathbf{x}_c(t))$ and $\mathbf{q}_1(t)$ are *colinear*. But by observing that $\phi(\mathbf{x}_c(t)) = t$ and using the chain rule of differentiation, we find

$$1 = \frac{d}{dt}\phi(\mathbf{x}_c(t)) = \nabla_\mathbf{x}^T\phi(\mathbf{x}_c(t))\mathbf{v}(t) = a\mathbf{q}_1^T(t)\mathbf{p}_1(t) = a, \tag{13.30}$$

so that $a = 1$.

Example 13.1 (Continued) Consider the oscillator (13.3). Because all trajectories rotate around the origin with the same angular speed, i.e., $d\theta/dt = 1$, independently of the initial point, we deduce that isochrons are formed by rays of the form $\theta = \phi_0$. These rays intersect the limit cycle $r = 1$

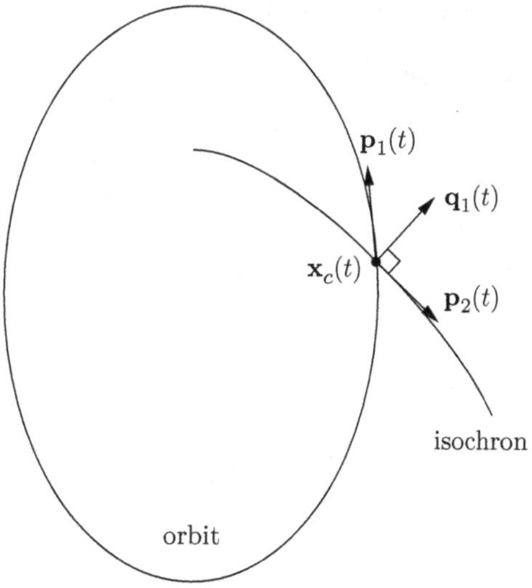

Fig. 13.4 Isochron geometry in the vicinity of orbit point $\mathbf{x}_c(t)$

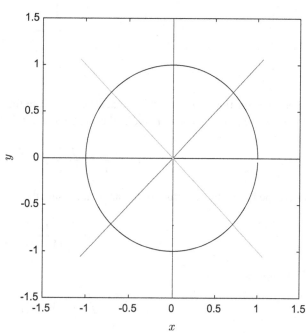

Fig. 13.5 Isochrons of oscillator (13.3) and oscillator orbit

perpendicularly as shown in Fig. 13.5, which is expected since in this case the second Floquet vector $\mathbf{p}_2(t)$ is perpendicular to $\mathbf{p}_1(t)$.

Example 13.3 (Continued) The Coram oscillator has the feature that its vector field is invariant under rotations. Accordingly, if ϕ_0 denotes the asymptotic phase corresponding to a trajectory starting from point $\mathbf{z} = (r, \theta)$, isochrons can be expressed as

$$\phi_0 = \theta + f(r),$$

Fig. 13.6 Isochrons of the
Coram oscillator and
oscillator orbit

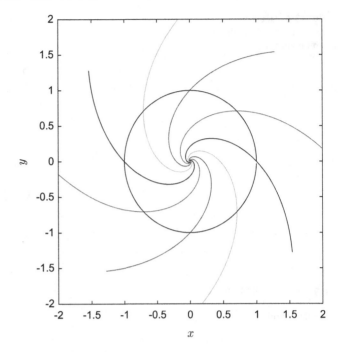

where the function $f(r)$ needs to be determined. Note that when $r = 1$, $\phi_0 = \theta$, so $f(1) = 0$. All points on the isochrons must move at the same angular rate to achieve the same asymptotic phase, so

$$\frac{d\theta}{dt} + \frac{df}{dr}\frac{dr}{dt} = \frac{d\phi_0}{dt} = 2,$$

where the last equality is due to the fact that the angular rate of the limit cycle trajectory is 2, so

$$\frac{df}{dr} = \frac{2 - (1 + r)}{r(1 - r)} = \frac{1}{r}.$$

This implies

$$f(r) = \ln(r) + C,$$

where $C = 0$ to ensure $f(1) = 0$. The isochrons are therefore given by

$$\phi_0 = \theta + \ln(r),$$

and isochron curves are plotted in Fig. 13.6.

13.3   Phase Noise Model

Consider a stable periodic oscillator subjected to random fluctuations, so that it satisfies a vector Ito stochastic differential equation of the form

$$d\mathbf{X} = \mathbf{f}(\mathbf{X})dt + \mathbf{G}(\mathbf{X})d\mathbf{W}, \tag{13.31}$$

where $G(\mathbf{X})$ is a $n \times m$ matrix nonlinear function, and where the vector process

$$\mathbf{W}(t) = \left[W_1(t) \ W_2(t) \ \dots \ W_m(t) \right]^T$$

is formed by m independent scalar Wiener processes $W_k(t)$, $1 \leq k \leq m$. In the absence of noise, the oscillator admits a stable orbit $\mathbf{x}_c(t)$ with period T.

13.3.1 General Model

The phase noise analysis proposed by Kaertner [4] and later refined by Demir, Mehrotra, and Roychowdhury [5] investigates the behavior of noisy oscillator trajectories in the vicinity of a stable orbit. It relies on two assumptions. The first one is that when the noise is small, because the oscillator is stable, random trajectory perturbations in directions tangent to the isochron surface passing through $\mathbf{X}(t)$ will never accumulate because the restoring force due to the characteristic exponents μ_i with $2 \leq i \leq n$ pushes back the trajectory towards the stable orbit since $\Re(\mu_i) < 0$ for $2 \leq i \leq n$. Roughly speaking, with high probability $\mathbf{x}(t)$ stays within a tube centered about the stable orbit $\mathbf{x}_c(\cdot)$, where the radius of the tube is proportional to the noise intensity. On the other hand, random fluctuations which are in the direction of the tangent to the stable orbit accumulate since they are never attenuated. This suggests therefore the approximation

$$\mathbf{X}(t) \approx \mathbf{x}_c(t + \alpha(t)) + \mathbf{\Delta}(t), \tag{13.32}$$

where $\mathbf{\Delta}(t)$ represents a small perturbation belonging to the tangent space to the isochron surface passing through $\mathbf{x}_c(t + \alpha(t))$. The approximation (13.32) implies that if $\phi(\mathbf{X}(t))$ is the asymptotic phase corresponding to trajectory point $\mathbf{X}(t)$, the phase noise

$$\alpha(t) = \phi(\mathbf{X}(t)) - t. \tag{13.33}$$

The second assumption, which was first stated explicitly in [6], is that isochron surfaces passing through orbit points $\mathbf{x}_c(t)$ are *locally flat*, i.e., the Hessian

$$\mathbf{H}(t) = \nabla_\mathbf{x} \nabla_\mathbf{x}^T \phi(\mathbf{x}) \mid_{\mathbf{x}=\mathbf{x}_c(t)} \approx \mathbf{0}. \tag{13.34}$$

By applying the vector version of Ito's rule of stochastic calculus (10.70) (see [18, Sec. 4.5]) we obtain

$$d\phi(\mathbf{X}) = \nabla_\mathbf{x}^T \phi(\mathbf{X})d\mathbf{X} + \frac{1}{2}\mathrm{tr}\,(G(\mathbf{X})G^T(\mathbf{X})\nabla_\mathbf{x}\nabla_\mathbf{x}^T\phi(\mathbf{X}))dt. \tag{13.35}$$

Then by using approximation (13.32) and the flatness assumption (13.34), we find

$$\begin{aligned}
d\alpha(t) &= d\phi(\mathbf{X}) - dt \\
&= \nabla_\mathbf{x}^T \phi(\mathbf{x}) \mid_{\mathbf{x}=\mathbf{x}_c(t+\alpha(t))} [\mathbf{f}(\mathbf{x}_c(t + \alpha(t))dt \\
&\quad + \mathbf{G}(\mathbf{x}_c(t + \alpha(t)))d\mathbf{W}(t)] - dt,
\end{aligned} \tag{13.36}$$

where by recognizing that

$$\nabla_\mathbf{x}\phi(\mathbf{x}) \mid_{\mathbf{x}=\mathbf{x}_c(t+\alpha(t))} = \mathbf{q}_1(t + \alpha(t))$$

$$f(\mathbf{x}_c(t + \alpha(t))) = \mathbf{p}_1(t + \alpha(t))$$

and noting that

$$\mathbf{q}_1^T(t + \alpha(t))\mathbf{p}_1(t + \alpha(t)) = 1, \tag{13.37}$$

we find that the phase noise $\alpha(t)$ satisfies the scalar Ito stochastic differential equation

$$d\alpha(t) = \mathbf{q}_1^T(t + \alpha(t))G(\mathbf{x}_c(t + \alpha(t)))d\mathbf{W}(t). \tag{13.38}$$

This equation models the *unwrapped* phase noise $\alpha(t)$, which takes values in $\mathbb{R}$. Since the oscillator is periodic with period T, the wrapped phase noise needs to be defined modulo T, and thus it is distributed over $[-T/2, T/2)$.

The phase noise model (13.38), which is due to Kaertner [4] and Demir et al. [5], is rather remarkable since it is scalar, even though the equation satisfied by the noisy oscillator $\mathbf{X}(t)$ is n-dimensional. However, it should be noted that approximation (13.32) and the isochron flatness approximation (13.34) are rather simplistic. Specifically, in the low noise case, even though for a fixed time t, $\mathbf{X}(t)$ stays usually close to the periodic orbit, with low probability the trajectory undergoes large excursions for which approximation (13.32) is wildly inaccurate. Likewise, it is not reasonable to assume that isochron surfaces have zero curvature at points located along the periodic orbit. These two sources of phase noise model errors have not escaped the attention of researchers, and several methods have been proposed to account for orbital fluctuations [19] and for isochron surface curvature [20] in phase noise models. These extensions introduce a significant increase in complexity to phase noise models, and since the purpose of our presentation is illustrative, we consider exclusively the baseline model (13.38), in spite of its obvious limitations.

If we introduce the m-dimensional periodic row vector function

$$\mathbf{v}^T(t) \overset{\triangle}{=} \mathbf{q}_1^T(t)\mathbf{G}(\mathbf{x}_c(t)), \tag{13.39}$$

the stochastic differential equation (13.38) can be written more compactly as

$$d\alpha(t) = \mathbf{v}^T(t + \alpha(t))d\mathbf{W}(t). \tag{13.40}$$

By recognizing that this equation models a scalar diffusion with diffusion coefficient $d(t + \alpha)$, where

$$d(t) = \mathbf{v}^T(t)\mathbf{v}(t) = ||\mathbf{v}(t)||^2, \tag{13.41}$$

this equation is equivalent to a stochastic differential equation of the form

$$d\alpha(t) = d^{1/2}(t + \alpha(t))dN(t), \tag{13.42}$$

where $N(t)$ is now a scalar Wiener process.

In this respect note that although the vector $\mathbf{v}(t)$ varies periodically, it is not uncommon for its length to be constant, in which case the differential equation (13.42) becomes linear, so that if we assume that the initial phase $\alpha(0) = 0$, the phase noise $\alpha(t) = d^{1/2}N(t)$ is just a scaled Wiener process.

Example 13.4 Consider the noisy Coram oscillator [14]

$$\begin{bmatrix} dR \\ d\Theta \end{bmatrix} = \begin{bmatrix} R - R^2 \\ 1 + R \end{bmatrix} dt + \sigma \begin{bmatrix} 1 \\ 0 \end{bmatrix} dW$$

where the noise dW is injected perpendicularly to the oscillator trajectory. For this oscillator, it was found in Example 3.3 that

$$\mathbf{P} = \begin{bmatrix} 0 & 1 \\ 1 & -1 \end{bmatrix}.$$

Then the inverse of $\mathbf{P}$ is

$$\mathbf{Q} = \begin{bmatrix} 1 & 1 \\ 1 & 0 \end{bmatrix},$$

so that the vector perpendicular to the isochron curve is

$$\mathbf{q}_1 = \begin{bmatrix} 1 \\ 1 \end{bmatrix}.$$

Thus although the noise has been injected perpendicularly to the oscillator trajectory, since

$$v^T = \sigma \mathbf{q}_1^T \begin{bmatrix} 1 \\ 0 \end{bmatrix} = \sigma \neq 0,$$

it has a component in the direction perpendicular to the isochron, and thus the phase noise

$$\alpha(t) = \sigma N(t)$$

is a scaled Wiener process. In this case, by applying the analysis presented in Example 10.5, it can be shown that the distribution of the wrapped phase noise

$$Z(t) = \alpha(t) \mod T$$

converges to the uniform distribution

$$f_G(z) = \begin{cases} \frac{1}{T} & -\frac{T}{2} \leq z < \frac{T}{2} \\ 0 & \text{otherwise} \end{cases}$$

as t tends to infinity.

13.3.2 Slow Time-Scale Model

The Ito stochastic differential equation (SDE) model (13.42) can be simplified further if we recognize that when the noise is small, i.e., when $d(t)$ is uniformly small for all t, the phase noise $\alpha(t)$ evolves on a time scale which is much slower than the rate of oscillation of the oscillator. For systems with multiple time scales, averaging or homogenization techniques are commonly used to construct a model of the slow time scale process whose coefficients are obtained by averaging or homogenization of the fast time scale coefficients. Bensoussan, Lions, and Papanicolaou [21] present a detailed account of homogenization methods for systems described by PDEs with periodic coefficients. Averaging and homogenization techniques for systems described by stochastic differential equations were proposed by Khasminskii [22] and Papanicolaou [7]. An introductory level presentation of multiple time scale analysis and averaging for ODEs, SDEs, and PDEs is given in Pavliotis and Stuart [8].

The homogenized phase noise model described below was first derived by Demir et al. [5] by using a rather intricate analysis, but a more insightful derivation can be obtained from a homogenization viewpoint. Consider an interval $[t, t + \tau]$ such that the interval length τ is large compared to the oscillator period T, i.e., $\tau / T \gg 1$. Then if we consider the integrated form

$$\alpha(t + \tau) - \alpha(t) = \int_t^{t+\tau} d^{1/2}(s + \alpha(s))dN(s) \tag{13.43}$$

of SDE (13.42), because the phase noise process evolves much more slowly than the oscillation period of the oscillator, for an interval length τ much larger than T, but still small enough, the phase noise process $\alpha(s)$ is almost constant over $[t, t+\tau]$, so that in (13.43) we can use the approximation $\alpha(s) \approx \alpha(t)$ for $t \leq s \leq t + \tau$. Then conditioned on $\alpha(t)$, the right-hand side of expression

$$\alpha(t + \tau) - \alpha(t) = \int_t^{t+\tau} d^{1/2}(s + \alpha(t))dN(s) \tag{13.44}$$

can be interpreted as a Wiener integral, since the function $d^{1/2}(\cdot, \alpha(t))$ is known. This implies that the conditional density of $\alpha(t+\tau) - \alpha(t)$ given $\alpha(t)$ is $N(0, v(\tau))$ distributed, where by using the Wiener isometry

$$v(\tau) = E[(\alpha(t + \tau) - \alpha(t))^2 | \alpha(t)] = \int_t^{t+\tau} d(s + \alpha(t))ds$$

$$= \int_{t-\alpha(t)}^{t-\alpha(t)+\tau} d(v)dv. \tag{13.45}$$

Let

$$<d> = \frac{1}{T} \int_0^T d(t)dt = \frac{1}{T} \int_0^T ||\mathbf{v}(t)||^2 dt \tag{13.46}$$

denote the average of periodic function $d(t)$ over a period. Since the length τ of the interval of integration in (13.45) is much larger than T, this integral can be approximated as $< d > \tau$. Thus we have shown that conditioned on $\alpha(t)$, the increment $\alpha(t+\tau) - \alpha(t)$ is $N(0, < d > \tau)$ distributed. Since this distribution does not depend on the value of $\alpha(t)$, the increment $\alpha(t + \tau) - \alpha(t)$ is independent of $\alpha(t)$. Because it is independent of t, it is also stationary. Since $\alpha(t)$ is Markov by construction, we deduce that it has stationary and Gaussian independent increments. If we assume $\alpha(0) = 0$, we deduce therefore that the homogenized phase noise model is given by

$$\alpha(t) = < d >^{1/2} N(t), \tag{13.47}$$

so that on a time scale much slower than the period T of oscillation, the phase noise is a scaled Wiener process with scaling parameter $< d >^{1/2}$. Thus quite remarkably, we have constructed a model of phase noise depending on the single parameter $< d >$.

The approximation (13.47) describes the distribution of the unwrapped phase noise for $t/T \gg 1$. As noted earlier in Example 10.5, this model implies that the wrapped phase noise

$$Z(t) = \alpha(t) \mod T \tag{13.48}$$

is uniformly distributed over $[-T/2, T/2)$ for $t/T \gg 1$.

13.4 Oscillator Autocorrelation and Spectrum

We are now in a position to compute the effect of the phase noise $\alpha(t)$ on the oscillator autocorrelation and spectrum. Since $\mathbf{x}_c(t)$ is periodic with period T, it admits a vector Fourier series representation

$$\mathbf{x}_c(t) = \sum_{k=-\infty}^{\infty} \mathbf{X}_k \exp(jk\omega_0 t), \tag{13.49}$$

where $\omega_0 = 2\pi/T$ denotes the fundamental frequency of oscillation and

$$\mathbf{X}_k = \frac{1}{T} \int_0^T \mathbf{x}_c(t) \exp(-jk\omega_0 t) dt \in \mathbb{R}^n \tag{13.50}$$

is the k-th Fourier coefficient vector of the oscillator orbit. Then assume that t/T is large enough to ensure that the wrapped phase (13.48) is uniformly distributed over $[-T/2, T/2)$. We seek to evaluate the autocorrelation matrix

$$\mathbf{R}_{\mathbf{X}}(\tau) = E[\mathbf{x}_c(t + \tau + \alpha(t + \tau))\mathbf{x}_c^H(t + \alpha(t))] \tag{13.51}$$

of the noisy oscillator. If we assume first that $\tau \geq 0$ and substitute the Fourier series representation (13.49) inside expression (13.51) for the autocorrelation, we find

$$R_{\mathbf{X}}(\tau) = \sum_{k=-\infty}^{\infty} \sum_{\ell=-\infty}^{\infty} \mathbf{X}_k \mathbf{X}_\ell^H \exp(-jk\omega_0 \tau) \exp(-j(k - \ell)\omega_0 t)$$

$$\times E[\exp(-jk\omega_0(\alpha(t + \tau) - \alpha(t))) \exp(-j(k - \ell)\omega_0(\alpha(t)))], \tag{13.52}$$

where the independence of $\alpha(t + \tau) - \alpha(t)$ and $\alpha(t)$ implies

$$E[\exp(-jk\omega_0(\alpha(t + \tau) - \alpha(t))) \exp(-j(k - \ell)\omega_0(\alpha(t)))]$$
$$= E[\exp(-jk\omega_0(\alpha(t + \tau) - \alpha(t)))]E[\exp(-j(k - \ell)\omega_0\alpha(t))]. \tag{13.53}$$

But since the wrapped phase is asymptotically uniformly distributed over $[-T/2, T/2)$, we have

$$E[\exp(-j(k - \ell)\omega_0\alpha(t))] = \frac{1}{T} \int_{T/2}^{T/2} \exp(-j(k - \ell)\omega_0) dz = \delta(k - \ell), \tag{13.54}$$

where the last equality is due to the fact that the functions

$$\{\phi_k(t) = T^{-1/2} \exp(-jk\omega_0 t), k \in \mathbb{Z}\}$$

form a complete orthonormal basis of $L^2[-T/2, T/2]$. By observing that the characteristic function of a $N(0, v)$ distributed random variable is

$$\Phi(u) = \exp(-u^2 v/2),$$

we find also

$$E[\exp(-jk\omega_0(\alpha(t+\tau) - \alpha(t)))] = \exp(-(k\omega_0)^2 <d> \tau/2), \tag{13.55}$$

and by substituting (13.53)–(13.55) inside (13.52) we obtain

$$\mathbf{R_X}(\tau) = \sum_{k=-\infty}^{\infty} \mathbf{X}_k \mathbf{X}_k^H \exp(-jk\omega_0\tau) \exp(-(k\omega_0)^2 <d> \tau/2) \tag{13.56}$$

for $\tau \geq 0$. The symmetry $\mathbf{R_X}(-\tau) = \mathbf{R_X}^H(\tau)$ then implies

$$\mathbf{R_X}(\tau) = \sum_{k=-\infty}^{\infty} \mathbf{X}_k \mathbf{X}_k^H \exp(-jk\omega_0\tau) \exp(-(k\omega_0)^2 <d> |\tau|/2), \tag{13.57}$$

so that the noisy oscillator $\mathbf{X}(t)$ is asymptotically WSS for $t/T \gg 1$.

By observing that the Fourier transform of $\exp(-a|\tau|)$ is

$$\frac{2a}{\omega^2 + a^2},$$

we deduce that the matrix power spectral density of $\mathbf{X}(t)$ is

$$\mathbf{S_X}(\omega) = \sum_{k=-\infty}^{\infty} \mathbf{X}_k \mathbf{X}_k^H \frac{(k\omega_0)^2 <d>}{(\omega + k\omega_0)^2 + (k\omega_0)^4 <d>^2}$$

$$= \sum_{k=-\infty}^{\infty} \mathbf{X}_k^* \mathbf{X}_k^T \frac{(k\omega_0)^2 <d>}{(\omega - k\omega_0)^2 + (k\omega_0)^4 <d>^2}, \tag{13.58}$$

where to go from the first to the second line we have replaced k by $-k$ and used the fact that since $\mathbf{x}_c(t)$ is real $\mathbf{X}_{-k} = \mathbf{X}_k^*$.

Remark Note that since the slow-scale model of the phase noise was used to compute both the autocorrelation matrix $\mathbf{R_X}(\tau)$ and the PSD matrix $\mathbf{S_X}(\omega)$, both of these expressions are only approximations. Specifically, the autocorrelation expression (13.57) holds only for $|\tau|/T \gg 1$, and the Lorentz spectrum expression

$$\frac{(k\omega_0)^2 <d>}{(\omega - k\omega_0)^2 + (k\omega_0)^4 <d>^2} \tag{13.59}$$

for the k-th term in the harmonic decomposition of the PSD $\mathbf{S_x}(\omega)$ is only valid as long as the frequency deviation

$$\Delta\omega = \omega - k\omega_0$$

from the k-th spectral line $k\omega_0$ is such that $|\Delta\omega| \ll \omega_0$. In other words (13.58) should be viewed as obtained by stitching together local approximations for each of the oscillator harmonic components. From expression (13.59) we see also that the 50% power bandwidth of the spectrum of the k-th harmonic is

$$2|\Delta\omega| = 2(k\omega_0)^2 <d>,$$

so that the scaling parameter $<d>$ of the Wiener process model of the slow time-scale phase noise controls the width of the spectra corresponding to each harmonic of the oscillator spectrum. In other words, the smaller $<d>$, the purest each of the oscillator tones. Thus, for the simplified phase noise model under consideration, the design of a low-noise oscillator reduces to the minimization of $<d>$.

Example 13.4 (Continued) In Cartesian coordinates, the limit cycle of the Coram oscillator is

$$\mathbf{x}_c(t) = \begin{bmatrix} \cos(2t) \\ \sin(2t) \end{bmatrix} = \mathbf{X}_1 \exp(j2t) + \mathbf{X}_{-1}\exp(-j2t),$$

where

$$\mathbf{X}_1 = \frac{1}{2}\begin{bmatrix} 1 \\ -j \end{bmatrix} \quad \text{and} \quad \mathbf{X}_{-1} = \frac{1}{2}\begin{bmatrix} 1 \\ j \end{bmatrix}.$$

The period and fundamental frequency of the oscillator are $T = \pi$ and $\omega_0 = 2$. When noise is injected perpendicularly to the oscillator trajectory, it was shown in the previous section that $d = \sigma^2$ is constant, so that $<d> = d = \sigma^2$ and the phase noise model $\alpha(t) = \sigma N(t)$ is valid at all time scales. The PSD matrix of the oscillator is therefore given by

$$S_{\mathbf{X}}(\omega) = \mathbf{X}_1^*\mathbf{X}_1^T \frac{4\sigma^2}{(\omega-2)^2 + 16\sigma^4} + \mathbf{X}_{-1}^*\mathbf{X}_{-1}^T \frac{4\sigma^2}{(\omega+2)^2 + 16\sigma^4}.$$

In this case the horizontal and vertical oscillation components have the same spectrum with

$$\mathbf{S}_{11}(\omega) = \mathbf{S}_{22}(\omega) \approx \frac{\sigma^2}{(\omega-2)^2 + 16\sigma^4}$$

for $\omega \geq 0$, which is plotted in Fig. 13.7 for $\sigma = 0.1$.

Remark Although the phase noise theory of Kaertner [4] and Demir et al. [5] predicts correctly the Lorentzian shape of the oscillator spectral components due to phase noise, it fails to account for a small spectral shift of the oscillator harmonics [23, 24]. This error is due to the neglect of amplitude fluctuations $\mathbf{\Delta}(t)$ of the noisy oscillator trajectory $\mathbf{X}(t)$ about the noiseless orbit in (13.32). Phase noise analyses which account for amplitude fluctuations have been proposed [19, 25] but at the expense of a significant increase in modeling complexity.

Fig. 13.7 PSD of the
horizontal and vertical
components of the noisy
Coram oscillator

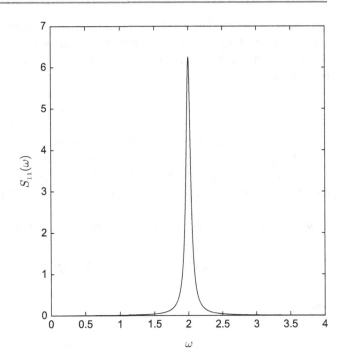

13.5   Bibliographical Notes

Noiseless oscillator dynamics and the Floquet theory are presented in detail in [10, 12]. Oscillator isochrons are studied in [15], and the relation between isochrons and phase noise computation is discussed in [6, 17]. The derivation of the phase noise model presented here relies primarily on the original articles of Kaertner [4] and Demir et al. [5]. While the averaging method employed to derive the slow time-scale model of phase noise can be best understood in the framework of modern time-averaging and homogenization methods, it is worth observing that a similar averaging idea was used by Lax [26] to study an early phase noise model.

13.6   Problems

13.1 Consider the Coram oscillator.

a) By rewriting its radial dynamics as

$$\left[\frac{1}{r} + \frac{1}{1-r}\right] dr = dt,$$

show that the radial coordinate of each trajectory can be expressed in closed form as

$$r(t) = \frac{1}{1 + (r_0^{-1} - 1)\exp(-t)},$$

where r_0 denotes the initial radial position of the oscillator. Then by integrating

$$\dot{\theta} = 1 + r(t),$$

show that

$$\theta(t) = \theta_0 + t + \ln\left[e^t + (r_0^{-1} - 1)\right] + \ln(r_0).$$

b) Use the closed form expressions obtained in part a) to prove that for $r_0 > 0$, all trajectories $(r(t), \theta(t))$ of the oscillator converge to the limit cycle given by $r_c(t) = 1$ and $\theta_c(t) = 2t + \theta_0 + \ln(r_0)$.

c) Conclude from the expression for $\theta_c(t)$ of part b) that the asymptotic phase of point (r, θ) is

$$\phi_0 = \theta + \ln(r),$$

which matches the isochron equation derived earlier for the Coram oscillator.

13.2 Consider the oscillator whose dynamics in polar coordinates are given by

$$\dot{r} = (1 - r)r^2$$
$$\dot{\theta} = r.$$

a) Use the Poincaré–Bendixson theorem to verify that

$$\begin{bmatrix} r_c(t) \\ \theta_c(t) \end{bmatrix} = \begin{bmatrix} 1 \\ t \quad \mathrm{mod}\ 2\pi \end{bmatrix},$$

is a stable limit-cycle of the oscillator, or equivalently in polar coordinates

$$\mathbf{x}_c(t) = \begin{bmatrix} \cos(t) \\ \sin(t) \end{bmatrix}.$$

b) By following the technique used in Example 3.3 to identify the isochrons of the Coram oscillator, assume that oscillator isochrons are parametrized as

$$\phi_0 = \theta + f(r),$$

where the function $f(r)$ is selected such that $f(1) = 0$, so that on the limit cycle

$$\phi_0 = \theta_c(t),$$

which ensures that

$$\frac{d}{dt}\phi_0 = 1.$$

Then by differentiating

$$\phi_0(t) = \theta(t) + f(r(t))$$

with respect to t verify that

$$\frac{df}{dr} = \frac{1 - \dot{\theta}}{\dot{r}}.$$

Use the oscillator dynamics and the constraint $f(1) = 0$ to find $f(r)$.

c) Plot the oscillator isochrons for $\phi_0 = k\pi/4$ where the integer k varies between 0 and 7.

13.3 Consider the oscillator with the polar coordinate dynamics

$$\dot{r} = \frac{1}{2}(1 - r^2) \tag{13.60}$$

$$\dot{\theta} = \omega_0 - \frac{v}{2}(1 - r^2). \tag{13.61}$$

a) Use the Poincaré–Bendixson theorem to show that

$$\begin{bmatrix} r_c(t) \\ \theta_c(t) \end{bmatrix} = \begin{bmatrix} 1 \\ \omega_0 t \quad \mathrm{mod}\ 2\pi \end{bmatrix},$$

is a stable limit-cycle of the oscillator, or equivalently in polar coordinates

$$\mathbf{x}_c(t) = \begin{bmatrix} \cos(\omega_0 t) \\ \sin(\omega_0 t) \end{bmatrix}.$$

b) An interesting feature of oscillator (13.60) and (13.61) is that by writing (13.60) as

$$\left[\frac{1}{1 + r} + \frac{1}{1 - r}\right] dr = dt$$

and observing that

$$\frac{d}{dt}(\theta + vr) = \omega_0,$$

its trajectories can be computed in closed form, parametrized by initial conditions (r_0, θ_0). Use this closed form to verify that independently of the choice of initial condition, all trajectories converge asymptotically to the limit cycle identified in a).

13.4 For the oscillator (13.60) and (13.61), evaluate the periodic matrix $\mathbf{A}(t) = \mathbf{J}(\mathbf{x}_c(t)$ parametrizing the linearized motion of small oscillator perturbations. Compute the Floquet basis $\mathbf{P}(t)$ and the characteristic exponents of the system. Verify that the first exponent $\mu_1 = 0$. Compute the conjugate Floquet basis $\mathbf{Q}(t)$.

13.5 For the oscillator (13.60) and (13.61), use the property

$$\frac{d}{dt}(\theta + vr) = \omega_0$$

of the oscillator dynamics to show that isochrons admit the parametrization

$$\theta + v(r - 1) = \phi_0,$$

where θ_0 denotes the asymptotic phase corresponding point (r, θ). Plot isochron curves for $v = \pi/4$ and $\phi_0 = k\pi/4$ with $0 \leq k \leq 7$. Verify that for $v \neq 0$, isochrons do not intersect perpendicularly to the limit cycle.

13.6 Consider now the noisy version [19]

$$dR = \frac{1}{2}(1 - R^2) + \sigma dW_1$$

$$d\Theta = \omega_0 - \frac{v}{2}(1 - R^2) + \frac{\sigma}{R}dW_2$$

of oscillator (13.60) and (13.61), where $W_1(t)$ and $W_2(t)$ are two independent Wiener processes. For this model, compute the vector

$$\mathbf{v}^T(t) = \mathbf{q}_1^T(t)\mathbf{G}(\mathbf{x}_c(t)),$$

where $\mathbf{q}_1^T(t)$ is the first vector of the conjugate Floquet basis computed in Problem 13.3. Evaluate the squared length $d(t)$ of $\mathbf{v}(t)$, and use it to specify the phase noise model

$$d\alpha(t) = d^{1/2}(t + \alpha(t))dN(t).$$

By observing that $d(t) = d$ is constant, conclude that $\alpha(t) = d^{1/2}N(t)$ where $N(t)$ is a Wiener process.

13.7 Compute the matrix autocorrelation and power spectral density of the noisy oscillator of Problem 13.6. Plot the diagonal components $\mathbf{S}_{11}(\omega)$ and $\mathbf{S}_{22}(\omega)$ of the PSD for $\omega_0 = 1$, $v = \pi/4$ and $\sigma = 0.1$.

13.8 Consider the noisy Coram oscillator

$$dR = (1 - R)Rdt + \sigma dW(t)$$

$$d\Theta = (1 + R)dt$$

considered in Example 13.4, where the noise is injected perpendicularly to the direction of motion. An interesting feature of these dynamics is that the first component obeys a Langevin diffusion with

$$(1 - r)r = -\frac{dU}{dr},$$

where the potential

$$U(r) = \frac{r^3}{3} - \frac{r^2}{2},$$

and diffusion coefficient $d = \sigma^2$.

a) Verify that $U(r)$ tends to infinity as $r \to \infty$, and that it has two local extrema. Show that $r = 1$ (the location of the limit cycle) is a global minimum.

b) By using the results of Chap. 10, show that the solution of the Fokker–Planck equation for this Langevin diffusion converges to the Gibbs distribution

$$f_R(r) = \frac{1}{Z(\beta)} \exp\left(-\beta U(r)\right) \tag{13.62}$$

with $\beta = 2/\sigma^2$ for $r \geq 0$, where the normalization constant

$$Z(\beta) = \int_0^\infty \exp\left(-\beta U(r)\right) dr .$$

c) The normalization constant $Z(\beta)$ cannot be evaluated in closed form, but for large β (small σ^2), an accurate approximation can be obtained by using Laplace's method [27, Sec. 6.4] for expanding integrals of the form

$$I(\beta) = \int_a^b g(r) \exp(-\beta U(r)) dr$$

for large values of β in the vicinity of a global minimum of $U(r)$. If the minimum is located at c with $a < c < b$, the first two terms of Laplace's expansion are given by

$$I(\beta) = \left(\frac{2\pi}{\beta \ddot{U}(c)}\right)^{1/2} \exp(-\beta U(c)) \Bigg[g(c) + \frac{1}{\beta} \Bigg(\frac{\ddot{g}(c)}{2\ddot{U}(c)} - \frac{g(c) U^{(4)}(c)}{8(\ddot{U}(c))^2}$$

$$- \frac{\dot{g}(c) U^{(3)}(c)}{2(\ddot{U}(c))^2} + \frac{5(U^{(3)}(c))^2 g(c)}{24(\ddot{U}(c))^2} \Bigg) \Bigg]. \tag{13.63}$$

Use the expansion (13.63) to approximate $Z(\beta)$ for large β.

13.9 The Gibbs distribution (13.62) of the amplitude of the noisy Coram oscillator can be used to obtain a precise estimate of the frequency shift of the noisy oscillator due to amplitude fluctuations [24]. The second oscillator equation implies that the instantaneous frequency

$$\Omega_i = \frac{d\Theta}{dt} = 1 + R,$$

so that the deviation of the average instantaneous frequency from the fundamental frequency $\omega_0 = 2\,\mathrm{rad/s}$ of the noiseless oscillator is

$$\Delta\omega = E[\Omega_i] - 2 = E[R] - 1,$$

where

$$E[R] = \int_0^\infty r f_R(r) dr = \frac{M(\beta)}{Z(\beta)}$$

with

$$M(\beta) \triangleq \int_0^\infty r \exp\left(-\beta U(r)\right) dr.$$

The expression $E[R] - 1$ for the oscillator frequency shift indicates it is due to amplitude fluctuations of the noisy oscillator trajectory about the limit cycle. By using Laplace's method to approximate both $M(\beta)$ and $Z(\beta)$, show that

$$\Delta\omega = -\beta^{-1} = -\frac{\sigma^2}{2},$$

so that for small noise, the oscillator frequency shift is proportional to σ^2.

References

1. D. Leeson, "A simple model of feedback oscillator noise spectrum," *Proc. IEEE*, vol. 54, pp. 329–330, Feb. 1966.
2. A. Hajimiri and T. H. Lee, "A general theory of phase noise in electrical oscillators," *IEEE J. Solid State Circuits*, vol. 33, pp. 179–194, 1998.
3. A. Hajimiri and T. H. Lee, *The Design of Low Noise Oscillators*. New York: Springer Verlag, 1999.
4. F. X. Kaertner, "Analysis of white and $f^{-\alpha}$ noise in oscillators," *Int. J. Circuit theory and Applications*, vol. 18, pp. 485–519, 1990.
5. A. Demir, A. Mehrotra, and J. Roychowdhury, "Phase noise in oscillators: a unifying theory and numerical methods for characterization," *IEEE Trans. Circuits Syst. I*, pp. 655–674, May 2000.
6. O. Şuvak and A. Demir, "On phase models for oscillators," *IEEE Trans. Computer-Aided Design of Integrated Circuits and Systems*, vol. 30, pp. 972–985, July 2011.
7. G. Papanicolaou, "Introduction to the asymptotic analysis of stochastic equations," in *Modern Modeling of Continuum Phenomena* (R. C. DiPrima, ed.), vol. 16 of *Lectures in Applied Mathematics*, pp. 109–147, Providence, RI: Amer. Math. Society, 1977.
8. G. A. Pavliotis and A. M. Stuart, *Multiscale Methods– Averaging and Homogenization*. New York: Springer, 2008.
9. L. Perko, *Differential Equations and Dynamical Systems*. New York: Springer Verlag, 1991.
10. M. Farkas, *Periodic Motions*. New York: Springer Verlag, 1994.
11. D. Wang, S. Zhou, and J. Yu, "The existence of closed trajectories in the van der Pol oscillator," in *Proc. IEEE Conf. on Communications, Circuits and Systems*, vol. 2, pp. 1629–1632, 2002.
12. E. A. Coddington and R. Carlson, *Linear Ordinary Differential Equations*. Philadelphia, PA: Soc. Industrial and Applied Mathematics, 1997.
13. R. W. Brockett, *Finite Dimensional Linear Systems*. New York: J. Wiley & Sons, 1970. Republished in 2015 by the Soc. for Indust. and Applied Math., Philadelphia, PA.
14. G. J. Coram, "A simple 2-D oscillator to determine the correct decomposition of perturbations into amplitude and phase noise," *IEEE Trans. Circuits Syst. I*, vol. 48, pp. 896–898, July 2001.
15. A. T. Winfree, *The Geometry of Biological Time, second edition*. Berlin: Springer Verlag, 2001.
16. K. Josic, E. T. Shea-Brown, and J. Moehlis, "Isochron," *Scholarpedia*, vol. 1, no. 8, p. 1361, 2006.
17. T. Djurhuus, V. Krozer, J. Vidkjaer, and T. K. Johansen, "Oscillator phase noise: a geometric approach," *IEEE Trans. Circuits Syst. I*, vol. 56, pp. 1373–1382, July 2009.
18. A. H. Jazwinski, *Stochastic Processes and Filtering Theory*. New York, NY: Academic Press, 1970. Reprinted in 2007 by Dover Publications, Mineola NY.
19. F. L. Traversa and F. F. Bonani, "Oscillator noise: a nonlinear perturbative theory including orbital fluctuations and phase-orbital correlation," *IEEE Trans. Circuits Syst. I*, vol. 58, pp. 2485–2497, Oct. 2011.
20. O. Şuvak and A. Demir, "Quadratic approximations for the isochrons of oscillators: a general theory, advanced numerical methods, and accurate phase computations," *IEEE Trans. Computer-Aided Design of Integrated Circuits and Systems*, vol. 29, pp. 1215–1228, Aug. 2010.
21. A. Bensoussan, J.-L. Lions, and G. Papanicolaou, *Asymptotic Analysis for Periodic Structures*. Amsterdam, The Netherlands: North-Holland, 1978.
22. R. Khasminskii, "On averaging principle for Ito stochastic differential equations," *Kybernetika*, vol. 4, pp. 260–279, 1968. in Russian.

23. R. S. Swain, J. P. Gleeson, and M. P. Kennedy, "Influence of noise intensity on the spectrum of an oscillator," *IEEE Trans. Circuits Syst. II*, vol. 52, pp. 789–793, Nov. 2005.
24. F. L. Traversa, M. Bonnin, F. Corinto, and F. Bonani, "Noise in oscillators: a review of state space decomposition approaches," *J. Comput. Electron.*, vol. 14, pp. 51–61, 2015.
25. M. Bonnin and F. Corinto, "Phase noise and noise induced frequency shift in stochastic nonlinear oscillators," *IEEE Trans. Circuits Syst. I*, vol. 60, pp. 2104–2115, Aug. 2013.
26. M. Lax, "Classical noise. V. Noise in self-sustained oscillators," *Phys. Review*, vol. 160, pp. 290–307, Aug. 1967.
27. C. M. Bender and S. A. Orzag, *Advanced Methods for Scientists and Engineers*. New York: McGraw-Hill, 1978.

Printed in the United States
By Bookmasters